s are to be returne on or before
the   low.

Springer Series in Optical Sciences   Volume 30

Edited by A. L. Schawlow

Springer Series in Optical Sciences

Editorial Board: J.M. Enoch D.L. MacAdam A.L. Schawlow K. Shimoda T. Tamir

Laser Spectroscopy V

Proceedings of the Fifth International Conference
Jasper Park Lodge, Alberta, Canada,
June 29 – July 3, 1981

Editors:
A.R.W. McKellar T. Oka B.P. Stoicheff

With 319 Figures

Springer-Verlag Berlin Heidelberg New York 1981

Dr. A. ROBERT W. MCKELLAR
Herzberg Institute of Astrophysics,
National Research Council of Canada,
Ottawa, Ontario, K1A OR6, Canada

Professor TAKESHI OKA
Department of Chemistry, of Astronomy,
and Astrophysics University of Chicago,
Chicago, IL 60637, USA

Professor BORIS P. STOICHEFF
Department of Physics, McLennan Physics
Laboratory, University of Toronto,
Toronto, Ontario, M5S 1A7, Canada

Conference Chairman: B.P. Stoicheff T.Oka

ISBN 3-540-10914-5 Springer-Verlag Berlin Heidelberg New York
ISBN 0-387-10914-5 Springer-Verlag New York Heidelberg Berlin

D
535,84
INT

We are pleased to dedicate this volume to our colleagues and friends

Bloembergen and Schawlow

On October 19th the Royal Academy of Science of Sweden announced that Professors Nicolaas Bloembergen of Harvard University and Arthur L. Schawlow of Stanford University would share half of the 1981 Nobel Prize in Physics for their contributions to Laser Spectroscopy.

Preface

The Fifth International Conference on Laser Spectroscopy or VICOLS, was held at Jasper Park Lodge, in Jasper, Canada, June 29 to July 3, 1981. Following the tradition of the previous conferences in Vail, Megève, Jackson Lake, and Rottach-Egern, it was hoped that VICOLS would provide an opportunity for active scientists to meet in an informal atmosphere for discussions of recent developments and applications in laser spectroscopy. The excellent conference facilities and remote location of Jasper Park Lodge in the heart of the Canadian Rockies, amply fulfilled these expectations.

The conference was truly international, with 230 scientists from 19 countries participating. The busy program of invited talks lasted four days, with two evening sessions, one a panel discussion on Rydberg state spectroscopy, the other a lively poster session of approximately 60 post-deadline papers.

We wish to thank all of the participants for their outstanding contributions and for preparation of their papers, now available to a wider audience. Our thanks go to the members of the International Steering Committee for their suggestions and recommendations. We are especially pleased to have held this conference under the auspices of the International Union of Pure and Applied Physics. VICOLS would not have been possible without the financial support of the Natural Sciences and Engineering Research Council of Canada, and the Office of Naval Research and Air Force Office of Scientific Research of the United States* of America. We are indebted to them for their willing assistance. Thanks are due to ten industrial sponsors for their informative exhibits, and for financing several excursions on the way to Jasper, and in the area on July 1st, as well as nutritional refreshments during intermissions. We are especially grateful to the National Research Council of Canada for administrative assistance through its Conference Office and the efficient Lois Baignée and her staff who helped enormously in the preparations for and during the conference. Finally, we thank members of the Canadian Organizing Committee for their welcome assistance.

July 1981

Bob McKellar
Takeshi Oka
Boris Stoicheff

---

*The work relates to Department of the Navy Grant N00014-81-G-0087 issued by the Office of Naval Research. The United States Government has a royalty-free licence throughout the world in all copyrightable material contained herein.

Contents

Part III. *Double Resonance*

Part IV. *Collision-Induced Phenomena*

Part V. *Nonlinear Processes*

Part VI. *Rydberg States (Panel Discussion)*

Part VII. *Methods of Studying Unstable Species*

Part VIII. *Cooling, Trapping and Control of Ions, Atoms, and Molecules*

Part IX. *Surface and Solid State*

Progress and Perspectives in Laser Spectroscopy

H. Walther

Sektion Physik der Universität München

and

Max-Planck-Institut für Quantenoptik
D-8046 Garching, Fed. Rep. of Germany

Introduction

Laser spectroscopy is one of the most lively fields in physics at the moment. Therefore a review of the present status in an introductory paper like this must be incomplete and can only be restricted to a few examples. The discussion of future perspectives is only possible on the basis of an extrapolation of long term developments and can therefore, of course, not contain the many surprising and unforseen findings which make the field interesting and lively.

A complete review of the field of laser spectroscopy would include the three main chapters: lasers, methods, and applications. This paper cannot follow this scheme: the discussion of lasers must be omitted; under "methods" only the development of high-resolution spectroscopy will be discussed, and as an example of "applications" the use of laser spectroscopy for the study of fundamental physical problems will be briefly tackled. So this introductory paper will by far not be complete. We are all aware of new interesting developments of the field in connection with the study of ultrafast phenomena, of the new results in the study of chemical processes in general, in photochemistry and photobiology, in the investigation of laser-induced collective processes, in nonlinear optics, and in many other parts of the field being covered during this conference. The selection of the topics for the following discussion is more or less arbitrary and largely influenced by the interests of the author.

Methods of High-Resolution Laser Spectroscopy: Present and Future

There are essentially two groups of methods available in laser spectroscopy allowing a resolution which may only be limited by the natural width. The first group includes the methods of optical radio-frequency spectroscopy, e.g. the optical double resonance method, the quantum beat method and others being listed in Table 1 on the right-hand side. Especially in atomic spectroscopy these methods have played a large role in the past years. The techniques can be applied using discharge lamps for excitation having a spectral distribution comparable to or larger than the Doppler width. Many precision measurements have been performed, especially of hyperfine and Zeeman splittings. With the advent of tunable lasers these methods did not lose their significance since the high spectral brightness of the laser allowed a much better population of the investigated levels, so that afterwards some of these classical methods had a remarkable renaissance, e.g. the quantum beat method. For this method it is essential to have short light pulses exciting the atoms or molecules since the Fourier-limited spectral distribution of the pulse has to be larger than the splitting of the levels under investigation so that a coherent population is guaranteed. Laser excitation is therefore more advantageous than classical light sources.

Table 1 Survey on the methods of high-resolution spectroscopy used for the study of electronically excited states.

Narrow-banded excitation		Broad-banded excitation		
Narrow absorption	Broad banded absorption	"Incoherent"population	Coherent population of two or more states	
atomic beam	saturated absorption	double resonance method	time integral time differential observation	
	two photon spectroscopy (stepwise excitation)	optical pumping	level crossing	quantum beats
	fluorescence line narrowing	anti-crossing		modulated excitation

The techniques listed on the left-hand side of the table are those which are only applicable with narrow-banded excitation as provided by monomode lasers. With these methods, especially the atomic beam scattering, the non-linear absorption and the two photon spectroscopy became much more important.

It is impossible to discuss in this introductory paper all developments in high-resolution laser spectroscopy in recent years; two examples have to be selected: saturation spectroscopy and the quantum beat method. Both are very suitable for demonstrating the continuous progress in the field.

a) Saturation Spectroscopy

Table 2 shows the different variants of saturation spectroscopy known at present. Saturation spectroscopy started when MC FARLANE et al [1] and SZÖKE et al. [2] demonstrated the Lamb dip caused by gain saturation at the middle of the tuning curve of a single-mode HeNe laser. Later with intracavity and external absorption cells the method was used for high resolution spectroscopy and for the frequency stabilization of lasers. In saturation spectroscopy a signal is observed when two counterpropagating laser beams of the same frequency can interact with the same atom in a nonlinear way. The various methods used differ by the approach by which this nonlinear interaction is detected, resulting in a different sensitivity and selectivity.

A first improvement of the detection sensitivity can be obtained if one of the beams is chopped and the modulation of the other beam traced via phase-sensitive detection [3,4]. Nonlinear spectroscopy can also be performed by observing fluorescence instead of absorption. When the absorption is reduced by saturation, so too is the fluorescence reduced. This alternative detection of the signal is especially useful at low pressures where the total absorption is small. Favourable signal detection may be obtained if both laser beams are chopped with different frequencies, and the fluorescence is detected at the sum frequency, as it is done in the inter-modulated fluorescence technique introduced by SOREM and SCHAWLOW [5]. Since the upper-state population is modulated at the sum frequency as well, the detection of the saturation signal can also be performed using optogalvanic [6] or optoacoustic techniques [7]. The latter are two important extensions of the intermodulated fluorescence method.

The nonlinear coupling of the counterpropagating beams can also be detected by observing either the dispersion of the medium in an interferometric set-up [8,9] or else the laser induced birefringence or dichroism, as is done in the latter case in polarization spectroscopy introduced by HÄNSCH and WIEMAN [10]. The latter method brings many advantages for the assignment of complex molecular spectra since the light-induced anisotropy can be used to label molecular levels so that all optical transitions sharing common levels can be easily identified [11]. Polarization spectroscopy is of particular interest for measurements with optically thin samples or of weak lines. The signal shape depends strongly on the choice of polarization for the saturating laser beam and on the ad-justment of the polarizers used to probe the light-induced birefringence and dichroism. Therefore unsymmetric lines may be obtained, being a draw-back of this method compared to intermodulated fluorescence. However, polari-zation spectroscopy has a big advantage compared to the other methods des-cribed so far: the broad signal background of the narrow line due to the collisional redistribution of the particle velocities is absent since the collisions change the light-induced alignment or orientation; therefore the particles which experienced a collision do not contribute to the signal.

A method which combines the advantages of intermodulated fluorescence and polarization spectroscopy has recently been proposed and used by HÄNSCH et al. [12]. The essence of this polarization intermodulated excitation (POLI-NEX) spectroscopy is that the polarization of one or both counterpropagating beams is modulated (instead of using amplitude modulation). When the combined absorption depends on the relative polarization of both beams, an intermodul-ation in the total rate of excitation is observed. The signal detection can therefore be performed either by observing the fluorescence or by the use of the indirect methods, e.g. the optogalvanic detection (Table 2).

Table 2 Development of saturation spectroscopy

Coupling of oppo-site laser beams	direct signal	modulation one beam	modulation two beams		heterodyne detection
Absorption	[1,2] 1963	[3,4] 1971	[5] [6]* [7]**	1972 1979 1979	[13] 1980
Dispersion	[8] 1973	[9] 1975			
Polarization		[10] 1976	[12] [12]*	1981*** 1981***	

*   optogalvanic detection
**   optoacoustic detection
*** polarizers are rotated

This new technique has another important advantage compared to inter-modulated fluorescence: the POLINEX-signal does not have to be detected on a strongly modulated background (modulated at fundamental frequencies), because neither beam alone can produce a modulated signal in an isotropic medium; therefore nonlinear mixing in the detection system cannot produce spurious signals. However, there may be a dependence of the signal on external magnetic fields since the polarized excitation can produce a coherent population of Zeeman sublevels being affected by the external field. Therefore the compensation of the earth's magnetic field is not only necessary for the sake of achieving the highest possible resolution, but also to obtain the highest possible signal.

Another new technique has been proposed since the last laser spectroscopy meeting: it is the heterodyne detection scheme of saturated absorption [13, 14]. The method uses resonant degenerate four-wave mixing with two close optical frequencies to perform high-frequency optically heterodyne saturation spectroscopy. The specific features of this method are that the phase delays in the heterodyne signal for the crossovers make it possible to measure the relaxation rates of lower and upper states separately. In this respect the technique resembles the phase shift method in modulated fluorescence. The technique also gives information on line assignment and is generally applicable to any nonlinear spectroscopic scheme, e.g. the polarization spectroscopy where it becomes possible to optimize the signal-to-noise ratio by an adequate choice of the heterodyne frequency. In particular the influence of amplitude fluctuations of the laser can thus be eliminated when the measurement is performed in a frequency range where the shot-noise limit can be reached. The technique can in addition be applied to Doppler-free two-photon spectroscopy (see post deadline papers of this meeting) and to Raman spectroscopy.

b) Quantum Beat Spectroscopy

The quantum beat method is essentially of the same nature as other methods based on coherent effects in fluorescence, such as the Hanle effect or level-crossing, modulated excitation, and also perturbed angular correlation in nuclei. The techniques require that the system be excited into a coherent superposition of substates and that afterwards the evolution of this coherence is monitored as a change either of the polarization properties or of the angular distribution of the reemitted radiation.

The quantum beat method is much easier to understand than the more sophisticated steady-state atomic coherence phenomenon seen in a Hanle or level-crossing experiment. The fact that the experimental demonstration came much later has technical reasons: until recently it has been very difficult to detect fast modulation signals and also to produce the short and intense pulses needed for an efficient excitation of the atomic system. So after the first demonstration of the quantum beat method with classical light sources [15,16] the method was not widely applied until the pulsed tunable lasers became available. Since then a large number of experiments have been performed, especially in connection with the investigation of Rydberg levels.

Table 3 Quantum beat spectroscopy

Excitation	Observation	References	Remarks
short light pulse	Fluorescence	[15,16]* [17]**	Time differential observation
	Absorption	[18]	Probing is delayed versus excitation pulse. Signal is measured as function of delay time.
	Birefringence and dichroism	[19]	
	Field ionization	[20]	
	Photoionization	[22-24]	

* Pulsed discharge lamp
** First laser experiment

In the standard quantum beat experiment light pulses are used to excite the coherent superposition of two closely spaced levels. Detection is performed by observing the temporal change of the fluorescence. Quantum beats can also be observed by means of stimulated transitions. In this case, the detection is performed by a second light pulse which measures, e.g. the absorption of the system starting from the coherently populated intermediate levels. The quantum beats are obtained by measuring the absorption as a function of the time delay between the exciting and probing light pulses. The first quantum beat experiment using this absorption method has been performed by DUCAS et al. [18]. Furthermore the quantums beats can be investigated in a corresponding setup observing the laser-induced birefringence and dichroism [19]. In comparison with the standard experiment these new methods have the advantage that the spontaneous lifetime of the levels investigated can be large, as is the case, e.g. for Rydberg states. However, the fact that a second resonant probing light pulse must be available complicates the experiment and a nonresonant probe has many advantages. In the case of highly excited Rydberg states field ionization is a very simple way to probe the quantum beats. This has been demonstrated for the first time by LEUCHS et al. [19]. Due to the long lifetime of the levels, large delay times could be used and a rather good accuracy has been obtained in the experiments. Another way for nonresonant probing of quantum beats is the use of photoionization. Here either the change in the total current of photoelectrons [23,24] or in the angular distribution can be measured [21,22].

If the atoms are ionized in an n-photon process using linearly polarized light, the photoelectron angular distribution can be described by a linear combination of even Legendre polynomials up to P_{2n} (cos θ), where θ is the angle between the polarization direction of the ionizing laser pulse and the direction of emission of the photoelectron. For two-photon ionization the angular distribution of photoelectrons is given by

$$I(\theta) = 1 + \beta_2 P_2(\cos \theta) + \beta_4 P_4(\cos \theta)$$

The angular distribution of photoelectrons in two-photon ionization of sodium via the $3^2P_{3/2}$ state and its dependence on the time delay between the first and the second light pulse is shown in Fig. 1. The quantum interference signal due to the hyperfine splitting of the $3^2P_{3/2}$ state can be obtained from this measurement by Fourier analysis [22].

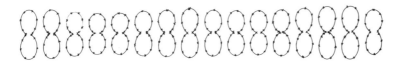

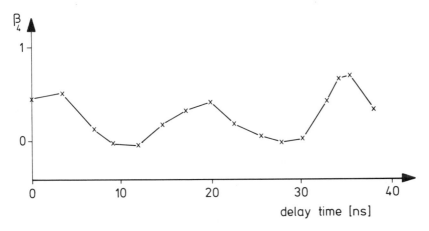

Fig. 1 Quantum beats in the shape of the angular distribution of photoelectrons resulting from photoionization of the $3^2P_{3/2}$ state of sodium. The signal is due to the hyperfine splitting of the $3^2P_{3/2}$ state.

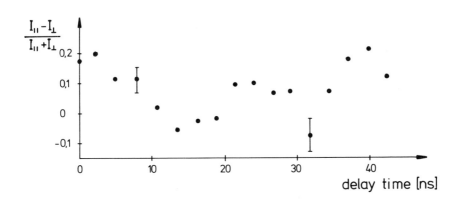

Fig. 2 Quantum beats in the total photoelectron current. The signal is due to the hyperfine splitting of the $3^2P_{3/2}$ state (see also Fig. 1).

For the measurement of the quantum beat signal in the total photoelectron current, the signals for parallel and perpendicular polarizations of the two laser pulses are evaluated; their contributions have opposite phase and considering the ratio $(I_{\parallel} - I_{\perp}) / (I_{\parallel} + I_{\perp})$ the intensity changes which were intruduced by an optical delay line were thereby eliminated (Fig. 2).

c) Ultimate Spectral Resolution

With the development of frequency stabilized single mode lasers and particularly tunable lasers, and with the introduction of methods eliminating the Doppler width, the natural linewidth set the limit for the resolution. This limit, however, is only achievable as long as the transition frequency to linewidth ratio is not larger than 10^{10}. Long lifetimes as associated with vibrational transitions in molecules or forbidden transitons in atoms allow, in principle, a much higher resolution. A limitation is that of the transit time broadening and the influence of the quadratic Doppler effect (Fig. 3). The transit time broadening can be reduced by the use of expanded laser beams. In this way it was possible to resolve the radiative recoil-induced doublets of the hyperfine components of the methane transitions at 3.39 μm. This high resolution (5 parts in 10^{10}), derived from an external absorption cell with a 30 cm aperture, nevertheless remained two orders of magnitude larger than the natural linewidth [25]. As larger cells and associated optics are difficult to realize, alternative schemes have to be found. An attractive solution is the multiple interaction in standing wave fields [26], the optical analogue to the well-known Ramsey technique routinely used in radio frequency spectroscopy of atomic and molecular beams. In these interference methods the resolution is limited by the travel time between radiation zones rather than by the transit time through one zone. The first experiments with this technique have been reported at the laser spectroscopy conferences in Jackson Lake Lodge 1977 and in Rottach-Egern 1979. The highest resolution has been obtained in the saturated absorption experiment of the Ca $^1S_0$ - $^3P_1$ intercombination line at 657 nm. The radiation beams were spatially separated by up to 3.5 cm [27]. The linewidth observed was 1 kHz. An important condition for the application of the Ramsey technique is that the phases of the standing wave fields are stationary, otherwise the fringes will wash away within the averaging time. Therefore this method also has technical limitations and it seems difficult at the moment to reduce the linewidth much further than 1 kHz.

The new technique of spectroscopy of trapped ions (being reviewed in this volume by DEHMELT), however, opens up the possibility of overcoming the present limit of ultimate resolution. So far two groups are exploring this new exciting method. In 1978 the Heidelberg group [28] succeeded in confining a single Ba^+-ion in a Paul radio frequency trap with the localization of about 2000 Å via laser side band cooling to about 10 mK [29]; furthermore the ion could visually and photoelectrically be observed. Recently also the Boulder group performed a mono-ion oscillator experiment using Mg^+. They have demonstrated a localization of an ion to $\leq$ 15 μm in their Penning trap [30]. These experiments prepare the ground for a transit time-free and second order Doppler-free spectroscopy and yet systems have been proposed which may make it possible to reach a resolution of one part in 10^{14} [31]. It is quite obvious that the new techniques open a new door in ultimate resolution spectroscopy and it is sure that before long new exciting results will be obtained in this connection.

Experiments with Ultimate Resolution

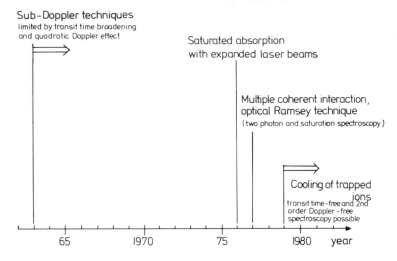

Fig 3 Development of the ultimate resolution in laser spectroscopy.

    Precision spectroscopy requires an accurate frequency or wavelength dif-
ference measurement. In the visible spectral region the laser spectroscopist
has to rely on an interferometric determination of the wavelength difference which allows accuracies in the range of several parts in 10^9 being
limited by the present definition of the meter. A possibility to measure
frequency differences would therefore be a big advantage for precision
spectroscopy and metrology. So far the direct and difference measurement
of the light frequency was limited to the infrared spectral region. The
most successful device, the metal-insulator-metal diode in point contact
configuration, was used for this purpose for more than ten years. However,
the application to visible laser light failed until recently. DANIEL et al.
[32] could measure frequency differences up to 170 GHz between a cw dye
laser and a krypton laser at 568 nm by mixing laser and microwave radiation
with a metal-insulator-metal point contact diode (Fig. 4). An extension to
higher frequency differences was successful up to 2.5 THz so far [33]. The
frequency roll-off for the signal-to-noise ratio observed in the latter
experiments is not too big so that one can hope that even larger frequency
differences are measurable. This is an important progress which gives some
hope that also the frequency chain for the absolute determination of laser
frequencies using the metal-insulator-metal diode can be extended into the
visible spectral range.

d) Spectral Resolution beyond the Natural Linewidth

Spectroscopists quite often have the problem of unravelling structures
with a splitting smaller than or comparable to the natural width of a
transition. In many cases a computer fit to the observed line profile
using the theoretical information on the structure can help to get the

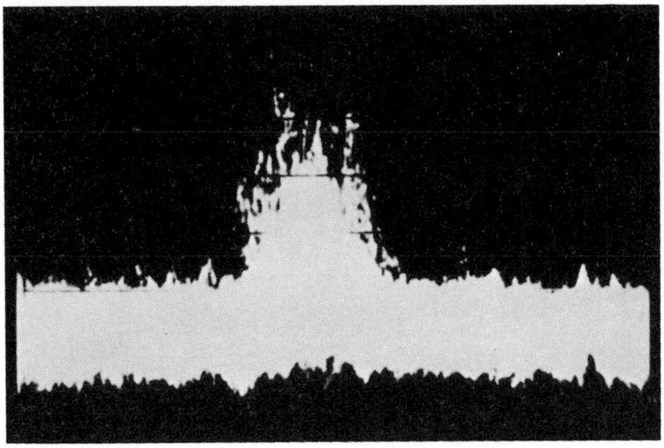

Fig. 4 Beat signal between light of a dye laser, a kryption laser, and a microwave frequency of 170 GHz obtained with a metal-insulator-metal point contact diode. For details see [32]

desired data. However, the experimentalist always feels better if more detailed a priori information is available. Therefore schemes have been proposed which make it possible to reach a "subnatural" resolution by limiting the observation to ensembles of atoms which are selected according to their time differential behaviour or, roughly speaking, by limiting the observation to "longer-lived" atoms. It is quite clear that in this way the signal of a certain number of atoms is disregarded. Therefore the observed signal decreases, resulting in longer averaging times. Methods of this type have been used in Mössbauer and optical spectroscopy (see for reviews [34,35]). Recently it was pointed out by MEYSTRE et al. [36] that in the transient behaviour of two levels with the decay rates γ_a and γ_b the line shape may be governed by $(\gamma_a - \gamma_b)/2$ instead of $(\gamma_a + \gamma_b)/2$, which is observed in the usual emission or absorption experiment. This fact enables one to perform spectroscopy beyond the natural linewidth.

The level scheme under consideration is shown in Fig. 5. Level b is initially excited by a laser pulse and the atoms interact with resonant radiation of frequency ν promoting them to level a. The decay from a is observed a time delay Θ after the excitation. With no time delay the usual Lorentzian line shape is observed. As Θ is increased, however, the width of the line measured as a function of ν decreases. The decrease of the linewidth is strongest for the smallest differences between γ_a and γ_b (Fig. 6).

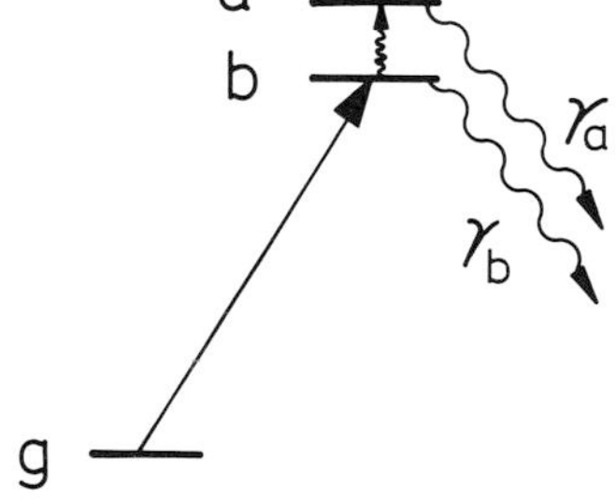

Fig. 5 Two-level system considered for transient line narrowing. The two levels a and b interact with a resonant electric field. Initially level b is excited by a laser pulse

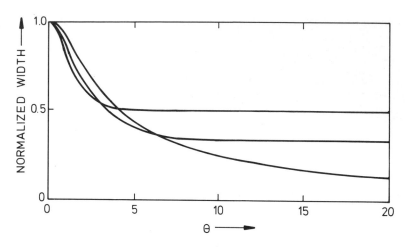

Fig. 6 Full width at half-maximum as a function of the delay θ normal-
        ized to the linewidth for no delay. γ_b = 1 in all cases and
        γ_a = 1.01, 2, and 3. The strongest linewidth reduction at large
        θ is observed for γ_a = 1.01 [36]

This partial cancellation of the excited-state width by the initial state
in transient experiments also occurs if the initial state is the ground state
atom interacting with a pulse of radiation showing an exponential time behaviour [37]. With this extension the new method becomes rather broadly applicable.

Application of Laser Spectroscopy to Fundamental Physical Problems

Laser spectroscopy also makes it possible to obtain many results which are
relevant to fundamental physics. A classical example is the spectroscopy of
the hydrogen atom: an improved value for the Rydberg has been obtained and
the Lamb shift of the ground state was determined [38]. There is the potential
for a further improvement of the accuracy of these data as soon as narrower
laser sources have been developed for the measurement of the 1s - 2s two-
photon transition. In this case it would also be possible to improve the
value for the electron/proton mass ratio by a precise determination of the
H-D isotope shift. (The improved electron/proton mass ratio is also of importance to obtain the highest possible precision of the Rydberg).

Another contribution of laser spectroscopy to fundamental physics is
expected in connection with the search for parity violation in atoms due
to the weak neutral coupling between electrons and nucleons predicted by
the Weinberg-Salam model. The experiments will show whether the theory of
the unification of weak and electromagnetic interactions is also valid under
atomic conditions. Several experiments are presently under way (see [39] for
a review) and recent data of two groups have been published [40,41] which are
consistent with each other and also agree with the result expected from the
Weinberg-Salam model.

Other examples of laser spectroscopy applications are the measurement of the relativistic Doppler shift [42] and an improved Michelson-

Morley experiment [43]. With increasing progress in laser stabilization and in the achievement of ever higher resolution it is also expected that experiments on the gravitational redshift or on the search for a secular drift of the frequency ratio of atomic clocks whose modes of operation are based on different physical principles will become feasible and worthwhile. It has also been proposed that "preferred frame" effects should manifest themselves in more accurate measurements of the relativistic redshift.

There are still more new challenges of laser spectroscopy coming up. During this meeting, the gravitational wave detection using laser interferometric techniques will be discussed. In addition to those experiments it is also possible to use narrow optical sources and highly stable lasers as detectors for gravitational waves [44]. Furthermore, the techniques of ring laser interferometry (see e.g. [45]) open up the possibility to perform tests of general relativity. The measurements of space curvature, Lense-Thirring effect, and preferred frame effects have been proposed using the new techniques.

The few examples I have given show that we are still at the beginning of the laser revolution of spectroscopy. New and important results will come up in the future and the development of the new techniques of ultrahigh resolution will make the field even more exciting than it has been so far.

References

1   R.A. Mc Farlane, W.R. Bennett, W.E. Lamb, Jr., Appl. Phys. Lett. 2, 189 (1963)
2   A. Szöke, A. Javan, Phys. Rev. Lett. 10, 521 (1963)
3   Ch. Bordé, C.R. Acad. Sc. Paris, 271, 371 (1970)
4   T.W. Hänsch, M.D. Levenson, A.L. Schawlow, Phys. Rev. Lett. 26, 946 (1971)
5   M.S. Sorem, A.L. Schawlow, Opt. Comm. 5, 148 (1972)
6   J.E. Lawler, A.J. Ferguson, J.E.M. Goldsmith, D.J. Jackson, A.L. Schawlow, Phys. Rev. Lett. 42, 1046 (1979)
7   E.E. Marinero, M. Stuke, Opt. Comm. 30, 349 (1979)
8   Ch. Bordé, G. Camy, B. Decombs, L. Pottier, C.R. Acad. So. 277, 381 (1973)
9   B. Couillaud, A. Ducasse, in Laser Spectroscopy, ed. by S. Haroche, J.C. Pebay-Peyroulu, T.W. Hänsch, S.E. Harris, Lecture Notes in Physics, Vol. 43 (Springer, Berlin, Heidelberg, New York 1975) p. 476
10 C. Wieman, T.W. Hänsch, Phys. Rev. Lett. 36, 1170 (1976)
11 R. Teets, R. Feinberg, T.W. Hänsch, A.L. Schawlow, Phys. Rev. Lett. 37 683 (1976)
12 T.W. Hänsch, D.R. Lyons, A.L. Schawlow, A. Siegel, Z - Y. Wang and G - Y. Yan, Opt. Comm. 37, 87 (1981)
13 R.K. Raj, D. Bloch, J.J. Snyder, G. Carmy, and M. Ducloy, Phys. Rev. Lett. 19, 1251 (1980)
14 D. Bloch, R.K. Raj, M. Ducloy, Opt. Comm. 37, 183 (1981)
15 E.B. Alexandrov, Opt. Spectrosc. 17, 957 (1964)
16 J.N. Dodd, R.D. Kaul, D.M. Warrington, Proc. Phys. Soc. (London) 84, 176 (1964)
17 S. Haroche, J.A. Paisner, A.L. Schawlow, Phys. Rev. Lett. 30, 948 (1973)

18   T.W. Ducas, M.G. Littman, M.L. Zimmerman, Phys. Rev. Lett. 35, 1752
     (1975)
19   W. Lange, J. Mlynek, Phys. Rev. Lett. 40, 1373 (1978)
20   G. Leuchs, H. Walther, in Laser Spectroscopy III, ed. by J.L. Hall,
     J.L. Carlsten, Springer Series in Optical Sciences, Vol. 7 (Springer,
     Berlin, Heidelberg, New York 1977) p. 299 and Z. Physik A 293, 93 (1979)
21   M.P. Strand, J. Hansen, R.L. Chien, R.S. Berry, Chem. Phys. Lett. 59,
     205 (1978)
22   G. Leuchs, S.J. Smith, E.J. Khawaja, H. Walther, Opt. Comm. 31, 313 (1979)
23   T. Hellmuth, G. Leuchs, S.J. Smith, H. Walther, Proceedings of the 2nd
     International Conference on Multiphoton Processes, Budapest, 14-18 April
     1980, Hungerian Academy of Sciences, Budapest
24   T. Hellmuth, G. Leuchs, S.J. Smith, H. Walther, in Lasers and Applications,
     ed. by W.O.N. Guimaraes, C.-T. Lin, A. Mooradian, Springer Series in
     Optical Sciences, Vol. 26 (Springer, Berlin, Heidelberg, New York 1981)
     p. 194
25   J.L. Hall, C.J. Bordé, K. Uehara, Phys. Rev. Lett. 37, 1339 (1976)
26   Ye. V. Baklanov, B. Ya. Dubetsky, V.P. Chebotayev, Appl. Phys. 9, 171
     (1976); 11, 201 (1976)
27   R.L. Barger, Opt. Lett. 6, 145 (1981)
28   W. Neuhauser, M. Hohenstatt, P.E. Toschek, H.G. Dehmelt, Phys. Rev.
     Lett. 41, 233 (1978)
29   W. Neuhauser, M. Hohenstatt, P.E. Toschek, H.G. Dehmelt, Phys. Rev.
     A22, 1137 (1980)
30   D.J. Wineland, W.M. Itano, Phys. Lett. 82 A, 75 (1981)
31   H.G. Dehmelt, Bull. Am. Phys. Soc. 18, 1571 (1973)
32   H.-U. Daniel, M. Steiner, H. Walther, Appl. Phys. 25, 7 (1981)
     and Appl. Phys. B 26, 19 (1981)
33   H.-U. Daniel, J.C. Bergquist, R.E. Drullinger, D.A. Jennings,
     K.M. Evenson, to be published
34   G. zu Putlitz, Comments Atom. Mol. Phys. 1, 74 (1969)
35   P.L. Knight, Comments Atom. Mol. Phys. 10, 241 (1981)
36   P. Meystre, M.O. Scully, H. Walther, Opt. Comm. 33, 153 (1980)
37   P.L. Knight, P.E. Coleman, J. Phys. B13, 4345 (1980)
38   A.I. Ferguson, J.E.M. Goldsmith, T.W. Hänsch, E.W. Weber, in
     Laser Spectroscopy IV, ed. by H. Walther, K.W. Rothe, Springer
     Series in Optical Sciences, Vol. 21 (Springer, Berlin, Heidelberg,
     New York 1979) p. 31
39   E. Commins, in Atomic Physics 7, ed. by D. Kleppner, F. Pipkin
     (Plenum Press, New York, London 1981)
40   P. Bucksbaum, E. Commins, L. Hunter, Phys. Rev. 46, 640 (1981)
41   J.H. Hollister, G.R. Apperson, L.L. Lewis, T.P. Emmons, T.G. Vold,
     E.N. Fortson, Phys. Rev. 46, 643 (1981)
42   J.J. Snyder, J.L. Hall, in Laser Spectroscopy, ed. by S. Haroche,
     J.C. Pebey-Peyroula, T.W. Hänsch, S.E. Harris, Lecture Notes in
     Physics, Vol. 43 (Springer, Berlin, Heidelberg, New York 1975) p. 6
43   A. Brillet, J.L. Hall, in Laser Spectroscopy IV, ed. by H. Walther,
     K.W. Rothe, Springer Series in Optical Sciences, Vol. 21 (Springer,
     Berlin, Heidelberg, New York 1979) p. 12
44   S.N. Bagayev, V.P. Chebotayev, A.S. Dychkov, V.G. Goldort,
     Appl. Phys. 25, 161 (1981)
45   M.P. Haugan, M.O. Scully, K. Just, Phys. Lett. 77A, 88 (1980)
     and M.S. Zubairy, M.O. Scully, K. Just, Opt. Comm. 36, 175 (1981)

Part I
Fundamental Applications
of Laser Spectroscopy

Precision Spectroscopy and Laser Frequency Control Using FM Sideband Optical Heterodyne Techniques

J.L. Hall*, T. Baer, L. Hollberg, and H.G. Robinson

Joint Institute for Laboratory Astrophysics
University of Colorado and National Bureau of Standards, 325 S.Broadway
Boulder, CO 80302, USA

In fundamental physical experiments using laser servolocking techniques,[1] in passive ring laser gyro experiments, and in precision atomic/molecular spectroscopic measurements as well, the two overriding experimental concerns are maximizing the signal-to-noise ratio and obtaining highly symmetrical resonance profiles to facilitate precise line splitting. In this paper we discuss the technique of FM sideband optical heterodyne spectroscopy,[2,3] which appears to be the experimentally optimum method for obtaining such high precision resonance profiles of maximal signal/noise ratio. We discuss the process in simple physical terms relative to stabilization of a laser to a resonant optical cavity, before turning to sub-Doppler resonance spectroscopy obtained by applying the sideband techniques to cw dye lasers and color center lasers. The final topic concerns our study of optical transients resulting from laser phase changes, studied with the optical heterodyne technique.

We begin with the observation that resonance absorption measurements in whatever frequency domain -- rf, microwave or laser -- are almost always limited in practice by residual noise of a technical nature present on the output of the coherent source. In the laser case this problem has led to the invention of polarization spectroscopy[4] and interference spectroscopy[5] as techniques to reduce the direct feedthrough of laser amplitude noise while preserving most of the signal. Recent progress has involved the use of high modulation frequencies[6] and four-wave mixing[7] to further improve the sensitivity. Although the technique of FM sideband optical spectroscopy recently has been invented independently at least twice,[2,8] and although its realization and application in the laser domain as optical heterodyne saturation spectroscopy will undoubtedly provide major advances in sensitive and precise spectroscopy, in fact the technique was first discussed in connection with NMR spectroscopy by SMALLER[9] and ACRIVOS.[10] For historical interest and perspective about the optical work, we briefly describe the several component ideas developed for rf and microwave spectroscopy.

To stabilize a microwave klystron oscillator to a stable microwave cavity, POUND[11] introduced the idea of a reflective modulator driven by an rf source, say at frequency ω, to produce sidebands on the carrier. A high Q reference cavity is used in a reflection-mode experimental setup. When the modulation frequency is high enough, the resulting FM sidebands lie well outside the passband of the resonant cavity and are fully reflected. However, the carrier, being in near resonance with the cavity, is reflected with a tuning-dependent phase and amplitude. When the reflected sidebands

---

*Staff Member, Quantum Physics Division, National Bureau of Standards.

and phase-shifted carrier are heterodyned in the detector, one has the cavity's resonance information converted to amplitude and phase changes of the output signal at the rf frequency ω. After rf amplification, subsequent phase-sensitive detection of this signal relative to the modulator's drive phase at ω leads to the desired "discriminator-type" response function. This description is totally appropriate also for the optical case.[8] POUND recognized the advantages of processing signal information at an rf frequency, thereby avoiding much of the intrinsic source noise and the detector's low frequency excess noise. TRELA and FAIRBANK,[12] in building superconducting cavity-stabilized oscillators, further improved the long-term stability with an experimental topology in which the FM sidebands and carrier leaving a phase modulator travel together along the same path to the resonant cavity and -- after reflection -- on to the detector. Mechanical perturbations thus develop phase-shifts proportional only to the <u>difference</u> in the k vectors.

STEIN and TURNEAURE,[13] further improved the locking precision by measuring in a separate detector, before interaction with the cavity, the residual AM component of the modulation sidebands. This AM information was then used to servo-control the dc bias on the modulator crystals to produce pure frequency modulation, free of any spurious AM. A locking stability σ_y ($\tau = 100$ sec, BW = 10 HZ) $\approx 6 \times 10^{-16}$ was obtained with a microwave cavity $Q \sim 10^{11}$ (optimum stability $\approx 10^{-4}$ cavity line widths).

The intrinsic resonance symmetry associated with pure FM sideband spectroscopy is a fundamental issue for high precision lineshape studies and precision servo locking. It appears that our optical studies may be the first to investigate this point. See below.

To choose the rf modulation frequency intelligently we consider the frequency distribution of the laser's amplitude noise, presented in Fig. 1 for our cw ring dye laser. Several observations may be made from the data.

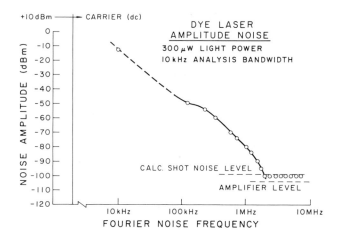

<u>Fig. 1.</u> Amplitude spectral noise density of our ring dye laser. It finally reaches the shot-noise floor at -148 dB/$\sqrt{Hz}$ relative to the carrier for Fourier frequencies above 2 MHz.

1. For sufficiently high Fourier frequencies, $\omega/2\pi > 2$ MHz, the dye laser output is indeed quiet enough to closely approach the fundamental photoelectron shot noise limit associated with the dc light level. [Not shown on this graph are some extremely noisy bands around 17 and 85 MHz that arise from argon laser mode-beating. It is impressive how strongly these frequencies propagate through the dye laser!]

2. At the usual ≈kilohertz modulation frequencies the laser noise spectral density is increased by ~80 dB (=10^8 power ratio) over the fundamental shot noise level. Under such conditions it is not surprising to find that over the years laser spectroscopists have developed a plethora of experimental techniques which aim to exclude laser noise from the measurement channel.

3. For this particular photodiode preamplifier, a light level of 300 μW is marginally sufficient for the shot noise level to exceed the noise floor set by the amplifier. A larger feedback resistor, R_f, could be used to reduce the amplifier noise and useful rf bandwidth proportionally (present $R_f = 3k$ gives $f_{3dB} = 17$ MHz). A further necessary criterion is $I_{dc} \times R_f >> 0.025$ V to mask thermal noise from the feedback resistor.

The use of optical FM sideband techniques has been suggested in relation to interferometric gravity-wave antennas.[14] From a physical discussion of the transient ringing behavior of the carrier field stored in the proposed high finesse resonant optical cavity, Drever[8] was led to suggest use of the cavity "leakage" field as a phase reference for phase locking a laser to the reference cavity. This surprising idea of phase locking a laser onto itself via a reference cavity does not appear to have been previously suggested although the stabilized microwave oscillator of Stein was in fact operating in this transient regime in view of the high Q of the superconducting cavity. POUND has shown[15] that this system exhibits a smooth crossover into his previously-described cavity frequency discriminator mode for times longer than the cavity ringing time.

It may be interesting to present some information about our experimental laser stabilization results[16] with these methods. To establish the reality of tight, phase-locking of lasers to a cavity, we used the Pound/Drever technique to lock a cw dye laser and a He/Ne laser to the same high finesse cavity. High stability of their beat could be possible only if they were individually well stabilized. Use of separate external modulator crystals, different rf frequencies, and orthogonal polarizations isolated the two systems effectively even though both oscillations were at 633 nm. In one series of experiments we locked the two systems on adjacent axial modes of the reference cavity (c/2L = 100 MHz, F = 300).

Small portions of each beam were mixed together via a separate beam splitter and avalanche diode receiver. The resulting 100 MHz beat was down converted in a balanced mixer by heterodyne with a stable frequency synthesizer. A surprisingly stable beat frequency was produced by tuning the synthesizer about 600 Hz away from the 100 MHz inter-laser beat. We could easily see that the rf heterodyne line width -- and hence the independent laser line widths -- were well below 100 Hz. A complete description will be published elsewhere,[16] and new experiments are being prepared.

We now turn to the use of FM sideband techniques for sub-Doppler spectroscopy. The two most striking features of this method are the high

signal-to-noise ratio and the excellent symmetry of the observed resonances. The physical origin of narrow resonances within the Doppler profile is most easily understood by representing the original pure phase-
modulated wave in terms of an optical carrier Ω plus FM sidebands spaced
by the modulation frequency ω. We have

$$E = E_0 \sin(\Omega t + \beta \sin \omega t)$$

$$= E_0 \left\{ \sum_{n=0}^{\infty} J_n (\beta) \sin(\Omega + n\omega)t + \sum_{n=1}^{\infty} (-1)^n J_n(\beta) \sin(\Omega - n\omega)t \right\} ,$$

where β is the modulation index and $J_n(\beta)$ is the Bessel function of order
n. To simplify the discussion we may assume a small modulation index $\beta < 1$,
so the full FM signal collapses to the carrier and its first order modulation sidebands. After passing through our absorption cell these three frequencies will be incident on the photodetector where they will produce,
in addition to the three terms at dc, two alternating currents at the rf modulation frequency ω. An essential point of the FM spectroscopy method, as
emphasized by BJORKLUND,[2] is that these two rf currents are equal but of
opposite signs. This cancellation, characteristic of pure phase modulation, can be affected by the optical medium by phase shift of any of the
spectral components or by differential attenuation of one of the sidebands.
BJORKLUND[2] emphasized the latter mode, especially where the modulation frequency was so high that only one of the sidebands interacted appreciably
with the absorbing medium. For precision spectroscopy, on the other hand,
we prefer to use sidebands spaced close together with respect to the
Doppler width and/or possible hyperfine structure. Of course the modulation frequency must be high enough to avoid the regime of laser technical
noise. See Fig. 1.

Figure 2 shows a schematic diagram of the apparatus employed in our spectroscopy experiments. The output beam of a cw ring dye laser beam passes
through a beam splitter, an external phase modulator, and a 50 cm I_2 absorption cell. This 1 mW probe beam of 3 mm diameter causes little saturation.
After passing an additional beam splitter, this information-bearing probe
is detected by a photoreceiver broadly tuned to the rf frequency ($\omega/2\pi \simeq$
15 MHz) driving the phase modulator. The resulting rf signal at ω is coherently detected in a doubly-balanced mixer whose phase reference comes from
the rf source at ω via a step-selected delay cable. The strong unmodulated
laser beam from the first beam splitter passes through an acousto-optic
modulator whose Bragg-deflected output beam is introduced via the second
beam splitter to become the counter-running saturating beam ($\sim$20 mW).[17]
The probe and saturating beams are coaxial, collimated, and precisely antiparallel. As noted by SNYDER et al.,[6] the frequency shift ($\Delta/2\pi = 80$ MHz)
associated with the acousto-optic Bragg-scattering prevents interferometric
noise problems due to optical scattering of the saturating beam back into
the probe channel direction: heterodyne terms with this radiation do not
lie at the information-bearing frequency ω. The sub-Doppler resonances are
efficiently isolated from any drifts or background by using phase sensitive
detection of the rf mixer output. See Fig. 2.

This apparatus responds to differences in the medium's refractive index
across the spectral interval 2ω. Since our modulation frequency, ω, is
much smaller than the Doppler line width, the probe beam alone picks up
only a very weak signal due to the overall Doppler profile. However, when
we supply a counter-running, saturating beam, we can affect the absorption/

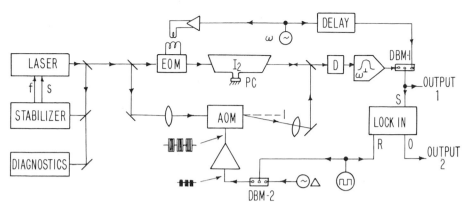

Fig. 2. Optical Heterodyne Saturation Spectrometer. The frequency-
stabilized single frequency laser output is divided into probe and saturating beams by a beam splitter. The probe beam is phase modulated by an
Electro-Optic Modulator and passed through the Iodine cell whose pressure
is controlled by the Peltier Cooler. The saturating beam, frequency offset
and chopped by the Acousto-Optic Modulator, is collimated and aligned coaxial and antiparallel to the probe beam in the I_2 cell. The signal-
bearing probe beam is detected by a fast photodiode (D) whose output is
filtered for the rf component at frequency ω and applied to the signal port
of an rf Doubly-Balanced Mixer. The rf reference signal is phase-shifted
by an adjustable delay line. The dc output of DBM-1 may be further pro-
cessed by a lockin amplifier to recover the signal (output 2) synchronous
with the saturation chopping.

dispersion on a velocity-resolved basis. For example, suppose the saturating beam modifies the population of the velocity group detuned by one
homogeneous line width on either side of the group resonant with the carrier frequency in the probe beam: these tuning conditions will produce the
maximum dispersive effect. (Tuning the saturating beam to resonate with
the same molecules as the probe carrier leads to zero dispersive effect.)
These small changes in the medium's index of refraction integrate over the
cell length to a resonant phase shift for the carrier which unbalances the
equality of the rf detector currents at the modulation frequency ω. When
the saturating beam influences the molecules nearly resonant with either
sideband field, an analogous dispersion-shaped feature will result, so
that two auxiliary resonances are obtained with $2(\omega/2)$ separation.[18] Under
suitable experimental conditions the entire signal will be purely odd symmetric around the central dispersion resonance.

To illustrate these ideas experimentally we have used the $^{127}I_2$ line at
589.2141 nm. Figure 3c shows a spectrum taken with the rf phase shift set
to recover the saturated dispersion signal. The signal-to-random-noise
ratio for the display is ~200 in a 10 kHz bandwidth. This signal-to-noise
ratio should permit laser stabilization to $<10^{-4}$ of a typical 1 MHz I_2 line
width in a 1 s averaging time.

Another type of signal may be obtained when the radio frequency phase
reference has been adjusted 90° from this present quadrature location. In

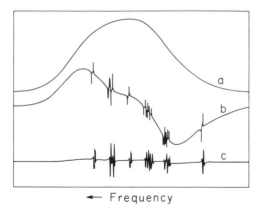

← Frequency

Fig. 3. Broad-scan spectral profiles. (a) Fluorescence from exciting the I_2 589.214 nm line by the probe beam alone. Overall sweep width about 2 GHz. (b) Output of rf doubly-balanced mixer with reference phase set to recover saturated absorption signals. (c) Same as (b) except rf phase set for dispersion. Note the absence of Doppler background even without saturation chopping.

the in-phase position we will be sensitive mainly to changes in the amplitude of either of the two sideband signals. This detection mode has recently been used by BJORKLUND and LEVENSON[19] for sub-Doppler spectroscopic experiments, with such a high FM frequency ω that only one sideband appreciably interacted with the Doppler distribution. In contrast, our probe frequency separations are much smaller than the Doppler width: in the absence of the saturating beam our linear absorption signal at ω approximates the derivative of the Doppler profile. With the saturating beam present, there are two resonant tuning conditions in which preferential bleaching occurs for one or the other of the two phase-modulation sidebands. Considering that the beat terms of these sidebands with the carrier have opposite signs, we again find a saturation resonance profile with odd symmetry around the center frequency. Figure 3b shows the in-phase saturation resonance signal at the rf balanced mixer output in which the large frequency scan clearly shows the Doppler (derivative absorption) background. For comparison, note that the rf dispersion signal (Fig. 3c) even without saturation chopping is almost perfectly free of additive Doppler background problems. The corresponding fluorescence signal is displayed in Fig. 3a. The optical frequency scale can be estimated from the observed spacing between the central and lower frequency singlets: (580 ± 10) MHz.

We have made a simple rate equation theory of this optical heterodyne saturation spectroscopy, extending the results of BJORKLUND[2] to include the effects of finite absorption and the nonlinear absorption/dispersion response of the resonant medium. We calculate the path-integrated phase and amplitude changes of the several spectral components of the probe beam due to the (nonlinear) medium's (complex) index of refraction. We follow the cross-terms of interest to recover rf currents at ω, phase-shifted and attenuated by the medium. The medium is modeled by an infinitely wide distribution of narrow, saturable velocity packets. Molecular transit time effects are lumped into the effective relaxation rate. After the Doppler convolution it is found that the sharp nonlinear resonances have pure odd symmetry[20] for all values of the rf phase shift, modulation index, and absorption coefficient. This absence of underlying opposite symmetry terms is fundamentally important in using these signals for laser frequency locking. See below. To facilitate presentation of our results, we introduce the notation

$$L^{\pm} \equiv \frac{\Gamma^2}{\Gamma^2 + (\omega^{\pm})^2} \quad , \quad D^{\pm} = \frac{\Gamma \cdot \omega^{\pm}}{\Gamma^2 + (\omega^{\pm})^2} \quad ,$$

with $\omega^{\pm} \equiv \Omega-\Omega^*\pm(\omega/2)$, $\omega^* \equiv \Omega-\Omega^*$, $\omega^{++} \equiv \Omega-\Omega^*+2(\omega/2)$ and $\omega^{--} \equiv \Omega-\Omega^*-2(\omega/2)$.[18] The natural center frequency, Ω_0, is shifted to $\Omega^* = \Omega_0 \pm \Delta/2$, depending on which acousto-optic sideband was chosen. Γ is the homogeneous width of a velocity packet, power-broadened by the average of the sideband and saturating intensities. The path-integrated absorption measure is $A = \alpha_0 L$, where α_0 is the saturated absorption coefficient and L the cell length. For the in-phase case, in the limit of modest modulation index $\beta<1$ and small absorption $A \ll 1$, we have $Sig_{sat,0°} = B \cdot A \cdot J_1[(J_0+J_2)(L^--L^+)-J_2(L^{++}-L^{--})]\cos \omega t$, which is clearly interpretable as the bleaching by the saturating beam of the opacity resonant with the four sidebands. The term at Ω^* vanishes because of equal but opposite contributions from each pair of sidebands.

The rf quadrature term in the same low absorption, modest modulation index limit is $Sig_{sat,90°} = -CAJ_1[(J_0-J_2)(D^+-2D^*+D^-)+J_2(D^{++}-2D^*+D^{--})]\sin \omega t$. At low saturation, the model gives $B \approx C$, which is well confirmed experimentally.[21] The most conspicuous effect of higher order terms in the absorption A is a small step near zero detuning in the in-phase signal.

To test the full theoretical function from which the above limiting forms were derived, we obtained high precision profiles of an isolated hfs component near the Doppler center of the I_2 line at 589.2141 nm. See Figs. 3 and 4. The dye laser was locked to a highly stable high finesse cavity which was itself locked to a HeNe local oscillator which was in turn scanned relative to a $^{129}I_2$ stabilized HeNe laser via a computer-driven frequency synthesizer. Laser line width was <10 kHz. The data are displayed as points in Fig. 4a (in-phase case) and Fig. 4c (dispersion phase). The full theory, separately least-squares fitted to the two sets of data, is plotted as solid lines in Fig. 4. Residuals of the fits, magnified five-fold, are shown in Figs. 4b and 4d. The agreement with the experimental data is seen to be remarkable! For example, the fitted displacement of the absorption phase components is (7.56±0.01) MHz versus 15.114 MHz/2 = 7.557 MHz expected. The dispersion phase gives (7.51±0.05) MHz displacement. The absorption phase gives a line width (HWHM) = (0.77±0.01) MHz; the dispersion phase also gives (0.77±0.01) MHz. Consideration of the spatial averaging of saturation further improves the fits. The I_2 pressure was ≈0.1 mbar.

The use of the dispersion-phase signal for laser locking is extremely attractive as no slight error (e.g.,in setting the rf phase-shift) can spoil the symmetry of the resonance (see Fig. 4). For example, our several tests on the dispersion-phase residuals, Fig. 4d, show no spurious term of even symmetry to within the noise level (<1/500 of the pp dispersion signal). Thus we can immediately see the prospect of laser stabilization to basically any of the iodine hfs lines with <u>kiloHertz accuracy</u>.

Another interesting aspect of these resonances concerns their excellent signal-to-noise ratio. If the laser is shot-noise-limited at the chosen modulation frequency, theory offers promise of an ultimate sensitivity limited only by quantum noise in the signal.[22]

A third interesting aspect of these resonances concerns their behavior under very rapid changes in the laser phase if the probe beam is intense enough to cause saturation. The carrier frequency component of the probe

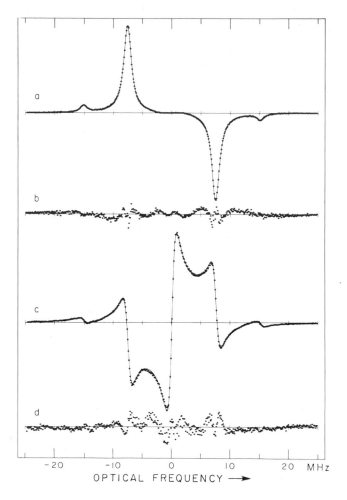

Fig. 4. Modulated saturation optical heterodyne spectrum of central hfs component of Fig. 2, with best-fitting theoretical representations. (a) Experimental absorption-phase profiles (o) and best-fitting theoretical shape (solid line). Frequency scale 100 kHz per channel. (b) Residuals of absorption-phase fit (5× expanded vertical scale). (c) Dispersion-phase profiles and best-fitting theoretical shape. (d) Residuals (×5) of dispersion fit. Note that the residuals have essentially pure odd symmetry.

OPTICAL FREQUENCY ⟶

beam will find its own velocity group with which a strong resonance is established, leading to a nonlinear macroscopic polarization at this frequency. This polarization cannot follow the rapid laser phase change and continues to radiate at the original optical phase. However, the sidebands, produced by modulation of the laser, carry its rapid phase shift. Heterodyne terms between the field radiated by the macroscopic polarization at the carrier frequency (with the original phase) and the now phase-shifted sidebands lead to a transient signal in the optical detector. Following a step laser phase jump of θ at $t = t'$, simple analysis leads to a transient signal $\propto E_0^2 J_0 J_1(\alpha L/2)(S/1+S) \, e^{-2\gamma(t-t')} \sin \omega t \sin \theta$ prior to rf demodulation. S is the saturation parameter. The sign of the transient signal is determined by the sign of the laser phase shift. Thus we have a "phase memory" provided by the molecules! This physical effect is closely related to the Stark switching[23] and frequency switching[24] coherent optical transients. These signals due to laser phase slewing are dramatically visible as a ~30 fold "degradation" of the quadrature phase signal-to-noise ratio relative to the in-phase case, if the laser stabilization is deactivated and if the probe beam is strong enough to cause appreciable saturation.[25]

The transient signals from the balanced mixer can be used for an extremely high speed phase-control loop of the dye laser anywhere within the Doppler profile. Experiments with both dye and color center lasers confirm this expectation. Interesting new effects at longer times are observed.

As a final remark we note that the necessary pure FM sidebands can be easily produced by a weakly-driven phase modulator crystal located inside the laser resonator. We used $LiNbO_3$ inside our color center laser both as the FM modulator and also as the fast phase/frequency feedback element in a frequency control loop. It also seems feasible to detect the residual AM on the laser output for use in centering the intracavity etalon. Figure 5 shows the resolved hfs of the P(2) line of HF near 2.6 μm. The full line width at half height is about 50 kHz, still importantly limited by transit broadening ($w_0 = 2.5$ mm) and laser line width ($\delta\nu \sim 25$ kHz).

In this paper we have described a new technique, optical heterodyne saturation spectroscopy, which can produce narrow peaks with high signal-to-noise ratios, limited only by the fundamental fluctuation in the signal itself. The method provides a resonance of essentially ideal symmetry, uncontaminated with offsets or terms of the opposite symmetry. It thus should allow unprecedented <u>accuracy</u> in locking to a quantum absorber's resonance. The method also provides a fast dispersion-phase signal which may be directly used for reducing the spectral width and frequency slewing rate of the laser source.

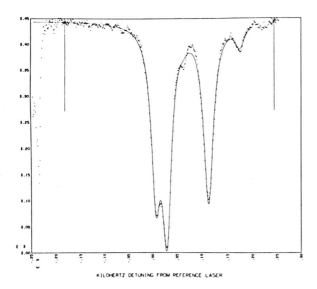

KILOHERTZ DETUNING FROM REFERENCE LASER

<u>Fig. 5.</u> Optical heterodyne saturated absorption spectrum of P(2) line of HF. Taken with color center laser at ~2.6 μm. Separation of two main hfs peaks ~200 kHz, line width ~50 kHz. Noise at left is artifact. Bold lines indicate fit limits.

We thank G.C. Bjorklund and M.D. Levenson for a preprint of their recent and closely-related work. One of us (JLH) is happy to acknowledge numerous stimulating conversations with R.W.P. Drever. We also thank J.H. Shirley for several useful discussions. H.G.R. is a JILA Visiting Fellow, 1980-81, on leave from Duke University. The work has been supported in part by the National Bureau of Standards under its program of precision measurement research for possible application to basic standards, and in part by the Office of Naval Research and the National Science Foundation through grants to the University of Colorado.

REFERENCES

1. J.L. Hall, in Atomic Physics VII, edited by D. Kleppner and F. Pipkin (Plenum, New York, 1981), pp. 267-296.
2. G.C. Bjorklund, Opt. Lett. 5, 15 (1980).
3. J.L. Hall, L. Hollberg, T. Baer and H.G. Robinson, Appl. Phys. Lett. to appear Nov. 1981.
4. C. Weiman and T.W. Hänsch, Phys. Rev. Lett. 36, 1170-1173 (1977).
5. F.V. Kowalski, W.T. Hill and A.L. Schawlow, Opt. Lett. 2, 122 (1978).
6. J.J. Snyder, R.K. Raj, D. Bloch and M. Ducloy, Opt. Lett. 5, 163 (1980).
7. R.K. Raj, D. Bloch, J.J. Snyder, G. Camy and M. Ducloy, Phys. Rev. Lett. 44, 1251 (1980).
8. R.W.P. Drever, private communication, August 1979.
9. B. Smaller, Phys. Rev. 83, 812 (1951).
10. J.V. Acrivos, J. Chem. Phys. 36, 1097 (1962).
11. R.V. Pound, Rev. Sci. Instr. 17, 490 (1946).
12. W.J. Trela, Thesis, Stanford Univ., pp. 36-41, unpublished, 1967 (under W.M. Fairbank).
13. S.R. Stein, Thesis, Stanford Univ., pp. 19-21, unpublished, 1974 (under J.P. Turneaure).
14. R. Weiss, unpublished.
15. R.V. Pound, private communication.
16. R.W.P. Drever, J.L. Hall, F.V. Kowalski, J. Hough, G.M. Ford and A.J. Munley, September 1979. Manuscript in preparation.
17. Polarizing beam dividers/combiners could be used with appropriate wave plates or Faraday devices to produce a more energy-efficient super-position of probe and saturating beams.
18. Note that only the probe beam has the sideband frequency offset but that both probe and saturation beam optical frequencies change with laser tuning: Thus the first-order sideband resonances are offset from Ω^* by $\pm\omega/2$. Line widths however are still given by $(\Omega-\Omega^*)/\Gamma = 1$.
19. G.C. Bjorklund and M.D. Levenson, Phys. Rev. A (in press).
20. Small symmetry departures can arise from multiplicative effects of the Doppler background.
21. We find $C/B = \left[1+(I_p/I_0)\right]^{1/2}/[1+(J_2/J_0)]$ where I_p is the probe beam intensity and I_0 is the saturation intensity.
22. M.D. Levenson and G.L. Eesley, Appl. Phys. 19, 1-17 (1979).
23. R.L. Shoemaker and R.G. Brewer, Phys. Rev. Lett. 27, 631 (1971).
24. J.L. Hall, in Atomic Physics III, edited by S.J .Smith and G.K. Walters (Plenum, New York, 1973), pp. 615-646.
25. The rf phase can be precisely set for the in-phase condition by tuning for a minimum of this noise from the balanced mixer. A calibrated phase change of 90° then produces accurate tuning for the dispersion-phase signal.

High Precision Laser Interferometry for Detection of Gravitational Radiation

K. Maischberger, A. Rüdiger, R. Schilling, L. Schnupp, W. Winkler, and
H. Billing

Max-Planck-Institut für Astrophysik, Karl-Schwarzschildstraße 1
D-8046 Garching, Fed. Rep. of Germany

1. Introduction

The "First Generation" gravitational wave experiments, carried out in the 1970s, were not able to confirm definitely the existence of gravitational radiation. The strain sensitivity of these detectors was limited to $\delta \ell / \ell \simeq 3 \cdot 10^{-17}$. Estimates by astrophysicists suggest that a sensitivity about four orders of magnitude higher would be required for a realistic chance to detect gravitational waves [1].

To achieve this goal, the "Second Generation" experiments follow mainly three lines: Weber-type resonant bars (high Q materials, low temperature), Doppler tracking by space craft, and laser interferometers, which will be the topic of this paper.

Two properties of gravitational radiation are favourable for a laser interferometer:

(a) Gravitational radiation causes free test masses to exhibit displacements $\delta \ell$ which are proportional to their distance ℓ. In an interferometer the distance ℓ between the mirrors, acting as test masses, can be made very large.

(b) Expected sources of gravitational radiation to be detectable on earth are catastrophic events in the universe (e.g. collapsing stars). The events are believed to have a typical duration of a few milliseconds. Thus, we require our interferometer to have low noise in the frequency range from, say, 500 Hz to a few kHz, only. This demand differs favourably from many other precision laser applications in which long term stability is of importance.

Interferometric techniques for the detection of gravitational radiation - all based on a Michelson configuration - are being investigated in several laboratories. In 1971 the first prototype detector was put in operation at Hughes Laboratories [2]. Pioneering work was also done at MIT [3]. Similar activities were started by groups in Munich [4, 5] and Glasgow [6], and recently at Caltech.

2. The Michelson Interferometer

2.1 Principle and Fundamental Limits

The principle of the Michelson interferometer is shown in Fig.1a: the light beam is split up into two beams which after reflection by the mirrors M_1 and

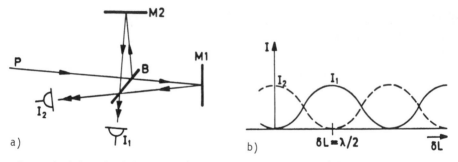

Fig.1a,b (a) Principle of Michelson interferometer, (b) Response of photo-currents I_1 and I_2 to pathlength difference δL

M2 are recombined at the beam-splitter B. The response of the interferometer to changes in pathlength difference δL for the interfering beams is indicated in Fig.1b.

One common way of using this set-up as a displacement sensor is to monitor the two outgoing beams on two photo diodes, choosing as operating point a crossing of the two response curves. The difference signal of the photocurrents, $\delta I = I_1 - I_2$, represents the interferometer output. An alternate nulling method is to choose the operating point at a minimum of one of the response curves, which can be achieved by the well known modulation technique.

The measurement is limited by two quantum mechanical noise sources. The photon counting noise (shot noise of the photo current) by far dominates the noise due to fluctuations of the radiation pressure on the mirrors. Thus, the photon counting noise can be considered our fundamental limit, leading to an equivalent pathlength difference δL expressed by the spectral density

$$\widetilde{\delta L}^2_{sh} = \frac{hc}{\pi} \cdot \frac{\lambda}{\eta P} \quad [m^2 / Hz] . \tag{1}$$

At present the Ar^+ laser ($\lambda = 0.514\ \mu m$) is the most suitable light source. For an effective light power ηP of 1 W for example (η representing the overall efficiency of the optical system), the shot noise equivalent value of δL is

$$\widetilde{\delta L}_{sh}(1W) \simeq 7 \cdot 10^{-17}\ m/\sqrt{Hz}.$$

2.2 The Interferometer as a Strain-Meter

According to (1) the minimum detectable path length difference is independent of the interferometer pathlength L. To convert the displacement sensor into a very sensitive strain meter, L has to be chosen as large as possible. The optimum pathlength is given by half the wavelength of the gravitational wave. For the expected events this optimum is at approximately 100 km, a distance which is far too big for terrestrial experiments.

But one can realize such an optical path also by the use of multiple reflection schemes. Two different methods are currently being investigated:

(a) an optical delay-line [7] with a discrete number of reflections, and

(b) an optical cavity of high finesse. The second method will be described

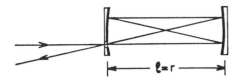

Fig.2 Optical delay line, shown
for N = 4 beams

elsewhere during this conference. We have adopted the delay-line technique,
which shall be explained now: two mirrors with radius of curvature r are
placed at a distance $\ell \simeq r$, as shown in Fig.2.

The beam, entering the left-hand mirror through an entrance hole will pass
this hole after four bounces if the mirror distance ℓ is equal to the radius
of curvature r (confocal case). The number of bounces can be increased by
choosing the distance $\ell \neq r$. For a desired number N, the distance ℓ has to be
adjusted according to the relation $\ell = r \cdot [1 + \sin(\pi/N)]$. For these distances
the so-called re-entrance condition is fulfilled, and the outgoing beam be-
haves as if reflected at the virtual entrance surface. This feature is of great
importance in our application, since it makes the interferometer insensitive
to possible lateral or angular motions of the far mirrors (neglecting second
order effects).

With the high reflection coatings available today ($R \geq 99.8\%$) the optical
losses can be kept small, and N > 300 bounces can easily be achieved. Then a
mirror distance ℓ of only 300 m is necessary to have the optimum path length
$L = N \cdot \ell$ of $\simeq$ 100 km. Using (1), with an effective laser power of 1 W and a
detection bandwidth of 1 kHz, the obtainable strain sensitivity is
$(\delta L/L)_{sh}$(1W, 1kHz) $\simeq 2 \cdot 10^{-20}$. To obtain a higher sensitivity, lasers with
higher power (or shorter wavelength) would be necessary.

It should be mentioned that all multiple reflection schemes have an increased response to mechanical disturbances, hence more effort has to be put
in the design. The noise due to radiation pressure fluctuations increases
by the square of the number of reflections, but it can still be neglected
for the laser powers currently available.

3. Noise Sources

In practice, we have to cope with many other noise contributions, which can
be investigated with our prototype interferometer of 3 m armlength, operated
currently at N = 138 reflections [5, 8]. Here, we will discuss only the most
prominent noise sources.

3.1 Mechanical Noise

One obvious noise contribution is due to relative mirror motions caused by
ground noise and acoustics. A sufficient isolation against acoustics is
achieved by evacuating the interferometer housing. The ground noise, drop-
ping at least with a $1/f^2$ law, has small amplitudes at the frequency window
of interest. With additional mechanical filtering - by suspending the mirrors
on thin wires for example - the ground noise can be kept below the shot noise
level. Of course, isolation via pendulums requires a rather complex servo
control system to avoid enhancement of the mirror motions at the various pendulum resonances (translations, torsion and tilts). This problem has been

solved by auxiliary lasers and position sensing photodiodes, monitoring the excursions of the mirrors relative to their local surroundings. The photodiode signals, appropriately amplified and filtered, are converted into magnetic forces which attenuate the pendulum motions of all mirrors. A more detailed description is given in [5] .

Another source of noise is given by motions of the optical components due to thermal excitation of their eigenmodes. The amplitude of Brownian motion of a mechanical oscillator at resonance is given by $\delta x^2(Br) = kT/m\omega^2$. For a mass m = 1 kg and a resonance frequency f = 1 kHz, for example, the mechanical oscillation amplitude is $\delta x \simeq 10^{-14}$m, a value far bigger than our shot noise goal. As cooling to low temperature is not considered, the only way out is to avoid mechanical resonances in the intended frequency window by a careful design of the mirror mounts. Though thermal noise still troubles us, we think that this problem can be solved.

At present, it seems that noise originating from the laser is the most severe noise source.

3.2 Laser Noise

Laser noise, generally, can only be converted into phase signals in a Michelson interferometer if the symmetry of the two arms is imperfect. Small deviations from symmetry lead to various noise contributions, in particular by three types of laser instabilities: power fluctuations, jitter of beam geometry, and – most severe – jitter of the laser frequency.

Power fluctuations can be sufficiently suppressed using a nulling method, i.e. choosing a suitable position of the operating point of the interferometer. If the differential photo current method is used, the operating point has to be chosen such that the two currents are made equal. Then, fluctuations of the laser power are cancelled. Deviations from this optimum position – as due to mechanical motions of the mirrors – can be kept small by installing an optical feedback loop by placing a Pockels cell in the light path. Excursions of the mirrors are then converted into corresponding changes in phase by the Pockels cell. Though the power fluctuations of an argon laser are relatively large, they can be sufficiently suppressed if the open loop gain of the circuit is made high enough.

The laser also shows fluctuations of the beam geometry. In particular, the lateral beam jitter (i.e. variations in position and direction) causes a signal if the interfering wave fronts are tilted with respect to each other (by a misalignment of the beam splitter, for example). A pulsation in beam width becomes disturbing if the wave fronts are not matched in curvature (due to poor quality of the mirror surfaces, e.g.).

We have found that the beam jitter of our Ar^+ laser is responsible for an excess interferometer signal, even after careful alignment of the beam splitter. Hence, a device for suppression of the fluctuations was necessary. The way we have chosen is to pass the laser beam through an optical cavity [9]. The geometric fluctuations of the beam, as viewed from the cavity, can be described as a superposition of a ground mode h_0 and higher order modes h_n with small time varying amplitudes. The most prominent modes h_1 and h_2, representing the lateral beam jitter and beam pulsation, respectively, are shown in Fig.3.

By proper choice of the cavity parameters, the resonator, tuned for maximum transmission of the ground mode, will not allow the higher order modes to pass. The suppression of these modes is a function of geometry and finesse

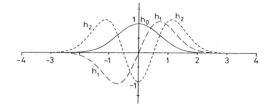

*Fig.3 Normalised cavity modes:
Gaussian ground mode h_0;
first and second order
modes h_1 and h_2*

of the cavity. In our present experiment, a suppression by one order of magnitude was already sufficient.

The <u>fluctuations of the laser frequency</u> cause phase error signals in a Michelson interferometer in two distinct ways: (a) by a static path length difference in the two interferometer arms, and (b) by a parasitic effect, in which scattered light interferes with the main beam.

If the path lengths L_1 and L_2 differ by $\Delta L = L_1 - L_2$, a frequency jitter $\delta\nu/\nu$ is converted into phase signals

$$\delta\phi = k\,\Delta L\,\delta\nu/\nu , \qquad \text{with } k = 2\pi/\lambda . \tag{2}$$

The inevitable tolerances in the radii of curvature of the mirrors lead to finite static pathlength differences ΔL, but they can be made small enough not to cause severe problems.

We are worried much more by the effect arising from scattered light. This shall be demonstrated in a simple example, as shown in Fig.4a. A fraction σ^2 of the main beam is scattered at an optical surface S and interferes with the outgoing beam, according to its phase delay $\Phi = k\Delta L$. Here, ΔL is the pathlength difference between main beam and scattered light. The phase error signal $\delta\phi$ due to frequency fluctuations $\delta\nu/\nu$ is

$$\delta\phi \simeq \sigma\,k\Delta L\,\cos(k\Delta L)\,\delta\nu/\nu \tag{3}$$

which can be easily derived using the relations given in Fig.4b. Though the fraction of scattered light may be small ($\sigma^2 \simeq 10^{-8}$, typically) it can become disturbing if the pathlength difference is large. In the example of Fig.4a ΔL is equal to L, the total pathlength of the delay-line. In practice, scattering also originates on the surfaces of the delay-line mirrors, leading to pathlength differences up to high multiples of the total round trip.

One straight-forward method to cope with this problem is an active frequency control of the laser. We have also investigated another method [10] which can be applied in addition to frequency stabilization: the phase error signal $\delta\phi$ in (3) would disappear for the special case $\Phi = \pi/2$ mod π. In practice this condition cannot be maintained, but it would be sufficient to make

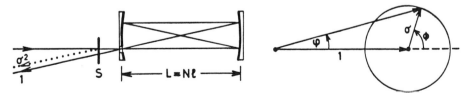

*Fig. 4a,b (a) Paths of main beam and of scattered light, (b) Scattered
light, with relative field amplitude σ, phase delay ϕ; resulting phase
error ϕ*

the time average of this signal equal to zero. If one could force the phase delay Φ to change in such a way that the vector σ points in its original direction for one half of a cycle, and in the opposite direction for the other half, it seems plausible that the time average would vanish. We have successfully tested a similar method in which we phase modulate the laser beam with special choices of phase swing and modulating frequency. The path-length difference ΔL in (3) is, in general, a function of time. In our case, the pendulum suspension can lead to large changes in phase delay Φ at low frequencies. If the swing of Φ is significantly above 2π, higher frequencies are generated in the phase error signal $\delta\phi$. In order to eliminate this effect, an armlength control with an auxiliary laser has been installed.

4. Experimental Set-up and Results

The principal components and control circuits of our present set-up are shown in Fig.5. With a mirror distance $\ell \simeq 3$ m and N = 138 reflections at present, the total pathlength L is about 400 m. The beam splitter and the near mirrors are mounted in one block, the far mirrors have separate mounts. For isolation against ground noise, all mirror mounts are suspended on thin steel wires of 0.7m in length. The servo loops, for the control of the various degrees of freedom of the mirror mounts are omitted in Fig.5 since attention should be focussed mainly on the control circuits for reduction of the laser noise.

The circuit (1) controls the operating point of the interferometer. The diode signal, representing the differential excursions of the mirrors, is fed back to keep the optical pathlength constant. This feedback is split up into two branches. The high pass filtered signal is applied to a Pockels cell to compensate the fast fluctuations in pathlength difference. This signal will later be analysed for possible gravitational wave events. At low frequencies a high gain mechanical control keeps the armlength difference stable within 10^{-9}m. Hence, even with 138 reflections the interferometer is kept well within one fringe.

The circuit (2) shows the mode selector with the servo loop to lock the cavity to maximum transmission. The cavity is composed of two spherical mirrors (R = 95%) with a radius of curvature of 10 cm, placed at a distance of 16 cm. The suppression of the first order mode was measured by introducing

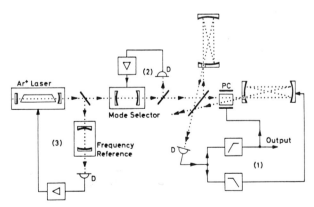

Fig.5 Interferometer with servo loops for the suppression of laser noise

an artificial lateral beam jitter. The result, a suppression by a factor of ≃ 30, was in good agreement with the theory.

The circuit (3), the servo loop for the frequency stabilisation, is built up in a conventional way, as shown in Fig.6. The reference to which the laser frequency is locked is a commercial optical cavity (TROPEL). It is essential to mention here that long term stability of the laser frequency is not required, and that only in the frequency range between 500 Hz and 5 kHz a strong suppression of the frequency fluctuations is necessary. Therefore, we can tolerate the inevitable thermal drift of a passive cavity, but it proved necessary to evacuate the cavity to eliminate the pressure fluctuations in the kHz-range.

We prefer to express frequency noise by its spectral density rather than by the Allan variance or the laser line width, both of which are integral measures of the fluctuations that include contributions even from the lowest frequencies. Typical frequency noise spectra of our Ar^+ laser are shown in Fig.7 with and without the active frequency control. Apart from small excess signals, mainly due to mechanical deficiencies of the reference cavity, the obtained frequency stability $\delta\nu \simeq 1$ Hz/$\sqrt{Hz}$ was limited by the shot noise of the photo current rather than by the open loop gain of the circuit. We hope that a further reduction of the frequency noise by approximately one order of magnitude can be achieved by improving the design of the reference cavity.

Finally, after having discussed the various noise sources and methods of their suppression, let us take a look at the interferometer sensitivity achieved. In Fig.8 a typical interferometer noise spectrum is shown. The signal is practically down to the shot noise limit at frequencies above 5 kHz. In the frequency window of interest between 500 Hz and 5 kHz, signals became apparent, originating from thermal noise. We have been successful in shifting most of the mechanical resonances to higher frequencies, but obviously an even more sophisticated mechanical design will be required to further "clean" this frequency window from mechanical oscillations. At frequencies below 1 kHz there is a steep increase of the signal. We have not yet analysed this frequency range thoroughly, but we think of two possible sources: a remainder of the ground noise and/or of the geometric beam fluctuations of the laser.

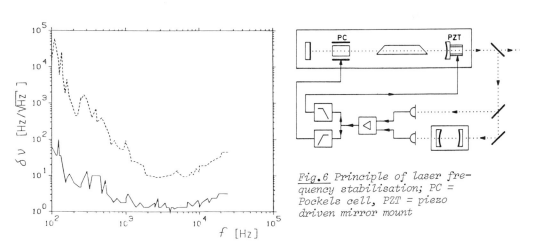

Fig.6 *Principle of laser frequency stabilisation; PC = Pockels cell, PZT = piezo driven mirror mount*

Fig.7 *Spectral density of laser frequency jitter δν (in linear measure Hz/√Hz), vs. spectral frequency f (in Hz). Dotted line: original jitter Solid line: with frequency control*

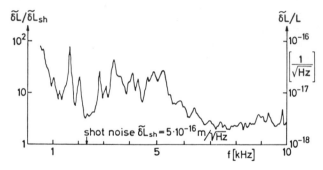

Fig.8 Spectral density of noise in our interferometer, with L = 400m, effective light power ηP = 50 mW. Vertical axis: Pathlength difference δL, normalised to the shot noise level δL_sh (left hand); strain sensitivity δL/L (right hand axis)

5. Conclusion

The main purpose of our prototype interferometer is to study the various noise phenomena. Even though we have practically reached the fundamental shot noise limit, our present pathlengths of only 138×3.0 m $\simeq 400$m and our low laser power of only 100 mW do not allow us to reach a strain sensitivity better than $3 \cdot 10^{-17}$ in a bandwidth of 1 kHz. This value is several, say four, orders of magnitude away from the final goal. Our next step towards this goal is to extend the geometric armlength of the interferometer to 30 m. We are aware of the fact that with increased pathlength the scattered light effect will become more severe. One possible solution of this problem is to improve the present conventional frequency control.

But we are also thinking of an additional control circuit that would operate independently of the present scheme. Here, one would use the interferometer as the frequency reference: Fractions of the outgoing beam and of the undelayed incoming beam are brought to interference. This interference is very sensitive to frequency fluctuations, and an appropriate compensating signal is fed back to a Pockels cell placed at the input of the interferometer. This Pockels cell locks the phase (and thus the frequency) of the incoming beam to the interferometer reference, at least for fast fluctuations.

We have tested this method recently and we were able to show that it works in principle. We hope that after having solved some technical problems an essential step in suppression of frequency noise can be achieved. Thus, we are looking forward with more optimism to a further increase of the interferometer sensitivity.

References

1. K.S.Thorne, Rev. Mod. Phys. 52, 285 (1980).
2. R.L.Forward, Phys. Rev. D 17, 379 (1978).
3. R.Weiss, Quarterly Progress Report, M.I.T. 105, 54 (1972).
4. W.Winkler, Proc. of the Intern. Sympos. on Experimental Gravitation, Pavia, Sept. 1976; Accad. Nazionale dei Lincei, Roma, 1977.
5. H.Billing, K.Maischberger, A.Rüdiger, R.Schilling, L.Schnupp, W.Winkler, J. Phys. E: Sci. Instr.12, 1043 (1979).
6. R.W.P.Drever, GR9, Jena (1980).
7. D.Herriot, H.Schulte, Appl. Opt. 4, 883 (1965).
8. K.Maischberger et al; Trieste, 1979: preprint MPI-PAE/Astro 209 (1979).
9. A.Rüdiger et al, Optica Acta 28, 641 (1981).
10. R.Schilling et al, J.Phys. E: Sci. Instr. 14, 65 (1981).

Optical Cavity Laser Interferometers for Gravitational Wave Detection

R.W.P. Drever[1][2], J. Hough[2], A.J. Munley[2], S.A. Lee[1], R. Spero[1], S.E. Whitcomb[1]
H. Ward[2], G.M. Ford[2], M. Hereld[1], N.A. Robertson[2], I. Kerr[2], J.R. Pugh[2],
G.P. Newton[2], B. Meers[2], E.D. Brooks III[1], and Y. Gursel[1]

[1]Department of Physics, California Institute of Technology, 130-33
Pasadena, CA 91125, USA
[2]University of Glasgow, Glasgow G12 8QQ, Scotland

1. Introduction

Most of the techniques being developed for detection of gravitational
radiation involve sensing the small strains in space associated with the
gravitational waves by looking for changes in the apparent distance
between two (or more) test masses. In many of the experimental searches
performed so far the detectors consisted of massive aluminium bars, the
metal near the ends of the bars acting as the test masses, and impulsive
strains induced in the bars were searched for. The strain sensitivity of
such experiments has been in the range 10^{-16} to 10^{-18} for pulses of
duration of order 1 millisecond, the limits usually being set by thermal
noise in the bar, and transducer and amplifier sensitivity. Current predictions of gravitational waves to be expected from various types of
astrophysical sources suggest that strain sensitivities some three orders
of magnitude better than these are likely to be required for detection of
gravitational wave bursts from known types of sources at a useful rate,
although indeed signals may be present over a wide frequency range - from
10 kHz to 10^{-4} Hz or lower. (A good summary is given in the proceedings
of a conference on "Sources of Gravitational Radiation" [1]). Work on bar
gravity wave detectors is continuing; but an alternative approach is to
use widely separated and nearly free test masses, and monitor changes in
their separation by optical interferometry techniques. This method shows
considerable promise for both high sensitivity and wide bandwidth and frequency coverage. At the sensitivity levels required absolute length
measurements would be difficult, but a comparison of two baselines perpendicular to one another, which may be affected in opposite senses by a
gravitational wave travelling in a suitable direction, provides a practical alternative. Early experiments of this type were carried out at
Hughes Laboratories [2] using a simple Michelson interferometer to monitor
separations between three test masses suspended in vacuum. The displacement sensitivity of such an arrangement may be improved by causing the
light in each arm of the interferometer to travel back and forth many
times between mirrors attached to the test masses, and a multireflection
system of this type using Herriott delay lines was proposed by R. Weiss [3].
Experimental work on multireflection Michelson interferometers for gravity
wave detection has been carried out at MIT, the Max-Planck Institute at
Munich, and the University of Glasgow.

Early work at Glasgow with a Michelson interferometer used a modified
White cell multireflection system, and the optical components were attached
to test masses from an old "split-bar" gravity wave detector [4]. These
experiments showed us that incoherent scattering of light from the surfaces

of the multireflection mirrors could be a serious noise source, for this could provide differing optical paths through the interferometer which could cause frequency fluctuations in the laser to give phase fluctuations at the final photodetector. This effect was also found independently by the Munich group, who subsequently developed a frequency modulation technique to minimise it [5]. At Glasgow the scattering problem led us to reconsider an earlier plan we had for use of Fabry-Perot interferometers instead of a multireflection Michelson. A Fabry-Perot cavity could be expected to be less affected by scattering than a system with many discrete reflections, for possible optical paths are more tightly constrained; and there is the practical advantage that the diameters of the mirrors can be smaller than for a Michelson interferometer with multiple discrete reflection points. The latter factor could be important with the large interferometers we are considering for the future, when with baselines of order a kilometre or more it may be useful to minimise the cost of the vacuum pipe between the test masses. There is, however, an obvious difficulty with a very long high finesse Fabry-Perot cavity: the very narrow transmission bandwidth requires use of a laser whose frequency is very well controlled. A new method of laser frequency stabilisation was devised to help overcome this problem, and seems to provide a satisfactory solution. It was therefore decided some three years ago to develop a Fabry-Perot cavity sensing system in a 10 metre gravity wave detector being built at Glasgow University, and it was subsequently also decided to investigate the same basic system in a project started more recently at California Institute of Technology.

2. Basic Arrangement of Optical Cavity Interferometer

There are many ways in which Fabry-Perot cavities might be used for detection of gravity waves, and indeed the idea that a gravity wave detector may be made by attaching the mirrors of an optical cavity to free test masses is not new [6]. The arrangement which we devised for our initial experimental work was chosen to be as simple to investigate as possible, and we will describe this first. Later we will mention some slightly more elaborate variants which we expect to give rather better performance [7].

A schematic diagram of the main parts of the system in its simplest form is shown in Fig.1. The circles represent three test masses, which in the current experiments at Glasgow are suspended to form two perpendicular baselines 10 metres long (labelled arm 1 and arm 2 in the Figure). A triangular optical cavity is set up along each arm between two plane mirrors and a concave mirror, as shown. We have chosen to use ring cavities at the present stage to reduce the need for isolators to prevent optical feedback to the laser[1]. The two cavities are illuminated by light from a single argon laser, and information on the relative motions of the masses is obtained by monitoring phase differences between the light from the laser and the light within each of the two cavities when they are in resonance. Feedback systems are used to maintain the resonance conditions in both arms, this being done in the present experiments by controlling the laser wavelength to be in resonance with the cavity in arm 1, and controlling the length of the cavity in arm 2 to be in resonance with the resulting laser wavelength.

---

[1]We are indebted to Alain Brillet for this suggestion.

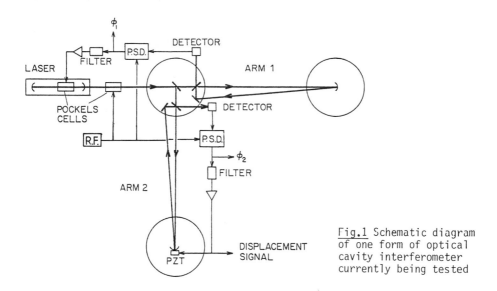

Fig.1 Schematic diagram of one form of optical cavity interferometer currently being tested

It may be noted at this stage that good stabilisation of the laser frequency is required to maintain these resonance conditions sufficiently well. With cavities 10 metres long, and only moderate finesse, control of the laser frequency to much better than 10 kHz is necessary, and the requirements become greater as the cavity length increases. Standard methods of locking the frequency of a laser to a cavity usually depend on imposing some frequency modulation on the laser, and using the resulting variation in the intensity of light transmitted by the cavity to adjust the laser wavelength. Such a feedback system has, however, a response time limited by the bandwidth of the cavity; and this would be quite inadequate for the present application.

The laser stabilisation system devised in this work depends on comparing the phase of stored light from the cavity with that of new light from the laser, a comparison which can be made in a time short compared with the cavity storage time. The response time of the feedback loop then becomes free of any limit from cavity bandwidth. One convenient way of implementing this idea is indicated in the Fig. 1. The light from the laser is phase modulated by an external Pockels cell driven at radio frequency (15 to 100MHz), and the external light from the input mirror of the cavity in arm 1 is detected by a fast photodiode. The light reaching the detector may be regarded as having two components, one being the directly reflected phase modulated light from the laser, and the other being steady stored light emerging from the cavity, with its modulation sidebands

removed. These two components interfere to give a signal whose amplitude modulation is a measure of the mean phase difference between them, so a phase sensitive detector connected to the photodiode gives an output which indicates the deviation of the laser frequency from the cavity resonance. By suitably feeding back this signal to a second phase modulating Pockels cell within the laser cavity itself the laser can be made to lock to the resonance of the cavity in arm 1.

It may be noted that this laser stabilising system operates as a phase lock for time scales short compared with the cavity storage time, as a frequency lock over longer time scales. It may also be pointed out that as the signal measurement is made at a high radio frequency any low frequency amplitude noise from the laser is largely rejected, and good signal to noise ratio can be obtained.

In the gravity wave detection system outlined in Fig. 1, the same phase modulation technique is used a second time to lock the cavity in arm 2 to the laser light. In this case the light from the cavity input mirror passes to a photodiode connected to a phase sensitive detector whose output adjusts the length of the cavity, by a piezo transducer on which one of the cavity mirrors is mounted, so the cavity locks itself into resonance. In an ideal case in which both this feedback loop and the laser control loop have such high loop gains that residual phase errors are negligible, then the ratio of the lengths of the two cavities would be maintained constant. If the test masses were initially at rest, a change in the separations induced by a gravity wave might then be measured by the change in voltage fed back to the piezo transducer in arm 2 which provides a displacement for the mirror, relative to the mass, which compensates for the motions.

If the gain in the laser stabilisation loop is not high enough to make the phase error ϕ_1 there negligible, then a more appropriate phase signal to apply to the feedback filter in arm 2 would be a suitable algebraic combination of ϕ_1 and ϕ_2. If, in addition, the loop gain in arm 2 is insufficient to make ϕ_2 negligible at the gravity wave frequencies of interest then the displacement signal can be further corrected by addition of a suitable function of this combination of ϕ_1 and ϕ_2.

The basic arrangement just described is probably the simplest for practical tests and is the one used in our experiments to date, but other variants have obvious advantages. For example it would be possible to reduce some noise sources by direct optical measurement of the phase difference between light in the cavities of arm 1 and arm 2 by combining output beams from the two cavities at a second beam splitter, and using differential high frequency phase modulation of these beams to remove low frequency amplitude noise. The output from this direct phase difference measurement would then be used to control the difference in length of the two cavities in place of the phase signal ϕ_2 shown in Fig. 1. In this arrangement it would be important that the two cavities have matched phase response to residual changes in laser frequency, and provision for fine trimming of the finesse of the cavities might be necessary to achieve this. There could be advantages also in making the system more symmetrical by stabilising the laser to the mean of the two cavity frequencies instead of one alone.

These variations are relatively minor ones, however, and would not be expected to significantly change the photon noise limited sensitivity of the system, which is essentially similar to the sensitivity of a multi-reflection Michelson interferometer detector with a number of reflections in each arm equal to the effective number of reflections in each cavity. It is assumed here that the reflectivities of the input and output mirrors in the Fabry-Perot cavities are optimally chosen, and absorption and scattering losses in these mirrors are negligible.

3. Practical Aspects of the Interferometer System

At the levels of displacement sensitivity we would like to achieve with this interferometer (better than 10^{-17} m/$\sqrt{Hz}$ near 1 kHz) there are many practical difficulties to be overcome before the photon shot noise limit to sensitivity is reached. Serious problems can arise in various ways from seismic motion, for ground motions can easily have amplitudes of order 10^{-6}m, or more, at frequencies near 1 Hz.

Initially we are concentrating on detection of gravity waves at frequencies of about 1 kHz, and here passive vibration isolation is fairly effective. Our test masses are suspended like pendulums by steel wires, and a 1 Hz pendulum in itself can give in principle good vibration attenuation at 1 kHz; although violin-mode vibrations of the suspension wires can give dangerous transmission bands in practice. A more difficult problem is noise at lower frequencies which may itself be emphasised by the pendulum resonance, for locking onto one fringe of the Fabry-Perot cavities is only possible if relative changes in optical lengths are kept to less than about 10^{-9}m. One way of tackling this problem is to actively damp the swinging of the test masses to the local ground at each end of the interferometer arms, an approach being used at Munich, and supplement this by overall feedback from the interferometer. We have preferred to avoid any displacement sensing relative to ground, and in our experimental work to date we have based all displacement damping and feedback on outputs from the main interferometer. In the system shown in Fig. 1, for example, low frequency components of the error signal ϕ_1 in the first feedback loop are applied to piezo transducers which cause the points from which the suspension wires are supported to move in such a way as to damp the swinging of the masses in arm 1 relative to one another and control the mean length of the cavity in this arm. Similarly, low frequency components of the error signal ϕ_2 are used to damp differential motions of the masses in the second arm and control its length. The complete feedback system is thus rather more complicated than indicated in Fig. 1, but care is taken to make the added feedback forces negligible at the gravity wave frequencies of interest in initial experiments.

Another evident practical problem is the control of orientation of the test masses required to keep the optical cavities well aligned. In our current experiments at Glasgow the orientation of all three test masses is controlled by active feedback systems which use small helium neon lasers and optical levers to sense rotation or tilts of each mass. Each mass is suspended by two or three separate wires, and moving-coil loudspeaker drives are used to move the suspension points to correct the orientation of the masses: the masses behave rather like puppets on strings with respect to orientation, but their horizontal motion in response to any gravity wave force is essentially free.

4. Experimental Investigations

The first part of the cavity interferometer system to be tested was the proposed laser stabilisation scheme, for wideband stabilisation of the argon laser was essential. At Glasgow we had no prior experience of laser stabilisation, and the first experimental tests were made in collaboration with

38

J.L. Hall and F.W. Kowalski at the Joint Institute for Laboratory Astrophysics, University of Colorado. The performance of the stabilisation system was investigated there by locking a helium-neon laser and a dye laser to adjacent modes of a single cavity, and studying the fluctuations in the beats between the two lasers[2]. Results were encouraging, showing that relative frequency fluctuations between these two lasers could be reduced to about a part in 10^{-14} over times of order 0.1 second. Development of the complete interferometer system at Glasgow soon led to satisfactory frequency locking of an argon laser to one of the 10 metre cavities. Some indication of the magnitude of the residual frequency fluctuations of the laser relative to the cavity may be obtained from the spectrum of the phase detector error signal, shown in Fig. 2. It should be stressed that this is not an independent measurement of total frequency fluctuations, but it does indicate that adequate frequency stabilisation has been achieved for the current stage of development of the interferometer.

At present, detailed investigations are being made at Glasgow of some of the noise sources in the interferometer system. Measurements have been made of the contributions to system noise of residual fluctuations in laser frequency, of fluctuations in laser intensity, and of fluctuations in laser beam direction. A typical output noise spectrum, expressed in terms of displacement of the end masses, is shown in Fig. 3. (Light intensity at the photodiodes was 10 mW here). The observed fluctuations are considerably larger than photon shot noise, and are indeed larger than would be expected from the noise sources just mentioned. There is growing evidence that a major component of this observed noise arises from effects of fluctuation in the path of the laser beam within the Pockels cell producing the R.F. phase modulation. This is currently under investigation, along with techniques to control it.

At Caltech, studies of noise sources in the optical sensing system of this type of gravity wave detector are also being made, using a pair of overlapping cavities between common masses, so that differential seismic noise is minimised. A spectrum of displacement noise observed, with a light

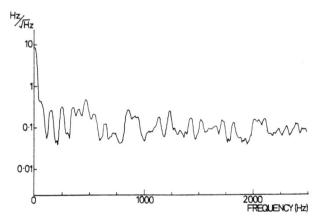

Fig.2 Spectrum of error signal in laser stabilising loop, expressed in terms of equivalent fluctuations in laser frequency

---

[2]Some of these laser stabilisation experiments and subsequent developments are discussed in a separate contribution to these proceedings by J.L. Hall.

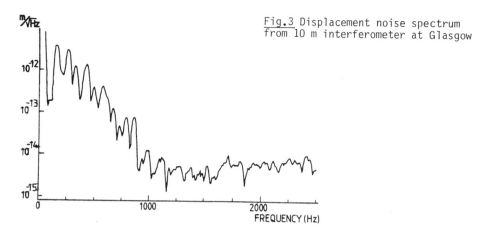

Fig.3 Displacement noise spectrum from 10 m interferometer at Glasgow

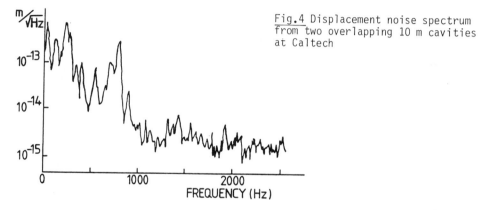

Fig.4 Displacement noise spectrum from two overlapping 10 m cavities at Caltech

intensity of 1 mW at the photodiodes, is shown in Fig. 4. This spectrum is surprisingly similar to that found in the rather different arrangement at Glasgow; and again there is evidence that noise associated with the R.F. Pockels cell is important.

In parallel with these noise studies, construction at Caltech of an optical cavity gravity wave detector is proceeding. This is essentially similar to, but slightly larger than, the one at Glasgow.

5. Conclusion

At the present stage in the development of these new optical instruments it is too early to assess their ultimate limits in sensitivity, and how successful they may be in detection of gravitational radiation. Progress so far is encouraging, however, and no serious difficulties have been encountered. Meanwhile the technical problems involved and the sensitivities required pose interesting challenges, whose solution may well have applications in widely differing fields.

Acknowledgements

We should like to acknowledge support for the work in the UK by the University of Glasgow and the Science Research Council, and in the USA by California Institute of Technology and the National Science Foundation (Grant PHY-7912305).

References

1. L. Smarr, editor, *Sources of Gravitational Radiation* (Cambridge University Press, Cambridge, 1979).
2. R.L. Forward, Phys. Rev. D 17, 379 (1978).
3. R. Weiss, Quarterly Progress Report, MIT 105, 54 (1972).
4. R.W.P. Drever, J. Hough, W.A. Edelstein, J.R. Pugh, W. Martin, Proc. of the Intern. Sympos. on Experimental Gravitation, Pavia 1976, ed. B. Bertotti (Accad. Nazionale dei Lincei, Rome, 1977) p. 365.
5. R. Schilling, L. Schnupp, W. Winkler, H. Billing, K. Maischberger and A. Rudiger, J. Phys. E; Sci. Inst. 14, 65 (1981).
6. V.B. Braginsky and A.B. Manukin, *Measurement of Small Forces in Physical Experiments* (Nauka, Moscow, 1974; University of Chicago Press, 1977).
7. R.W.P. Drever, G.M. Ford, J. Hough, I.M. Kerr, A.J. Munley, J.R. Pugh, N.A. Robertson and H. Ward, GR9, 9th International Conference on General Relativity and Gravitation, Jena 1980.

A Proposed Two Photon Correlation Experiment as a Test of Hidden Variable Theories*

M.O. Scully

Max-Planck-Institut für Quantenoptik
D-8046 Garching bei München, Fed. Rep. of Germany, and

Institute für Modern Optics, Department of Physics and Astronomy
University of New Mexico, Albuquerque, NM 87131, USA

According to hidden variable theories of quantum mechanics the various parameters of interest (e.g. a particle's position <u>and</u> momentum) are determinate but just not known to us. This is the logic of classical statistical mechanics wherein we settle for probabilities of certain configurations rather than the knowable (but unknown) complete specification of the state of the system. BELL's theorem [1] renders the hidden variable idea susceptible to experimental test and several experiments in this direction have been carried out. However, according to CLAUSER and SHIMONY the experimental case for or against hidden variables is not air-tight. To quote from their paper [1] (which is enthusiastically recommended):

> "Following Bell's (1965) results, many readers believed that local realistic [hidden variables] theories were *ipso facto* discredited, because quantum mechanics has been so abundantly confirmed in a variety of experimental situations. ...However, upon careful examination, one finds that situations exhibiting the disagreement discovered by Bell are rather rare, and none had ever been experimentally realised... In view of the consequences of Bell's theorem it is thus important to design experiments to test explicitly the predictions made for local realistic theories via Bell's theorem."

In what follows we discuss a two-atom correlation experiment which has the basic ingredients necessary for testing BELL's inequality.

In recent work [2] we have been interested in the question of photon correlations between γ and ϕ photons emitted from atoms localized at sites 1 and 2 as in Fig.1. As depicted in that figure, a weak incident pulse excites atoms to the upper state $|c\rangle$. Atoms so excited may decay to state $|b\rangle$ with the emission of a photon γ_1 or γ_2. Then the state of the system is given by

$$|\psi\rangle = |b_1 c_2\rangle |\gamma_1\rangle + |c_1 b_2\rangle |\gamma_2\rangle \ . \tag{1}$$

Subsequent decay from $|b\rangle$ to $|c\rangle$ with emission of a photon ϕ_1 or ϕ_2 takes us to the state

$$|\psi\rangle = |c_1 c_2\rangle [|\gamma_1 \phi_1\rangle + |\gamma_2 \phi_2\rangle]. \tag{2}$$

* Research supported by the Max-Planck-Gesellschaft zur Förderung der Wissenschaften, München; and the Alexander Von Humboldt-Stiftung, Bonn.

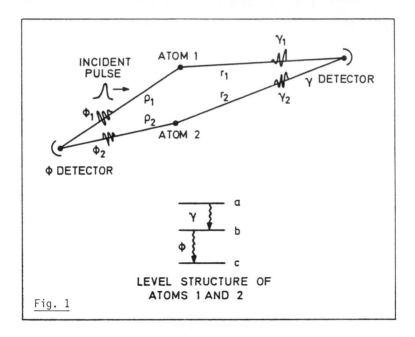

Fig. 1

The joint count probability $P_{\gamma,\phi}$ for detection of a photon in both the γ and ϕ counters can be calculated via the second order correlation function [3] given by (11) of Ref.2, which reads

$$G^{(2)}(\vec{r}t;\vec{\rho},\tau) = \left\langle\gamma_1|\hat{E}_\gamma^{(-)}(\vec{r},t)\hat{E}_\gamma^{(+)}(\vec{r},t)|\gamma_1\right\rangle\left\langle\phi_1|\hat{E}_\phi^{(-)}(\vec{\rho},\tau)\hat{E}_\phi^{(+)}(\vec{\rho},\tau)|\phi_1\right\rangle$$
$$+ \left\langle\gamma_1|\hat{E}_\gamma^{(-)}(\vec{r},t)\hat{E}_\gamma^{(+)}(\vec{r},t)|\gamma_2\right\rangle\left\langle\phi_1|\hat{E}_\phi^{(-)}(\vec{\rho},\tau)E_\phi^{(+)}(\vec{\rho},\tau)|\phi^2\right\rangle + \text{same } 1{\leftrightarrow}2. \quad (3)$$

The joint probability $P_{\gamma,\phi}$ is obtained from (3) by carrying out the appropriate time integrations. In this way we find•(to a good approximation)

$$P_{\gamma\phi}(\theta) = \kappa\,[\,1 + \cos\theta\,], \tag{4}$$

where κ is an uninteresting "constant" and

$$\theta = k_\gamma(r_1 - r_2) + k_\phi(\rho_1 - \rho_2). \tag{5}$$

The above notation is (hopefully) self-evident and is also illustrated in Fig.1.

That the present configuration could provide an experimental test of BELL's theorem is most easily established by making contact with the polarization correlation experiments. In these photon cascade experiments they generated a "polarization correlated photon state" such as

$$|\psi\rangle = |\gamma_x\phi_x\rangle + |\gamma_y\phi_y\rangle, \tag{6}$$

see Fig.2. They then measure (infer) the joint count distribution $p_{\gamma\phi}$ as a function of the angle between analyzers 1 and 2. By properly orienting

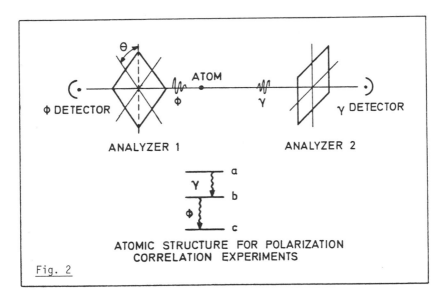

ATOMIC STRUCTURE FOR POLARIZATION
CORRELATION EXPERIMENTS

Fig. 2

the analyzers they are able to make contact with BELL's theorem in a simple fashion. Explicitly they consider the quantity $S(\theta)$ which is defined as

$$S(\theta) = \frac{3P_{\gamma\phi}(\theta) - P_{\gamma\phi}(3\theta)}{P_\gamma + P_\phi} .\tag{7}$$

In (7) P_γ and P_ϕ are the single γ and ϕ count probabilities. The comparison between $S(\theta)$ as predicted by quantum theory and BELL's Theorem is sketched in Fig.3[4]

Connections between the present photon correlation experiment and that of CLAUSER et al. are contained in Table 1.

Polarization Experiments	This work
$\lvert\psi\rangle = \lvert\gamma_s\phi_s\rangle + \lvert\gamma_y\phi_y\rangle$ Fld	$\lvert\psi\rangle = \lvert\gamma_1\phi_2\rangle + \lvert\gamma_2\phi_2\rangle$ Fld
One atom, two polarizations	two atoms, one polarization
$\theta \to$ angle between analyzers	$\theta \to k_\gamma \Delta r + k_\phi \Delta\rho$

Thus it is clear that there is a direct correspondence between the two problems and that the analysis of their experiment would apply, with obvious changes, to our problem. Thus, correlations involving the two-photon state given by (2) could provide a new test of hidden variable theories via BELL's theorem.

The presently envisioned experiment has a few points in its favor which recommend it for further study. For example there are no filters (analyzers) between the atoms and the detectors. Furthermore the present "two-photon, two-atom" set-up would involve spatial (rather than spin or polarization) correlations and would thus be a different type of "hidden vari-

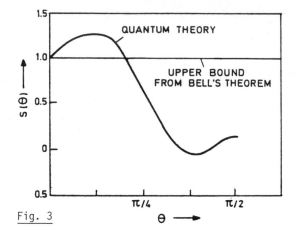

Fig. 3

ables" experiment. Difficulties associated with this set-up are likewise apparent, e.g. holding a few (it doesn't have to be one) atoms in a small volume at sites 1 and 2, etc. However there are several ways to deal with this and other envisioned difficulties.

Acknowledgment. It is a pleasure to thank Dr. J. Clausen for stimulating and helpful discussions.

REFERENCES

1. J. Clauser and A. Shimony, Rep. Prog. Phys. 41, 1881 (1978)
2. M. Scully and K. Drühl, to be published
3. See for example R. Glauber in "Quantum Optics and Electronics",
   Ed. B. De Witt, A. Blandin, C. Cohen-Tannoudji, Gordon and Breach N.Y. (1964)
4. This Figure derives from ref. 1.

Lamb Shift Studies of Cl^{16+}

O.R. Wood II, C.K.N. Patel, D.E. Murnick, E.T. Nelson, and M. Leventhal
Bell Laboratories, Murray Hill, NJ 07974, USA

H.W. Kugel* and Y.Niv**
Rutgers University, Piscataway, NJ 08854, USA

1. Introduction

This paper describes laser resonance spectroscopy of the Lamb Shift splitting in hydrogenic chlorine [1]. The experiment involved a fast ($v/c \approx 0.1$) Cl 17+ beam from the Brookhaven National Laboratory double MP Tandem accelerator facility that had been partially promoted into the excited $2s_{1/2}$ metastable state of the one electron ion (Cl 16+). The $2s_{1/2}$ - $2p_{1/2}$ Lamb Shift transition in this ion was driven by the E-field from a powerful CO_2 laser. The resonance condition was monitored by detecting the subsequent Lyman-α x-ray emission at 2.96 keV as a function of CO_2 laser frequency. Using this technique we have obtained the value, 31.19 ± 0.22 THz, for the Lamb Shift in this hydrogenic ion.

2. Experimental Arrangement

A fast Cl^{14+} beam is momentum analyzed and post-stripped to 17+ by passing it through 50 - 70 $\mu g/cm^2$ carbon foils. The beam then passes through a thin adder carbon foil (0.5 - 3 $\mu g/cm^2$) and is brought to a focus at the focal

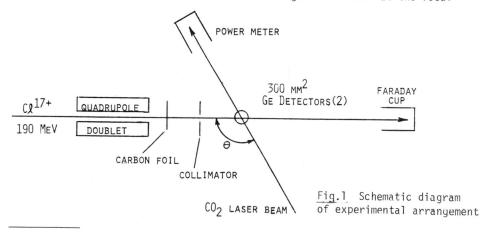

POWER METER

Cl^{17+}

QUADRUPOLE

190 MeV

DOUBLET

CARBON FOIL

COLLIMATOR

θ

300 MM^2
Ge DETECTORS(2)

FARADAY CUP

CO_2 LASER BEAM

Fig.1 Schematic diagram of experimental arrangement

* Present Address: Princeton Plasma Laboratory; Princeton, NJ
**Dr. Chain Weizmann Fellow; Resident Visitor at Bell Laboratories

point of a high power CO_2 laser system, see Fig.1. Although an equilibrium charge state distribution was not achieved, a fraction of the ions capture an electron and are excited into the $2s_{1/2}$ metastable state. After emerging from this focus the particle beam passes onto a Faraday Cup for normalization purposes. The CO_2 laser beam enters the interaction chamber, forming an angle θ with respect to the particle beam, is focused to a waist coincident with the particle beam waist and is re-imaged onto the surface of a water cooled power meter for laser power normalization. Two 300 mm^2 x 7 mm intrinsic Ge x-ray spectrometers are mounted perpendicular to the intersection plane formed by the particle and laser beams. The measured x-ray background, consisting of spontaneous M1 and 2E1 decays of the $2s_{1/2}$ metastable state as well as x-rays from components of the beam other than Cl^{16+}, was of the order of 70 kHz, whereas, the laser induced transition rate was typically of the order of 10 Hz for typical beam currents of 2 particle nanoamperes.

The CO_2 laser system consisted of an oscillator stage with intracavity diffraction grating and an amplifier state providing 13 m of amplification path. The CO_2 oscillator was operated as a waveguide laser [2] to ensure good transverse mode control. 'The device combined Q-switching and current pulsing and, consequently, produced infrared output pulses with higher peak power than could be produced with either technique alone. The average output power from the oscillator when operated at 480 Hz repetition rate was about 10 Watts. This corresponds to a peak output power of ≃ 175 Watts in pulses of 120 μsec duration. The oscillator output was mode matched into the 13m long amplifier that provided a saturated gain of 14. The amplified laser beam had good spatial coherence and was focused with f/3.8 optics to a 150 μm spot coincident with the particle beam.

The output of the CO_2 oscillator/amplifier system could be discretely tuned over the 4 bands of frequencies shown in Fig.2. The complete laser frequency scan, ≃ 4 THz, unfortunately covered only one HWHM of the Lamb Shift resonance. Hence, in order to achieve a more complete coverage of the resonance, several different Doppler shifts (using different laser beam/particle beam intersection angles and particle beam velocities) were used.

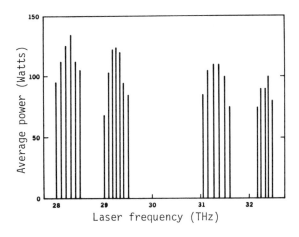

Fig.2   Average CO_2 laser power versus frequency

3. Results

The measured transverse intensity profile of the laser beam at the point of intersection with the particle beam is shown as the solid curve in Fig.3. The measured transverse profile of the particle beam at the point of intersection with the laser beam is shown in Fig.3 as a dashed curve. Also shown in Fig.3 is the result of a measurement of the overlap between the particle beam and the laser beam (shown as solid points labeled "laser induced count rate"). Notice that the overlap is slightly larger than the particle beam ≈ 0.5 mm and the laser beam ≈ 0.15 mm.

An example of our resonance data is shown in Fig.4. The intersection angle between laser and particle beams was 130°. The particle beam energy was 190 MeV. The data presented (as solid triangles) was the result of 16 passes through 8 different laser lines chosen from among the 4 THz available range and represents approximately 8 hours of data taking. Also shown in Fig.4 (as a solid curve) is the best-fit of a relativistically corrected Lorentzian absorption profile to the data.

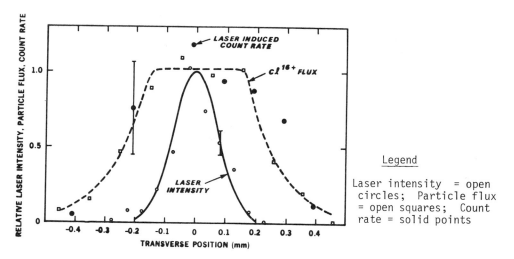

Legend

Laser intensity = open circles; Particle flux = open squares; Count rate = solid points

Fig.3   Measured overlap between laser and particle beams in horizontal plane

4. Conclusions

Our experimental value for the Lamb shift in hydrogenic chlorine, the composite best-fit to more than 50 hours of data taking, is 31.19 ± 0.22 THz. Our result is in agreement with the calculation of MOHR [3] and with the series expansion in powers of $(Z\alpha)$ given in the review article by TAYLOR et al. [4]. On the other hand, our result is three standard deviations below the calculation of ERICKSON [5]. Scaling the discrepancy recently reported for the Z = 1 Lamb shift [6] with an order higher than Z^5 is in disagreement with our result and suggests that the inconsistency in the

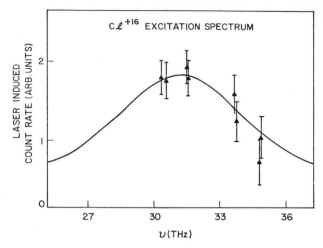

Fig.4   Typical Lamb shift resonance data (solid traingles) together with best-fit of line shape function (solid curve)

hydrogen case is due to low order corrections other than the QED terms, possibly involving proton size and structure effects or involving corrections to the relativistic recoil terms in the theory.

5.   References

1.   D. E. Murnick, C. K. N. Patel, M. Leventhal, O. R. Wood, II and H. W. Kugel; J. de Physique. 40, C1(1979)

2.   P. W. Smith: Appl. Phys. Letters. 19, 132(1971)

3.   P. J. Mohr: Phys. Rev. Letters. 34, 1050(1975)

4.   B. N. Taylor, W. H. Parker and D. N. Langenberg: Rev. Mod. Phys. 41, 375(1969)

5.   G. W. Erickson: Phys. Rev. Letters. 27, 780(1971)

6.   S. R. Lundeen and F. M. Pipkin: Phys. Rev. Letters. 46, 232(1981)

Part II
Laser Spectroscopic
Applications

Selective Spectrum Simplification by Laser Level Labeling

N.W. Carlson[1], K.M. Jones, G.P. Morgan, A.L. Schawlow, A.J. Taylor[2],
H.-R. Xia[3], and G.-Y. Yan[3]
Department of Physics, Stanford University
Stanford, CA 94305, USA

1. Introduction

Even small molecules produce spectra complicated enough so that their
analysis is complex and difficult. In Na_2, for instance, only six excited
singlet electronic states have been identified by conventional
spectroscopy. We have, therefore, studied several methods of using lasers
to identify a selected energy level, and thereby simplify the spectrum.

2. Population Labeling

Even before the advent of lasers, monochromatic light was used to excite a
few chosen upper levels. The subsequent fluorescence spectrum was then
much simpler, and quickly revealed the structure of the lower levels. This
method is an important one for molecules of sufficient complexity, but the
lower levels which it can explore are the ones most likely to be already
known.

We can instead probe the absorption spectrum from molecules in a
particular state whose population can be either increased or decreased by
the laser pumping. Thus KAMINSKY et al.[1] used a chopped beam from an
argon laser to modulate the population of one vibration-rotation level of
the ground electronic state of Na_2. Absorption lines originating on the
chosen level thus decreased in strength whenever the pumping beam reduced
the population, and returned when the beam was blocked and population was
restored by relaxation. A tunable c.w. dye laser was used to explore part
of the A ← X band. For each vibrational level of the A state, the
selection rules permit only two lines, with $\Delta J = + 1$ and $\Delta J = - 1$. These
lines were observed for all values of v from 19 to 33, including those of
v = 22, which were found to be displaced by nearly 1 cm^{-1} through
perturbations by triplet levels.

3. Polarization Labeling

Searching for unknown lines is facilitated by the polarization labeling
technique [2], shown in Fig.1. The pumping laser is polarized, either
linearly or circularly, so that it selectively excites molecules with some

*Work supported by the National Sience Foundation, Grant PHY80-10689
[1]Fannie & John Hertz Predoctoral Fellow
[2]Present address: Bell Laboratories, Murray Hill, New Jersey
[3]Permanent address: East China Normal University, Shanghai, People's
Republic of China

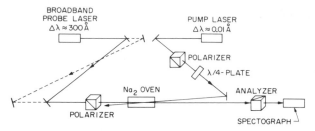

Schematic diagram of apparatus for polarization labeling

particular orientation, and leaves the remainder with a complementary orientation. Then the medium becomes optically anisotropic at all wavelengths corresponding to absorption from the labeled level, and so can depolarize light at those wavelengths.

The probe beam passes through crossed polarizers, before and after the sample region, and so is blocked from reaching the spectrograph except at the wavelengths where the medium is anisotropic. Thus the absorption lines from the labeled level appear as bright lines at the spectrograph's output (Fig.2).

With either population or polarization labeling, we may observe several kinds of signals obtained through secondary processes, as in Fig.3. Most important so far are the transitions upward from the level to which the pump beam excites the molecules, to still higher states. The lifetime of the intermediate level is generally short, but the two-step transitions can be observed if the probe pulse follows the pump pulse within a few nanoseconds. Moreover, with a polarized pump pulse and a broadband probe, the polarization labeling can show these upward transitions, and thus can reveal new molecular states of the same parity as the ground state [3,4,5].

By this method, we have detected a number of excited gerade electronic states in Na_2. Polarization selection rules permit us to distinguish the different values of the projection of the angular momentum on the molecular axis, so h the states can be identified as $^1\Sigma$, $^1\Pi$ or $^1\Delta$. In general, the

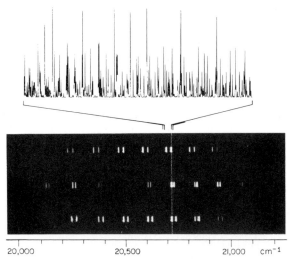

(a)

(b)

A small section of the Na_2 spectrum revealed by conventional spectroscopy and by polarization labeling

20,000   20,500   21,000   cm^{-1}

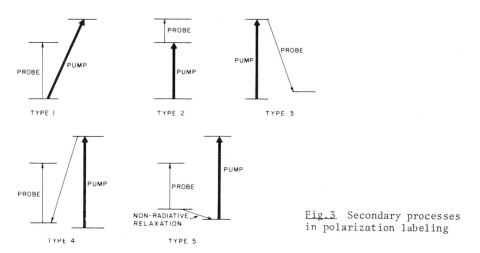

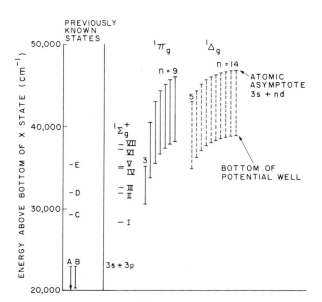

Fig.3 Secondary processes
in polarization labeling

observed electronic states are molecular Rydberg states, with one electron
far outside the Na_2^+ ion core. These states differ from atomic Rydberg
states in that the core is not spherical, and that the outer electron has
some effect on the molecular bond. For high values of the principal
quantum number, n, the outer electron contributes little to the bonding and
so the depth of the potential well does not change much with n. Thus we
are able to set up a correspondence between the atomic sodium levels and
the molecular levels, as is shown in Fig.4. As n decreases, the well
becomes deeper (Fig.5), and the bond length decreases (rotational constant
increases, Fig.6) for the Δ states. On the other hand, the outer electron
of the Π states is antibonding, so that when it comes closer to the core,
the well depth is decreased and the bond lengthened. From either sequence

Fig.4 Diagram of excited
states of Na_2

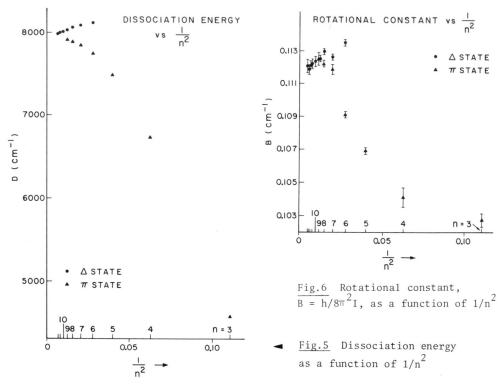

Fig.6 Rotational constant,
$\overline{B = h/8\pi^2 I}$, as a function of $1/n^2$

◄ Fig.5 Dissociation energy
as a function of $1/n^2$

of states, we can plot the molecular constants against $1/n^2$, extrapolate to
$n = \infty$ and so obtain good values for the ionization energy and for the
molecular constants of the Na_2^+ ion [5] ground state.

Seven Σ states have been observed, of which five have been analyzed [6].
They can come from configurations involving excited s or d electrons and,
in at least one case, from a doubly excited 3p-3p configuration.

4. C.W. Two-photon Transitions in Na_2

HARVEY [7] and WOERDMAN [8] have observed c.w. two-photon transitions in
Na_2. They may be comparable in strength to the atomic two-photon
transitions, which was at first surprising, because of the much smaller
number of molecules than atoms, and because there would be fewer still in
any one molecular lower state. They are, nevertheless, easily observable
because they are enhanced by accidental near-coincidence of the two-photon
transition with some line in the A ← X band of the molecule. WOERDMAN
recognized that the enhancement would occur only when the intermediate
state was such that selection rules would permit real transitions between
it and each of the upper and lower states. That is, the upper level can
differ from the enhancing level by not more than one unit of the angular
momentum, J. Thus, if the intermediate level is identified from the

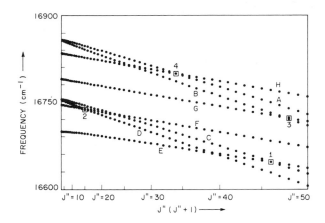

Fig.7 Transition frequency vs. J'(J''+1) for one-photon P and R branch transitions from v=0 in the X state to V=19 and 20 in the A state, and for two-photon Q and S branch transitions from the same v=0 levels of the X state to v=15, 16, 17, and 18 in the $^1\Sigma_g^+$ (IV) state. The numbered data refer to the assigned transitions indicated in Table 1.
Key: (v, transition branch, upper electronic state)
A: (20,R,A); B: (20,P,A); C: (19,R,A); D: (19,P,A);
E: (15,S,Σ); F: (17,Q,Σ); G: (16,Q,Σ); H: (18,Q,Σ)

constants of the A state, we can come close to knowing the angular momentum of the unknown upper level.

We have recently observed about 70 such enhanced c.w. two-photon transitions in the 16600 - 17600 cm^{-1} region [9]. The way these occur is illustrated by Fig.7, in which the one-photon and two-photon frequencies are plotted against J for several vibronic levels of one of the recently analyzed Σ states, which lie in the right region. Where these curves cross, we may perhaps expect to find strongly enhanced two-photon transitions, although some of the levels are displaced from the expected positions by perturbations. Table 1 shows some of the two-photon transititions, and their intermediate enhancing levels, that have been identified.

The offset, Δ, between the one-photon and two-photon transition frequencies, has been measured for some of these lines by doing simultaneous one-photon, and two-photon Doppler-free spectroscopy [10]. They are generally consistent with the expected values, confirming the constants measured by TAYLOR and JONES for the Σ states. The measured offsets range from 3 GHz down to as little as 34 MHz. That is, most of them are less than the Doppler width of 1 GHz, and a few are even less than the lifetime width of about 40 MHz. Under these circumstances, the line shape of the two-photon transition depends strongly on Δ (Fig.8). This occurs because the offset is quite different for molecules which have different Doppler shifts because of their different velocities. When $\Delta=0$, molecules with nearly zero velocity component along the direction of the laser beams have more strongly enhanced transitions than those with higher velocities. Thus the Doppler-free line comes mostly from those nearly stationary molecules, whereas when Δ is large all molecules contribute equally to the Doppler-free

Table 1 Enhanced two-photon transitions from $X^1\Sigma_g^+$ to $^1\Sigma_g^+$ (II) and (IV)

KEY	LASER FREQ. (cm^{-1})	UPPER STATE: $^1\Sigma_g^+$ (IV) (V'',J'')	GROUND STATE: $X^1\Sigma_g^+$ (V,K)	ENHANCING STATE: $A^1\Sigma_u^+$ (V',J')	$\Delta=\|\nu_{2\phi}-\nu_{X\leftarrow A}\|$ CALCULATED (GHz)	UPPER STATE ENERGY (cm^{-1}) PREDICTED*	OBSERVED
1	16647.22	(15,48)	(0,46)	(19,47)	0.75	33706.82	33704.65
2	16738.02	(16,15)	(0,15)	(19,14)	1.2	33591.85	33592.40
	16635.59	(16,38)	(1,36)	(20,37)	0.9	33712.40	33711.50
	16611.07	(17,27)	(2,27)	(21,28)	2.1	33731.52	33730.34
3	16720.47	(17,48)	(0,48)	(20,47)	1.2	33885.26	33879.96
4	16799.37	(18,34)	(0,34)	(20,35)	2.7	33860.93	33860.88
	16704.17	(18,46)	(1,44)	(21,45)	0.09	33953.84	33946.82
	16601.82	(18,50)	(2,50)	(22,49)	0.72	33990.92	33981.97
	16785.42	(19,19)	(1,19)	(21,20)	0.3	33870.00	33866.08
		$^1\Sigma_g^+$ (II)					
	16724.84	(8,19)	(0,21)	(19,20)	1.5	33599.27	33600.20
	16641.98	(11,27)	(3,29)	(23,28)	0.27	33962.64	33963.39
	16648.41	(13,43)	(4,43)	(25,44)	2.4	34281.52	34281.06
	16672.64	(17,17)	(7,15)	(29,16)	1.62	34532.68	34532.74

*Preliminary molecular constants

line. The Doppler-broadened background, on the other hand, comes from molecules in resonance with the Doppler-shifted beam in one direction, and so all parts of it are equally enhanced. Thus the ratio of the Doppler-free peak to the Doppler-broadened pedestal is less for Δ = 0 than for large values of Δ. This ratio has a maximum when the offset is about equal to the half-width of the Doppler-broadened line. This is both shown by theoretical calculations and confirmed by experimental measurements. The transition is more easily saturated near the center than in the wings, so that care is needed to avoid serious distortion of the ratio of the peak to the pedestal.

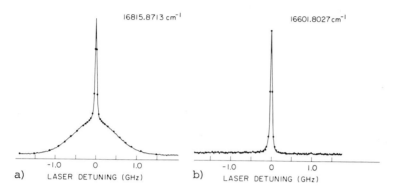

Fig.8 Experimental (solid curve) and theoretical (dots) two-photon line shapes (a) Δ = 34 MHz; (b) Δ = 960 MHz

5. More Complex Spectra

Labeling techniques can be very useful for polyatomic molecules, with their even higher complexity. TEETS et al.[11] used polarization labeling to study NO_2. Although the intensity of the available laser was not enough to label many of the lines with low oscillator strengths, they were able to measure some rotational constants and band origins in the $^2T_2$ state.

As the molecular complexity is further increased, eventually the spectral lines will be so dense that an accidental overlap of lines within the natural line width will occur. Then, a laser tuned to one line cannot help exciting several unconnected levels, so that several levels would be labeled where only one is desired. However, we could use two pumping lasers, tuned to different wavelengths that can be absorbed by molecules in one of the levels, and modulated at different frequencies f1 and f2. Then the common lower level population would be modulated at frequency f1 + f2, while the accidentally overlapping lines would be modulated only at f1 or f2. In such ways, we should be able to extend labeling methods to much more complicated molecules.

6. Stimulated Emission Enhanced Polarization Labeling

BRAND et al.[12] have recently shown that when oriented molecules are put into a level of an excited electronic state, they can be stimulated to emit to unoccupied levels of the ground state, providing strongly enhanced polarization labeling signals. They were thus able to see high resolution labeled spectra in NO_2 despite the weakness of the NO_2 lines.

7. Conclusions

Labeling chosen levels, by using laser light to change the population or orientation of the molecules in them, is a powerful technique for simplifying complicated spectra. It can be extended to more complex molecules, and to other spectral regions if suitable probes can be found. The probe beam does not need to be a laser, if it is intense enough to provide substantial light through the pumped region of the sample. Multi-step exciation [13] can also be used in labeling studies, with either optical or ionization detection.

References

1. M.E.Kaminsky, R.T.Hawkins, F.V.Kowalski, and A.L.Schawlow,
     Phys. Rev. Letters 36, 671 (1976).
2. R.E.Teets, R.Feinberg, T.W.Hansch, and A.L.Schawlow,
     Phys. Rev. Letters 37, 683 (1976).
3. N.W.Carlson, F.V.Kowalski, R.E.Teets, and A.L.Schawlow,
     Opt. Comm. 29, 302 (1979).
4. N.W.Carlson, A.J.Taylor, and A.L.Schawlow,
     Phys. Rev. Letter 45, 18 (180).
5. N.W.Carlson, A.J.Taylor, K.M.Jones, and A.L.Schawlow,
     Phys. Rev. A. (accepted for publication 1981).
6. A.J.Taylor, K.M.Jones, and A.L.Schawlow,
     Opt. Comm. (submitted for publication, 1981).
7. K.C.Harvey, Ph.D. Thesis, Stanford University (1975).
8. J.P.Woerdman, Chem. Phys. Letters 43, 279 (1976).
9. G.P.Morgan, H.-R.Xia, and A.L.Schawlow,
     Opt. Comm. (submitted for publication, 1981).
10. H.-R.Xia, G.-Y.Yan, and A.L.Schawlow, Opt. Comm.
     (submitted for publication, 1981).
11. R.E.Teets, N.W.Carlson, and A.L.Schawlow,
     J. Mol. Spectro. 78, 415 (1979).
12. J.C.D.Brand, K.J.Cross, N.P.Ernsting, and A.B.Yamashita,
     Opt. Comm. 37, 178 (1981).
13. E.F.Worden, R.W.Solarz, J.A.Paisner, and J.G.Conway,
     J. Opt. Soc. Am. 68, 52 (1978).

Resonance Ionization Spectroscopy: Counting Noble Gas Atoms*

G.S. Hurst, M.G. Payne, C.H. Chen, R.D. Willis[1], B.E. Lehmann[2], and S.D. Kramer

Chemical Physics Section, Oak Ridge National Laboratory, P.O. Box X
Oak Ridge, TN 37830, USA

1. Introduction

From its beginning, Resonance Ionization Spectroscopy (RIS) [1] has been recognized as a method for the ionization of matter which is spectroscopically selective and has one-atom sensitivity. Prior to the development of RIS for sensitive analytical purposes, similar laser ionization schemes were proposed [2] for the separation of isotopes. Two-step selective photoionization of Rb atoms was reported [3] by AMBARTSUMYAN, KALININ, and LETOKHOV in 1971. For further information on the history of RIS and the use of various laser methods for sensitive detection of atoms, the reader is referred to recent reviews [4-6].

With five simple laser schemes [5,6], RIS can be used to selectively ionize each element in the periodic table except He and Ne. Most of the work reported thus far on one-atom detection has involved the detection of single electrons created by a laser in a proportional counter. For example, the first such experiment [1] detected single atoms which diffused from a warm Cs metal and migrated by chance into a region of the space of a proportional counter where a laser could be pulsed. In another case [7], Cs atoms were detected following their photodissociation from the chemically stable CsI molecule. Single atoms of Cs have also been found [8] in individual ionization tracks created by the fission decay of ^{252}Cf. Recently [9] a practical application of RIS was made by using laser beams in proportional counters to detect Na impurity atoms in electronics grade Si.

The purpose of this paper is to describe new work on the counting of noble gas atoms, using lasers for the selective ionization and detectors for counting individual particles (electrons or positive ions). When positive ions are counted, various kinds of mass analyzers (magnetic, quadrupole, or time-of-flight) can be incorporated to provide A selectivity. We show that a variety of interesting and important applications can be made with atom-counting techniques which are both atomic number (Z) and mass number (A) selective.

*Research sponsored by the Office of Health and Environmental Research, U.S. Dept. of Energy under contract W-7405-eng-26 with the Union Carbide Corp.

[1] On leave from Scripps Institution of Oceanography and supported in part by Scripps.

[2] Postdoctoral Research Appointment through Western Kentucky University; supported in part by the Swiss National Foundation for Scientific Research.

2. RIS of Noble Gases (Z Selection)

A laser scheme to demonstrate [10] RIS of Xe is the following. A Nd-Yag
laser beam was frequency doubled and used to pump a red dye to obtain tunable
light between 645 and 675 nm with a pulse duration of 5 ns. The dye laser
output was doubled and then mixed with 1.06 μm to yield tunable light from
246 to 258 nm. At a two-photon resonance in Xe (i.e., 252.6 or 249.6 nm),
the energy per pulse was ∼1 mJ and the bandwidth was estimated to be ∼0.1 Å.
During the experiment the laser beam was focused down to 0.1 mm at the center
of an ionization chamber which contained Xe at pressures ranging from 10^{-8} to
10^{-6} Torr.

The ionization line shape for the three-photon ionization of Xe near the
two-photon resonances with $Xe[1/2]_0(5p^56p)$ and $Xe[3/2]_2(5p^56p)$ states is
shown in Fig.1. We observed that the two-photon transition occurred at a
rate of $2 \times 10^{-9} I^2$ (sec^{-1}) when I is W/cm^2 and could be saturated for laser
power densities higher than 6×10^8 W/cm^2. The ratio of background ions from
residual gases to the saturated resonance ionization signal is ∼10^{-2} when
the Xe pressure is 1×10^{-6} Torr. When a quadrupole mass filter was introduced, the background was reduced to such a low value that the detection of
single Xe atoms of a selected isotope was possible.

Following the experimental demonstration of RIS with Xe, calculations [11]
were made to show how to generalize the results to the other noble gases with
commercial lasers (Kr and Ar). We describe here estimates of volumes of
ionization, ΔV, for RIS beginning with the following two-photon resonance
excitation schemes:

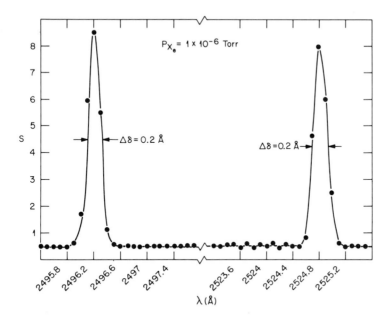

$P_{Xe} = 1 \times 10^{-6}$ Torr

$\Delta\delta = 0.2$ Å

$\Delta\delta = 0.2$ Å

λ (Å)

<u>Fig.1</u> Tuning curve for Xe atoms involving two-photon excitation followed
with one-photon ionization at the two indicated wavelengths

$$Ar(3p^6) + \hbar\omega_1(\lambda = 1933 \text{ Å}) + \hbar\omega_2(\lambda = 1807 \text{ Å}) \rightarrow Ar(3p^54p[1/2]_0) \, ,$$

$$Ar(3p^6) + \hbar\omega_1(\lambda = 1576.3 \text{ Å}) + \hbar\omega_2(\lambda = 2292.8 \text{ Å}) \rightarrow Ar(3p^54p[1/2]_0) \, ,$$

$$Kr(4p^6) + 2\hbar\omega(\lambda = 1933 \text{ Å}) \rightarrow Kr(4p^56p[3/2]_2) \, ,$$

$$Kr(4p^6) + \hbar\omega_1(\lambda = 1576.3 \text{ Å}) + \hbar\omega_2(\lambda = 3262 \text{ Å}) \rightarrow Kr(4p^55p[1/2]_0) \, ,$$

$$Xe(5p^6) + \hbar\omega_1(\lambda = 1576.3 \text{ Å}) + \hbar\omega_2(\lambda = 5995 \text{ Å}) \rightarrow Xe(5p^56p[1/2]_0) \, ,$$

$$Xe(5p^6) + 2\hbar\omega(\lambda = 2496 \text{ Å}) \rightarrow Xe(5p^56p[1/2]_0) \, .$$

The lasers assumed here are the ArF excimer laser (1933 Å), the F_2 laser (1576.3 Å), and other laser outputs that can be generated with commercial dye laser systems pumped by harmonics of the Nd-Yag. Thus far, the use of the F_2 laser has been limited due to the fact that no major effort has been devoted to developing a F_2 laser with good beam quality and with gas mixtures having long lifetimes.

The effective volume for ionization, ΔV, was calculated by using a rate analysis and assuming a Gaussian radial intensity profile. The two-photon excitation rates were calculated [11] with reasonable self-consistency by several approximation schemes, and the photoionization rates for the upper state were based on the results of DUZY and HYMAN [12]. Numerical values for the effective ionization volumes were obtained for a beam divergence of 10^{-3} rad and assuming a lens of 10-cm focal length. If the F_2 laser can be made to work reliably, ΔV can be at least 3×10^{-4} cm^3 for Ar, Kr, or Xe.

Recently, there have been important advances in the development of tunable light sources in the vacuum ultraviolet spectral region where Ar and Kr have strong, one-photon allowed transitions [13]. Using a two-photon resonance, four-wave mixing process in Hg vapor, $\sim$1 kW of peak power in a bandwidth of 0.04 cm^{-1} was generated at the Kr transition line at 1235.8 Å [14]. By frequency tripling a dye laser in Xe gas, ZAPKA [15] obtained tunable output with a maximum peak power of almost a watt in the range from 1061 to 1068 Å. Using a more powerful dye laser, a peak power of 100 W at the 1066.7 Å Ar transition is possible. Doppler-free, two-photon RIS with isotopic selectivity has been demonstrated in both Xe and Kr [16]. This method is particularly useful in achieving separation between even isotopes and those odd isotopes that have large hyperfine shifts; however, these separations are too small for use in the technique described here.

3. Laser-Mass Spectrometer Techniques for Noble Gas Detection

While a substantial amount of work on RIS has been done with proportional counters, it is clear that evacuated particle detectors also have unique applications with RIS. Proportional counters are excellent analogue devices and can be used to count just one electron (or one atom using RIS), however they require gases of selected properties for their operation. Fortunately, other particle detectors are available which can be used to count individual charged particles (electrons, negative ions, or positive ions) with nearly 100% efficiency; and they work at pressures below 10^{-5} Torr. Positive ion detectors can be substituted for electron detectors, which eliminates any noise problems associated with photoelectrons produced by laser interactions with window materials.

Detection of positive ions also make possible mass analysis (A selectivity) prior to counting. The combination of Z selection (made possible by the laser) and A selection (with a mass spectrometer) can be put to great advantage in practical applications. In particular, it is a natural combination for the isotopically selective counting of individual noble gas atoms as discussed below. More generally, however, all of the standard forms of mass spectrometers play their own special roles when combined with RIS. Magnetic deflection offers both a high transmission (approaching 100% of the ions) and good resolution (> 10^6 discrimination between A and A $\pm$ 1). Time-of-flight mass spectrometers have much less resolution but retain high transmission and offer the advantage of simultaneous recording of a range of A values. Quadrupole mass filters have low transmission and medium resolution, but they have the unique feature that the ions being analyzed are kept at a very low energy. In the experiment of CHEN et al. [10] a simple version of a quadrupole mass filter was used to selectively detect Xe atoms. With the laser linewidth greater than isotopic splittings, it was found that Xe isotopes were detected by amounts according to their normal abundances.

4. Maxwell's Sorting Demon

4.1 The Concept

In 1871, JAMES CLERK MAXWELL visualized a demon which could see and identify atoms and open and close doors to sort atoms into well-defined compartments. We consider the system shown in Fig.2 to be a practical realization of Maxwell's demon for noble gases. The ultraviolet laser ionizes all noble gas atoms of a particular type (Z selection) in a small volume; the quadrupole mass filter separates ions of a particular isotope (A selection), which are then implanted into a CuBe target. The electron multiplier counts secondary electrons emitted from the activated CuBe surface.

4.2 The Atom Buncher

Since the effective volume of the high-vacuum chamber is rather large ($\sim$2.5 liters) and the volumes that can be saturated in a two-photon process are quite small (3 x 10^{-4} cm^3), it is necessary in a practical application of the demon concept to find a way to increase the chance of a noble gas atom being in the beam at the time the laser is fired. In our case, the sample will be frozen onto the liquid helium cold finger. When the light of a pulsed CO_2 laser strikes the cold finger, the noble gas atoms will be released instantly and travel a few millimeters before the time-delayed ultraviolet laser is pulsed just above the cold finger and ionizes a significant fraction of the sample. The ionization probability for a single atom of a selected isotope can be adjusted to approximately 10% in order to achieve true digital counting.

4.3 Implantation and Counting of Single Ions

Ions of 10-keV energy have been implanted at an average depth of $\sim$100 Å in Al or glass [17]. The diffusion of atoms in solids at room temperature is extremely slow; thus, ions which are implanted into a metal foil such as CuBe remain stored in the foil for months without significant escape.

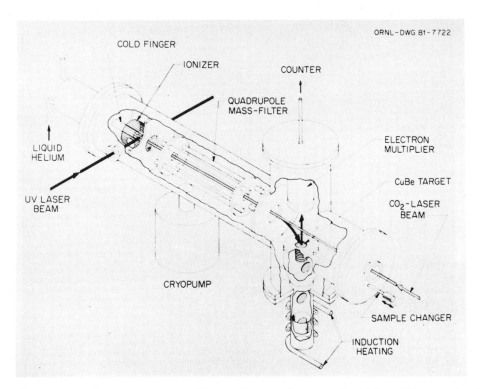

ORNL-DWG 81-7722

COLD FINGER

IONIZER

COUNTER

QUADRUPOLE
MASS-FILTER

LIQUID
HELIUM

UV LASER
BEAM

ELECTRON
MULTIPLIER

CuBe TARGET

CO₂-LASER
BEAM

CRYOPUMP

SAMPLE CHANGER

INDUCTION
HEATING

Fig.2 Artist's conception of "Maxwell's sorting demon" which utilizes pulsed lasers and a quadrupole mass spectrometer to count Z- and A-selected noble gas atoms.

The ability to sort Xe atoms in a small volume by means of selective laser ionization and ion implantation was demonstrated in our laboratory [10]. Fig.3 shows the depletion of Xe atoms in volumes of 50 and 15 cm^3 [curves (a) and (b), respectively], using a pulsed Nd-Yag laser to produce Xe^+ ions followed by implantation of the ions into a channeltron detector. The measured Xe pump-out rates were consistent with an implantation probability of nearly 100%, in agreement with measurements by KORNELSEN [18]. When ions are implanted, they can be detected by observing secondary electron emission (see Fig.3). BAUMHÄKEL [19] measured electron yields from activated CuBe surfaces bombarded with noble gas ions. For 3-keV Ar^+, Kr^+, and Xe^+ ions, the electron yields are 2 to 3.5 secondary electrons per incident ion. For ion energies of 10 keV, electron yields of 6 to 8 electrons per ion are expected.

4.4. Recovery of Implanted Atoms, Recycling

Since the abundance ratio of the quadrupole mass spectrometer is only 10^3, many atoms of neighboring isotopes will be implanted even for samples where the selected isotope occurs only in a very small fraction of the sample,

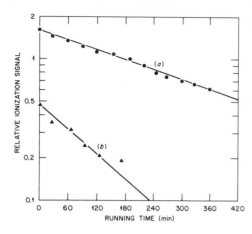

Fig.3 Selective pumping of Xe atoms using a pulsed laser in the RIS process in gas volumes of 50 cm^3(curve a) and 15 cm^3 (curve b)

e.g., 1:10^{15} (see ^{39}Ar applications, section 5.2). The system, therefore, is designed to allow for a target to be heated or melted inside the high vacuum in order to release the implanted atoms, which will then be collected again on the liquid He cold finger. After a small number, n, of these cycles, an enrichment of 10^{3n} will be achieved and the selected atoms will be counted by implantation into the final fresh CuBe target.

Some of these pre-enrichment cycles may be done without lasers by using the electron ionizer of the quadrupole. In a 2.5-liter system, with the mass spectrometer tuned to mass 84 and operating with a throughput efficiency of 2 mA/Torr, the time required to reduce the ratio of ^{84}Kr to total Kr concentration by one e-fold was found [10] to be 82 min.

4.5 Summary of Expected Characteristics of the Machine

The laser/mass spectrometer realization of Maxwell's demon should enable one to count small numbers (i.e., 100–1000) of atoms of any selected isotope of Ar, Kr, Xe, or Rn in isotopic backgrounds of up to 10^{15}.

The final counting, after any necessary pre-enrichment steps, should only take a few minutes, so that interferences with noble gas atoms degassing from the walls of the vacuum system can be minimized.

5. Application of Noble Gas Counting

5.1 Solar Neutrinos

Many very timely problems in physics can now be considered seriously for the first time because Maxwell's demon for counting noble gas atoms is becoming available. One of these applications involves the work of RAY DAVIS and his group at Brookhaven National Laboratory. In a 100,000-gallon tank filled with perchloroethylene, the enormous flux of neutrinos from the sun that pass through the earth can convert a few atoms of ^{37}Cl to ^{37}Ar via the process ^{37}Cl(ν,e$^-$)^{37}Ar. After exposure to neutrinos from the sun for about 6 months, the 100 or so ^{37}Ar atoms are removed and counted as they decay in a proportional counter. Another interesting measurement could be made by using ^{81}Br$(\nu,e^-)^{81}$Kr. Here the "direct" counting of ^{81}Kr atoms with Maxwell's demon is essential because more than 10^5 yr are required for the 100 atoms of ^{81}Kr to decay in a proportional counter! The very new idea in weak interaction physics that neutrinos could have a finite rest mass and hence oscillate between neutrinos of different types (e.g., electron, muon, and tau neutrinos) can also be tested by using these new detectors.

Another interesting test of weak interaction physics involves the measurement of double beta $(\beta^-\beta^-)$ decay. For example,

$$^{82}Se \rightarrow\ ^{82}Kr + \beta^-\beta^-$$

can occur even though ^{82}Se is stable against β^- decay. The $\beta^-\beta^-$ half-life is uncertain by more than an order of magnitude but is about 10^{19} or 10^{20} yr. Direct measurement of ^{82}Kr with a RIS technique is quite feasible and is the basis of a collaboration between ORNL, BNL, and JACK ULLMAN of the Herbert H. Lehman College in New York. PETER ROSEN and WICK HAXTON at Purdue University are interested in extending these studies to the $\beta^-\beta^-$ decay of Te to Xe.

5.2 Oceanography

The radioisotope ^{39}Ar, produced by cosmic rays in the upper atmosphere and dispersed through the oceans by oceanic transport processes, is potentially one of the most valuable tracers available for studying the pattern and rates of circulation in the deep ocean and the processes by which the ocean is internally mixed. The half-life of ^{39}Ar (270 yr) is ideally matched to the characteristic time scales of ocean mixing and circulation, while the inert nature of ^{39}Ar ensures that its distribution in the oceans is determined only by advection, mixing, and radioactive decay.

The application of ^{39}Ar to ocean studies has been thwarted by the extremely low ^{39}Ar activities expected in the ocean (0.01 to 0.1 dpm/liter). Such low levels presently necessitate the collection and processing of at least 2000 liters of ocean water per sample for conventional decay counting. This requirement is absolutely prohibitive for routine application. With Maxwell's sorting demon, however, the sample size requirement will be reduced to only 10 liters, thus enabling our collaborators at the Scripps Institution of Oceanography to routinely make use of this powerful and long-awaited tracer.

5.3 Other Applications

Other applications of direct counting of noble gases which have been considered in some detail are cosmic-ray histories (^{39}Ar and ^{81}Kr), radon transport (^{222}Rn), polar ice cap dating (^{81}Kr), aquifer ages (^{81}Kr), earthquake prediction (noble gases), transuranic waste isolation (Xe fission product), diagnoses of bone diseases (^{37}Ar), and neutron dosimetry (^{37}Ar).

Acknowledgements

The authors wish to acknowledge the technical support of R. C. Phillips and S. L. Allman and the suggestions and information received from J. W. T. Dabbs and G. D. Alton.

References

1. G. S. Hurst, M. G. Payne, M. H. Nayfeh, J. P. Judish, and E. B. Wagner, Phys. Rev. Lett. 35, 82 (1975); M. G. Payne, G. S. Hurst, M. H. Nayfeh, J. P. Judish, C. H. Chen, E. B. Wagner, and J. P. Young, Phys. Rev. Lett. 35, 1154 (1975); G. S. Hurst, M. H. Nayfeh, and J. P. Young, Appl. Phys. Lett. 30, 229 (1977); G. S. Hurst, M. H. Nayfeh, and J. P. Young, Phys. Rev. A 15, 2283 (1977).

2. J. Robieux and J. M. Auclair, French Patent No. 1,391,738 (Compagnie Général d'Electricité), 1965.

3. R. V. Ambartsumyan, V. N. Kalinin, and V. S. Letokhov, Zh. Eksp. Teor. Fiz. Pis'ma Red. 13(6), 305 (1971).

4. W. M. Fairbank, Jr. and C. Y. She, Optics News 5, 4 (1979).

5. G. S. Hurst, M. G. Payne, S. D. Kramer, and J. P. Young, Rev. Mod. Phys. 51, 767 (1979).

6. G. S. Hurst, M. G. Payne, S. D. Kramer, and C. H. Chen, Phys. Today 33, 24 (September 1980).

7. L. W. Grossman, G. S. Hurst, M. G. Payne, and S. L. Allman, Chem. Phys. Lett. 50, 70 (1977).

8. S. D. Kramer, C. E. Bemis, Jr., J. P. Young, and G. S. Hurst, Optics Lett. 3, 16 (1978).

9. S. Mayo, T. B. Lucatorto, and G. G. Luther, "Laser Ablation and Resonance Ionization Spectroscopy (LARIS) for Trace Analysis of Solids," to be submitted for publication.

10. C. H. Chen, G. S. Hurst, and M. G. Payne, Chem. Phys. Lett. 25, 473 (1980).

11. M. G. Payne, C. H. Chen, G. S. Hurst, S. D. Kramer, W. R. Garrett, and M. Pindzola, Chem. Phys. Lett. 79, 142 (1981).

12. C. Duzy and H. A. Hyman, Phys. Rev. A 22, 1878 (1980).

13. J. Reintjes, Appl. Optics 19, 3889 (1980).

14. F. S. Tomkins and R. Mahon, Optics Lett. 6, 179 (1981); F. S. Tomkins, private communication.

15. W. Zapka, D. Cotter, and U. Brackmann, Optics Comm. 36, 79 (1981).

16. T. J. Whitaker and B. A. Bushaw, J. Opt. Soc. Am. 70, 1413 (1980); T. J. Whitaker, private communication.

17. R. G. Wilson and G. R. Brewer, ION BEAMS (Wiley, New York, 1973).

18. E. V. Kornelsen, Can. J. Phys. 42, 364 (1964).

19. R. Baumhäkel, Z. Phys. 199, 41 (1967).

Linewidth Characteristics of (GaA1)As Semiconductor Diode Lasers*

A. Mooradian, D. Welford, and M.W. Fleming**

Lincoln Laboratory, Massachusetts Institute of Technology, Box 73
Lexington, MA 02173, USA

Use of semiconductor diode lasers as high resolution spectral sources requires an understanding of the linewidth characteristics of these devices. Reported here is a study of the power and temperature dependence of the fundamental linewidth of continuously operating (GaAl)As single mode diode lasers. The linewidths were observed to decrease linearly with reciprocal output power with a slope significantly greater than the calculated Schawlow-Townes linewidth at room temperature. At liquid nitrogen temperature, within experimental error, the slope decreased to the Schawlow-Townes limit and had a finite contribution which is attributed to index fluctuations arising from electron density fluctuations in the small gain volume of these devices.

The experiments were carried out on single frequency channel-substrate-planar (CSP) Hitachi and transverse-junction-stripe (TJS) Mitsubishi diode lasers. The devices were thermally isolated in a Dewar to avoid temperature fluctuations and measurements were carried out as previously described [1], with the exception that the data taking and reduction was performed by a computer. The Mitsubishi TJS diode lasers usually worked well down to liquid helium temperature in contrast to the Hitachi CSP lasers which usually operated poorly or not at all at 77 K and below. The lineshape remained Lorentzian at all times and was carefully checked by heterodyning against a narrow linewidth and stable external cavity (GaAl)As diode laser. Figure 1 shows such a heterodyne spectrum which is consistent with similar measurements taken on lead salt diode lasers by HINKLEY and FREED [2].

The linewidth due to quantum phase fluctuations is given by the modified Schawlow-Townes relation [3]

$$2\Gamma = (h\nu/8\pi P_0) (c/n\ell)^2(\ln R-\alpha\ell) (\ln R)n_{sp} \qquad (1)$$

where 2Γ is the full-width at half-maximum of the emission line at frequency ν, P_0 is the single-ended output power, n is the refractive index, ℓ is the cavity length, α is the material absorption coefficient, R is the facet reflectivity, and n_{sp}, the spontaneous emission factor, is the ratio of the spontaneous emission rate per mode to the stimulated emission rate per laser photon. In a semiconductor laser, the spontaneous photon emission factor n_{sp} may be written as [4]

*This work was sponsored by the Department of the Air Force.
**Present address: Central Research Laboratories, 3M Center, St. Paul, Minnesota 55144.

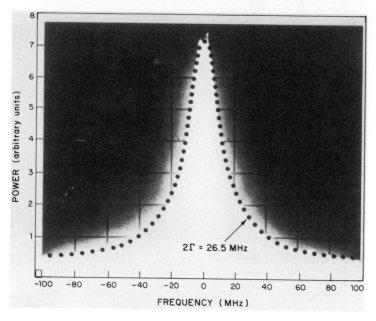

<u>Fig.1</u> Heterodyne spectrum of Hitachi CSP (GaAl)As diode laser at room
temperature. Output power of 4.2 mW. Dotted curve is a theoretical fit.

$$n_{sp} = \{1 - \exp[(h\nu + E_{Fv} - E_{Fc})/kT]\}^{-1} \qquad , \qquad (2)$$

where E_{Fc} and E_{Fv} are the conduction- and valence-band quasi-Fermi levels,
k is the Boltzmann constant, and T is the temperature. In most lasers, n_{sp}
is nearly unity, but for semiconductor lasers with nondegenerate carrier
distribution functions, n_{sp} is greater than unity at room temperature.
For these devices, n_{sp} is about 2.3 at room temperature and becomes unity
below 77 K.

    Figure 2 shows the linewidth plotted as a function of reciprocal single-
ended output power for several Mitsubishi TJS diode lasers at ice temperature together with the calculated linewidth dependence using (1). This discrepancy with theory has been reported [1] previously for the Hitachi CSP
diode lasers at room temperature. The difference between the experimental
and theoretical slopes is a factor of seven for the Mitsubishi TJS diodes
and a factor of twenty for the Hitachi CSP lasers. This discrepancy between
the two types of laser may be due to the difference in their construction.
The difference between experimental and theoretical slopes, however, which
nearly disappears at low temperature must be explained by some other process.

    Figure 3 shows the linewidths measured at 77 K for the Mitsubishi diodes.
Within experimental error, the observed slope has a value much closer to
that calculated from the Schawlow-Townes expression but with the unexpected
result of a finite intercept of 13 MHz. This finite intercept for $1/P = 0$
occurs at all temperatures, increasing from about 1-3 MHz at room temperature to as much as 30-40 MHz at liquid helium temperature. This fundamental limit to the laser linewidth is attributed to fluctuations in refractive

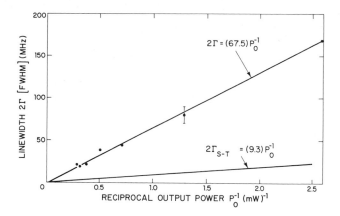

<u>Fig.2</u> Linewidth of Mitsubishi TJS lasers at 273 K compared to calculated Schawlow-Townes width.

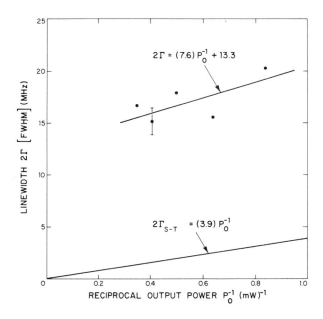

Fig.3 Linewidth vs recipro cal power for Mitsubishi diode lasers at 77 K compared with calculated Schawlow-Townes width.

index in the gain volume of the laser. Such fluctuations are significant for the small gain volumes in these devices. The fluctuation in the laser frequency is given by

$$\delta\nu = \frac{\nu \sqrt{N}}{n} (\delta n/\delta N)_{\omega_\ell}$$ (3)

where N is the number of electrons in the gain volume, and $(\partial n/\partial N)_{\omega_\ell}$ is the change in refractive index with electron number at the laser frequency. For the Mitsubishi TJS diode lasers at 77 K, $N \cong 10^7$ electrons which is

estimated from the threshold current, carrier lifetime and a gain volume
of 2 μm x 0.1 μm x 200 μm. The factor $(\partial n/\partial N)\omega_\ell$ is obtained from the room
temperature value of the change in refractive index with electron density
(at a density of 1 x 10^{18} cm^{-3} which is approximately the same as our density of 2.5 x 10^{17} cm^{-3}) as reported by ITO and KIMURA [5]. The result is
a calculated residual linewidth of about 30 MHz which is surprisingly close
considering the approximations made. The variation of this residual linewidth with temperature is also consistent with the reduced threshold currents.

In order to decrease the linewidth for useful spectroscopic applications, an external cavity with its much higher cavity Q can be used. This
is demonstrated in Fig. 4 where two single-frequency external cavity
lasers are shown heterodyned together. A linewidth of 15 kHz limited by
the spectrum analyzer resolution is shown. The expected linewidth for
these devices having an output power of 0.4 mW is about 400 Hz.

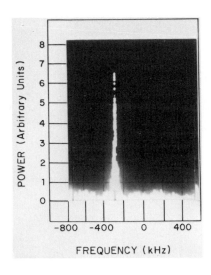

Fig.4 Heterodyne beat spectrum
between two external cavity
(GaAl)As CW diode lasers with
output power of 0.4 and 0.34 mW,
respectively.

REFERENCES

1. M. W. Fleming and A. Mooradian, Appl. Phys. Lett. 38, 511 (1981).
2. E. D. Hinkley and C. Freed, Phys. Rev. 23, 277 (1969).
3. M. Lax, in Physics of Quantum Electronics, edited by P. L. Kelley,
   B. Lax, and P. E. Tannenwald (McGraw-Hill, New York, 1966), p. 375.
4. H. Haug and H. Haken, Z. Phys. 204, 262 (1967).
5. M. Ito and T. Kimura, IEEE J. Quantum Electron. 16, 910 (1980).

High Sensitivity Spectroscopy with Tunable Diode Lasers – Detection of O_2 Quadrupole Transitions and [14]C

J. Reid

Department of Engineering Physics, McMaster University
Hamilton, L8S 4M1 Ontario, Canada

1. Introduction

In recent years, tunable lead-salt diode lasers (TDLs) have found widespread application in all fields of infrared spectroscopy. However, most applications of TDLs utilise only the tunability and high resolution of these devices, and few experiments have employed the ability of the TDL to detect very small absorption coefficients. We have developed a laser absorption spectrometer (LAS) which can detect absorption coefficients as small as 10^{-6} to 10^{-7} m^{-1}, while retaining the full tunability and resolution of the TDL. This instrument has been used as a point monitoring system for many trace gases of atmospheric significance [1-3]. In this paper, we describe two additional applications of the LAS: (i) the detection of very weak transitions such as quadrupole lines in oxygen, and (ii) the detection of rare isotopes, with ^{14}C in CO_2 as an example. Details are given in the following sections.

2. Experimental Apparatus

The LAS basically consists of a TDL and a multipass optical cell. Two f/1 Ge Lenses are used to focus the TDL beam into the multipass cell, and to refocus the output onto a HgCdTe detector. A detailed description of the apparatus is given in previous publications [1-3]. Initially, a monochromator is placed in the laser beam to characterise the TDL behaviour, and to select operating conditions which give regions of good single mode tunability. The monochromator is then removed as reflections from the input and exit slits seriously degrade the instrument performance.

Typical results obtained with a TDL operating near 1600 cm^{-1} are shown in Fig. 1. The diode tunes in a single mode until a mode hop occurs at 924 mA. For the lower trace, second harmonic detection was used to enhance sensitivity [1-3]. The two additional lines which appear in the air sample are identified as HDO lines from the AFGL compilation [4]. By tilting all optical components to avoid feedback, carefully optimising the optical alignment, and adding a second modulation to minimise the effect of optical fringes, the LAS noise level can be reduced until it is equivalent to an absorption coefficient of $\sim 2 \times 10^{-7}$ m^{-1}, or less than 100 ppt of NO_2 [3].

The NO_2 lines shown in Fig. 1 are conventional electric dipole transitions and have substantial linestrength. As an alternative to detecting NO_2 at low concentrations, one can clearly use the LAS to search for very weak transitions in a pure gas. An example is given in the next section.

72

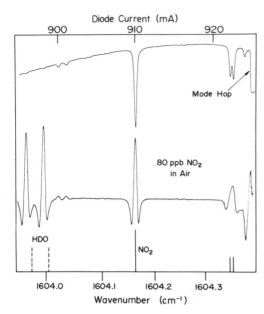

Fig. 1 Spectra of pure NO_2 and NO_2 in air. For the upper trace a 10-cm cell containing NO_2 at low pressure was placed in the laser beam, and conventional amplitude detection was employed. The lower trace was taken using second harmonic detection with the laser beam traversing a path length of 40 m in air at 30 Torr and a small calibration cell giving the equivalent of 80-ppb NO_2. The predicted HDO and NO_2 spectra are shown at the bottom of the figure [4,5].

3. Observation of Electric Quadrupole Transitions in O_2

The TDL used to produce the traces in Fig. 1 is ideally suited to study the $v=1 \leftarrow 0$ vibration-rotation quadrupole band of O_2. At room temperature one of the strongest features in this band is the S(7) quadrupole line, which is expected to lie near $1603 \cdot 8$ cm^{-1} [6]. A further incentive to search for a weak O_2 transition in this region was provided by the recent observation of an unidentified line triplet at $1603 \cdot 8$ cm^{-1} in a long path infrared atmospheric spectrum reported by NIPLE et al. [7].

For the initial experiments, we used an NO_2 reference cell in addition to the 1m base pathlength White Cell, and set the TDL to reproduce the upper trace of Fig. 1. The diode heat sink temperature was then adjusted very slightly to enable the TDL to tune from $1603 \cdot 5$ cm^{-1} to $1604 \cdot 1$ cm^{-1} in a single mode. Results taken with conventional amplitude detection are shown in Fig. 2. The NO_2 reference cell was then removed, and a high sensitivity scan taken through 40 m of pure O_2 at 40 Torr using second harmonic detection. A distinct triplet appeared above the noise level, as shown in Fig. 3. We have unambiguously identified this triplet as the $J=10 \leftarrow 8$, $9 \leftarrow 7$, and $8 \leftarrow 6$ spin components of the S(7) (N=9$\leftarrow$7) quadrupole transition of $^{16}O_2$.

We have made careful measurements of the positions and line splittings of the O_2 quadrupole lines, and found good agreement between our results and previous measurements on the Raman and microwave spectra of O_2 [6]. The laboratory measurements also enabled us to confirm that the triplet observed by NIPLE et al. [7] is in fact the S(7) quadrupole transition of atmospheric oxygen. Similar agreement between laboratory and solar spectra was obtained for the S(5) quadrupole transition [6], and several quadrupole lines of N_2 have recently been observed in long-path atmospheric spectra.

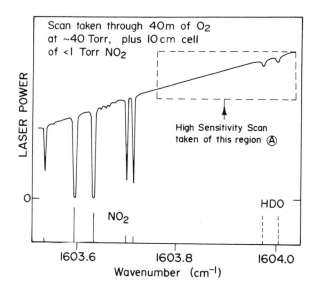

Fig.2 Diode laser scan of the 1603·8 cm⁻1 region using conventional ampli--tude detection. Calcula--ted line positions, shown at the bottom of the figure, are taken from the AFGL tapes [4,5].

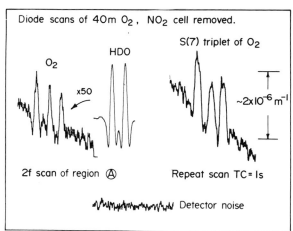

Fig.3 High sensitivity scans of the S(7) triplet of O_2 in region A of Fig.2. The sensitivity is close to detector noise limited due to the relatively low power a-vailable in this laser mode.

4. Detection of Radiocarbon

The LAS has been used to examine the feasibility of dating carbon samples using infrared spectroscopy. We propose to convert the carbon sample to CO_2, and then measure the absorption coefficients of the strong vibration-rotation lines in the ν_3 band of $^{14}CO_2$ near 4.5 μm. Modern samples of carbon have a $^{14}C/^{12}C$ ratio of $\sim 10^{-12}$, and hence one must be able to detect less than 1 ppt ^{14}C in the presence of the stable isotopes of carbon and oxygen. This requires a careful choice of $^{14}CO_2$ absorption line to minimise the background interference from the lines of the stable isotopic species.

We carried out a line-by-line evaluation of the P-branch of $^{14}CO_2$, and determined that the P(20) transition is one of the best lines for detection of $^{14}CO_2$ [8]. The upper trace of Fig. 4 shows a TDL scan made with a 32 m pathlength of normal CO_2 plus a small cell containing a sample of $^{14}CO_2$ to locate the P(20) line. Second harmonic detection is used for the lower trace, and expanded views of the P(20) region are shown in Fig. 5. Clearly, line B interferes with the $^{14}CO_2$ P(20) line, and poses a severe problem for radiocarbon dating, unless steps are taken to suppress the interference.

Fortunately, the ratio of interference linestrength to $^{14}CO_2$ linestrength shown in Fig. 5 only holds at room temperature. All weak lines in CO_2 in the 2209 cm^{-1} region are expected to be hot band lines, and should undergo a dramatic reduction in intensity if the CO_2 sample is cooled. To estimate the reduction in linestrength of line B upon cooling, one must know the energy of its lower level, E_L. We determined E_L by heating a sample of CO_2 and comparing the increase of absorption on line B with increased absorption

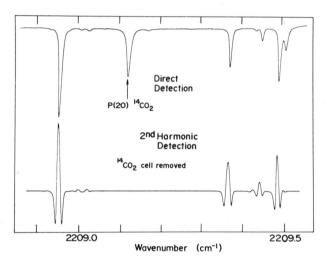

Fig. 4 Diode laser scans of the wavenumber region near the P(20) line of $^{14}CO_2$. For the upper trace the laser beam passes through a 32 m pathlength of normal CO_2 at 20 Torr, followed by a small cell containing $^{14}CO_2$.

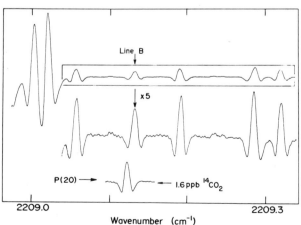

Fig. 5 More sensitive scans of the region around the P(20) $^{14}CO_2$ line. The weak line B is present in normal CO_2 and overlaps the P(20) $^{14}CO_2$ line which is represented by the trace at the bottom of the figure

on several lines of known E_L. In this manner E_L for line B was determined to be 3430 ± 150 cm^{-1}. This high value of E_L ensures that there will be a very substantial reduction in the interference on cooling the CO_2 sample. Thus, we are confident that the detection of $^{14}CO_2$ in a cooled multipass cell will not be limited by interferences from the stable isotopic forms of CO_2.

If infrared spectroscopy is to provide an alternative technique for radiocarbon dating, the experimental sensitivity must be sufficient to detect $^{14}CO_2$ at concentrations much less than 1 ppt. At present our noise level of 2×10^{-7} m^{-1} is equivalent to 10 ppt $^{14}CO_2$, but our results have been confined to 1 m multipass cells with time constants of 1s. Under these conditions, the LAS is more sensitive to ^{14}C than conventional disintegration counters. However, it remains to be seen whether the LAS results can be extrapolated to very long time constants and longer cells [8]. A combination of signal averaging and background subtraction techniques will be required to approach the sensitivities needed for practical radiocarbon dating.

References

1. J. Reid, J. Shewchun, B.K. Garside, and E.A. Ballik: Appl. Opt. 17, 300-307 (1978).
2. J. Reid, B.K. Garside, and J. Shewchun: Opt. and Quantum Elect. 11 385-391 (1979).
3. J. Reid, M. El-Sherbiny, B.K. Garside, and E.A. Ballik: Appl. Opt. 19, 3349-3354 (1980).
4. L.S. Rothman: Appl. Opt. 20, 791-795 (1981).
5. L.S. Rothman, S.A. Clough, R.A. McClatchey, L.G. Young, D.E. Snider, and A. Goldman: Appl. Opt. 17, 507 (1978).
6. J. Reid, R.L. Sinclair, A.M. Robinson, and A.R.W. McKellar: Phys. Rev. A, in press.
7. E. Niple, W.G. Mankin, A. Goldman, D.C. Murcray, and F.J. Murcray: Geophys. Res. Lett. 7, 489-492 (1980).
8. D. Labrie and J. Reid: Appl. Phys. 24, 381-386 (1981).

New Techniques and Applications for High Resolution Tunable Submillimeter Spectroscopy

H.R. Fetterman and W.A.M. Blumberg*

Lincoln Laboratory, Massachusetts Institute of Technology, P.O. Box 73
Lexington, MA 02173, USA

Recent advances in submillimeter detectors [1] and sources [2] have opened up new possibilities for high resolution spectroscopy in the 300 to 3000 GHz regime (10 to 100 cm^{-1}). The techniques described here are based on the development of GaAs Schottky diode detectors mounted in a corner-cube-configuration, and upon new optically pumped cw submillimeter lasers. The diodes, in conjunction with the lasers have been used to make high sensitivity heterodyne radiometers and receivers suitable for high resolution spectroscopy. In addition, the same Schottky diode system has been utilized for efficient sideband modulation of laser signals. This use of the diode permits tunable radiation to be generated $\pm$ 1 cm^{-1} about any submillimeter laser line.

Several types of tunable experiments are feasible with this high sensitivity radiometer:

a) Use of the sideband generator and heterodyne detector for high resolution absorption spectroscopy.

b) The study of molecular absorption using a black body continuum source and a heterodyne detector with a tunable IF.

c) Molecular emission studies of relatively hot sources using a filter bank.

An application of tunable sideband generation is the study of double resonance in CH$_3$F[3]. This experiment, in which the rotational levels in the excited state were examined, demonstrated the sensitivity of this technique. It did not, however, realize the high resolution intrinsic to this approach. A subsequent experiment [4], discussed below, used a molecular beam formed by capillary arrays to observe sub-Doppler linewidth rotational transitions in D$_2$O and CH$_3$F. The linewidth of the molecular absorption is reduced below the Doppler limit by observing the molecular beam transverse to its direction of flow.

Figure 1 shows the configuration of source, sample and detector. The tunable submillimeter radiation is both generated and detected using the

---

*Present Address: Air Force Geophysical Laboratory, Hanscom Air Force Base, Bedford, MA 01731, USA

MOLECULAR BEAM SPECTROMETER

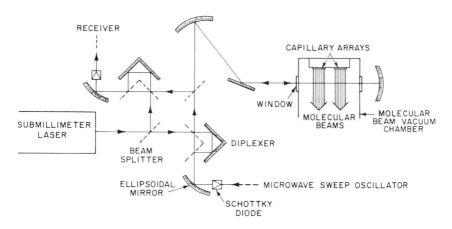

Fig.1 Schematic diagramm of the molecular beam apparatus and tunable sideband
generator and receiver

corner-cube-mounted GaAs Schottky diodes. A beam splitter directs half of
the output of a submillimeter laser into the sideband generator and half
into the heterodyne receiver for use as a local oscillator. An optical di-
plexer and ellipsoidal mirror couple the radiation into the antenna structure
of the diodes. The radiation from the sideband generating diode is directed
through a second beam splitter to the molecular beam where it is double
passed to increase its absorption length. The return is then heterodyne-
detected in the receiver diode, where typical IF signal to noise ratios are
approximately 40 dB. Weak absorptions of the sideband radiation are observed by frequency modulating the sideband signal and observing the second
derivative.

The molecular beam is produced from an array of parallel 5 μm diameter
capillary tubes. The resolution limits are then determined by the pressure required to get adequate absorption lengths. A demonstration of this
technique is shown in Fig.2. Observing the $(J,K) = (13,0) \rightarrow (14,0)$ and
$(13,1) \rightarrow (14,1)$ rotational transitions in CH_3F we find that this pair is
split by 12 MHz. In the bottom of the figure the absorption is Doppler
limited, while the top shows the sub-Doppler spectra with a linewidth reduction of more than an order of magnitude. This sub-Doppler spectroscopy
technique has a number of applications, and experiments are now underway
to examine the use of atomic beams for submillimeter frequency standards.

The receiver system shown in Fig.1 can of course also be used with a
continuum source. In this case the IF output can be sent either to a tunable
filter or to a filter bank, for data analysis. An experiment of this type
has been reported on water absorption from nozzles[5]. This has particular significance for remote sensing applications.

Perhaps the most dramatic application of this system is to look at
emission spectra outside the laboratory. In Fig.3 we show the experimental set up developed for linking the radiometer to a telescope in order to

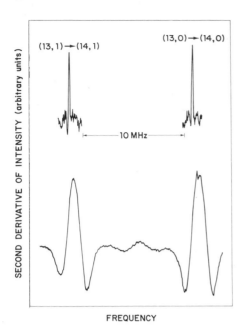

SECOND DERIVATIVE OF INTENSITY (arbitrary units)

$(13,1) \rightarrow (14,1)$

$(13,0) \rightarrow (14,0)$

⟵ 10 MHz ⟶

FREQUENCY

Fig.2 Second derivatives of sub-Doppler linewidth absorptions due to the (J,K) = $(13,0) \rightarrow (14,0)$ and $(13,1) \rightarrow (14,1)$ rotational transitions in the ground state of CH_3F obtained with the molecular beam spectrometer. The driving pressure was 0.50 Torr.

2(bottom) The same transitions observed with a static cell.

carry out the first laser heterodyne submillimeter astronomical measurements. The first experiment [6] last year successfully demonstrated the feasibility of making ground-based measurements at frequencies as high as 691 GHz--namely, the detection of the $J = 6 \rightarrow 5$ transition of CO using the NASA Infrared Telescope Facility (IRTF) located at an altitude of 4200 m on Mauna Kea, Hawaii. The system with a filter bank of 64 discrete 5 MHz filters gave a resolution corresponding to $\lambda/\Delta\lambda = 1.4 \times 10^5$ at the 434 μm.

More important, however, than the frequency resolution, in this case was the spatial resolution of 35" associated with the relatively short wavelength. This permitted the central region of the Orion nebula to be mapped out as a function of frequency (Doppler velocity of the CO molecule). In addition to the broad plateau region previously observed, it was possible to observe other regions with extremely hot, narrow line emission corresponding to different excitation mechanisms within the clouds. A portion of the data is shown in Fig.4. Observations of other CO astronomical sources, as well as a detailed analysis of the Orion data, are now in preparation[7]. However, this experiment demonstrates the sensitivity and potential applications of this type of laser radiometer.

Since the pioneering work of Gordy and his collaborators [8] some 20 years ago, the development of new semiconductor devices and laser sources have significantly changed submillimeter spectroscopy. Sensitivities and resolutions are now achievable which make possible whole new classes of both laboratory and fieldable experiments.

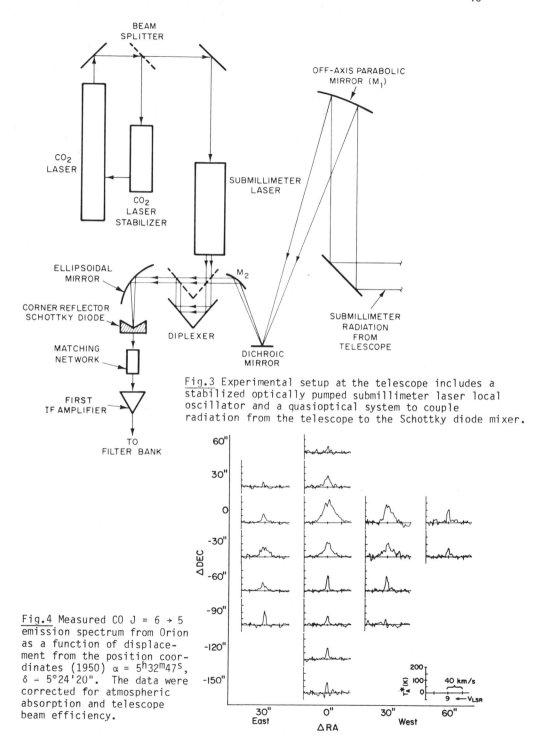

Fig.3 Experimental setup at the telescope includes a stabilized optically pumped submillimeter laser local oscillator and a quasioptical system to couple radiation from the telescope to the Schottky diode mixer.

Fig.4 Measured CO J = 6 → 5 emission spectrum from Orion as a function of displacement from the position coordinates (1950) $\alpha = 5^h32^m47^s$, $\delta = -5°24'20"$. The data were corrected for atmospheric absorption and telescope beam efficiency.

REFERENCES

1. H. R. Fetterman, P. E. Tannenwald, B. J. Clifton, C. D. Parker, W. D. Fitzgerald and N. R. Erickson, Appl. Phys. Lett. 33, 151 (1978).

2. G. Koepf, H. R. Fetterman, Nelson McAvoy, Int. J. of Infrared and Millimeter Waves 1, 597 (1980).

3. W. A. M. Blumberg, H. R. Fetterman, D. D. Peck and P. F. Goldsmith, Appl. Phys. Lett. 35, 582 (1979).

4. W. A. M. Blumberg, D. D. Peck, H. R. Fetterman, Appl. Phys. Lett. to be published.

5. G. F. Dionne, J. F. Fitzgerald, T. S. Chang, M. M. Litvak and H. R. Fetterman, Int. J. of Infrared and Millimeter Waves 1, 581 (1980).

6. H. R. Fetterman, G. A. Koepf, P. F. Goldsmith, B. J. Clifton, D. Buhl, N. Erickson, D. D. Peck, N. McAvoy and P. E. Tannenwald, Science 211, 580 (1981).

7. G. A. Koepf et al., to be published.

8. See for example Frank C. DeLucia in Molecular Spectroscopy: Modern Research, edited by K. N. Rao, (Academic Press, New York 1976) p. 69.

Infrared Heterodyne Spectroscopy of Ammonia and Ethylene in Stars

A.L. Betz

Space Sciences Laboratory, University of California
Berkeley, CA 94720, USA

1. Astronomical Background

The study of molecules in stars has greatly expanded in scope during the past decade, principally because of the development of high resolution heterodyne spectrometers for radio astronomy and the subsequent discovery of maser emission from OH, H_2O, and SiO. Although infrared frequencies have a fundamental advantage for the detection of new and more complicated molecules, the development of infrared spectrometers with the required spectral resolution and sensitivity has heretofore lagged behind equivalent advances in microwave instrumentation. In stellar sources, the linewidths of molecular transitions are generally dominated by large scale mass motions rather than simple thermal or pressure-broadening effects. Nevertheless, Doppler-broadened linewidths as narrow as 1 km/s (100 MHz at 30 THz) can still be expected. Generally, the line profiles are asymmetrical, and the details of the shapes made visible by high resolution are of critical importance for unraveling the dynamics of the stellar environment. Fortunately, developments in laser and photodiode technology now bring the advantages of coherent heterodyne detection to the infrared, with the result that the vibrational transitions of molecules can now be observed with the same Doppler-limited resolution commonly used for observations of the rotational, hyperfine, and inversion transitions which occur at radio frequencies. The importance of laser heterodyne spectroscopy for astronomical observations has been demonstrated by the first detection of ammonia [1,2] and ethylene [3] in stars. In heterodyne spectroscopy, the signal radiation collected through a telescope is mixed with a local oscillator (LO) beam from a stabilized laser in a high speed photodetector. The resulting difference frequencies are then amplified over a broad radio-frequency band and analyzed in a contiguous set of radio-frequency filters. The actual resolution is determined by the chosen widths of the filters, but the ultimate achievable resolution is limited only by the spectral linewidth of the laser.

Stellar molecules appear to be most abundant around the older "supergiant" stars with low photospheric temperatures (T < 2000 K). (The term "giant" is appropriate because these stars span a distance larger than that between the Earth and the Sun.) In their relatively late stage of life, the "red" supergiants seem to be shedding considerable amounts of material from their outer layers in a "mass-loss" process akin to the solar wind. Some of the more durable molecules, such as the metal oxides and the basic hydrocarbons (e.g. acetylene, methane, and ethylene) can form in the photosphere of the star from which they are subsequently expelled. (The exact mechanisms responsible for this initial expulsion of material are not as yet known, but some theoretical models invoke the propagation of shock waves in the photosphere.) Other more fragile species such as ammonia do not appear in much abundance until the temperature of the expanding "cloud" of expelled gas drops below ~700 K. The formation of ammonia and some other polyatomic species may be catalytically enhanced by the previous formation of "dust-grain" condensates in the cooling circumstellar material. In stars rich in oxygen the condensates appear to be silicates whereas in those stars richer in carbon the grains are thought to be silicon carbide or graphite. Atoms and radicals absorbed on the surfaces of these grains can easily combine to form more complicated molecules which can subsequently be liberated from the grain surface by photo-desorption . In addition, stellar radiation pressure on these grains accelerates them even faster away from the star, and the grains in turn tend to sweep the gas along in a radially outward expansion of cooling material. The expansion quickly carries the gas to a terminal velocity ranging from 5 to 30 km/s . In some stars, the mass-loss rate appears to have been relatively constant for at least thousands of years, while in others the process may be modulated by the 1 to 2 year periodic variations in stellar activity.

These astronomical details are relevant for the spectroscopic detection of circumstellar molecules in that the Doppler shift from the expansion velocity field (and its gradients) will govern both the observed frequency and shape of the line profile. Conversely, observations of the positions and widths of the usually asymmetric line profiles will tell us a lot about the dynamics of the circumstellar environment as well as the chemical abundances of the observed species. Transitions from different states of the molecule can be used to measure the temperature and density of the line-forming region. In an accelerating and cooling gas cloud, lines originating from levels requiring higher excitation will predominantly form closer toward the star, and the differing Doppler shifts of various transitions can be used to "map" the radial velocity gradient of the circumstellar gas. These and other observable details are of interest to the astronomer. In this context molecular spectroscopy is an applied science used as a tool to probe the physical conditions existing near stars hundreds and thousands of light years away.

At radio frequencies, molecular transitions detected in the circumstellar gas around evolved stars have been usually restricted to either those undergoing maser amplification or those from relatively abundant linear molecules such as CO, CS, HCN, HCCCN, etc. Nonlinear polyatomic molecules in relatively warm stellar environments have a larger partition function and fractionally fewer molecules in any particular state; consequently, they are more difficult to detect. Fortunately, the higher line strengths of vibrational transitions help to offset the lower population per level of polyatomics and thus enhance their chance for detection. Furthermore, in the infrared, the diffraction-limited field of view of a large telescope is better matched to the small angular sizes of the stars (<1 arcsec). Individual radio telescopes typically have main-beam-spreads (fields of view) larger than $\sim\frac{1}{2}$ arcmin and are only able to detect molecules (aside from those that are masers) which remain excited in the outer parts of the circumstellar "cloud". Another factor favoring the infrared is that many chemically important molecules have a symmetrical structure (eg. C_2H_2, C_2H_4, C_2H_6, etc.) and no permanent dipole moment. Consequently, these symmetric molecules when undistorted lack transitions available for radio-frequency observations. The perpendicular-band vibration-rotation transitions of the symmetric hydrocarbons, on the other hand, can be very intense and readily detected in the infrared.

The technology of infrared heterodyne spectroscopy has developed principally around 30 THz because of the availability of stable CO_2 lasers and the better than 90% transparency of the atmosphere in this frequency range. It is particularly fortunate that one of the most important molecules, ammonia, has its strong ν_2 vibration-rotation band interlaced with the 25 to 33 THz laser bands of CO_2 and N_2O. Frequency coincidences of <3 GHz between laser lines and ammonia transitions are quite common. The same may be said for the ν_7 band of ethylene centered near 28.5 THz. Both molecules were expected to be detectable in the infrared, and the development of the laser spectrometer was oriented toward their observation. An earlier version of the instrument had been used for the study of CO_2 in planetary atmospheres [4], but improvements in both the sensitivity and bandwidth of the system were required to accommodate the wider line profiles expected from circumstellar molecules.

2. Instrumentation

Figure 1 shows a simplified schematic of the laser spectrometer. The light collected from the star by the telescope enters on the left, comes to a focus, and then passes through a coated NaCl beamsplitter. (The telescope used for this work is the 1.5-m McMath Solar Telescope of Kitt Peak National Observatory near Tucson, Arizona.) The vertically polarized output from a

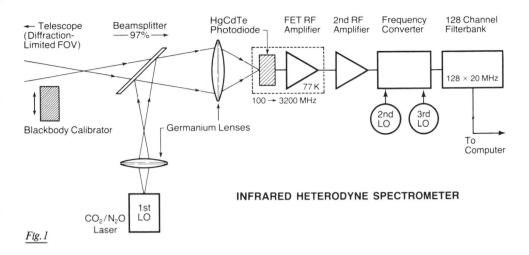

Telescope (Diffraction-Limited FOV) · Blackbody Calibrator · Beamsplitter 97% · Germanium Lenses · CO_2/N_2O Laser · 1st LO · HgCdTe Photodiode · FET RF Amplifier · 77 K · 100 → 3200 MHz · 2nd RF Amplifier · Frequency Converter · 2nd LO · 3rd LO · 128 Channel Filterbank · 128 × 20 MHz · To Computer

INFRARED HETERODYNE SPECTROMETER

Fig. 1

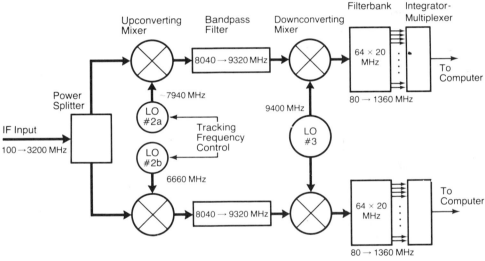

Upconverting Mixer · Bandpass Filter · Downconverting Mixer · Filterbank · Integrator-Multiplexer · Power Splitter · IF Input 100 → 3200 MHz · LO #2a ~7940 MHz · LO #2b 6660 MHz · Tracking Frequency Control · 8040 → 9320 MHz · 9400 MHz · LO #3 · 64 × 20 MHz · 80 → 1360 MHz · To Computer

Fig. 2   **INTERMEDIATE FREQUENCY CONVERTER**

CO_2/N_2O laser is first matched to the divergence of the telescope beam and then partially reflected by the beamsplitter. By expanding the Gaussian laser beam before the beamsplitter and then selecting only the central portion, a nearly uniform transverse intensity distribution can be obtained to match that of the telescope beam. In practice, the "lens" at the output of the laser is actually a pair of small mirrors which perform the equivalent beam-matching function illustrated in the diagram. The second germanium lens just before the photomixer is as shown, however, and is used at approximately f/3 in order to achieve a small spot size of ~60 μm at an optimum point on the 150 μm diameter photomixer, a HgCdTe photodiode developed by D. Spears at Lincoln Laboratory. With the cooled GaAs-FET amplifier mounted next to the detector as shown, quantum-noise limited detection performance can be achieved over a 3 GHz intermediate frequency (IF) bandwidth, even though the -3dB output rolloff of the mixer itself is only 1500 MHz. The amplified IF output is then directed through frequency conversion electronics and into a pair of 64x20 MHz filter banks. Details of the converter electronics are illustrated in Fig. 2. Since the laser oscillates at a fixed frequency, changes in the Doppler shift of the star due to both the orbital and rotational motions of the Earth must be tracked at intermediate frequencies. Depending on the direction of the star, the shift can be as large as ~ 10 MHz/hr. The 128 detected-power outputs from the filter-bank channels are simultaneously integrated in a following set of analog integrators and then sampled by a computer. Integrations are alternated between the star and a laboratory-blackbody calibrator for periods as long as several hours. Not shown in Fig. 1 are the sky-chopper fore-optics which alternate the telescope field-of-view on and off the star at 150 Hz. The chopped signals from the filter bank are synchronously demodulated at this frequency prior to integration. This differential method of observation is necessary in the thermal infrared, where radiation from the telescope structure protruding into the field of view can be orders of magnitude stronger than that from the star.

All our observational work to date has been with fixed-frequency gas lasers, which are step-tunable to discrete vibration-rotation transitions of N_2O and various CO_2 isotopes. The particular model now in use is based on the Lincoln Laboratory design of C. Freed in using a semi-sealed discharge tube, invar stabilizing rods, and grating-tuning with zero-order output coupling [5]. This design maximizes the Q of the 1.5-m laser cavity and permits CW TEM_{00} oscillation on J-values up to ~ 60 on the stronger isotopes of CO_2. Aside from continuous tunability, this laser offers all the features desired in a local oscillator: more than adequate power in a clean mode, freedom from AM and FM perturbations, reasonable physical size and efficiency, and accurate absolute frequency calibration for all the discrete oscillation frequencies [7]. In fact, the lack of continuous tunability, and hence frequency uncertainty, may be viewed as a positive feature for many spectroscopic applications, in that the laser frequency is auto-calibrated either to better than 1 part in 10^7 accuracy using the discharge impedance sensing technique [8], or to better than 1 part in 10^9 absolute accuracy with the saturated resonance-fluorescence technique [9]. Consequently, a separate involved calibration mechanism is not necessary, as would be the case with tunable diode lasers, for example. The completeness of spectral coverage over the 25 to 33 THz band may be estimated by counting the total number of laser lines available from isotopic CO_2 and N_2O lasers in the sequence and "hot" bands as well as the more conventional laser bands. The average spacing between lines is ~ 3 GHz, which is the same as the IF output bandwidth available from the photomixer. One more factor aiding the practicality of a fixed-frequency LO is that the Doppler shift from the orbital motion of the Earth with respect to the star can produce as much as ± 3 GHz of fine-frequency tuning. Once again the exact amount depends on the direction of the star, but the shift is readily calculable. Two other measurements are first required, though, before observations can be attempted. First the intrinsic velocity of the star relative to the Sun must be determined. It can usually be estimated with sufficient accuracy of 5 to 10 km/s from other types of astronomical observations. Secondly, the offset frequency between the transition of the target molecule and the LO frequency must be measured in the laboratory, preferably to an accuracy better than the 20 MHz width of a single channel. Aside from heterodyne spectroscopy done in our own laboratory on selected transitions of NH_3, C_2H_4, OCS, SiH_4, and C_2H_6, both heterodyne and an number of other laser-related techniques have been used to measure line frequencies of astrophysically important molecules with ~ 3 MHz accuracy [10-12].

3. Observations

Figure 3 shows absorption by ammonia in the circumstellar gas expanding from the supergiant star IRC +10216. Even though the star is about 900 light years away, it is the brightest stellar object in the mid-infrared. Observations of ammonia in IRC +10216 and several other stars were done in collaboration with R. McLaren of the University of Toronto [13]. The aR(0,0) line is formed by predominantly cooler gas (<200 K) which has been swept a considerable distance from the star. Most of this ammonia is no closer to the star than 100 to 1000 A.U. (1 A.U = 1 astronomical unit = the average Earth-Sun distance). The linewidth (FWHM) is

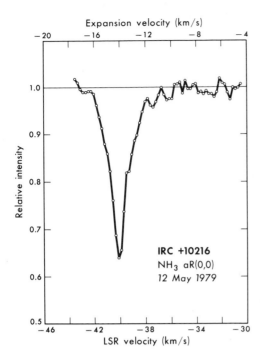

Fig.3 The aR(0,0) line of ammonia in IRC +10216. The integration time was 80 min., and the expansion velocity scale assumes an intrinsic stellar velocity of -26 km/s. The rest frequency for this transition was measured to be 28.533534 THz. A small uniform slope in the "continuum" has been removed from the data.

only 120 MHz, which is much too narrow to be resolved with other spectroscopic techniques currently used in astronomy. This vibration-rotation line is the only available probe for ground-state ammonia, since the (0,0) lower level does not have a 24 GHz inversion transition, and the pure rotational transition falls in the far infrared where the Earth's atmosphere is opaque. The horizontal scale in the diagram expresses the frequency scale in terms of the Doppler velocity shift of the observed line from its rest frequency. This style of frequency presentation is commonly used to display astronomical spectra when lines of widely differing frequency are to be compared. The Doppler velocity is measured relative to the center of velocity of the Sun and a group of nearby stars, which leads to the term : local standard of rest (LSR). At the time these observations were done, only one bank of 64 filters had been constructed. The rest frequency for the aR(0,0) line was measured in the laboratory with the heterodyne spectrometer to be 28.533534 THz, about 753 MHz below the P(20) laser line of $^{12}C^{18}O_2$. Because of the widespread abundance of cold ammonia, this particular laser line has become the "workhorse" LO frequency for the detection of NH_3 in new sources.

From microwave observations of CO rotational transitions in IRC +10216, the intrinsic stellar velocity has been measured to be -26 km/s. (In astronomical convention, a negative velocity implies approaching motion.) The velocity of the circumstellar ammonia relative to the star can thus be determined and is indicated across the top scale of the figure. The gas in the line of sight toward the star is seen to be approaching us 14 km/s faster than the star itself. Observations of this "metastable" line (J=K) and two other lines originating from non-metastable rotational levels are shown in Fig. 4. The vertical scale of the aQ(3,2) line has been expanded by a factor of 5 and that of the sP(1,0) line by 1.5 to illustrate the small but measurable acceleration between the hotter ammonia seen in the aQ(3,2) line and the colder gas picked up by the aR(0,0) line. Populations of the non-metastable levels are maintained against far-infrared radiative decay mainly by collisional excitation from molecular hydrogen, which is the principal constituent of the circumstellar gas. The non-metastable lines of ammonia are therefore good probes for the gas density. Because the populations of rotational levels are sensitive to the change in excitation conditions with distance from the star, absorption lines from

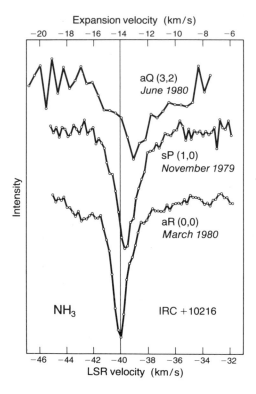

Expansion velocity (km/s)

−20 −18 −16 −14 −12 −10 −8 −6

aQ (3,2)
June 1980

sP (1,0)
November 1979

aR (0,0)
March 1980

NH₃

IRC +10216

−46 −44 −42 −40 −38 −36 −34 −32

LSR velocity (km/s)

Fig.4 Comparison spectra of 3 ammonia lines in IRC +10216 which show the radial acceleration of the gas from hotter to cooler regions. Adjacent 20 MHz channels in the aQ(3,2) line have been averaged to reduce the peak-to-peak fluctuations. The noise level in each trace can be estimated from the scatter in the 5 leftmost channels. The aR(0,0) line shown here was observed 10 months after that illustrated in Fig.3 and does not have any continuum slope removed.

different lower levels can be expected to show any gradient in the radial velocity field. In addition, it was found that lines originating from rotational levels much higher than (J,K) = (6,6) do not show any absorption at all, which indicates that ammonia does not exist in any appreciable abundance at temperatures higher than 700 K.

Ethylene, on the other hand, is stable at temperatures as high as 1200 K and thus is a better probe of the inner parts of the circumstellar region. The two lines (of admittedly low signal to noise ratio) shown in Fig. 5 were detected in IRC +10216 by using an N_2O laser [3]. Molecular constants derived from recent laser spectroscopy of ethylene in the laboratory were important for calculating line frequencies for the astronomical observations [14]. Since ethylene is an asymmetric rotor, and a relatively large number of levels are populated at high temperatures, there are fewer molecules to detect in any one state. Consequently, the lines tend to be weaker than those of ammonia, even though the fractional abundances of the two molecules are both about 10^{-7} relative to molecular hydrogen. On the other hand, the lack of a permanent dipole moment in ethylene means that there are no "permitted" rotational transitions, and the rotational levels are more likely to remain in LTE even at the low densities encountered in circumstellar gas $(\rho < 10^{12})$. Ethylene is thus a better probe of the temperature structure of the expanding gas cloud.

4. Prospects

Although the 30 THz band is currently the most favorable spectral region for infrared heterodyne spectroscopy, if practical considerations of atmospheric transparency and laser technology are considered, the 60 THz band is also attractive because of the astrophysically important bands of carbon monoxide. HgCdTe photomixers function at these higher frequencies, and isotopic CO lasers appear to be the obvious choice for the LO. On the other hand, more complete

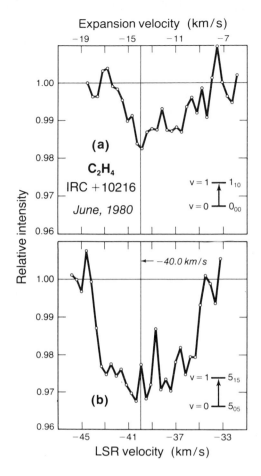

Fig.5 Two ethylene lines observed in IRC +10216. The vertical line at -40 km/s marks the velocity of peak absorption for ammonia lines seen in this star. Adjacent 20 MHz channels have been averaged to give 40 MHz resolution per data point. Integration times are 3 hours for (a) and 1.5 hours for (b).

spectral coverage around 60 THz may come from frequency-doubled CO_2 and N_2O lasers. Developmental work on a high frequency system is now being pursued at Berkeley. For the longer term, difference-frequency generation between CO and CO_2 lasers is an interesting technique for the synthesis of accurate LO frequencies beyond the ranges of the lasers themselves. This emphasis on gas laser technology is not meant to exclude other viable methods for LO generation. Following the work at other institutions, a smaller effort is now underway in our laboratory to evaluate tunable diode lasers as local oscillators for frequencies unobtainable with simple gas lasers. Much of our work, however, will concentrate on refining heterodyne spectroscopy in the 30 THz region. The spectrometer has recently been installed at the 3-m Shane telescope of Lick Observatory. This larger telescope will enhance the observations of fainter objects, and permit a more statistical survey of the abundances of polyatomic molecules in late-type stars. Initial tests indicate that phase fluctuations caused by atmospheric turbulence across the 3-m telescope aperture are often small enough in the infrared that the angular resolution of the telescope is diffraction-limited. Such diffraction-limited performance is required by the heterodyne spectrometer in order to realize the full sensitivity of the telescope on spatially unresolved stellar sources.

This work was supported in part by NSF grant AST 7920845 and NASA grant NGR 05-003-452.

References

1. A.L. Betz, R.A. McLaren, D.L. Spears: *Astrophys. J. Lett.* **229**, L97 (1979)
2. R.A. McLaren, A.L. Betz: *Astrophys. J. Lett.* **240**, L159 (1980)
3. A.L. Betz: *Astrophys. J. Lett.* **244**, L103 (1981)
4. A.L. Betz: in *Laser Spectroscopy III*, eds. J.S. Hall and J.L. Carlsten (Springer-Verlag, 1977), p.31
5. C. Freed: in *Proc. Frequency Standards and Metrology Seminar* (University Laval, Quebec, Canada, 1971), p.226
6. C. Freed, L.C. Bradley, R.G. O'Donnell: *IEEE J. Quant. Electr.* **QE-16**, 1195 (1980)
7. B.G. Whitford, K.J. Siemsen, H.D. Riccius, G.R. Hanes: *Opt. Commun.* **14**, 70 (1975)
8. M. Skolnick: *IEEE J. Quant. Electr.* **QE-6**, 139 (1970)
9. C. Freed, A. Javan: *Appl. Phys. Lett.* **17**, 53 (1970)
10. S.M. Freund, T. Oka: *Phys. Rev. A.* **13**, 2178 (1976)
11. Y. Ueda, K. Shimoda: in *Laser Spectroscopy II,* ed. S. Haroche (Springer-Verlag, 1975), p.186
12. J.J. Hillman, T. Kostiuk, D. Buhl, J.L. Faris, J.C. Novaco, M.J. Mumma: *Optics Lett.* **1**, 81 (1977)
13. A.L. Betz, R.A. McLaren: in *Proc. IAU Symposium 87, Interstellar Molecules,* ed. B.H. Andrew (D. Reidel, 1980), p.503
14. Ch. Lambeau, A. Fayt, J.L. Duncan, T. Nakagawa: *J. Molec. Spectrosc.* **81**, 227 (1980)

Stark Modulation Spectroscopy in the Visible Region

K. Uehara

Department of Physics, Keio University
3-14-1 Hiyoshi, Kohoku-ku, Yokohama 223, Japan

1. Introduction

The idea of Stark modulation spectroscopy in the visible region is a natural extension of the classical technique which has been used for many years in microwave spectroscopy and recently in infrared spectroscopy. By means of the Stark modulation technique very high sensitivities can be achieved and, at the same time, important clues for the assignment of absorption lines are obtainable. Although the potential usefulness of this technique using a cw dye laser as a light source has been demonstrated in a previous work [1] on the NO_2 spectrum, little other application of the method to visible spectra has been reported. In the present work selected parts of the visible absorption spectra of two diatomic molecules, ICl and I_2, are reexamined by Stark modulation spectroscopy.

2. Experimental

The Stark modulation spectrometer used here is essentially the same as that in [1]. A commercial cw dye laser (Spectra-Physics Model 580, dye cell type) employs rhodamine 6G as a dye and is tunable over 400 GHz at any wavelength between 555 and 600 nm. The Stark electrodes made of chromium coated glass strips are 60 or 40 cm in length with 0.5- or 0.1-mm spaces between them. A sinusoidal voltage at 100 kHz is superposed upon a dc voltage applied across the Stark electrodes. The output of the dye laser is split into two beams, one of which is focused on a detector behind the Stark absorption cell and the other of which is focused directly on the second detector. The difference between the signals from the two detectors is fed into a lock-in amplifier. An adjustment is made so that the two signals balance and give the lowest noise level. The natural isotopic mixture of ICl is prepared by evaporation of solid ICl_3 at room temperature. Frequency calibration is done by simultaneous recording of the I_2 absorption lines whose accurate wavenumbers are known [2].

3. Stark Modulation Spectrum of ICl Around 5760 Å

The spectrum of ICl around 5760 Å ($17360\ cm^{-1}$) is of particular interest because the transitions from the ground vibronic level to vibrational levels close to the lowest dissociation limit appear in this region. HULTHÉN and

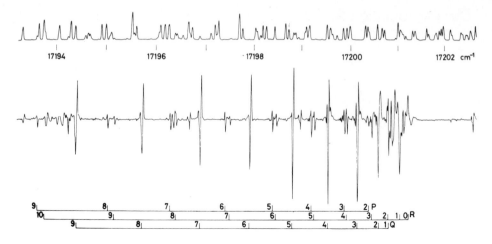

Fig.1 Portion of the Stark modulation spectrum of ICl (lower trace). The low-J rotational structure of the 28-0 band of the $A\,^3\Pi_1 \leftarrow X\,^1\Sigma^+$ system of I ^{35}Cl is resolved. Pressure,1.2 Torr; dc field,10 kV/cm; ac field,10 kV/cm peak to peak; E(optical) parallel to E(Stark); scan time,7 min. The upper trace is the absorption spectrum of I$_2$ as a wavelength reference

coworkers [3] analyzed the rotational structure of the vibrational levels in the $A\,^3\Pi_1$ state up to v' = 35 for I ^{35}Cl and v' = 32 for I ^{37}Cl. In addition, they observed that in I ^{35}Cl there are five anomalous vibrational levels (X_1, X_2, $\cdots$, X_5) between the v' = 35 level and the dissociation limit. Similar structure in I ^{37}Cl was identified by KING and McFADDEN [4]. Recently a precise analysis of the $A\,^3\Pi_1 \leftarrow X\,^1\Sigma^+$ system was performed by COXON and co-workers [5] to get more reliable molecular constants.

In the $A\,^3\Pi_1$ state each rotational level splits into two nearby sublevels (e and f levels) due to Ω-doubling. The dipole moment operator has a nonzero matrix element between these two levels and, therefore, appreciable Stark shifts are expected in the $A\,^3\Pi_1$ state even at a medium strength of the electric field. On the other hand, the Stark shift in the ground electronic state is much smaller except for very low-J levels. Since the e levels are associated with the P and R branches while the f levels with the Q branch, the Stark shift is opposite in sign in the P and R branches, and the Q branch. Such behavior is clearly seen in the 28 - 0 band of I ^{35}Cl shown in Fig.1. The shift of the Q branch lines in the high-frequency direction indicates that the f levels lie above the e levels in the v' = 28 state. The band head structure can be easily recognized because the low-J lines are enhanced by their larger Stark effect.

When an applied electric field is so strong that the wave functions of the e and f levels are mixed appreciably, the forbidden transitions become allowed and, therefore, the symmetric line shapes appear as shown in Fig.2. From the Stark splittings of the low-J lines as those in Fig.2, the permanent dipole moments in the v' = 22 and 23 vibrational levels of I ^{35}Cl were estimated as 0.9$_5$ and 0.9$_1$ Debye, respectively, which are consistent with a previous measurement by CUMMINGS and KLEMPERER [6].

In the spectral range from 17200 to 17370 cm^{-1} many new bands were found in addition to those observed by previous workers. The band origin, ν_0, and the difference of rotational constants, B" – B$_{vf}$, could be determined from the Q branch lines. Preliminary values of these constants of the new bands are summarized in Table 1. Five of the new bands were identified as those

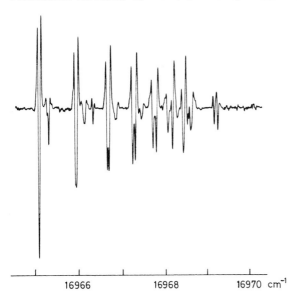

16966   16968   16970 cm^{-1}

Fig.2 Stark modulation spectrum around the band origin of the 23–0 band of the A $^3\Pi_1$ ← X $^1\Sigma^+$ system of I ^{35}Cl. dc field, 20 kV/cm; ac field, 10 kV/cm peak to peak; E(optical) parallel to E(Stark)

Table 1 Molecular constants of new bands of ICl between 17200 and 17370 cm^{-1}

Band	Band origin[cm^{-1}]	B"–B$_{vf}$[cm^{-1}]
	17221.30	0.0791
	17246.00	0.0811
	17249.75	0.0762
X$_{-4}$–0	17268.42	0.0811
	17272.67	0.0762
	17274	
X$_{-3}$–0	17288.51	0.0831
	17293.15	0.0785
	17294	
X$_{-2}$–0	17306.13	0.0848
	17310.93	0.0801
X$_{-1}$–0	17321.16	0.0871
	17326.05	0.0842
X$_0$ –0	17333.59	0.0897
	17338.49	0.0855
	17348.5	
I^{35}Cl 36–0?	17349.48	
	17364.87	
	17367.75	0.0994

associated with new members of the anomalous "X"-levels mentioned above. This assignment is supported by the fact that ν_0 and B" – B$_{vf}$ can be approximated by smooth functions of relative vibrational quantum number. The new levels are denoted by X$_{-4}$,····,X$_0$ in Table 1. In all the ten "X"–0 bands the Q branch lines show the Stark shifts in the high-frequency direction, indicating that the f levels are located above the e levels. This conclusion is not in agreement with the previous analysis [5].

A very weak band at 17349.48 cm^{-1} is probably the 36–0 band of I ^{35}Cl. Two bands were found beyond the X$_5$–0 band at 17363.34 cm^{-1}. Some of the unidentified bands appear to form a smooth progression which converges to around 17365 cm^{-1}.

4. Stark Effect in I_2

The Stark shift in molecular iodine has been observed by the same experimental setup [7]. So far as the author knows it is the first observation of the Stark shift in the visible absorption spectrum of any homonuclear diatomic molecule. A portion of the Stark modulation spectrum of I_2 is shown in Fig. 3. All lines in the region studied exhibited the second-order Stark shift of nearly the same amount in the low-frequency direction.

Recently LAYER and coworkers [8] reported a significant power-dependent frequency shift of a 633-nm I_2-stabilized He-Ne laser. Their measurement revealed that the output frequency of the laser is shifted by an amount which is proportional to the output power. This behavior is qualitatively the same as the dc Stark effect observed in I_2, if the power-dependent shift is assumed to arise from the Stark effect of I_2 caused by the optical field. However, this assumption leads to a very large value of the polarizability (assumed to be isotropic) $\alpha \sim 3 \times 10^3 \ \mathring{A}^3$, which is to be compared with 10 $\mathring{A}^3$ for the ground electronic state of I_2 determined from the dc Stark effect in hyperfine transitions [9].

An accurate measurement of the dc Stark effect in the visible transitions of I_2 can be made by monitoring the beat frequency of two lasers stabilized on the I_2 lines by the méthod proposed in [7].

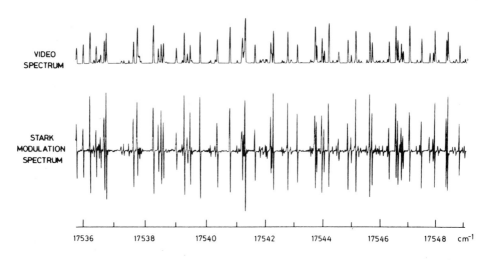

Fig.3 Video and Stark modulation spectra of I_2 around 17540 cm^{-1}; dc field, 50 kV/cm; ac field,10 kV/cm peak to peak; E(optical) parallel to E(Stark)

References

1. K.Uehara and K.Shimoda, Jpn. J. Appl. Phys. 16, 633 (1977)
2. S.Gerstenkorn and P.Luc, Atlas du spectre d'absorption de la molécule d'iode entre 14800-20000 cm^{-1} (Editions du C.N.R.S., Paris, 1978)
3. E.Hulthén, N.Johansson, and U.Pilsäter, Ark. Fys. 14, 31 (1958); E.Hulthén, N.Järlsäter, and L.Koffman, Ark. Fys. 18, 479 (1960)
4. G.W.King and R.G.McFadden, Chem. Phys. Letters 58, 119 (1978)
5. J.A.Coxon, R.M.Gordon, and M.A.Wickramaaratchi, J. Mol. Spectrosc. 79, 363 (1980); J.A.Coxon and M.A.Wickramaaratchi, J. Mol. Spectrosc. 79, 380 (1980)
6. F.E.Cummings and W.Klemperer, J. Chem. Phys. 60, 2035 (1974)
7. K.Uehara, Optics Letters 6, 191 (1981)
8. H.P.Layer, W.R.C.Rowley, and B.R.Marx, Optics Letters 6, 188 (1981)
9. D.W.Callahan, A.Yokozeki, and J.S.Muenter, J. Chem. Phys. 72, 4791 (1980)

Alignment Effects in Optogalvanic Spectroscopy

P. Hannaford* and G.W. Series

J.J. Thomson Physical Laboratory, The University of Reading
Whiteknights, Reading RG6 2AF, UK

1. Introduction

The principles of optogalvanic spectroscopy were recalled for us by Dr.
Lawler at Rottach-Egern [1].   More recently, he has published an analysis
of the mechanism for a discharge in helium [2].   The essentials of the
technique are recalled in fig. 1.   The current in a gas discharge is monitored while the discharge is irradiated with light.   As the frequency of
the light is tuned across some characteristic frequency of the atoms of the
gas, changes may be detected in the discharge current.   In favourable cases
these changes provide a very sensitive monitor of the resonance, so that
spectra may be plotted by recording the current as a function of the frequency
of the light.

A number of mechanisms, all ultimately leading to changes in ionization,
have been invoked to explain the optogalvanic effect, depending on the particular type of transition studied.   But in all cases the primary mechanism
has been the resonant transfer of atoms or ions between the states connected
by the radiation, thus changing the distribution of population over the
states, and hence the electrical characteristics of the discharge.

We report here a series of experiments which show that it may be possible
to gain useful spectroscopic information by concerning ourselves with the
polarization of the light rather than its frequency.   We are interested in

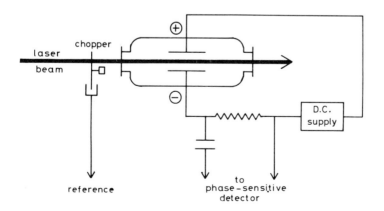

Fig. 1
Arrangement
for studying
the opto-
galvanic
effect

* On leave from CSIRO, Division of Chemical Physics, Clayton 3168, Australia

the frequency solely in order to enable us to excite some specified transition. By concentrating on the polarization we demonstrate level-crossing effects which, as is well known, can provide gτ values for excited states (zero-field crossings) and hyperfine interaction constants (finite-field crossings).

We began our experiments in order to investigate a proposal [3] to detect level-crossings through a mechanism depending solely on changes in the alignment of atoms with respect to the current in the discharge, not on changes in the overall population of states of a given angular momentum. We have not yet tested that proposal in suitable conditions. Preliminary studies revealed a level-crossing effect which has the characteristics of a mechanism proposed by Beverini and Inguscio [4]. It is non-linear in the intensity of laser light and arises from the changes of population which occur when, owing to degeneracy of levels, more than one transition in an atom is in resonance with the radiation. A detailed analysis in the context of calculating laser gain and fluorescence intensity in the side-light has been given by a number of authors [5, 6]. The name 'saturation level-crossing resonance' has been used for this situation.

2. Linear level-crossing resonance

Before we describe the non-linear effects we give a brief account of the proposed alignment effects, noticing particularly that they should manifest themselves in the linear region.

Irradiation of a gas discharge with linearly polarized light resonant with some transition will, in general, result in the alignment of the angular momenta of atoms, either directly through the radiation interaction, or indirectly by optical pumping. Now, aligned atoms will surely present different cross-sections for whatever collisional processes lead to ionization, according to their alignment with respect to the current in the discharge (Fig. 2). To be specific, we will consider ionization by electron impact. Changes of alignment - induced, for example, by means of a magnetic field - should lead to changes of current. Thus, it should be possible to detect the Hanle effect (or level-crossings at finite fields, or quantum beats) by an optogalvanic signal.

The validity of this argument depends on the ratio of directed to random motion of the electrons in the discharge; the corresponding velocity ratio is typically one to ten, but one would expect to improve this figure by going to low pressure discharges. Among the configurations we have considered are hot-cathode-sustained discharges, and r-f excited discharges in sealed cells. Such experiments at low gas pressures would then be in some

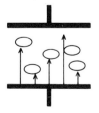 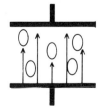

Fig. 2   Change of current arising from atomic re-alignment

sense the inverse of a technique which has been used on many occasions for studies of level-crossings and quantum beats: the creation of alignment in atoms by a directed beam of electrons and the monitoring of the alignment by observation of the fluorescent light [7]. The new proposal was to create the alignment by light and to monitor it by directed electron impact.

A specific geometrical configuration for observation of the opto galvanic Hanle effect would be: magnetic field perpendicular to discharge current, polarization vector of the light in the plane perpendicular to the field – either parallel or perpendicular to the current. It had not escaped our notice that the magnetic field would perturb the charge carriers in the current. Such perturbation was seen and allowed for in the electron-impact Hanle experiments, and could be allowed for in opto galvanic experiments.

3. Observation of alignment

Our first attempt to detect the Hanle effect by optogalvanic means was in a neon discharge, at pressures of 1 torr and below, using the transitions based on metastable states. These are known to give strong optogalvanic signals. We were unable to detect changes of current which could not be dismissed as the direct effect of the magnetic field on the current. It was concluded that these transitions were unsuitable because the optogalvanic effect is almost entirely due to perturbation of the population of metastable atoms, whose effect on the discharge, though strong, is very indirect.

The successful experiments were carried out on zirconium atoms, sputtered from a cold cathode of zirconium metal into a discharge maintained in neon at pressures in the range 0.5-3 torr. The transitions studied are shown in Fig. 3. It is to be noticed that the lower states are sufficiently close to the ground state to be heavily populated at gas-discharge temperatures, and that the group of transitions falls conveniently inside the tuning range of rhodamine 6G. The radiative lifetimes of the upper levels of most of these transitions have recently been determined [8]. The optogalvanic signals for these transitions are found to be always positive, that is to say, the discharge current is enhanced when the laser is tuned to resonance. This is consistent with the behaviour expected for signals which originate from population changes in the upper level of a transition.

The experimental arrangement for investigating alignment effects is similar to that shown in Fig. 1, but with the addition of a pair of Helmholtz coils to provide a magnetic field and a polarizer to generate the desired alignment in the upper levels. The geometrical configuration used for most of the experiments is shown in Fig. 4, where the direction of the polarization vector is arbitrarily chosen to be perpendicular to the discharge current. The cathode of the discharge was actually in the form of a sharp-edged rectangular strip (2 x 0.3 cm) and the laser beam was directed along the sharp edge. The laser source is a c.w. dye laser running multi-mode over a bandwidth about 10 GHz.

We find that the optogalvanic signal for some of the zirconium transitions exhibits a strong, sharp resonance as the magnetic field is swept through zero (Fig. 5). The recording shown in Fig. 5 was obtained for the zirconium transition 6127 Å (a^3F_4-$z^3F_4^o$), with the laser light polarized in the direc-

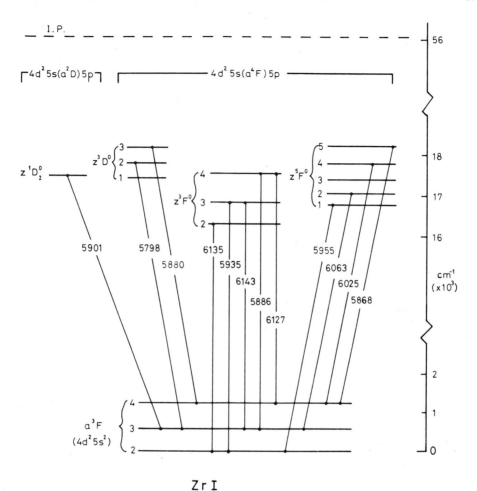

ZrI

Fig. 3   Transitions studied in Zr I

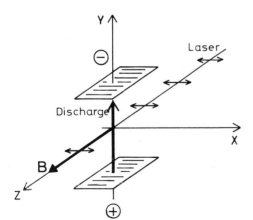

Fig. 4   Generation and monitoring of alignment

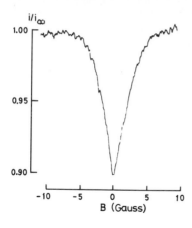

<u>Fig. 5</u> Fractional opto-galvanic signal, i/i$_\infty$, as function of magnetic field B. (0.8 torr neon. Laser power density: 6 W cm^{-2})

tion of the discharge current and with a discharge operated at 0.8 torr of neon and 5 mA current. The geometrical features of the resonance are summarised in Table 1.

The observed resonance is found to have some of the characteristics of a zero-field level-crossing signal: it is suppressed when the magnetic field is applied in the direction of polarization of the laser light, or in the case of circularly polarized light, when the magnetic field is along the axis of rotation of the electric vector, and it has an approximately Lorentzian-like shape with a width (FWHM 3.7 G) very much smaller than the Doppler width (400 G at 300 K). The resonance evidently represents a destruction of the atomic alignment by the applied magnetic field. There are

Table 1

Summary of geometrical features of observed resonance in the Zr 6127 $\overset{\text{o}}{\text{A}}$ optogalvanic signal (0.8 torr Ne, laser power density: 6 W cm^{-2}). Directions are defined in Fig. 4.

Polarization vector	Direction of Magnetic field	Discharge current	Sign and fractional amplitude of resonance
Linear Y	Z	Y	+0.10
X	Z	Y	+0.09
Y	X	Y	+0.10
X	X	Y	<0.005
Y	Y	Y	<0.005
X	Y	Y	+0.09
Y	Z	X	+0.10
X	Z	X	+0.09
Circular (left or right)	Z	Y	<0.003
	X	Y	+0.06

several features, however, that distinguish this resonance from the normal linear Hanle effect that might be observed in an optogalvanic experiment.

(i) For an axial magnetic field (along Z), the sign and shape of the resonance remain fixed for any direction of the polarization vector. In particular, the resonance does not invert when the polarization vector is rotated by $\pi/2$. There is a small systematic difference of about 10% in the amplitude of the resonance when the polarization is nominally changed from vertical (Y) to horizontal (X), but this we believe to be some polarisation artefact, as it persists when the discharge is rotated so that the current is in the horizontal polarization direction (X). Indeed, over the range of pressures used (0.5-3 torr) the ratio of the amplitudes for horizontal and vertical polarizations is invariant for any direction of the discharge current and for any position in the discharge. We conclude that the discharge current at these relatively high gas pressures is behaving as an isotropic probe.

(ii) The width of the resonance shown in Fig. 5 is a factor of about five greater than the width of the Hanle signal (0.7 G) measured in fluorescence under the same conditions. The fact that it is possible to observe the Hanle effect in fluorescence indicates that the alignment does survive in the discharge long enough to be monitored by fluorescent light. The width, sign and fractional amplitude (0.20) of the observed fluorescence Hanle signals are consistent with theoretical signals calculated from the known radiative lifetime of the $z^3F_4^o$ level (520 ns), assuming a collisional depopulation cross section of about 20 $\AA^2$. Superimposed on the sharp Hanle signals, however, is a weak, broad resonance which does not invert when the polarization vector is rotated by $\pi/2$ and which is not suppressed in the π fluorescence (analyser parallel to magnetic field). We identify this broad signal as the same resonance found in the optogalvanic experiment and conclude that it results from a change in the total population of the Zeeman sublevels when they become degenerate, rather than from the occurrence of an interference between the sublevels near zero field.

(iii) The resonance is found with different strengths in different transitions, most strongly in the transitions 6127, 6135 and 6143 $\AA$ (Table 2). The variation in strength over the different transitions is larger than can be explained by differences in the vector-coupling factors of the transitions. (The strengths of the calculated Hanle signals in fluorescent light are shown in the fifth column of Table 2.) In particular, the fractional amplitudes for transitions linked to the same upper level, such as the 6127 $\AA$ and 5886 $\AA$ transitions, can vary by more than an order of magnitude. From the results of Table 2 it is evident that the amplitude of the resonance is large when the Landé g-factors of the upper and lower levels are closely matched. This result was confirmed by extending our measurements to an additional zirconium transition at 6122 $\AA$ (a^1G_4-$y^1F_3^o$), for which the quantum numbers of the upper and lower levels are different, but for which S = 0, L = J and hence g_{LS} = 1.00 for both levels (Table 2).

(iv) The amplitude of the resonance is found to be strongly dependent upon the power density of the laser light and approaches zero at low power densities (Fig. 6). The reduction in amplitude with decreasing power density is accompanied by some narrowing of the resonance. The power densities of the laser are in the régime where some saturation in the population of the upper level is known to occur. It is clear that the resonance must be a non-linear effect which depends on partial saturation of the upper level by the laser light.

Table 2

Spectroscopic data and observed fractional amplitudes of resonance (at laser power densities of 2 and 6 W cm^{-2}) for thirteen transitions in Zr I.

λ(Å)	Transition	τ_2 [a] (ns)	$\Gamma_2^{(o)}$ [b] (G)	R [c]	g_1 [d]	g_2 [d]	Fractional amplitude of resonance (2 W cm^{-2})	(6 W cm^{-2})
6135	$a^3F_2 - z^3F_2^o$	500	0.34	0.45	0.66	0.67	0.066	0.089
5935	$a^3F_2 - z^3F_3^o$	410	0.26	0.32	0.66	1.08	0.004	0.015
6143	$a^3F_3 - z^3F_3^o$	410	0.26	0.47	1.06	1.08	0.076	0.097
5886	$a^3F_3 - z^3F_4^o$	520	0.18	0.27	1.06	1.23	0.004	0.018
6127	$a^3F_4 - z^3F_4^o$	520	0.18	0.48	1.24	1.23	0.072	0.100
5955	$a^3F_2 - z^5F_1^o$	228	1.66	0.06	0.66	0.30	0.022	0.038
6063	$a^3F_3 - z^5F_2^o$	288	0.42	0.04	1.06	0.95	0.011	0.015
6025	$a^3F_4 - z^5F_4^o$	255	0.33	0.48	1.24	1.35	0.008	0.015
5868	$a^3F_4 - z^5F_5^o$	220	0.37	0.24	1.24	1.40	<0.002	0.011
5798	$a^3F_3 - z^3D_2^o$			0.04	1.06	1.09	0.041	0.061
5880	$a^3F_4 - z^3D_3^o$	267	0.32	0.06	1.24	1.32	0.020	0.035
5901	$a^3F_3 - z^1D_2^o$	540	0.22	0.04	1.06	0.96	0.003	0.014
6122	$a^1G_4 - y^1F_3^o$			0.06	1.00	1.01	0.077	0.106

[a] Radiative lifetimes from ref. 8 and unpublished work of these authors.

[b] Width of Hanle signal corresponding to τ_2.

[c] Strength of Hanle signal in fluorescent light, as a fraction of incoherent light, calculated for the J-values involved in the transition, and assuming no collisions.

[d] Landé g-factors from ref. 9.

4. Discussion

The fact that the signals are observed under isotropic monitoring indicates that we are dealing with a change of total population in the excited state. The fact that the effects are non-linear in the intensity of the light indicates that stimulated emission plays a significant role in the process. The analysis of Feld et al. [5] would seem to describe the situation. Further evidence in support of Feld's analysis is that the observed resonances are more pointed than true Lorentzians, and could be interpreted as the superposition of Lorentzians of widths corresponding to the gT values of the states concerned.

An important feature of the resonances is the strong dependence of signal strength on the relative g-factors of the upper and lower levels. Conditions

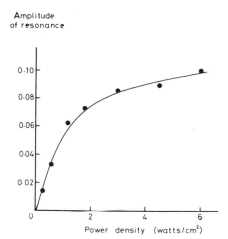

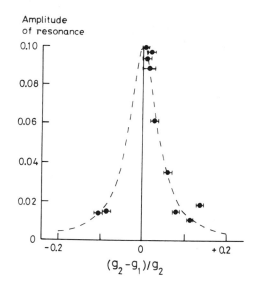

$$(g_2 - g_1)/g_2$$

Fig. 6 Amplitude of resonance
(fraction of optogalvanic
signal) as function of laser
power density. Conditions as
in Fig. 5

Fig. 7 Amplitude of resonance
as function of g-mismatch for 11
transitions in Zr I. Conditions
as in Fig. 5. The broken line
is a Lorentzian with width parameter 0.08. The horizontal
bars derive from rounding errors
in quoted g-values

which favour strong signals are that the Larmor precessional frequencies in
the upper and lower levels shall be as closely matched as possible: mismatch of the g-factors militates against efficient transfer of coherence
round the optical pumping cycle.

The use of zirconium allowed a quantitative investigation of the significance of g-mismatch. In Fig. 7 we show the signal strength, for fixed laser
power, for 11 of the transitions investigated, as a function of $(g_2-g_1)/g_2$,
where the suffixes 2 and 1 refer to upper and lower levels, respectively.
The empirical evidence suggests a Lorentzian-like dependence. We have
sketched out a theory which suggests a relationship proportional to

$\left\{ \left[(g_2-g_1)/g_2 \right]^2 + (T_{12}/T_1)^2 \right\}^{-1}$, which effectively compares the relative dephasing $(\omega_2-\omega_1)T_1$ in the lifetime of the lower state with the phase evolution
$\omega_2 T_{12}$ in the lifetime of the electric di   The linewidth parameter of the
Lorentzian shown in Fig. 7 (the broken line) would then be interpreted as
T_2/T_1. The value obtained is not unreasonable. In support of this interpretation is adduced the fact that the experimental plot narrowed when the
laser power density was reduced. Use of the data in Table 2 leads to the
result that reduction of power by a factor 3 leads to a narrowing by 20%.

Finally we emphasise that the occurrence of phenomena attributable to
isotropic monitoring of alignment reported in the latter part of this paper
does not preclude the existence of the effect suggested in the first part.

It is our intention to continue our investigations into the régime where directed motion of the electrons may be expected to predominate.

It is a pleasure to acknowledge the contribution made by Dr. S. Nakayama in the early stages of this work.

References

1. J.E. Lawler: In Laser Spectroscopy IV, ed. by H. Walther, K.W. Rothe, Springer Series in Optical Sciences, Vol. 21 (Springer Berlin, Heidelberg, New York 1979)
2. J.E. Lawler: Phys. Rev. A22, 1025 (1980)
3. G.W. Series: Comments At. Mol. Phys. X, 199 (1981)
4. N. Beverini and M. Inguscio: Nuov. Cim. Lett. 20, 10 (1980)
5. M.S. Feld, A. Sanchez, A. Javan and B.J. Feldman: 'Spectroscopie sans largeur Doppler...', CNRS No.217, Paris (1974)
6. B. Decomps, M. Dumont and M. Ducloy, in Laser Spectroscopy of Atoms and Molecules, ed. by H. Walther, Topics in Applied Physics, Vol. 2 (Springer Berlin, Heidelberg, New York 1976)
7. 'Excitation Electronique d'une Vapeur Atomique', CNRS No. 162, Paris (1976)
8. E. Biémont, N. Grevesse, P. Hannaford and R.M. Lowe, Astrophys. J., In press (1981)
9. C.E. Moore, 'Atomic Energy Levels', NSRDS-NBS 35 (1971)

Spin Polarization of Atoms by Laser Driven Two Photon and Excimer Transitions

N.H. Tran, N.D. Bhaskar, and W. Happer

Department of Physics, Princeton University, Jadcoin Hall, P.O.Box 708
Princeton, NJ 08544, USA

Spin polarization of atoms by one-photon optical pumping has been an important tool ever since its discovery by Brossel and Kastler [1] and by Hawkins and Dicke [2]. In some situations the use of two-photon optical pumping to produce spin polarization would be advantageous. For example, one might use two-photon optical pumping with wavelengths between 2000 Å and 3000 Å to spin polarize Xe^{129}, Xe^{131}, Kr^{83} and other noble gases for which direct one-photon optical pumping of the ground state necessitates the use of vacuum ultraviolet photons. The localization of the spin polarization produced by nonlinear two-photon pumping near the focus of the pumping beam would have important advantages for remote sensing applications.

The basic scheme of two-photon optical pumping is similar to that of conventional one-photon optical pumping as illustrated in Fig. 1.

The transition rates out of the spin down and spin up ground state sublevels are different and atoms accumulate in the weakly absorbing sublevels. If the atoms deexcite with equal probability to both ground state sublevels one can easily show that the mean atomic spin $\langle \vec{J}_g \rangle$ produced by photons of mean spin $\vec{s}$ is

$$\langle \vec{J}_g \rangle = \frac{n\vec{s}}{2} \tag{1}$$

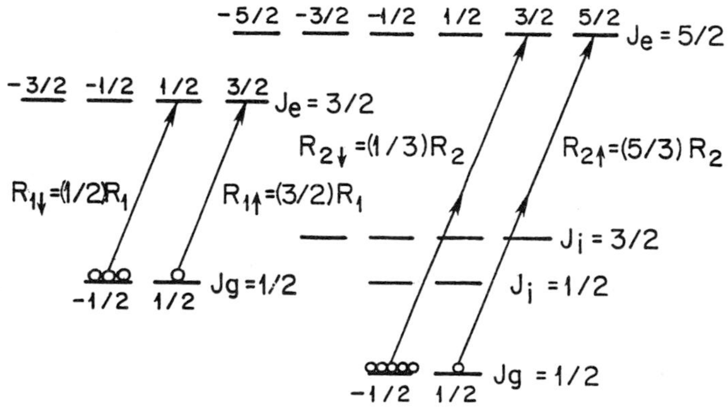

Fig.1 Spin polarization by one-photon and two-photon optical pumping with σ_+ light (R_1 = one-photon pumping rate; R_2 = two-photon pumping rate)

when the efficiency is

$$\eta = \frac{R\!\downarrow - R\!\uparrow}{R\!\downarrow + R\!\uparrow} \quad . \tag{2}$$

To test some of the basic ideas involved in two-photon optical pumping we have carried out a series of experiments with rubidium atoms, which can be excited with a laser at our disposal. The low-lying energy levels of rubidium are sketched in Fig. 2.

We have chosen to excite the 5^2D state with the $5^2S_{1/2} \to 5^2D_{5/2}$ transition with a resonant wavelength of 7781 Å for which the mean excitation rate is calculated to be

$$R_2 = \phi^2\, 3.6 \times 10^{-4} \text{ cm}^4 \text{ sec} \tag{3}$$

where ϕ is the flux of monochromatic 7781 Å photons incident on rubidium atoms with a collision broadened two-photon linewidth of 10^{10} sec^{-1}. For diffraction limited 500 mW laser beams of an initial diameter of 0.5 mm we expect $R_2 \simeq 10^6$ sec^{-1} at the focus of a lens of 16 cm focal length and correspondingly higher rates for shorter focal length lenses. Diffusion of the polarized spins out of the laser focus and other relaxation rates should be small compared to the two-photon pumping rates for the conditions of our experiments, and we therefore expect two-photon optical pumping to produce large spin polarizations.

The experimental apparatus used in our work is shown in Fig. 3. The sample is a small aluminosilicate glass cell filled with rubidium metal and various amounts of nitrogen buffer gas (typically a few tens of torr) to prevent multiple scattering of radiation and helium or argon (typically several hundred torr) to hinder diffusion of the spin polarized atoms out of the laser focus. A krypton-ion laser (Spectra Physics 171-01) with 5 watts of power in the red lines is used to pump a jet-stream dye laser with oxazine dye. We obtain multimode ($\sim$ 50 GHz linewidth) dye laser output powers of 500 mW at the two-photon wavelength, 7781 Å. The incident light is circularly po-

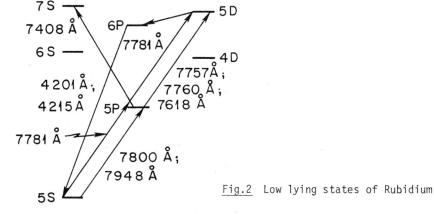

Fig.2 Low lying states of Rubidium

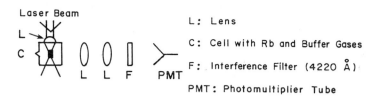

L: Lens

C: Cell with Rb and Buffer Gases

F: Interference Filter (4220 Å)

PMT: Photomultiplier Tube

Fig.3 Experimental apparatus

larized and is focussed into the cell with a short focal length lens (f = 1 cm),
optically joined to the cell with index matching fluid. A magnetic field of
several hundred gauss can be applied to the cell at any angle with respect
to the laser beam. When the field is parallel to the beam it will have no
effect on the spin polarization but when the field is perpendicular to the
beam the spins will be depolarized. We note that exceptionally large magnetic fields are required to depolarize the atoms because of the light shift
of the pumping light, which is equivalent to longitudinal fields of several
hundred gauss. As indicated in Fig. 3, we observe the 4201; 4215 Å fluorescence from the 6^2P state which can be populated by cascading from the $5^2D_{5/2}$

upper state of the two-photon transition. This eliminates interference from
any instrumental scattering of the laser beam, since highly effective optical
filters can be used to isolate the blue light.

A plot of the blue fluorescence versus the laser wavelength is shown in
Fig. 4. Note that in addition to the 7781Å two-photon peak, which is clearly
visible, there are even larger additional peaks at 7757 Å, 7618 Å and 7408 Å
and a continuum background of blue fluorescence is observed at all intermediate
wavelengths. We believe that this continuum and the peaks at 7757 Å, 7618 Å
and 7408 Å are due to excimer [3] transitions in the Rb-He molecule like those

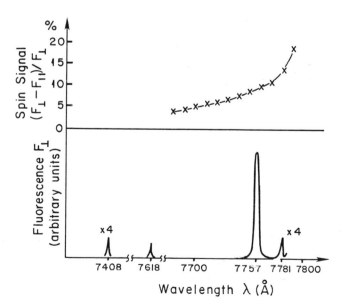

Fig.4 Fluorescence and
spin signal versus laser
wavelength (the peaks at
7757 Å and 7760 Å are
not resolved)

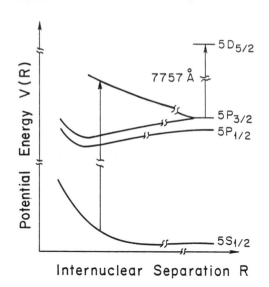

Fig.5 Rubidium-noble gas excimer transitions

sketched in Fig. 5. Support for this idea comes from the fact that the continuum background and the peaks at 7757 Å, 7618 Å and 7408 Å become much weaker relative to the 2-photon transition at 7781 Å when less buffer gas is placed in the cell. Any wavelength shorter than the 7800 Å D_2 line (but longer than any blue satellite of the line) will excite the Rb-He molecule at an appropriate internuclear separation, and the 5P atoms produced can be further excited, either resonantly to give the peaks at 7757 Å, 7618 Å and 7408 Å, or by an as yet undetermined non-resonant mechanism to produce the blue continuum.

The spin polarization is determined by measuring the difference in the blue fluorescence $F_\perp$ for a transverse magnetic field and the blue fluorescence $F_\parallel$ for a longitudinal magnetic field. Under reasonable assumptions about the optical pumping mechanism we expect to find

$$\frac{F_\perp - F_\parallel}{F_\perp} = \eta^2 \tag{4}$$

where the optical pumping efficiency η was defined in (2). The spin signal $\frac{F_\perp - F_\parallel}{F_\perp}$ is plotted in Fig. 4 versus wavelength. Note that the spin signal decreases monotonically as the laser is detuned from the D_2 line. There may be a slight enhancement of the spin signal at the two-photon wavelength but the effect is hardly significant with the present noise level.

These experiments prove that the excimer transitions are strongly spin dependent and they can be used for spin production by optical pumping. According to (4) we may regard the spin signal $\frac{F_\perp - F_\parallel}{F_\perp}$ as a measure of the efficiency for excimer pumping and we see that η decreases for shorter wavelengths, i.e. for smaller internuclear separations between the rubidium atom

and the buffer gas atom. This is to be expected because the upper state of
the excimer transition will be a $^2\Sigma_{1/2}$ state for sufficiently small internuclear separations and one can easily show that η must be zero for transitions to a $^2\Sigma$ state. Because of the efficient optical pumping on the excimer transition, we cannot yet be sure that we have seen spin polarization
produced by two-photon pumping.

References

1. J. Brossel and A. Kastler, Compt. Rend. 229, 1213 (1949).

2. W.B. Hawkins and R.H. Dicke, Phys. Rev. 91, 1008 (1953).

3. R.E.M. Hedges, D.L. Drummond and A. Gallagher, Phys. Rev. A6, 1519 (1972);
    D.L. Drummond and A. Gallagher, J. Chem. Phys. 60, 3426 (1974).

High Power Tunable Mid IR NH$_3$-N$_2$ Laser and Its Application for Selective Interaction with Multiatomic Molecules

A.Z. Grasiuk, A.P. Dyad'kin, A.N. Sukhanov, B.I. Vasil'ev, and
A.B. Yastrebkov

P.N. Lebedev Physical Institute Acad. Sci. USSR, Leninsky prospekt 53
Moscow 117924, USSR

1. Introduction

Since discovering multiphoton IR dissociation of polyatomic molecules a CO_2 TEA laser was for several years practically the only device which was used for such experiments.

However there are many polyatomic molecules of interest which can be effectively dissociated only by using radiation with lower frequency than that of a CO_2 laser. Therefore the laser community has been making efforts to develop high power tunable mid IR laser sources in the frequency range approximately between 900 and 500 cm^{-1}. Such efforts have already resulted in the creation of several types of new mid IR lasers, in particular using resonant laser pumping [1,2]. In the family of such optically (laser) pumped mid IR lasers the NH$_3$ laser using a CO_2 TEA laser as the pump source (and originally reported in [3]) is among the most promising.

The paper reported some new results of work on development and application of the NH$_3$ laser. The work which we has been performing since 1976 in the Laboratory of Quantum Radiophysics of P.N. Lebedev Physical Institute, Moscow, has resulted in creating a laser of a new type based on a NH$_3$-N$_2$ mixture. In such a NH$_3$-N$_2$ laser the pulse energy, specific energy, peakpower, efficiency, and tunability (between 745 and 890 cm^{-1}) are comparable with those of the CO_2 TEA laser of the same active volume [2,4-7]. There are two important problems in the development of the NH$_3$-N$_2$ laser. The first is shortening the pulse duration. The second problem is achieving a smooth (stepless) frequency tuning in addition to a stepped one (from line to line [2,5,7]. Possible approaches for solving the problems are, respectively, mode locking and increasing the operation pressure of the NH$_3$-N$_2$ mixture.

2. Mode Locking

We obtained mode locking in the NH_3-N_2 laser using a saturable absorber placed inside the NH_3-N_2 laser resonator [8]. Figure 1 shows the setup. A cell containing NH_3 is placed in the part of the resonator which has no pump radiation. Mode locking was achieved at several frequencies (796 cm^{-1}, 816 cm^{-1}, 753.6 cm^{-1} etc.) by rotating the grating G_3. The time history of the mode locking laser emission involved a train of pulses with pulsewidths from 6 to 8 ns. The time interval between the pulses was about 27 ns and corresponded to the resonator length of 4 m. Various absorbing cell lengths and various gas pressures were tested. Figure 2 shows a typical dependence of the NH_3-N_2 emission percentage modulation on the NH_3 pressure in the absorbing cell. Inspite of some decrease in the mode locking pulse train energy (by 1.5 times) as compared to the conventional regime the peak power increased by 3 times.

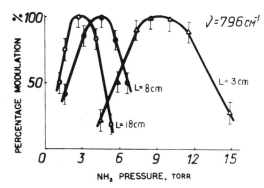

Fig.1. Optical scheme of the mode locking NH_3-N_2 laser

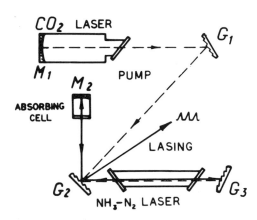

Fig.2. Dependence of percentage modulation of the NH_3-N_2 laser emission on the NH_3 pressure in the absorbing cell

3. High Pressure NH$_3$-N$_2$ Laser

Optimum laser action is possible when $I_p \gtrsim I_s$ where I_p and I_s are pump and saturation intensities, respectively. So laser action at high NH$_3$-N$_2$ mixture pressure P_{NH3-N2} is possible only at high pump intensities because $I_s \sim P^2_{NH3-N2}$ [5]. The high pump intensity must be combined with spatial homogeneity. To meet such requirements we has been using special optical schemes including a focusing raster in combination with a lightguide [2,6,7]. In the given experiments the lightguide was 18 cm long and had a cross section of 0.8×0.8 cm^2.

Using the raster-lightguide technique we previously [6] achieved efficient laser action at the NH$_3$-N$_2$ pressure of 230 torr and observed lasing at the NH$_3$-N$_2$ pressures of up to 1 atm. Optimization of the NH$_3$-N$_2$ mixture and improvement of the optical scheme resulted in obtaining lasing at pressures of up to 3 atm. Figure 3 shows a dependence of the laser output energy on the N$_2$ pressure at various values of the NH$_3$ pressure in the NH$_3$-N$_2$ mixture at the pump fluence of about 5 J·cm^{-2}. Our estimates show that lasing at pressure of about 3 atm permits smooth frequency tuning in the range of about 0.3 cm^{-1} in addition to the line-to-line tuning from 745 to 890 cm^{-1} [2,5,7]. Moreover, such a high pressure operation regime also permits further shortening of the mode-locked pulses up to subnanosecond values.

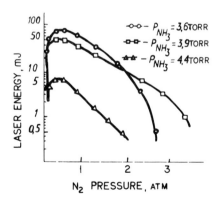

Fig.3. Dependence of the NH$_3$-N$_2$ laser output pulse energy on the N$_2$ pressure at various pressures of NH$_3$ in the operating NH$_3$-N$_2$ mixture

4. Isotopically Selective Dissociation of CCl_4 Molecules

In [9] we applied a NH_3 laser at the frequency of 780.5 cm^{-1} for isotopically selective dissociation (ISD) of $^{12}CCl_4$. The tunability of the NH_3-N_2 laser allowed us to produce ISD of either $^{12}CCl_4$ and $^{13}CCl_4$ molecules.

Figure 4 shows some results of the experiments in which we used the tunable NH_3-N_2 laser at the frequency of 753.6 cm^{-1} to affect the $^{13}CCl_4$ molecules. The experiments showed that using the mode-locked pulse train increases the isotopic selectivity as compared to the convention laser pulse.

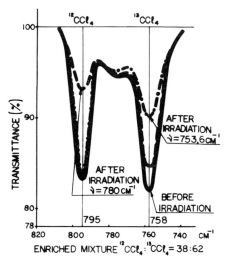

Fig.4. Isotopically selective dissociation of $^{12}CCl_4$ and $^{13}CCl_4$ using the tunable NH_3-N_2 laser emission
- - - - $\nu = 780$ cm^{-1} without mode locking [9]
- · - $\nu = 753.6$ cm^{-1} mode locking pulse train

5. Multiphoton Dissociation of UF_6 by Excitation of the $\nu_3 + \nu_5$ Combination Mode

Multiphoton IR dissociation of UF_6 has so far been produced by excitation of the ν_3 (~ 625 cm^{-1}) fundamental mode using an optically pumped CF_4 laser [10-12]. In our experiments [13] we excited $\nu_3 + \nu_5 = 824$ cm^{-1} combination mode using several NH_3-N_2 laser lines at frequencies of 828.0 cm^{-1}, 816 cm^{-1}, 812 cm^{-1} and 798 cm^{-1} with an energy of up to 0.5 J in the ~ 1 μs pulse.

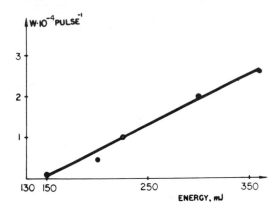

Fig.5. Dependence of the dissociation yield of UF$_6$ on the NH$_3$-N$_2$ laser energy at ν = 816 cm^{-1} exciting the $\nu_3 + \nu_5$ combination mode

Figure 5 shows a dependence of the dissociation yield W of UF$_6$ on the NH$_3$-N$_2$ laser energy at the frequency of 816 cm^{-1}. The threshold dissociation energy was 150 mJ. It corresponds to the fluence of 10 J·cm^{-2} which was only several times more than the threshold fluence observed at the excitation of the ν_3 fundamental band with the CF$_4$ laser.

Figure 6 shows a dependence of the dissociation yield W on the NH$_3$-N$_2$ laser frequency at the pulse energy of 0.3 J. One can see that maximum yield is achieved at the frequencies which are shifted by 7.5 cm^{-1} to the "red" spectral region.

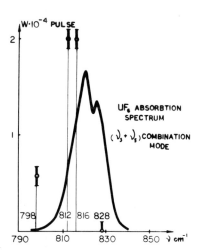

Fig.6. Dissociation yield of UF$_6$ as a function of the NH$_3$-N$_2$ laser frequency exciting the $\nu_3 + \nu_5$ combination mode of the UF$_6$ molecule

References

1. T.Y. Chang: In *Nonlinear Infrared Generation*, ed. by Y.-R. Shen, Topics in Applied Physics, Vol.16 (Springer, Berlin, Heidelberg, New York 1977)
2. A.Z. Grasiuk, U.S. Letokhov, V.V. Lobko: Prog. Quantum Electron. *6*, 245 (1980)
3. T.Y. Chang, J.D. McGee: Appl. Phys. Lett. *28*, 526 (1976)
4. B.I. Vasil'ev, A.Z. Grasiuk, A.P. Dyad'kin: Kvantovaya Elektron. *4*, 1085 (1977)
5. B.I. Vasil'ev, A.Z. Grasiuk, A.P. Dyad'kin, A.N. Sukhanov, A.B. Yastrebkov: Kvantovaya Elektron. *7*, 116 (1980)
6. B.I. Vasil'ev, A.Z. Grasiuk, S.V. Efimovsky, V.G. Smirnov, A.B. Yastrebkov: Kvantovaya Elektron. *6*, 648 (1979)
7. A.Z. Grasiuk: Appl. Phys. *21*, 173 (1980)
8. B.I. Vasil'ev, Sh.A. Mamedov: Pis'ma Zh. Eksp. Teor. Fiz. *6*, 201 (1980)
9. R.V. Ambartzumian, A.Z. Grasiuk, B.I. Vasil'ev et al.: Appl. Phys. *15*, 27 (1978)
10. J.J. Tee, C. Wittig: Opt. Commun. *27*, 377 (1978)
11. A. Kaldor et al.: Opt. Commun. *27*, 381 (1978)
12. R.V. Ambartzumian, N.G. Basov, A.Z. Grasiuk et al.: Kvantovaya Elektron. *6*, 2612 (1979)
13. B.I. Vasil'ev, A.P. Dyad'kin, A.N. Sukhanov: Pis'ma Zh. Eksp. Teor. Fiz. *6*, 311 (1980)

Part III
Double Resonance

Infrared-Ultraviolet Double Resonance Studies of Molecular Energy Transfer

B.J. Orr

School of Chemistry, University of New South Wales
Sydney, NSW 2033, Australia

1. Introduction

Of the various optical double resonance techniques available to laser spectroscopists, that involving the combination of infrared and ultraviolet
radiation has until recently received relatively little attention [1]. We
have developed a time-resolved infrared-ultraviolet double resonance (IRUVDR)
technique [2] and have applied it extensively to spectroscopic and energy
transfer studies of the molecules D_2CO and HDCO [3-5].

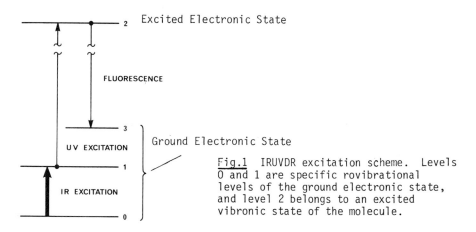

Fig.1 IRUVDR excitation scheme. Levels
0 and 1 are specific rovibrational
levels of the ground electronic state,
and level 2 belongs to an excited
vibronic state of the molecule.

Our method is shown schematically in Fig.1. Sequential pulsed excitation
by an infrared CO_2 laser and a tunable ultraviolet dye laser is detected by
time-integrated visible fluorescence. It should be noted that use of two
monochromatic pulsed lasers provides sufficient time- and wavelength-resolu-
tion to enable a specific molecular rovibrational state (1 in Fig.1) to be
probed without any elaborate sample-handling or detection apparatus [3]. The
minimum resolvable product $p\tau$ of pressure and IR-UV delay is ~0.5 ns Torr,
compared to $p\tau$=100 ns Torr for D_2CO-D_2CO gas-kinetic collisions, so that
effectively collison-free detection of single molecular rovibrational
states is feasible. If the pressure or the IR-UV delay is increased, it is
possible to study relaxation to other rovibrational states.

The areas in which our IRUVDR technique has so far been applied are three-fold: simplification and assignment of congested molecular spectra; collision-induced rotational and rovibrational relaxation rate determination; mechanistic studies of relevance to infrared multiple photon (IRMP) excitation and dissociation schemes. These will be reviewed in turn.

2. A Spectroscopic Example

Accidental coincidences between CO_2 laser lines and rovibrational transitions of D_2CO and HDCO have now been characterised by the IRUVDR method in considerably more cases than previously reported [3]. A good example is provided by that between the CO_2 10P16 laser line and the $12_{3,10} \leftarrow 12_{2,10}$ transition in the ν_4 band of D_2CO. Excitation of the $\nu_4=1$, $12_{3,10}$ rovibrational level by this means results in a great simplification of the rovibronic fluorescence excitation spectrum, as shown in Fig.2. The upper trace shows the incompletely resolved 4_1^0 band spectrum. Below it is the corresponding IRUVDR difference spectrum, in which the correlated set of six rovibronic transitions derived from the $\nu_4=1$, $12_{3,10}$ level is evident.

3. Rotational Relaxation

Figure 3 illustrates the effect of changing the IR-UV delay (and hence the product $p\tau$), in the case of the $10P16/4^0$ $PQ_3^0(12)$ IRUVDR feature of D_2CO.

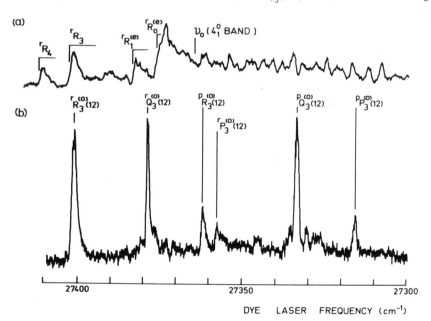

Fig.2 (a) Background 365-nm 4_1^0 band excitation spectrum of D_2CO. (b) IRUVDR difference spectrum for D_2CO pumped by the CO_2 10P16 laser line, recorded with 500-ns IR-UV delay. For both spectra the D_2CO pressure is 5 mTorr.

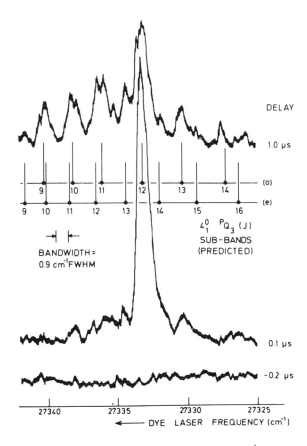

DELAY

1.0 μs

(o)

(e)

4^0_1 $^PQ_3$ (J)
SUB-BANDS
(PREDICTED)

BANDWIDTH =
0.9 cm^{-1}FWHM

0.1 μs

-0.2 μs

| | | | |
| 27340 | 27335 | 27330 | 27325 |

◄——— DYE LASER FREQUENCY (cm^{-1})

Fig.3 IRUVDR spectra for 20 mTorr of D_2CO pumped by the CO_2 10P16 laser line and probed near the 4^0_1 band $_pQ_0$ (12) transition, with IR–UV delays as indicated

A null effect is observed when the UV pulse precedes the IR pulse, a single strong 'parent' IRUVDR peak occurs with $p\tau$=2 ns Torr, and a cluster of collision-induced satellites appears as the delay is further increased. The satellites represent states adjacent to the v_4=1, $12_{3,10}$ directly pumped rovibrational level and arise from rotational relaxation which follows well established [6] collisional selection rules.

By monitoring each spectral feature in turn and scanning the IR-UV delay, a family of time-evolution curves is generated as shown in Fig.4. These have been analysed [4] in terms of a linear rate equation model to yield the rate constants governing rotational relaxation in which $\Delta J=\pm1$ and $\Delta J=0$ (that is, across an asymmetry doublet). The corresponding values of $p\tau$ (ns Torr) are 20±5 and 40±15, respectively, demonstrating the long-range nature of the collisions responsible for rotational relaxation. Our IRUVDR technique provides absolute estimates of rate constants in contrast to many other techniques [6,1], from which only relative estimates are available.

We have also monitored rotational relaxation involving a change in K_a, by observing (for example) the migration of intensity from the parent $^rR_3$ head to other ^rR heads in the IRUVDR spectrum. Such relaxation in D_2CO typically requires a single gas-kinetic collision ($p\tau$=100 ns Torr) and follows the anticipated [6] ΔK_a=±2 selection rules.

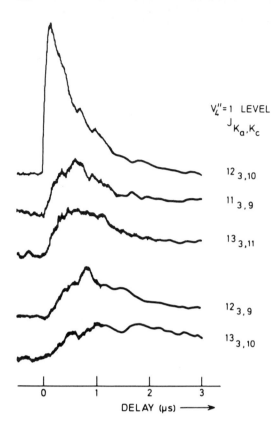

$V_i'' = 1$ LEVEL

J_{K_a, K_c}

$12_{3,10}$

$11_{3,9}$

$13_{3,11}$

$12_{3,9}$

$13_{3,10}$

0   1   2   3

DELAY (μs) ⟶

<u>Fig.4</u> Time evolution of selected IRUVDR spectral features, corresponding to population of specific D_2CO rovibrational levels by pumping with the CO_2 10P16 laser line

4. Rovibrational relaxation

The proximity of the ν_4 and ν_6 modes in D_2CO (and of ν_6 and ν_5 in HDCO) yields the possibility that vibrational energy transfer between these modes can be relatively rapid and, moreover, that it can preserve some degree of rotational specificity. To cite an example [5], when D_2CO is pumped with the 10P14 CO_2 laser line, no trace of IRUVDR excitation is found in the 4_1^0 band when $p\tau$ is small. For larger values of $p\tau$, however, intensity is seen to grow in the IRUVDR 4_1^0 band spectrum, favouring $K_a=3$ and 5 in the $\nu_4=1$ level. This is attributed to rotationally specific vibrational energy transfer which follows direct IR excitation in the ν_6 band, requires only 2 or 3 gas-kinetic collisions, and presumably is quasi-resonant.

5. Infrared Multiple Photon Excitation

We have recently adapted the IRUVDR approach to examine the specific rovibrational states involved in IRMP excitation of D_2CO and HDCO by a given CO_2 laser line [5]; such a process is known to lead to isotopically-selective dissociation. For D_2CO, it is relatively straightforward to monitor population in $v_4=2$ rovibrational states by probing the 4^1_2 band and to characterise the IR excitation mechanism involved. An interesting example is that of excitation by the 10P14 CO_2 laser line, in which a particularly complicated IRUVDR 4^1_2 band parent spectrum is attributed to specific excitation of the $6_{2,4}$, $8_{4,5}$ and $9_{4,5}$ states in the $v_4=2$ level. The experimental evidence [5] favours single-photon excitation in the $2v_4-v_4$ hot band, rather than two-photon excitation in the $2v_4$ band. As $p\tau$ is increased the types of relaxation described above occur, leading to highly selective population growth in $v_4=2$ rovibrational states with even K_a (particularly 4 and 6). This information has enabled us to devise a prototype IRMP excitation scheme consisting of a collision-assisted sequence of single-photon transitions within the (v_4,v_6) vibrational ladder of D_2CO and to demonstrate that collisions do not necessarily destroy specify in the IRMP dissociation of small polyatomic molecules such as formaldehyde.

Acknowledgments

The work reported has benefitted greatly from the efforts of John Haub, Gary Nutt, Jim Steward, and Os Vozzo. Support from the Australian Research Grants Committee is gratefully acknowledged.

References

1. J.I. Steinfeld and P.L. Houston in *Laser and Coherence Spectroscopy*, ed. J.I. Steinfeld, Plenum Press, New York 1978, pp.1-123.

2. B.J. Orr and G.F. Nutt, Opt.Lett. 5, 12-14(1980).

3. B.J. Orr and G.F. Nutt, J.Mol.Spectrosc. 84, 272-287(1980).

4. B.J. Orr, J.G. Haub, G.F. Nutt, J.L. Steward, and O. Vozzo, Chem.Phys.Lett. 78, 621-625(1981).

5. B.J. Orr and J.G. Haub, Opt.Lett. 6, 236-238(1981).

6. T. Oka, Advan.At.Mol.Phys. 9, 127-206(1973).

Pulsed Optical-Optical Double Resonance Spectroscopy of $^7Li_2$

R.A. Bernheim, L.P. Gold, P.B. ·Kelly, T. Tiptón, C.A. Tomczyk, and D.K. Veirs
Department of Chemistry, 152 Davey Laboratory
The Pennsylvania State University
University Park, PA 16802, USA

One of the more fascinating applications of laser spectroscopy is the use of multiphoton phenomena to explore molecular electronic states which cannot be excited by one photon electric dipole transitions. Such studies are particularly rewarding when applied to diatomic molecular spectroscopy where a comparison between theory and experiment is often possible and of fundamental interest. The electronic state structure of Li_2, the least complex stable homonuclear diatomic besides molecular hydrogen, is an important example. Recently, we initiated OODR experiments on Li_2 with the intention of making a comprehensive spectroscopic investigation of the one-photon "forbidden" states [1-4]. A pulsed OODR technique was developed which is extremely useful for the characterization of states in this molecule that have been previously unobserved. The pulsed approach to OODR yields clean, simple spectra which are easily and unambiguously assigned. The enormous simplification of the OODR spectrum arises because (1) a single transition between the ground and intermediate state can be selectively pumped by one laser while a second laser is scanned to produce excitation from the populated intermediate level to the final states; and (2) the amount of relaxation within the intermediate state can be controlled by adjusting the delay time separating the pump laser pulse from the scanned laser pulse.

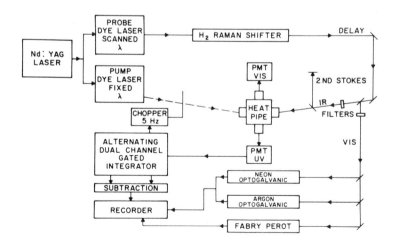

Fig.1 Schematic diagram of the pulsed OODR experiment on Li_2 in which one tunable dye laser is used to produce 1st Stokes stimulated Raman scattering for tunable radiation in the near infrared

In the initial experiments the output of two tunable dye lasers, simultaneously pumped by a nitrogen gaser, were used to pump and probe the first and second steps of the OODR transition [1-4]. It soon emerged that the probe laser would have to be scanned in the near infrared region in order to obtain all of the required data. Rather than use the inconveniently dangerous near-infrared dyes, the tunable near-infrared radiation was produced by Stokes shifting the output of the scanned probe dye laser with a high pressure (200 psi) H_2 gas filled cell. To obtain the threshold power required to generate the spontaneous Raman scattering the pair of dye lasers was pumped by the frequency doubled output of a Nd:YAG laser as shown in Fig.1. The pump dye laser was operated with 0.5-1.2 mJ/pulse output and the probe laser with up to 0.5 mJ/pulse output at 10 Hz. The unshifted probe dye laser radiation, anti-Stokes, and 2nd Stokes radiation were eliminated with filters and dielectric coated mirrors. The remaining 1st Stokes radiation was time delayed and overlapped with the pump laser beam position in the heat pipe cell. The remaining experimental details are similar to those described previously [1-4].

Either visible or uv Li_2 fluorescence is used to detect the spectral excitation. In order to distinguish the desired OODR transitions from resonantly enhanced two-photon excitation produced by the probe laser alone, a subtraction technique was applied. Both dye lasers were excited at 10 Hz. The pump laser was additionally interrupted at 5 Hz. All signals that originated from the probe laser alone at 10 Hz could be subtracted from the total output of the gated integrator, leaving only those signals at 5 Hz which were dependent on both lasers.

The experiments were conducted on lithium vapor (99.99% ^{7}Li) in a cruciform heat pipe cell operated at approximately 800°C where rough estimates of the partial pressures of the species present are 70 Torr He, 0.1 Torr Li_2 and 2.6 Torr Li. Calibration of the OODR spectra was achieved by simultaneously recording the spectra of neon and argon produced in optogalvanic cells and the fringes produced when part of the probe laser radiation was sampled by a Fabry-Perot interferometer. Both pump and probe dye lasers

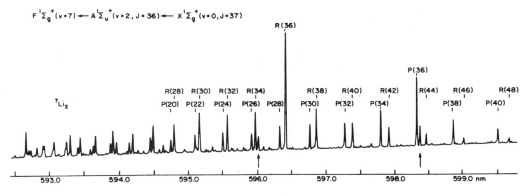

$F\,^1\Sigma_g^+\,(v=7)\leftarrow A\,^1\Sigma_u^+\,(v=2,J=36)\leftarrow X\,^1\Sigma_g^+\,(v=0,J=37)$

Fig.2 One of the bands observed in a pulsed optical-optical double resonance experiment on Li_2 showing an example of a strong perturbation of the J=35 level of the excited $^1\Sigma_g^+$ state. The extra lines occur near R(34) and P(36), marked by arrows. The second pulse in the optical optical double resonance experiment has been delayed by about 5 nsec after the first pulse producing rotational relaxation in the $A\,^1\Sigma_u^+$ state and giving rise to all the transitions other than R(36) and P(36)

were operated without etalon tuning elements, and both produced radiation
linewidths of 0.3 cm^{-1}. The wavelength measurements of the OODR transitions
could be made with an estimated accuracy of 0.2 cm^{-1} over the experimentally
covered wavelength range of 380-775 nm.

The observed spectra consist of vibrational bands in which the rotational
components originating from the pumped level in the intermediate state have
a pronounced intensity as shown in Fig.1. The transitions which require a
rotational relaxation in the intermediate state are weaker. Occasionally,
strong rotational perturbations produce extra lines in the spectrum as
indicated by the arrows in Fig.2. In this case, these are likely due to
neighboring triplet states and offer a means of access to the triplet state
manifold for spectroscopic study. As a result of differing Franck-Condon
factors, different groups of final state vibrational bands are produced
depending upon which vibrational level in the intermediate state is pumped.
This makes it possible to obtain a data field that covers a major portion of
the potential energy curve. Comparison of calculated Franck-Condon factors
with observed band intensities as well as missing bands yields the vibrational numbering assignment. The data were fit by a Dunham coefficient
expansion to give molecular constants which were then used to generate an
RKR potential. Franck-Condon factors were determined for the range of
observed vibrational levels.

The spectroscopic results have shown that the lower 60% of the $G^1\Pi_g$ and
$F^1\Sigma_g^+$ states are now known with reasonable accuracy. The experiments with
the Nd:YAG pumped dye lasers have resulted in an accurate determination of
the lower 30% of the $E^1\Sigma_g^+$ state. These three states are shown in Fig.3.

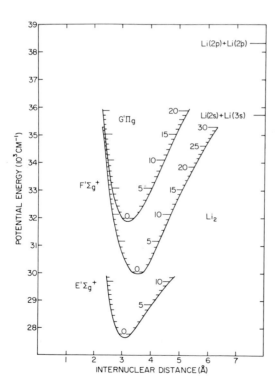

Fig.3 The potential energy
curves of three of the new
gerade states of $^7Li_2$ observed
for the first time by the
pulsed OODR techniques. The
curves are RKR potentials
constructed from the Dunham
molecular constants determined in this work

The E state is thought to dissociate into the $(2^2S)+(3^2S)$ lithium atomic states while the F and G states are thought to dissociate into the $(2^2P) + (2^2P)$ atomic states. Among the interesting features associated with these results are a homogeneous perturbation of the $F^1\Sigma_g^+$ state, probably by the $E^1\Sigma_g^+$ state, and evidence for the possibility of a double minimum in the $E^1\Sigma_g^+$ state. This latter behavior is analogous to similar findings in molecular hydrogen and has its origin in the (+,-) ion pair potential for the molecule [5]. A summary of selected molecular constants for $^7Li_2$ is given in Table 1.

Table 1 Summary of selected molecular constants for $^7Li_2$

State	$T_e(cm^{-1})$	$\omega_e(cm^{-1})$	$B_e(cm^{-1})$	$r_e(Å)$
$G^1\Pi_g$	31868.45	229.26	0.46887	3.2014
$F^1\Sigma_g^+$	29975.12	227.25	0.38177	3.5479
$E^1\Sigma_g^+$	27410.21	245.88	0.50469	3.0857
$D^1\Pi_u$	34518	201.68	0.4628	3.222
$C^1\Pi_u$	30550.6	237.9	0.5075	3.077
$B^1\Pi_u$	20436.33	270.66	0.55737	2.9363
$A^1\Sigma_u^+$	14068.31	255.47	0.49742	3.1082
$X^1\Sigma_g^+$	0	351.42	0.67245	2.6733

This research was supported by the National Science Foundation, the Donors of the Petroleum Research Fund administered by the American Chemical Society, and the U.S. Naval Sea System Command under contract with The Pennsylvania State University Applied Research Laboratory. The measurements with the Nd:YAG laser were performed at the Regional Laser Laboratory located at the University of Pennsylvania.

References

1. R.A. Bernheim, L.P. Gold, P.B. Kelly, C. Kittrell and D.K. Veirs, Phys. Rev. Lett. 43, 123 (1979).
2. R.A. Bernheim, L.P. Gold, P.B. Kelly, C. Kittrell, and D.K. Veirs, Chem. Phys. Lett. 70, 104 (1980).
3. R.A. Bernheim, L.P. Gold, P.B. Kelly, T. Tipton, and D.K. Veirs, J. Chem. Phys. 74, 2749 (1981).
4. R.A. Bernheim, L.P. Gold, P.B. Kelly, C. Tomczyk, and D.K. Veirs, J. Chem. Phys. 74, 3249 (1981).
5. D.D. Konowalow, private communication.

Double-Resonance Polarization Spectroscopy of the Caesium – Dimer (CS_2)

M. Raab and W. Demtröder

Fachbereich Physik der Universität Kaiserslautern, Postfach 3049
D-6750 Kaiserslautern, Fed. Rep of Germany

1. Introduction

The application of Doppler-free polarization spectroscopy [1] to the investigation of complex molecular spectra has several definite advantages compared with other Doppler-free techniques. Besides its high resolution and
extreme sensitivity this technique allows to distinguish between transitions
with $\Delta J = 0$ and P- or R-lines with $\Delta J = \pm 1$. According to the polarization
mode of the pump beam (circular or linear polarization), Q-lines in the polarization spectrum can be either suppressed or enhanced [2] . This facili-
tates the assignment of dense spectra considerably.

In spite of these advantages the assignment of a perturbed complex molecular spectrum may still be tedious. In order to reach an unambiguous J-num-
bering of individual rotational lines it may be necessary to record many
lines in each band. In such cases the combination of polarization spectroscopy with optical double resonance techniques is helpful and time-saving for
the assignment because it reduces the numerous lines of a spectrum to a few
double resonance signals [3] .

2. Double-Resonance Schemes

For optical-optical double resonance (OODR) molecular spectroscopy two different lasers are used. While the "pump laser" is stabilized onto a selected
molecular transition $(v_i^!, J_i^!) \leftarrow (v_k'', J_k'')$ the "probe-laser" is tuned through
the spectral range of interest. Chopping of the pump laser intensity at a
frequency f_1 results in a corresponding modulation of the population densities N_i, N_K of lower and upper level. Any time the probe laser is tuned to
a transition sharing either the lower level (v_k'', J_k'') or the upper level
$(v_i^!, J_i^!)$ with the pump, a modulated probe signal is detected with a modulation phase that is opposite for both cases. The pump may either affect the
absorption of the probe level (level schemes b) and c) in Fig.1) or the induced emission (scheme a)).

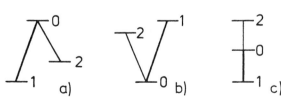

Fig.1 Different transition
schemes for double-resonance
spectroscopy

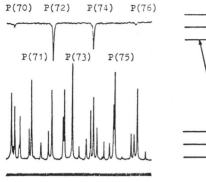
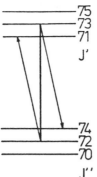

Fig.2 Illustration of the simplification of the polarization spectrum (lower trace trace) by double resonance spectroscopy. The two OODR-signals are due to the V- and Λ-scheme

In OODR-polarization spectroscopy the selective pumping of different M-sublevels influences the state of polarization of the probe wave. Pumping on a P- or R-transition yields OODR-signals which differ in sign for P and for R-probe transitions while Q-transitions can be either completely suppressed or greatly enhanced according to the pump polarization.

The significant simplification of a "normal" polarization spectrum by the OODR-technique is illustrated in Fig.2, which shows two OODR-signals generated at the same pump transition where the probe shared either the lower level (V-scheme) or the upper level (Λ -scheme) with the pump laser. It is worth noting, that for copropagating pump and probe beams the parametric interaction of the two waves with the molecule generate stimulated resonance Raman scattering. In case of the Λ -scheme of Fig.1a this brings about an OODR-signal with a linewidth which equals approximately the sum $\gamma_1 + \gamma_2$ of the levelwidths of the two nonshared levels [4] . If these are rotational levels of the electronic groundstate with long lifetimes, the linewidth of the OODR-signal may become smaller than the natural linewidth of the optical transition. Such subnatural linewidths have been indeed found by us in Cs_2 where the natural linewidth is much larger than in I_2, where this effect has been first observed [5] .

3. Experimental Setup

Figure 3 shows a schematic diagram of the experimental arrangement. Two actively stabilized tunable single mode cw dye lasers (a "homemade" and a commercial CR 599) are used. While the pump beam is sent unfocussed through the heat pipe containing the caesium vapor, the probe beam is slightly focussed by a f = 200 cm lens. The vapor pressure was kept at about 10 millitorr. A small fraction of the pump beam was split off and used as an extra probe. This allows to record a normal polarization spectrum with the pump laser and to stabilize the pump laser onto the center of the desired molecular line. Chopping of the probe laser at a frequency f_2 and of the pump laser at f_1 allows to monitor the OODR-signals through a lock in at the sum frequency $f_1 + f_2$ while the polarization spectrum of the pump laser can be detected at the frequency f_1. This "intermodulated polarization spectroscopy" is analoguous to the intermodulated fluorescence technique [6] . The wavenumbers of the molecular transitions were determined by stabilizing the probe laser

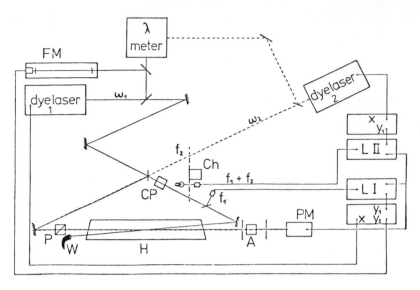

Fig.3 Experimental arrangement

onto the center of a molecular line and measuring its wavelength in a Michelson-type wavemeter [7] with an iodine-stabilized He-Ne-laser as reference. Frequency marks for interpolation of wavenumbers between absolutely measured lines were provided by a long Fabry-Perot-interferometer with a free spectral range of 63 MHz.

4. Applications to Molecular Spectroscopy of Cs_2

Polarization spectroscopy and OODR were used to measure and unambiguously assign about 2400 lines in the $C^1\pi_u \leftarrow X^1\Sigma_g^+$-system of Cs_2. Together with 450 lines obtained by Höning et al.[8] from laser induced fluorescence measurements these lines were used in a weighted least squares fit to get accurate molecular constants for the X-groundstate and the $C^1\pi_u$-state [9] . These constants reproduce the measured lines within $10^{-3}cm^{-1}$ for transitions up to v' = 10. Transitions to higher vibrational levels showed increasing deviations which can be explained by configuration interaction with a higher lying $D^1\Sigma_u^+$-state. This state which is also responsible for the Λ-doubling in the $C^1\pi_u$-state, has not been assigned previously although COLLINS et al. [10] have postulated its existence to explain their measured laser photolytic cross sections. Using polarization spectroscopy and OODR we could measure and identify some bands and extended vibrational progressions in the $D^1\Sigma_u^+ \leftarrow X^1\Sigma_g^+$ system. The missing Q-lines identify the state to be a Σ_u^+-state. From the vibrational spacings and the measured rotational constants, the shape of the potential curves and its minimum distance could be determined (see Fig.4). At larger internuclear distances the two potential curves approach each other, both merging into the 5^2D+5^2S-atomic states. This narrow approach causes a deformation of the $C^1\pi_u$-potential curve and is responsible for the sudden decrease of the vibrational spacings for v' > 10 in this state.

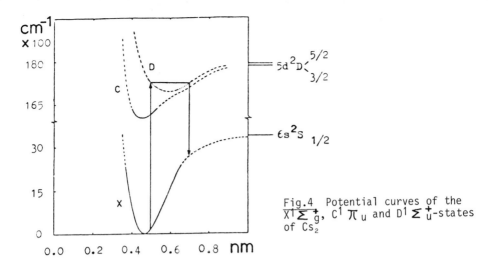

Fig.4 Potential curves of the $X^1\Sigma_g^+$, $C^1\Pi_u$ and $D^1\Sigma_u^+$-states of Cs_2

Populating selected levels in this D-state by optical pumping offers in-teresting possibilities for spectroscopic applications. Since the Franck-Condon factors for transitions from these levels to high lying levels in the X-ground state, close to the dissociation energy, are large, an OODR-experiment, based on the Λ-scheme of Fig.1a will yield an accurate deter-mination of the ground-state dissociation energy, which is still in question.

Since these levels can be effectively depopulated by collision induced dissociation, the possibility to maintain inversion between these levels and optically pumped (v', J')-levels in the $D^1\Sigma_u^+$-state is certainly worth-while to investigate. Because of the large level density the realisation of an optically pumped Cs_2-dimer laser in this spectral range would offer a quasi-continuous tunable source around 700 nm.

References

1. C. Wieman, Th.W. Hänsch, Phys. Rev. Lett. 36, 1170 (1976)
2. R.E. Teets, F.V. Kowalski, W.T. Hill, N. Carlson, T.W. Hänsch, Proc. Soc. Photoopt. Instr. Eng. 113, Advances in Laser Spectroscopy, San Diego (1977), p. 80
3. R.E. Teets, R. Feinberg, T.W. Hänsch, A.L. Schawlow, Phys. Rev. Lett. 37, 683 (1976)
4. V.P. Chebotayev, V.S. Letokhov, Nonlinear Laser Spectroscopy, Springer Series in Optical Sciences, Vol. 4 (Springer Berlin, Heidelberg, New York 1977)
5. R.P. Hackel and S. Ezekiel, Phys. Rev. Lett. 42, 1736 (1979)
6. M.S. Sorem and A.L. Schawlow, Opt. Commun. 5, 148 (1972)
7. F.V. Kowalski, R.E. Teets, W. Demtröder, A.L. Schawlow, J. Opt. Soc. Ann. 68, 1611 (1978)
8. G. Höning, M. Czajkowski, M. Stock, W. Demtröder, J. Chem. Phys. 71, 2138 (1979)
9. M. Raab, G. Höning, W. Demtröder, C.R. Vidal, to be published
10. C.B. Collins, F.W. Lee, J. A. Anderson, P.A. Viacharelli, D. Popescu and I. Popescu, J. Chem. Phys. 74, 1067 (1981)

Laser-Microwave Spectroscopy of Li⁺ Ions*

R. Neumann, J. Kowalski, F. Mayer, S. Noehte, R. Schwarzwald, H. Suhr,
K. Winkler, and G. zu Putlitz*

Physikalisches Institut der Universität Heidelberg*
D-6900 Heidelberg, and
Gesellschaft für Schwerionenforschung
D-6100 Darmstadt, Fed. Rep. of Germany

The Li^+ ion, a two-electron system like He is of fundamental interest as a
testing ground for atomic structure theory. This paper reports on combined
laser-microwave measurements of hyperfine structure (hfs) splittings of the
metastable ($\tau \approx 50$ sec) $1s2s\ ^3S_1$ state in $^{6,7}Li^+$ and of the short-lived
($\tau = 43$ nsec) $1s2p\ ^3P$ multiplet in $^7Li^+$. The precise data from this experiment should stimulate improved ab initio calculations of the $2\ ^3S_1$ hfs not
available yet. It is also an aim of this experiment to test the very precise
theoretical hfs [1] and fine structure [2] values existing for the $2\ ^3P$
states. For very accurate $2\ ^3P$ fine structure measurements in ^4He see [3].

In Li^+ a resonant transition of $\lambda = 548.5$ nm connects the $2\ ^3S_1$ and $2\ ^3P$
terms (Fig.1). The $2\ ^3S_1$ hfs has been measured for $^6Li^+$ (nuclear spin I=1)
and $^7Li^+$ (I=3/2) by means of the optical pumping method. The laser-microwave
spectrometer (Fig.2), including the ion beam production and signal processing, has been described in more detail in a recent publication [4]. Fig. 3 a

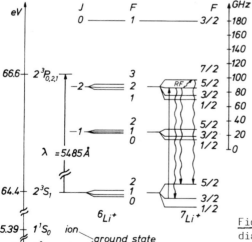

Fig. 1 Part of $^{6,7}Li^+$ energy level
diagram with $2\ ^3S_1$ state and $2\ ^3P$ multiplet (the isotope shift is omitted)

*This work is sponsored by the Deutsche Forschungsgemeinschaft.

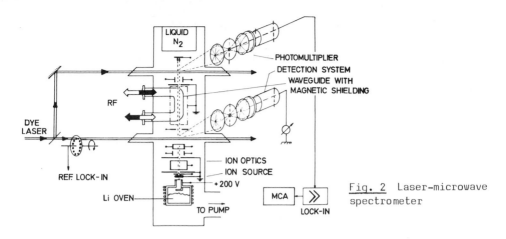

Fig. 2 Laser-microwave spectrometer

and b give examples of microwave resonance signals. From the splittings the magnetic hfs interaction constants A $(^6Li^+)$ and A $(^7Li^+)$, and small depressions $\delta(^6Li^+)$ and $\delta(^7Li^+)$ of the F=I hfs sublevels were extracted (Table 1). The hfs anomaly $\Delta^{6,7}_{ion}$, defined by the equation $\Delta^{6,7} = {^6A}/{^7A} \cdot {^7g}/{^6g} - 1$, with $^{6,7}g$ being the nuclear g-factors, agrees well (Table 2) with $\Delta^{6,7}_{atom}$ obtained from the atomic $2\,^2S_{1/2}$ ground state A-factors [5]. Obviously the atomic $1s^2 2s\,^2S_{1/2}$ configuration and the ionic $1s2s\,^3S_1$ configuration are sensitive to the same distribution of the nuclear magnetism. This behaviour has also been observed for the $1s\,^2S_{1/2}$ and $2s\,^2S_{1/2}$ of hydrogen and deuterium [6]. With the equation for the energy splitting between two hfs sublevels F and F', as given in [7],

$$\Delta E = \frac{8}{3} \alpha^2 \frac{m_e}{m_p} g_I\, Z^3\, [(1+\epsilon) - 3\,\frac{me}{M} + \frac{\alpha}{2\pi}\,]\, R_\infty\, (Y_F - Y_{F'})\,,$$

theoretical A values have been calculated (Table 1). The equation corrects for the presence of the second electron (Breit-Doermann correction ϵ [8]), for the anomalous magnetic moment of the electron in lowest order ($\sim \alpha/2\pi$), and for the reduced mass. The uncertainty of ϵ dominates the error of A_{theor}.

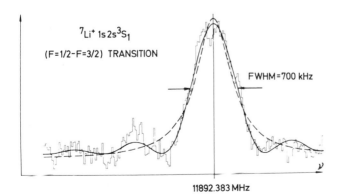

Fig. 3a $2\,^3S_1$ (F=1/2 – F=3/2) microwave resonance signal in $^7Li^+$ (with fit curves discussed in [4])

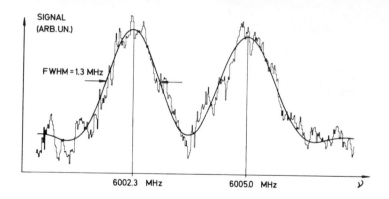

SIGNAL
(ARB.UN.)

FWHM = 1.3 MHz

6002.3 MHz   6005.0 MHz ν

Fig. 3b $2\ ^3S_1$ (F=1 - F=2) transition in $^6Li^+$ with two Doppler shifted signal curves caused by microwave reflection

Higher order QED and relativistic corrections are needed in order to cover the difference between experimental and theoretical values. We know from ^{3}He, that there should also be a significant nuclear structure effect [9]. The depression δ (Fig. 4), within the present error bars significant only for $^7Li^+$, is caused by $2\ ^1S_0 - 2\ ^3S_1$ mixing via hfs interaction [10]. Table 1 also lists theoretical values for δ , calculated with the hfs theory of two-electron systems, given in [11]. Improved calculations, using better wave-functions are in progress [12].

The laser-rf method has been extended to hfs measurements in the $2\ ^3P$ multiplet. Microwave transitions inside a microwave cavity between two ^3P hyperfine levels change the population of the $2\ ^3S_1$ hfs substate from which the laser excitation starts. This rf resonance induced optical pumping is monitored via the fluorescence light intensity change in a second intersection zone of the laser beam and the ion beam far outside the cavity. The laser-rf resonance cycle is drawn in Fig. 1 for the $2\ ^3P_2$(F=5/2 - F=7/2) rf transition in $^7Li^+$. Fig. 5 shows a microwave resonance signal. Besides a rigorous test of fine and hfs theory a complete laser-microwave measurement of the $2\ ^3P$ energy splittings could provide an improved value of the electric nuclear quadrupole moment Q of ^7Li, till now known to 15% [13,14].

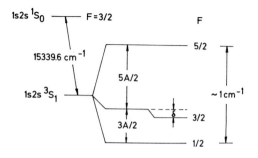

1s2s $^1S_0$ ——— F = 3/2   F

15339.6 cm^{-1} 5/2

5A/2

1s2s $^3S_1$ ~1cm^{-1}

                                      δ 3/2

3A/2

                                      1/2

Fig. 4 Scheme of $2\ ^3S_1$ hfs multiplet of $^7Li^+$ including the F=3/2 depression δ (enlarged)

Table 1. Experimental and theoretical values (in MHz) of the $2\,^3S_1$ hfs
splittings in $^6Li^+$ (a) and $^7Li^+$ (b).

a.

F=0 – F=1 splitting	3001.780 (50)
F=1 – F=2 splitting	6003.600 (50)
$A(^6Li^+,\,2\,^3S_1)_{exp.}$	3001.793 (17)
$A(^6Li^+,\,2\,^3S_1)_{theor.}$	3000.81 (10)
F=1 depression $\delta(^6Li^+,\,2\,^3S_1)_{exp.}$	0.013 (37)
$\delta(^6Li^+,\,2\,^3S_1)_{theor.}$	0.039

b.

F=1/2 – F=3/2 splitting	11890.018 (40)
F=3/2 – F=5/2 splitting	19817.673 (40)
$A(^7Li^+,\,2\,^3S_1)_{exp.}$	7926.923 (14)
$A(^7Li^+,\,2\,^3S_1)_{theor.}$	7925.15 (23)
F=3/2 depression $\delta(^7Li^+,\,2\,^3S_1)_{exp.}$	0.366 (29)
$\delta(^7Li^+,\,2\,^3S_1)_{theor.}$	0.511

Table 2. Hyperfine Structure Anomaly $\Delta^{6,7}$

Li^+ ion $(2\,^3S_1)$	Li atom $(2\,^2S_{1/2})$
$6.76(60)\cdot10^{-5}$	$6.806(63)\cdot10^{-5}$ [5]

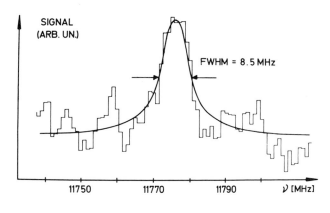

SIGNAL
(ARB. UN.)

FWHM = 8.5 MHz

11750   11770   11790 ν [MHz]

Fig. 5 Microwave resonance
signal of the $2\,^3P_2$ (F=5/2 –
F=7/2) transition in $^7Li^+$

134

References

[1] N.A. Jette, T. Lee, T.P. Das: Phys. Rev. A $\underline{9}$, 2337 (1974)
[2] B. Schiff, Y. Akkad, C.L. Pekeris: Phys. Rev. A $\underline{1}$, 1837 (1970)
[3] W.E. Frieze, E.A. Hinds, V.W. Hughes, F.M. Pichanick: Phys. Lett. $\underline{78}$
     A, 322 (1980)
[4] U. Kötz, J. Kowalski, R. Neumann, S. Noehte, H. Suhr, K. Winkler,
     G. zu Putlitz: Z. Physik A $\underline{300}$, 25 (1981)
[5] A. Beckmann, K.D. Böklen, D. Elke: Z. Physik $\underline{270}$, 173 (1974)
[6] H.A. Reich, J.W. Heberle, P. Kusch: Phys. Rev. $\underline{104}$, 1585 (1956)
[7] H.A. Bethe, E.E. Salpeter: Quantum Mechanics of One- and Two-Electron Atoms
     (Springer Berlin, Göttingen, Heidelberg 1957)
[8] P.J. Luke, R.E. Meyerott, W.W. Clendenin: Phys. Rev. $\underline{85}$, 401 (1952)
[9] S.D. Rosner, F.M. Pipkin: Phys. Rev. A $\underline{1}$, 571 (1970)
[10] M.M. Sternheim: Phys. Rev. Lett. $\underline{15}$, 545 (1965)
[11] A. Lurio, M. Mandel, R. Novick: Phys. Rev. $\underline{126}$, 1758 (1962)
[12] R. Herman, J. Kowalski, R. Neumann, G. zu Putlitz: to be published
[13] H. Orth, H. Ackermann, E.W. Otten: Z. Physik A $\underline{273}$, 221 (1975)
[14] P. Egelhof, W. Dreves, K.-H. Möbius, E. Steffens, G. Tungate,
     P. Zupranski, D. Fick, R. Böttger, F. Roesel: Phys. Rev. Lett. $\underline{44}$,
     1380 (1980)

Laser-Induced Coherence Effects in Molecular Beam Optical-RF Double Resonance

S.D. Rosner, A.G. Adam, T.D. Gaily, and R.A. Holt

Department of Physics, The University of Western Ontario
London, Ont. N6A 3K7, Canada

1. Introduction

Optical pumping, from its earliest days, has been used with rf magnetic or electric resonance to make precise measurements of atomic and molecular fine structure, hyperfine structure (hfs), and g-factors [1]. Many such experiments have been performed in cells, where the molecules interact simultaneously with the light, the rf field, the walls, a buffer gas, each other, and in some cases free electrons. The resulting alteration of the rf resonance line profile must be accounted for in extracting the constants appropriate to an isolated molecule.

Most of these undesirable effects can be eliminated by replacing the vapor cell with a molecular beam. In such experiments [2,3,4], a molecule interacts with the light only *before* and *after* it undergoes rf resonance and so it might be expected that only the rf field itself would influence the rf resonance line profile. We report here observations and calculations which demonstrate that the rf resonance line profile is a strong function of light intensity and polarization when the light is coherent, and the rf levels are optically unresolved.

2. Experiment and Results

This work arose from an earlier experiment [2] in which the hfs constants of a single rovibronic level of Na_2 were measured using laser/rf double resonance on a molecular beam. Many details of the apparatus and method which are unchanged may be found in [2].

The apparatus consists of a Na_2 supersonic beam source, two regions A and B where the molecular beam is crossed at $90°$ by a linearly polarized laser beam, and an intermediate region C where the magnetic resonance takes place in flattened solenoid 36.2 cm long. A frequency-stabilized single-mode Ar^+ laser is tuned to 476.5 nm and couples the $v''=0$, $J''=28$ level of the $^1\Sigma_g^+$ ground state to the $v'=6$, $J'=27$ level of the $B^1\Pi_u$ excited state. The spontaneously emitted fluorescence at the B region is detected by a photomultiplier, and the change in fluorescent light level when the rf magnetic field is on or off provides the signal. For this work, the Doppler width of the optical absorption was made approximately equal to the natural width of 25 MHz. Also the laser beam was expanded to illuminate the Na_2 beam more uniformly. Ambient static magnetic fields were cancelled with 3 orthogonal pairs of Helmholtz coils to $<0.5\mu T$.

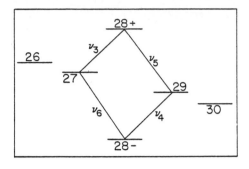

Fig.1 The hfs of v"=0,J"=28,$^1\Sigma_g^+$. The energy levels are labelled by F" and the relevant transitions are shown as solid diagonal lines labelled ν_3, ν_4, ν_5, ν_6.

The hfs of the v"=0, J"=28 level arising from the electric quadrupole and much smaller spin-rotation hf interactions [2] is shown in Fig. 1. The transitions ν_3, ν_4, ν_5, ν_6, were chosen for study because they are strong and effectively free of Bloch-Siegert shifts [5]. With a transit-time linewidth of a few kHz, the differences $|\nu_3-\nu_4|$ and $|\nu_5-\nu_6|$ are too small to be observed. The rf spectrum of these transitions from 100 to 132 kHz was measured at different rf magnetic field strenghts (B_1), laser powers (P_A,P_B), and laser polarizations. Figs. 2 and 3 show the dramatic change in the spectrum when the relative (linear) polarization of the laser and the rf magnetic fields is changed from parallel (P) to perpendicular (S). While both spectra show the expected two principal peaks, their sign, width and size are markedly different. The peak positions are also slightly shifted away from the central region in the S case, relative to the P case.

In the P case, increasing B_1 to 1.38 mT splits each peak into two, with a minimum near the resonance frequency, amplifies the subsidiary peaks, but does not change the sign of the signal. In the S case, increasing B_1 produces oscillatory changes in signal amplitude at any given frequency which at B_1=865 μT results in an inverted version of Fig.3. In fact, at the intermediate value B_1= 623 μT, the signal at all frequencies is virtually obscured by the noise.

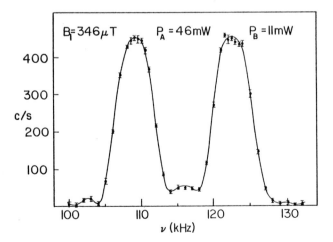

Fig.2 Rf spectrum of transition $\nu_3-\nu_6$ in the P polarization case

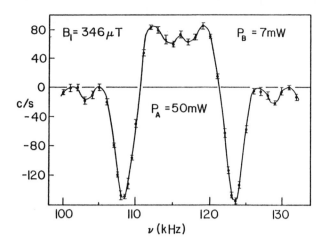

Fig.3 Same as Fig.2
except laser polariza-
tion is rotated by 90^0
(S-case)

Increasing the laser power over the range 25-65 mW increases the signal
amplitude and, in the S case, amplifies the subsidiary peaks and deepens
the valley between the two principal peaks.

3. Discussion and Conclusions

It can be shown [6] that any model of the A-C-B regions which considers
only populations, and the probabilities of their transfer among states by
the optical and rf fields, must necessarily produce *positive* signals. For
any pair of hf states, the population of the state that absorbs the light
more strongly will be depopulated relative to the other; a resonant rf
field will therefore always transfer population to the strongly absorbing
state, producing an increase in fluorescence at B. Rate equation models
are thus ruled out by the data, and one must explicitly consider the field-
induced phase coherence between states.

A molecule which absorbs a laser photon has only a 3% chance of returning
to the initial rovibronic level by *spontaneous* emission. Neglecting this
process and the effect of the counter-rotating light field allows the time-
dependent Schrödinger equation to be reduced to a set of coupled equations
for the amplitudes of the states. This can be easily transformed to a time-
independent form which is exactly soluble by standard matrix techniques. The
important physical effect is that any pair of ground state hf levels which
are coupled to the same excited state level by the processes of absorption
and stimulated emission will emerge from the A region with a definite phase
relation between their amplitudes. These amplitudes become the initial conditions to the C-region Schrödinger equation which is handled in an identical manner (again with the justifiable neglect of the counter-rotating rf
field component). Finally the fluorescence is obtained as the difference
in each state population before and after the B optical pumping, summed over
the hf states of the v"=0, J"=28 level. The signal is then the difference
in this fluorescence with the rf on and off. It is very important to average
this signal over the velocity distribution of the molecular beam [7,8], and
over the phase of the rf field seen by a molecule as it enters the field;

this averaging removes many 'high-frequency' terms which have a significant influence on the shape of the model spectrum.

The $v''=0$, $J''=28$ level has 342 hf states counting all possible (F'',M_F'') values. If the axis of quantization is chosen parallel to the electric field vector of the light, only $\Delta M_F=0$ optical transitions are allowed and the optical pumping problem subdivides into 31 much smaller problems, one for each possible $|M_F|$ value. In the P case discussed earlier, this same simplification occurs for the rf problem. In addition, consideration of the M_F-dependence of the optical matrix elements shows that only the few largest M_F's contribute to the signal: molecules in lower M_F states are completely pumped away to other vibrational levels after the A region. In the S case, however, rf transitions with $\Delta M_F'' = \pm 1$ are allowed, coupling all 342 states together; most of these must be excluded to obtain a tractable model in the C-region.

Figure 4 shows a model calculation for the P case which includes all states with $M_F'' \geq 25$. While the overall shape compares very well with experiment, the linewidth is about 2/3 of the experimental value. The model shows that for the dominant $M_F'' = 28$ and 27 contributions to the signal, the phase and velocity averaging removes almost all coherence terms, giving nearly the same result as a population model. The S case model of Fig. 5, based on 8 states with large M_F'', shows the dominant, negative peaks but has the wrong sign in the central region. The outward shift of the peaks is present. The oscillatory behaviour of the signal vs B_1 at a fixed frequency is reproduced by the model. Positions of the extrema agree to within 8% in the P case and 20% in the S case.

Besides the limitation on the number of states used, a possible reason for the remaining disagreement between experiment and theory is that several effects have been omitted from the model which could enhance the contribution of states with smaller M_F'' values. These include averaging over the geometry of the laser/Na_2 interaction regions (Doppler shifts, transit time

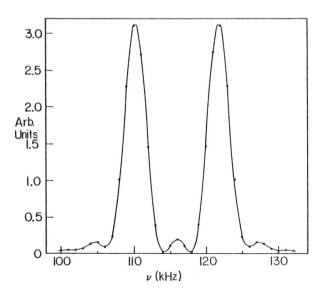

Fig.4 Model of the P-case data of Fig.2

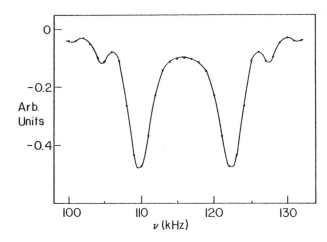

Fig.5 Model of the S-case data of Fig.3

variations), and anisotropic angular distribution of the fluorescence at the B region. An increased contribution from smaller $M_F"$ states in the P case would broaden the line because the matrix elements of the rf Hamiltonian get larger with small $M_F"$.

In addition to further study of these and other omitted effects, we plan to observe hf transitions in a level of much lower J" in order to reduce the number of states involved. This should lead to a complete understanding of the laser-induced coherence in double resonance experiments so as to permit high precision determinations of atomic and molecular constants by this powerful technique.

References

1. W. Happer: Rev. Mod. Phys. 44, 169 (1972).
2. S.D. Rosner, R.A. Holt, and T.D. Gaily: Phys. Rev. Lett.35,785-9(1975).
3. W. Ertmer and B. Hofer: Z. Phys. A276, 9 (1976).
4. W.J. Childs and L.S. Goodman: Phys. Rev. A21, 1216 (1980).
5. N.F. Ramsey: Molecular Beams (Clarendon Press, Oxford 1956).
6. A.G. Adam, T.D. Gaily, R.A. Holt, and S.D. Rosner: to be published.
7. T.D. Gaily, S.D. Rosner, and R.A. Holt: Rev. Sci. Instr. 47,143-5(1975)
8. K. Bergmann, U. Hefter, and P. Hering: J. Chem. Phys. 65, 488-90 (1976).

Probing the Dressed Molecule Energy Levels by Infrared-Radiofrequency Double Resonance

A. Jacques and P. Glorieux
Laboratoire de Spectroscopie Hertzieme, Associé au CNRS
Université de Lille I, F-59655 Villeneuve D'Asco Cédex, France

I. Introduction

The highly sensitive method of radiofrequency spectroscopy inside the cavity of a laser which was discussed at TICOLS [1] has been developed towards two fields of application, (i) spectroscopic studies [2,3,4] and (ii) multiphoton processes [5,6,7]. Such a radiofrequency spectrometer is a tool very well suited for observation of these processes not only because of its high sensitivity but also because of the high IR power density experienced by molecules since they are located inside the cavity.

The idea of considering the dressed molecule has already proved useful to give a simple picture of most IR - RF multiphoton processes. It has been used in a previous work on IR spectroscopy of molecules dressed by a RF field which may be either resonant or non-resonant with hyperfine splittings [8]. While in IR spectroscopy, the IR power is monitored as function of IR frequency, in RF double resonance spectroscopy the same quantity is monitored versus RF frequency. Unfortunately in that situation the coupling between molecules and RF may depend on RF frequency. *The basic idea of this study is to use two different* RF *fields*, the first one (pump or dressing field) is used to dress the molecule, its frequency ω_p is fixed; the second one (probe field) probes the transition of the dressed molecule by an IR - RF double resonance method, its frequency ω is swept and its power has to be strong enough to allow an efficient IR - RF double resonance. Thus in usual conditions, the proble field is weakly saturating.

Depending whether the levels of the bare molecule are single or double parity, the selection rules are markedly different and give RF spectra with different patterns. The case of levels with definite parity is first considered. Double parity levels are considered in another section. Many new resonances have been observed. They may be put into two classes depending on the strength of the RF probe field: (i) if this field is weak enough to involve RF probe interaction to the lowest order, this field acts only as a probe; (ii) at higher field strength, it induces multiphoton processes in the dressed molecule.

2. Single Parity Levels Weak Probe Field

A typical spectrum obtained in the low power limit for the probe is reported on Figure 1. Resonances are observed as soon as the probe frequency ω goes through the resonance frequency of a transition allowed in the dressed molecule and directly gives the rf absorption spectrum of the "molecule + rf field" quantum system. Our observation reveals two kinds of resonances at frequencies $(2n+1)$ ω_p and $2n\omega_p \pm \bar{\omega}_0$ with intensities decreasing as n increases. ω_p is the frequency of the fixed field and $\bar{\omega}_0$ is the resonance frequency of the molecule (in this experiment the K doubling frequency) corrected for high frequency Stark effect and Bloch-Siegert shift caused by the strong "dressing" field. Resonances at $2n\omega_p \pm \bar{\omega}_0$ appear as usual Lorentzians. The peculiar shape of the other class of resonances at $(2n+1)\omega_p$ is explained by the use of lock in detection which provides a signal proportional to RF-induced emission change.

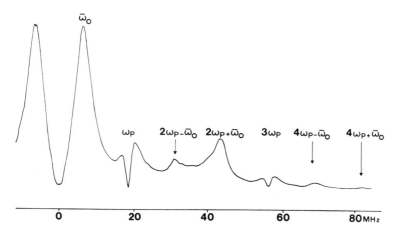

Fig.1. Example of RF spectrum of a molecule with single parity levels dressed by an RF field. The coincidence of the 9.66 μm P(32) line of CO_2 with the ν_6 $7_{4,3} \leftarrow 0,6_{3,4}$ line of D_2 CO has been used ω_p = 18.8 MHz. Pump power 18 W. Probe power 50 mW. Note that the lineshape of the signal at $\bar{\omega}_0$ is altered by the zero-field signal

The fact that only transitions following the parity selection rule, i.e., in which an odd number of photons are involved is easily understood if one considers the eigenstates of the perturbed Hamiltonian.

Using perturbation theory, the eigenstates of the perturbed dressed molecule $|\pm,n>$ may be expressed in terms of those of the dressed molecule neglecting the coupling between the bare molecule and the dressing field, $|\pm,n>$ where $|\pm>$ denotes the molecule state and $|n>$ the photon occupation number. It is easily shown that the state $|+,n>$ is a superposition of $|+,n+2p>$ and $|-, n+2p+1>$ levels. It is connected to the latter at order $(2p+1)$ via $2p$ intermediate states and to the former at order $2p$ via $(2p-1)$ intermediate states.

Figure 2 gives the energy level diagram of the dressed molecule. For sake of clarity, it has been drawn in the case $\omega_p \ll \omega_o$. It shows the different transitions which have been recorded on Figure 1. Note that the dressed molecule picture is not necessary to interpret all these transitions, but the semiclassical picture is much more complicated [9]. Especially as a high number of photons is considered, one single transition in the dressed molecule summarizes many different processes in the bare molecule interacting with RF field [8].

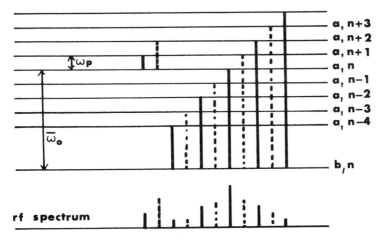

Fig.2. Sketch of energy diagrams of the molecule dressed by an RF field a and b are purely molecular states and n is the number of RF photons. With single parity levels only transitions in full lines are observed

3. Single Parity Levels High Probe Power

Observation with higher probe power (~1 Watt) reveals additional features
in the lower frequency range $\omega \lesssim \omega_p$. As shown on Figure 3, many sharp resonances are detected. Most of them appear at frequencies $\frac{m\omega_p}{n}$ while weaker
ones appear at $(2\omega_p \pm \omega_0) / 3 \ldots$. They correspond to multiphoton absorption
of the probe radiation. Since in that particular case, the probe field is
strongly saturating as can be seen from the width of the "one probe photon"
line, this field can induce multiphoton transitions respecting the parity
selection rule, i.e., such that an odd number of photons are required for a
$\pm \leftrightarrow \mp$ transition while an even number of photons correspond to $\pm \leftrightarrow \pm$
transitions that are forbidden in the weak probe field limit. Similar multi-
photon resonances have been observed by WINTER in magnetic resonance experiments [10].

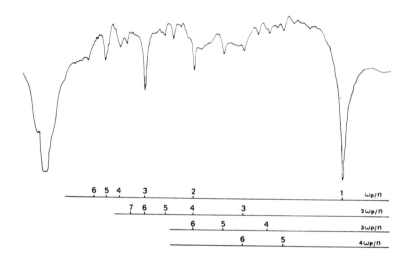

Fig.3. Exemple of multiphoton transitions at subharmonic frequencies. Same
laser line as Fig.1 ω_p = 23.4 MHz. Pump power 8 W "Probe" power 4 W

4. Double Parity Levels

A typical spectrum of the dressed molecule with double parity levels is given
on Figure 4. The probe radiation induces transitions at $n\omega_p$ and $\bar{\omega}_0 \pm n\omega_p$
where n is an integer. The parity selection rule now breaks out and any
transition is allowed. However, because of the phenomenon of destructive

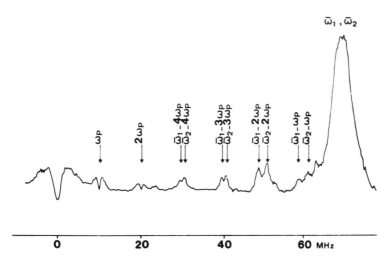

Fig.4. Exemple of RF spectrum of a dressed molecule with double parity levels. The coincidence of the 10.48 µm P(8) line of CO_2 with the $^rR(9,8)$ $\nu_6 \leftarrow 0$ absorption line of CH_3I has been used. $\omega_P = 10.1$ MHz. $\bar{\omega}_1$ and $\bar{\omega}_2$ are frequencies of RF transitions between hyperfine sublevels

interference the efficiency of some processes is strongly decreased. For instance, transitions at frequency ω_p should normally be forbidden because they occur between states of the dressed molecule corresponding to the same purely molecular state. However, because of a small contribution of second order Stark effect, there is a state mixing through which they gain some intensity.

5. Conclusion

In summary, radiofrequency spectrosocpy inside the cavity of an infrared laser may also be applied to molecules dressed by a radiofrequency field. Multiphoton resonances similar to overtone ($n\omega_p$) combination ($n\omega_p + \bar{\omega}_0$) and difference ($n\omega_p - \bar{\omega}_0$) absorption resonances have been observed. With a high probe power, "standard" multiphoton transitions ($^{n\omega}_m P$) have also been observed and interpreted with the help of the dressed molecule picture.

References

1. E. Arimondo, P. Glorieux, I. Oka: In *Laser Spectroscopy III*, ed. by J.L. Hall, J.L. Carlsten, Springer Series in Optical Sciences, Vol.7 (Springer Berlin, Heidelberg, New York 1977) p.278
2. R.F. Curl, T. Oka, D.S. Smith: J. Mol. Spectrosc. *46*, 518 (1973)
3. E. Arimondo, P. Glorieux, T. Oka: Phys. Rev. A *17*, 1735 (1978)
4. E. Arimondo, J.G. Baker, P. Glorieux, T. Oka, J. Sakai: J. Mol. Spectrosc. *82*, 54 (1980)
5. S.M. Freund, M. Romheld, T. Oka: Phys. Rev. Lett. *35*, 1487 (1975)
6. T. Oka: *Frontiers in Laser Spectroscopy*, Vol.2 (North-Holland Publishing Co 1977) p.531
7. J. Reid, T. Oka: Phys. Rev. Lett. *38*, 67 (1977)
8. E. Arimondo, P. Glorieux: Phys. Rev. A *19*, 1067 (1979)
9. F. Shimizu: Phys. Rev. A *10*, 950 (1974)
10. J.M. Winter: Ann. de Phys. (Paris) *4*, 745 (1959)

Infrared-Microwave Double Resonance Spectroscopy of Tetrahedral Molecules Using a Tunable Diode Laser

M. Takami

The Institute of Physical and Chemical Research
Wako, Saitama 351, Japan

A tetrahedral molecule has a vibrationally induced dipole moment in its triply degenerate vibrational state [1,2]. This dipole moment allows a pure rotational transition among the rotational manifold of the triply degenerate vibration. Such transitions have been observed in CH_4 [3], SiH_4 [4] and GeH_4 [5] by infrared-microwave double resonance. Because the lasers used in these previous works could not be tuned over a wide frequency range, however, the experiments had to rely upon accidental coincidences of the molecular absorption lines and laser frequencies.

For more extensive study of the pure rotational spectra of tetrahedral molecules, it is highly desirable to use a frequency tunable infrared laser for the pumping radiation source of double resonance. Along this line, IR-MW double resonance using a frequency tunable semiconductor diode laser (TDL) was developed in this work. The technique was first demonstrated in NH_3 [6], and then applied to the study of pure rotational spectra in $^{12}CF_4$ (ν_3) [7], SiH_4 (ν_3), $^{13}CF_4$ (ν_3), and $^{12}CF_4$ ($\nu_2 + \nu_3$). This paper reports briefly the results of TDL-pumped IR-MW double resonance spectroscopy of tetrahedral molecules.

An energy level diagram of the $CF_4 \nu_3$ state is shown in Fig. 1. In a triply degenerate vibrational state of a tetrahedral molecule, the vibrationally induced angular momentum ℓ couples with the pure rotational angular momentum R, to give the resultant total angular momentum J. For the first excited vibrational state of a triply degenerate vibration, therefore, a given J state splits into three Coriolis sublevels (R = J, J±1). Each Coriolis sublevel further splits into the well known tetrahedral fine structure by higher order vibration-rotation interactions. Each rovibrational state is labeled by its tetrahedral symmetry, F_1, F_2, E, A_1 and A_2. The selection rules for the pure rotational transitions are $\Delta J = 0$, ±1, and $F_1 \leftrightarrow F_2$, $E \leftrightarrow E$, $A_1 \leftrightarrow A_2$. In the ν_3 state of CF_4, the pure rotational transitions within a tetrahedral fine structure fall into a radiofrequency region, while those between the different Coriolis sublevels fall into a microwave and millimeter wave region.

Figure 2 shows a schematic diagram of the experimental setup. The excited state pure rotational transitions are observed based

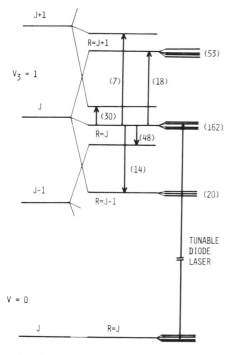

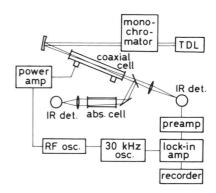

Fig.2 A schematic diagram of the IR-MW double reson- ance experiment. The laser beam is focused into the cell with a spherical mirror. The absorption cell is for monitoring weak absorption lines.

Fig.1 Energy levels of the $CF_4\nu_3$ state. The arrows indicate the types of observed microwave transitions. The numbers of measured pure rotational transitions are shown in parentheses.

on the principle that, when a vibration-rotation transition is saturated by a laser, simultaneous saturation of the rotational transition increases the infrared absorption. In order to saturate the infrared absorption line effectively with a small output power from a TDL, the laser beam was focused into a double resonance cell. A low sample pressure, typically from 0.1 to 2 Pa (0.8 to 15 mTorr), was also necessary for the effective pumping. The pure rotational transitions were saturated by an amplitude-molulated radiofrequency field or frequency-modulated microwave field. The resultant increase in the infrared absorption was detected phase sensitively.

A typical trace of the double resonance signal is shown in Fig. 3. The linewidth of the signal is about 100 kHz at HWHM when the RF field intensity is reduced to avoid saturation broadening. This narrow linewidth allows us to measure the transition frequencies with the accuracy of a few tens of kilohertz.

The advantage of using a tunable laser in double resonance was remarkable. In the ν_3 state of $^{12}CF_4$, nearly 400 RF and MW transitions were measured. In Fig. 1 the number of observed lines is indicated in the parentheses for each branch. The

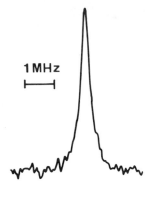

1 MHz

<u>Fig.3</u> A trace of the double resonance signal observed by scanning the radio-frequency. The observed transition is $^{12}CF_4 \nu_3$ 29_{30} Al(3)-A2(3) at 5.11 MHz. The time constant is 0.3s.

observed spectrum shows that the lowest Coriolis sublevels, $R = J - 1$, are largely perturbed by $2\nu_4$ which lies about 22 cm^{-1} below ν_3. Tentative molecular constants have, therefore, been determined from the transitions in the $R = J$ and $J + 1$ states.

The successful observation of the double resonance signals made it possible to analyse the badly congested ν_3 Q-branch spectrum. Each time a pure rotational transition was observed with a Q-branch vibration-rotation line, the infrared absorption and the double resonance signals were recorded simultaneously by scanning the TDL while the radiofrequency was fixed at resonance. One such trace is shown in Fig. 4. By this method, many infrared transitions were assigned in the Q-branch spectra of the $^{12}CF_4$ and $^{13}CF_4\nu_3$ bands.

The Fermi interaction observed in the pure rotational spectrum of the $CF_4\nu_3$ state motivated us to examine ν_3 of $^{13}CF_4$, which lies about 15 cm^{-1} *below* $2\nu_4$. The double resonance signals were observed with $^{13}CF_4$ in natural abundance (1.1%). The pressure of the sample was increased by an order of magnitude (about 2 Pa). About 100 pure rotational lines, most of which were in the $R = J$ state, were measured with a good S/N ratio (see Fig. 4). As expected,

IR ABSORPTION

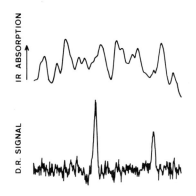

D.R. SIGNAL

<u>Fig.4</u> Simultaneous recording of the infrared absorption and double resonance signals observed by scanning the laser frequency over the $^{13}CF_4$ ν_3 Q-branch region. The radio-frequency is fixed at 869.20 MHz corresponding to the $38_{38}A2(3)-A1(3)$ transition.

large frequency anomalies were observed in the RF transitions of the R = J + 1 highest Coriolis sublevels, which were nearest to the $2\nu_4$ state.

The good S/N ratio of the $^{13}CF_4$ double resonance signals indicates the possibility of exploring the combination states using hot band transitions. If double resonance spectroscopy is possible for the vibrational levels which have more than 1% of the ground state population, the $\nu_2 + \nu_3$, $\nu_4 + \nu_3$, $\nu_1 + \nu_3$ and $2\nu_2 + \nu_3$ states of $^{12}CF_4$ should be accessible by this method. Double resonance spectroscopy with hot band transitions was examined in $\nu_2 + \nu_3$. About 30 pure rotational transitions were observed. This kind of measurements will give us very accurate informations on the combination states of tetrahedral molecules.

The double resonance spectroscopy was also applied successfully to the $SiH_4\nu_3$ state. Although the measurement was suspended due to a sudden breakdown of the laser diode, the observed three transition frequencies verified high reliability of the analysis of the $SiH_4\nu_3$ band by CABANA *et al.*[8].

I am grateful to Dr. E. Hirota and Dr. S. Saito of Institute for Molecular Science for their great assistance in this work. My gratitudes are also to Dr. J.T. Hougen for his valuable comments and discussions on the theoretical part of the work, and to Dr. R.G. Robiette for useful information on the CF_4 vibrational energies and computor programming. This work was supported by IMS Joint Research Program from 1979 to 1981.

References

1. M. Mizushima and P. Venkateswarlu, J. Chem. Phys. **21**, 705 (1953).
2. K. Uehara, K. Sakurai and K. Shimoda, J. Phys. Soc. Jpn. **26**, 1018 (1969).
3. R.F. Curl, Jr. and T. Oka, J. Chem. Phys. **58**, 4908 (1973); M. Takami, K. Uehara and K. Shimoda, Jpn. J. Appl. Phys. **12**, 924 (1973).
4. W.A. Kreiner and T. Oka, Can. J. Phys. **53**, 2000 (1975); W.A. Kreiner, T. Oka and R.G. Robiette, J. Chem. Phys. **68**, 3236 (1978).
5. W.A. Kreiner, B.J. Orr, U. Andresen and T. Oka, Phys. Rev. **A15**, 2298 (1977); W.A. Kreiner, U. Andresen and T. Oka, J. Chem. Phys. **66**, 4662 (1977).
6. M. Takami, Appl. Phys. Lett. **34**, 682 (1979).
7. M. Takami, J. Chem. Phys. **71**, 4164 (1979); ibid. **73**, 2665 (1980); ibid. **74**, 4276 (1981).
8. A. Cabana, D.L. Gray, A.G. Robiette and G. Pierre, Mol. Phys. **36**, 1503 (1978).

Molecular-Beam, Laser-RF, Double Resonance Studies of Calcium Monohalide Radicals

W.J. Childs, D.R. Cok, and L.S. Goodman

Physics Division 203-F105, Argonne National Laboratory
Argonne, IL 60439, USA

1. The Molecular-Beam, Laser-RF, Double-Resonance Technique

The molecular-beam, laser-rf, double-resonance technique has been described [1-3] a number of times. In essence, the occurrence of a radiofrequency (rf) transition in the electronic ground state of the molecule under study is detected by an increase in the laser-induced fluorescence of the molecular beam when the rf is on resonance. The technique makes it possible to measure small energy splittings (normally spin-rotational or hyperfine) in the electronic ground state of a molecule to an absolute precision of 1 kHz. The sensitivity of the technique is high because even a very small increase in fluorescence can be easily seen if the rf is swept repeatedly and digital data-handling techniques are used. The technique is useful for ionic [4] as well as for neutral atoms and molecules.

2. The Calcium Monohalide Radicals

The calcium monohalide radicals, Ca^+X^-, are a highly ionic species, and to an excellent approximation consist of a p^6 halide ion X^- and a single electron localized near a closed p-shell Ca^{2+} ion. Because most of their properties are determined by the single, nonbonding electron outside the closed-shell ions, these radicals are closely analogous to alkali atoms. Since ^{40}Ca has no nuclear spin ($I = 0$), the dipole hfs arises primarily from the halide dipole moment sensing the unpaired spin density of the outer electron. The extremely small observed dipole hfs of the CaX radicals (<1% of the atomic value) indicates the extreme localization of the outer electron at the Ca site.

The very small observed hfs has been difficult to account for theoretically; most recently, BERNATH, PINCHEMEL, and FIELD [5] have proposed that much of it arises from spin-polarization of the halide orbitals by the outer electron centered on the Ca site. Development of the theory has so far been severely handicapped by a nearly total lack of accurate hfs measurements, and the present series of experiments is designed to address this problem.

3. Fluorescence Spectroscopy of the Molecular Beam

If the pump laser beam is blocked and the fluorescence induced by the probe beam is recorded as a function of laser wavelength, a spectrum like that of

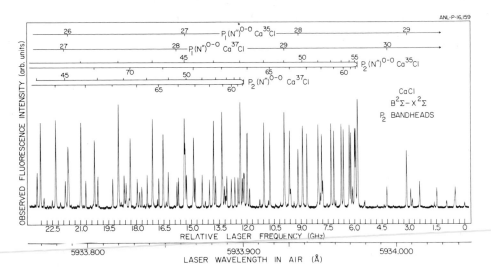

ANL-P-16,159

Fig.1 Fluorescence spectrum of a CaCl molecular beam near 5934 Å. The 37 MHz linewidth is due to unresolved hyperfine structure.

Fig. 1 is obtained. It shows the P_2 bandhead of the B $^2\Sigma \leftrightarrow$ X $^2\Sigma$, v = 0 $\leftrightarrow$ 0 transition in both Ca^{35}Cl and the weaker Ca^{37}Cl. Each of the individual lines of the P_2(J") branch contains four unresolved hfs components (I = 3/2 for both ^{35}Cl and ^{37}Cl).

4. Laser-RF Double Resonance

In the cases of CaF [3] and CaBr [6], the hfs components of the optical lines used were well resolved, and the double-resonance method was applied as indicated in section (1) above. Even in the case of CaCl, however, where (as shown in Fig. 1) the hfs of the optical lines is unresolved, the resolution

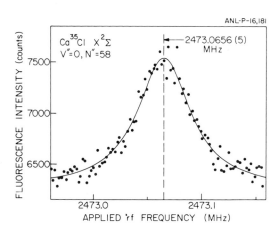

Fig.2 Laser-rf double resonance in Ca^{35}Cl. The laser-induced fluorescence increases about 20% when the rf is on resonance. The transition shown is characterized by J = 57.5, F = 59 $\leftrightarrow$ J = 58.5, F = 60 within the v = 0, N = 58, X $^2\Sigma$ molecular ground state.

was achieved by the use of double-resonance techniques. Fig. 2 shows the appearance of a typical rf transition for $Ca^{35}Cl$. An increase in fluorescence of about 20% is found on resonance. Observations of this kind have been made [3,5] for many values of N in the X $^2\Sigma$ ground states of $Ca^{19}F$, $Ca^{79}Br$, $Ca^{81}Br$, $Ca^{35}Cl$, and $Ca^{37}Cl$.

5. Interpretation

An interpretation of the double-resonance measurements exemplified by Fig. 3 is achieved with the help of the standard Hamiltonian [7,8] for a $^2\Sigma$ state of a diatomic molecule,

$$H = H_{el} + H_{vib} + H_{rot} + \gamma_N \underset{\sim}{S} \cdot N + b_N \underset{\sim}{I} \cdot \underset{\sim}{S} + c_N I_z S_z + \tag{1}$$
$$+ C_{IN} \underset{\sim}{I} \cdot \underset{\sim}{N} + (eqQ)_N \left[3I_z^2 - I(I+1) \right] \Big/ 4I(2I-1),$$

where γ is the spin-rotation interaction constant, b and c are the Frosch-Foley [7] hfs parameters, C_I is the (very small) coefficient for the weak $\underset{\sim}{I} \cdot \underset{\sim}{N}$ hfs interaction [9], and eqQ measures the strength of the electric-quadrupole hfs interaction. The subscripts N show that each parameter X is a function of the rotational state, and we express this as

$$X_N = \sum_j X_j N^j (N+1)^j. \tag{2}$$

In adjusting the Hamiltonian parameters to fit the observations, one uses as few of the parameters X_j as possible. The isotopic dependence determined from independent fits to the $Ca^{79}Br$ and $Ca^{81}Br$ data is in good agreement with the theory.

Ab initio calculations by G. L. GOODMAN [3] give reasonable ($\sim$20%) agreement for the lower order (in N) parameters in CaF, but fail to reproduce the higher order parameters satisfactorily. Calculations (including spin-polarization effects) are so far unsatisfactory for CaBr, and the spin-orbit interaction may have to be included [10].

Table I Hfs parameter values measured for calcium monohalide radicals by laser-rf double resonance. The bottom section of the table shows the results after removal of strictly nuclear effects.

hfs parameter	parameter value (MHz)		
	$^{40}Ca^{19}F$	$^{40}Ca^{35}Cl$	$^{40}Ca^{79}Br$
(b + c/3)	122.025	23.453	121.202
c	40.647	12.455	77.620
C_I	0.029	0.0017	0.0041
eqQ	0	-1.002	20.015
(b + c/3)/g_I	23.209	42.808	86.347
c/g_I	7.731	22.733	55.298
C_I/g_I	0.0056	0.0031	0.0029
eqQ/Q	0/0	10.0	54.1

Table I summarizes the experimental results so far. No other technique has yet been able to determine any of these hfs parameters, all vital for further development of the theory. The bottom section of the table shows that when nuclear effects are taken account of, a smooth dependence is found for all hfs parameters as one moves to heavier halides.

6. Acknowledgement

This research was supported by the U.S. Department of Energy, Office of Basic Energy Sciences, under Contract W-31-109-Eng-38.

References

1. S. D. Rosner, T. D. Gaily, and R. A. Holt, Phys. Rev. Lett. 35, 785 (1975).

2. W. Ertmer and B. Hofer, Z. Physik A276, 9 (1976).

3. W. J. Childs, G. L. Goodman, and L. S. Goodman, J. Mol. Spectrosc. 86, 365 (1981).

4. U. Kotz, J. Kowalski, R. Neumann, S. Noethe, H. Suhr, K. Winkler, and G. zu Pultlitz, Seventh International Conference on Atomic Physics, Abstracts, p. 162 (1980).

5. P. F. Bernath, B. Pinchemel, and R. W. Field, J. Chem. Phys. (in press).

6. W. J. Childs, David R. Cok, G. L. Goodman, and L. S. Goodman, J. Chem. Phys. (in press).

7. R. S. Frosch and H. M. Foley, Phys. Rev. 88, 1337 (1952).

8. J. L. Dunham, Phys. Rev. 41, 721 (1932).

9. K. F. Freed, J. Chem. Phys. 45, 4214 (1966).

10. G. L. Goodman, private communication (1981).

Part IV
Collision-Induced Phenomena

Collision-Induced Coherence in Four Wave Light Mixing*

N. Bloembergen, A.R. Bogdan, and M.W. Downer

Division of Applied Sciencesm Harvard University
Cambridge, MA 02138, USA

1. Introduction

The evolution of the density matrix of a material system under the influence
of one or more traveling electromagnetic waves is central in determining the
linear and nonlinear optical response. The role of spontaneous emission and
collisional processes may, under certain conditions which apply to a rather
large variety of experimental situations, be represented by phenomenological
damping terms in the equations of motion for the diagonal and off-diagonal
elements of the density matrix [1]. General expressions for the third order
nonlinear susceptibility have been given by many authors [2-4]. The book-
keeping of the many terms occurring in third order perturbation theory may
be systematized by the use of double-sided Feynman type diagrams [5,6,7].
It is important to consider explicitly the evolution of both bra vectors
$\langle\psi|$ and ket vectors $|\psi\rangle$ in the presence of damping. Authors who use only
one sided diagrams irretrievably lose some terms which have physical significance. One must not calculate the evolution in the absence of damping,
and then formally add the damping by considering some frequencies as complex
quantities. In the latter case one always ends up with resonant denominators
which contain only separations ω_{gn} from the initially occupied state,
$\rho_{gg}^{(o)} = 1$, to states $|n\rangle$. The correct expression also contains denominators
with frequency separations $\omega_{nn'}$, between two initially *unoccupied* excited
states.

It was recognized quite early in the development of nonlinear optics that
such extra resonances disappear if the damping terms satisfy the relation [8]
$\Gamma_{nn'} - \Gamma_{ng} - \Gamma_{n'g} = 0$. This relation is, for example, satisfied if $|g\rangle$ is
the ground state, and the Γ's represent natural widths due to spontaneous
emission from $|n\rangle$ and $|n'\rangle$. It was used by HANSCH and TOSCHEK [9] in
their treatment of saturation spectroscopy in three level systems. In the
presence of collisions these extra resonances should be present. LYNCH et
al. [2] have given explicit expressions for these extra resonances in $\chi^{(3)}$.
They have been confirmed by DRUET and TARAN [7], who have prepared an excellent up-to-date review of this question. It has an important bearing on the
detailed line shapes of resonant CARS (coherent Anti-Stokes Raman scattering)
and CSRS (coherent Stokes Raman scattering) signals. Some statements in the
recent literature are not quite correct [10,11], and details are critically
reviewed by DRUET and TARAN [7]. Furthermore the important details of the
role of diagonal elements in the density matrix calculation of $\chi^{(3)}$ have
been presented by OUDAR and SHEN [4]. References to many earlier theoretical contributions are quoted in these papers.

---

* This research was supported by the Joint Services Electronics Program
under Contract N00014-75-C-0648.

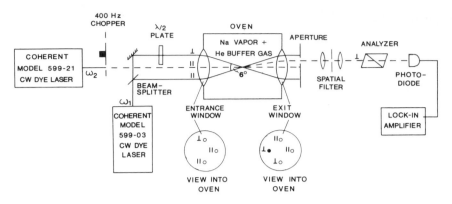

<u>Fig. 1</u> Schematic diagram of experimental arrangement to detect collisionally induced resonances. The symbols "$\perp$" and "$\parallel$" denote mutually orthogonal polarization directions.

Although an experimental effort to detect the extra resonant terms was mounted several years ago and was discussed at the preceding FICOLS conference [12], experimental confirmation was achieved only recently [13]. It is the purpose of this contribution to discuss the experimental evidence and the characteristics of these collisionally induced coherent signals [13-16]. The system of Na atoms with He buffer gas was chosen as all the pertinent matrix elements and energy levels are known for the Na atom. The Rhodamine 6G dye lasers can be conveniently tuned in the vicinity of the yellow resonance doublet. Furthermore the collisional damping contributions of Na-Na and Na-He collisions are known. In the experimental arrangement shown in fig. 1, two dye laser beams at frequency ω_1 have wave vectors in the vertical plane. The frequency $\omega_1 = \omega_{3p,3s} + \Delta$ is fixed. The off-set Δ from one of the yellow Na-resonance lines is chosen so that $\hbar^{-1} |\mu| |E| << \Delta << t_c^{-1}$. Since the Rabi frequency is small compared to the detuning, the perturbation approach is valid. Since the inverse of the duration of a collision is large compared to the detuning, the impact approximation is valid. The use of phenomenological damping terms leading to Lorentzian line shapes is justified. A third incident beam at a variable frequency ω_2 has a wave vector in the horizontal plane.

A new wave in a new direction $\tilde{k}_1' + \tilde{k}_1 - \tilde{k}_2$ in the horizontal plane is created at the frequency $2\omega_1 - \omega_2$. The angles between the near-forward moving beams are about 6°. The advantage of this "folded boxcars" momentum matching scheme [17] is that resonance can be detected with ω_2 in the immediate vicinity of ω_1. Figure 2 shows seven resonances that occur as ω_2 is varied. The resonances 2,3,5 and 6 are very strong resonant CARS signals not of interest for the present purposes. The resonances 1 and 7 are several orders of magnitude weaker and disappear when the helium pressure goes to zero. They are the pressure induced extra resonances discussed above, when $|\omega_1 - \omega_2|$ equals the frequency separation between two initially unpopulated excited states $3^2P_{1/2}$ and $3^2P_{3/2}$. While the original experimental detection was made with pulsed dye lasers, much better resolution has recently been obtained with cw dye laser beams [15]. The detailed results will be discussed further in the final section 4 of this paper.

A degenerate frequency resonance was found [14,16] for $\omega_1 = \omega_2$. The physical interpretation of this signal which also decreases in strength with decreasing helium pressure is more straightforward and is presented next.

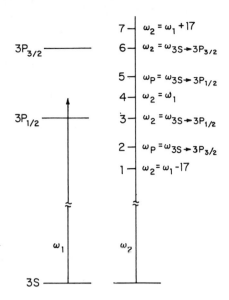

Fig. 2 The frequency ω_1 is chosen at a fixed off-set from a Na resonance line. As the frequency ω_2 is varied, seven resonances in the four wave mixing signal are observed.

2. Degenerate Frequency Resonance, Induced by Collisions [14,16]

Figure 3 shows the resonances observed in the intensity of the beam in the new fourth direction when ω_2 is varied in the immediate vicinity of ω_1. In this section we discuss the central component, whose width is essentially independent of helium pressure up to p_{He} = 500 torr, and whose intensity varies as p_{He}^2. The experimental width (FWHM) of about MHz arises principally from the spectral width of the beam at ω_1. Active stabilization of this frequency is expected to reveal a substantially narrower width, on the order of the natural line width for spontaneous emission.

The physical origin of the resonance is a population grating, rather similar to that observed for backward phase conjugate mixing [18]. The following differences should, however, be noted:

1. The frequencies are sufficiently far removed from resonance, so that no strong linear absorption and no saturation effects in the Doppler profile take place.

2. The population grating is due to collisions whose rate can be controlled.

3. The degenerate resonance can be observed in detail as no phase mismatch occurs when ω_2 is detuned from ω_1.

Suppose two beams at ω_1 and ω_2 have equal amplitude. Their combined influence may be represented by an amplitude modulated wave

A $\cos \frac{1}{2}(\omega_1 - \omega_2)t \cos \frac{1}{2}(\omega_1 + \omega_2)t$. In spite of the detuning of $\frac{1}{2}(\omega_1 + \omega_2)$ from resonance, some collision induced transitions from the 3s ground state

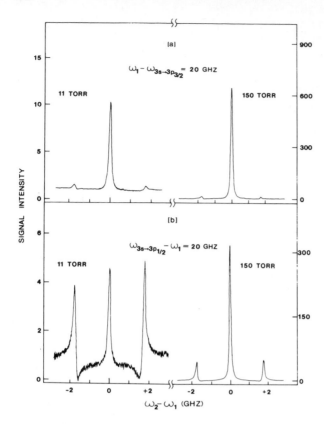

Fig. 3 Experimental scans of degenerate and nearly degenerate four wave mixing signals as a function of $\omega_1 - \omega_2$. The central component is discussed in section 2, the satellites in section 3.

$|g\rangle$ to the excited state $|n\rangle$ which stands for $3P_{1/2,3/2}$ will take place. The transition rate is modulated at the difference frequency $\omega_1 - \omega_2$. When this difference is small compared to the population relaxation rate, $T_{1,eff}^{-1}$ a modulated grating will be created, varying in time and space as $\exp[i(\underset{\sim}{k}_1 - \underset{\sim}{k}_2) \cdot \underset{\sim}{r} - i(\omega_1 - \omega_2)t]$. The other beam at ω_1 is diffracted by this population grating to yield the new beam. Second order perturbation theory for a two level model yields

$$\rho_{nn}^{(2)}(\omega_1 - \omega_2) = \frac{\hbar^{-2}|\mu_{gn}|^2 E_1 E_2^*}{(\omega_{ng} - \omega_1 - i\Gamma_{ng})(\omega_{ng} - \omega_2 + i\Gamma_{ng})} \left[1 + \frac{i(\Gamma_{ng} + \Gamma_{gn} - \Gamma_{nn})}{\omega_1 - \omega_2 + i\Gamma_{nn}} \right] \quad (1)$$

The intensity of the new diffracted beam is proportional to

$$|\chi^{(3)}|^2 \propto \left| 1 + \frac{i(\Gamma_{ng} + \Gamma_{gn} - \Gamma_{nn})}{\omega_1 - \omega_2 + i\Gamma_{nn}} \right|^2 \quad (2)$$

Here $\Gamma_{nn} = T_{1,eff}^{-1}$ and $\Gamma_{ng} = \Gamma_{gn} = T_2^{-1}$. In the absence of collisions, we have the characteristic cancellation $2\Gamma_{nn} - \Gamma_{gn} - \Gamma_{ng} = 0$, discussed in the introduction. The degenerate resonance is a collisionally induced effect. The effect of pressure on Γ_{nn} is small, as collisions cannot cause radiationless de-excitation of the Na atoms. They can produce fine structure changing collisions, so that population of the $3^2P_{\frac{1}{2}}$ state may be transferred to and relaxed via the $3^2P_{3/2}$ state [19]. For tuning near the former state one has

$$T_{1,\text{eff}}^{-1} = T_1^{-1} \frac{1 + 1.5\gamma_t p_{\text{He}} T_1}{1 + 0.5\gamma_t p_{\text{He}} T_1} \qquad (3)$$

where T_1 is the lifetime for spontaneous emission. For tuning near the $3^2P_{3/2}$ state, the coefficient 0.5 in the denominator should be replaced by 1.0. The off-diagonal terms have a strong pressure dependence from dephasing during collisions, $\Gamma_{ng} + \Gamma_{gn} = \Gamma_{nn} + \beta p_{\text{He}} + \beta' p_{\text{Na}}$. The co-efficients β and β' are known [20]. The two level model on which (Eq.1) is based, can be extended to take account of the various hyperfine levels. In such a multilevel system additional contributions to the nonresonant background arise. The theoretical peak to background ratio can be put in the form [21].

$$\frac{(\beta p_{\text{He}} + \beta' p_{\text{Na}})^2}{a^2 T_{1,\text{eff}}^{-1} \Gamma_{\text{exp}}} + \frac{2(\beta p_{\text{He}} + \beta' p_{\text{Na}})}{a\Gamma_{\text{exp}}} \qquad (4)$$

where $a = 2.53$ for tuning near $3^2P_{3/2}$ resonance, and $a = 2.82$ for tuning near the $3^2P_{\frac{1}{2}}$ resonance. Experimental results fit this expression with an accuracey of 30 percent in the range $p_{\text{He}} \sim 10 - 500$ torr.

The physical arguments for the origin of the collision induced degenerate frequency resonance remain valid for larger detunings from resonance $\Delta t_c \gg 1$. The impact approximation must then be replaced by the quasi-static treatment of the collisions. The collision induced degenerate resonance should remain observable for detunings $\Delta > 20$ cm^{-1} and it may serve as a probe for collisions in this regime, far in the wings of the line profile.

3. Collision Induced Resonance between States with Initially Equal Populations

The satellite resonances shown in fig. 3 occur when $\omega - \omega = \omega_{g',g} = 1.7$GHz. This corresponds to the hyperfine splitting of the 3s ^{1}ground state of the Na atom. We denote the levels with $F = 1$ and 2 by $|g>$, and $|g'>$ respectively. The following features are apparent: 1) The satellites are more pronounced for tuning ω_1 in the vicinity of the $3 P_{\frac{1}{2}}$ excited states. 2) The satellite line shape has a striking asymmetry at low helium pressure. 3) The intensity of the satellites is an increasing function of p_{He}.

The resonances may be interpreted as a Raman type of coherence induced between the levels $|g>$ and $|g'>$. These levels, when properly weighted for degeneracy, are equally populated, and the conventional theory of stimulated Raman scattering yields a vanishing result for equally populated states. This is, however, no longer precisely true in the presence of damping. Standard second order perturbation theory gives for the induced coherence

$$\rho_{g'g}^{(2)}(\omega_1 - \omega_2) = \frac{\hbar^{-2}\mu_{g'n}\mu_{ng}E_1E_2^*}{\omega_{g'g}-(\omega_1-\omega_2)-i\Gamma_{g'g}}\left(\frac{\rho_{gg}^{(o)}}{\omega_{ng}-\omega_1-i\Gamma_{ng}} - \frac{\rho_{g'g'}^{(o)}}{\omega_{ng'}-\omega_2+i\Gamma_{ng'}}\right). \quad (5)$$

We evaluate this expression, which is proportional to $\chi^{(3)}$, with the condition $\omega_{ng} - \omega_1 = \Delta \gg \Gamma_{ng} = \Gamma_{ng'}$ and $\omega_{ng'} - \omega_2 = \Delta + \delta$. Thus, ω_2, or δ, is variable, and the satellite resonance occurs for $\delta = 0$. Furthermore $\rho_{gg}^{(o)} = \frac{1}{2} + \varepsilon$ and $\rho_{g'g'}^{(o)} = \frac{1}{2} - \varepsilon$, where we allow for a small deviation from equality of population. This may be caused by optical pumping by preferential depopulation, as one hyperfine level of the ground state is somewhat less detuned from one photon resonance than the other. After some algebra, one finds that the intensity of the four wave mixing signal is proportional to

$$|\rho_{g'g}^{(2)}|^2 = \left| 1 + \frac{i(2\Gamma_{ng} - \Gamma_{g'g})}{\delta + i\Gamma_{g'g}} + \frac{4\Delta + 2\delta}{\delta + i\Gamma_{g'g}}\varepsilon \right|^2. \quad (6)$$

The nonresonant background term has again to be scaled by a correction factor due to contributions from other resonances. For $\varepsilon \neq 0$, the last term describes a CARS type resonance. This term is real, changes sign with δ and interferes with the nonresonant part to yield a characteristic asymmetric profile. For $\varepsilon = 0$, the middle term gives a resonance induced by the damping processes. As Γ_{ng} increases linearly with p_{He} this term dominates at higher helium pressure. It is 90° out-of-phase with the other terms and it alone would lead to a symmetric line profile. This new effect is the subject of this section. The ratio of the satellite intensity to the central component is very different for tuning near the $^2P_{\frac{1}{2}}$ and $^2P_{3/2}$ levels. From simple selection rules it is evident that one can only get contributions to a Raman process between the F = 1 and 2 ground levels, if one uses in the virtual intermediate state the F = 1 and 2 levels of the excited P states. The F = 3 and F = 0 level of the $^2P_{3/2}$ state do not contribute to the satellite resonances, but they make a contribution with considerable oscillator strength to the central components. The experimental curves shown in fig. 3 can be fitted [21] rather well by a computer program combining eqs. (2) and (6), if one assumes $\varepsilon = 0.007$ and $\Gamma_{gg'} = 40$ MHz. The amount of optical pumping is reasonable, but the value of $\Gamma_{gg'}$ is considerably larger than one would expect from spin-exchange collisions between Na atoms, which are largely responsible for transitions between the two hyperfine states. Further experimental checks with higher resolution obtainable from better stabilization of ω_1 are necessary. The systematic dependence on the detuning Δ, on the sodium density p_{Na}, on the other polarization configurations of the incident beams, and on the application of an external magnetic field must be investigated.

4. Collision Induced Extra Resonance

The pressure induced extra resonance in four wave light mixing (PIER 4) was mentioned in the introduction. One dye laser at ω_1 is tuned 15 GHz below the $3^2P_{1/2}$ level. The other frequency ω_2 is swept through the frequency which lies 15 GHz below the $3^2P_{3/2}$ level. Figure 4 shows the observed PIER 4 signal for three different helium pressures. The nonresonant background is independent of pressure and serves as a calibration point. The peak intensity, line width and integrated intensity are plotted as a function of p_{He} in fig. 5. This behavior is in agreement with the theory of $\chi^{(3)}$. This quantity contains a term proportional to the coherence induced between the two $^2P_{3/2,1/2}$ states, denoted by $|n>$ and $|n'>$, respectively,

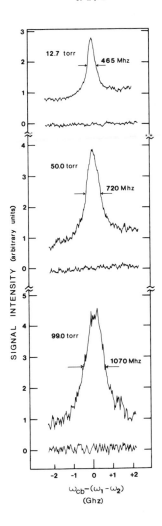

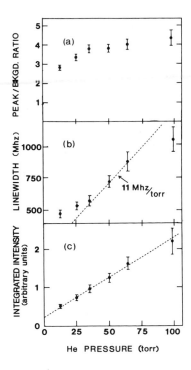

Fig. 5 The PIER 4 signal characteristics as a function of buffer gas pressure. (a) Ratio of peak height to nonresonant signal, (b) line width (FWHM), and (c) integrated intensity of the resonant signal.

Fig. 4 Experimental scans of the intensity of the four wave mixing signal for three buffer gas pressures. The lower scan at each pressure was taken with the input beams at ω_1 blocked. Note the nonresonant signal, independent of pressure, and the pressure induced resonance.

$$\rho_{nn'}^{(2)} (\omega_1 - \omega_2) = \frac{\hbar^{-2} \mu_{ng} \mu_{gn'} E_1 E_2^* \rho_{gg}^{(o)}}{(\omega_{ng} - \omega_1 - i\Gamma_{ng})(\omega_{n'g} - \omega_2 + i\Gamma_{n'g})} \left\{ 1 + \frac{\Gamma_{nn'} - \Gamma_{ng} - \Gamma_{n'g}}{\omega_{nn'} - (\omega_1 - \omega_2) - i\Gamma_{nn'}} \right\}. \quad (7)$$

Note the resonant term for $\omega_{nn'} - (\omega_1 - \omega_2) = 0$, which vanishes in the absence of collisions, because $\Gamma_{nn'}^{sp} = \Gamma_{ng}^{sp} + \Gamma_{n'g}^{sp}$ for spontaneous emission. The line width is pressure dependent as $\Gamma_{nn'} = \Gamma_{nn'}^{sp} + \gamma_{nn'} p_{He}$. With $\gamma_{nn'} = \gamma_{ng} = \gamma_{n'g} = \gamma = 5.5 \text{MHz/torr}$ as deduced from independent experiments [20] , one finds a signal proportional to

$$|\chi^{(3)}|^2 \propto N_g^2 I_1^2 I_2 \left\{ 1 + \frac{3(\gamma p_{He})^2 + 2\gamma p_{He} \Gamma_{nn'}^{sp}}{(\omega_{nn'} - \omega_1 + \omega_2)^2 + (\Gamma_{nn'}^{sp} + \gamma p H_e)^2} \right\} \quad (8)$$

in good agreement with the observations at high p_{He}. In the limiting behavior for $p_{He} \to 0$ the theoretical expression must be modified to take account of the following three factors. 1) The hyperfine splittings of the $^2P_{3/2,\frac{1}{2}}$ states, 2) the spectral width of the laser frequency ω_1, 3) the influence of Na-Na collisions.

The overall agreement between theory and experiment is satisfactory. In particular, the proportionality of the signal to $I_1^2 I_2$, characteristic of a $\chi^{(3)}$ process was verified in detail. This is significant, because it distinguishes the PIER 4 term from more mundane Raman-type resonances due to a build-up of real populations in the excited states. Of course, $\rho_{nn'}^{(2)}$ can only be nonvanishing, if at the same time $\rho_{nn}^{(2)} \neq 0$ and $\rho_{n'n'}^{(2)} \neq 0$. In fact, we have already used such induced population changes in section 2. There will be an ordinary Raman-type resonance in four wave mixing for $\omega_1 - \omega_2 = \omega_{nn'}$, proportional to $\rho_{nn}^{(2)} - \rho_{n'n'}^{(2)}$. Note that this signal would be proportional to a fifth power combination of the beam intensities I_1 and I_2. These Raman resonances in populated excited states have been investigated by other authors [22-24]. However, the intrinsic theoretical interest here is in a third order contribution to $\chi^{(3)}$ starting with $\rho_{gg}^{(o)} = 1$. At no stage in the calculation of Eqs. (7) and (8) do the populations ρ_{nn} and $\rho_{n'n'}$ occur.

There are several ways to give a physical rationale for the paradoxical emergence of a coherent new light beam solely in the presence of collisions. It is apparent from Eqs. (2), (6) and (8) that certain combinations of Feynman diagrams lead to a vanishing result in the absence of collisions. Thus collisions destroy the destructive interference of several coherent Feynman pathways. GRYNBERG [25] has adopted an alternative, very elegant description in terms of the dressed atom picture, which leads to identical results as the density matrix formulation. It has the advantage that it automatically includes higher order effects in the light field amplitudes and that it is not restricted to the impact approximation.

In conclusion, the observation of collision induced resonances has confirmed some subtle features of damping processes in nonlinear four wave light mixing in the vicinity of one and two photon resonances. These features are irretrievably lost in treatments with one-sided Feynman diagrams or equivalent approximations. The collision-induced four wave mixing signals may be used to determine the damping coefficients of diagonal and off-diagonal terms in the density matrix, and to verify its evolution in hyperfine structure manifolds of both ground and excited states.

References

1. N. Bloembergen and Y.R. Shen, Phys. Rev. 133, A37 (1964).

2. N. Bloembergen, H.Lotem and R.T. Lynch, Indian J. Pure Appl. Phys. 16, 151 (1978).

3. S.A.J. Druet, B. Attal, T.K. Gustafson and J.P. Taran, Phys. Rev. A18, 1529 (1978).

4. J.L. Oudar and Y.R. Shen, Phys. Rev. A22, 1141 (1980).

5. C.J. Bordé, C.R. Acad. Sc. Paris, 282B, 341 (1976).

6. S.Y. Yee and T.K. Gustafson, Phys. Rev. A18, 1597 (1978).

7. S.A.J. Druet and J.P.E. Taran, in Progress in Quantum Electronics, ed. by J.H. Sanders and S. Stenholm, Pergamon Press, Oxford, to be published.

8. N. Bloembergen, Nonlinear Optics, Benjamin, New York, 1965, p. 29

9. T.W. Hänsch and P. Toschek, Zeits. Phys. 236, 213 (1970). See also T.W. Hänsch, in Nonlinear Spectroscopy, edited by N. Bloembergen, p. 17, North-Holland Pub. Amsterdam, 1977.

10. L.A. Carreira, L.P. Goss and Th.B. Malloy, J. Chem. Phys. 69, 855 (1978).

11. G.L. Eesley, Coherent Raman Spectroscopy, Pergamon Press, Oxford, 1981. Some statements at the end of Appendix 2 (pp.115-116) are questionable.

12. N. Bloembergen, in Laser Spectroscopy IV, ed. by H. Walther, K.W. Rothe Springer Series in Optical Sciences, Vol. 21 (Springer Berlin, Heidelberg, New York 1979) p. 340.

13. Y. Prior, A.R. Bogdan, M. Dagenais and N. Bloembergen, Phys. Rev. Lett. 46, 111 (1981).

14. A.R.Bogdan, Y. Prior and N. Bloembergen, Opt. Lett. 6, 82 (1981).

15. A.R. Bogdan, M. Downer and N. Bloembergen, Phys. Rev. A24,   (1981).

16. A.R. Bogdan, M.W. Downer and N. Bloembergen, Opt. Lett. 6,   (1981).

17. Y. Prior, Appt. Opt. 19, 1741 (1980).
    J.A. Shirley, R.J. Hall and A.C. Eckbreth, Opt. Lett. 5, 380 (1980).
    S. Chandra, A. Compaan and E. Wiener-Avnean, Appl. Phys. Lett. 33, 867 (1978).

18. J.P. Woerdman and M.F.H. Schuurmans, Opt. Lett. 6, 239 (1981), and references quoted therein.

19. P.F.Liao, J.E. Bjorkholm and P.R. Berman, Phys. Rev. A20, 1489 (1979).

20. T.W. Mossberg, F. Whittaker, R. Kachru and S.R. Hartmann, Phys. Rev. A22, 1962 (1980).

21. A.R. Bogdan, Ph.D. thesis, Harvard University, 1981 (unpublished).

22. J. Herman and M. Landmann, Opt. Comm. 29, 172 (1979).

23. A. Lau, R. König and M. Pfeiffer, Opt. Comm. 32, 75 (1980).

24. M. Dagenais, Phys. Rev. A   (1981).

25. M. Grynberg, J. de Physique, Paris, (1981). See also this volume.

Collisional Effects in Resonance Fluorescence

S. Reynaud and C. Cohen-Tannoudji

Ecole Normale Supérieure and Collège de France,24, rue Lhomond
F-75231 Paris Cedex 05, France

1. Introduction

The collisional redistribution of near resonant scattered light has been
extensively studied both experimentally [1] and theoretically [2 to 16]. The
usual perturbative picture given for such a redistribution is sketched on
Fig. 1. In the absence of collisions and at the lowest order in the laser
intensity, the fluorescence spectrum is given by the elastic Rayleigh scattering process of Fig. 1a. Collisions are responsible for the appearance of
a new fluorescence line around ω_0 which can be interpreted as due to collision induced transitions populating the excited atomic level e from the
"virtual" level reached after the absorption of a laser photon (dotted
line of Fig. 1b).

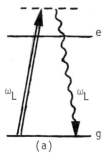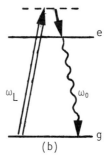

$$(a) \qquad (b)$$

Fig. 1 Perturbative interpretation of Rayleigh scattering (a) and
collisional redistribution (b)

Saturation effects associated with an increase of the laser intensity have
been investigated by different methods. One of them is based on a non-
perturbative solution of the "optical Bloch equations"

$$\frac{d}{dt} \sigma_a = - i[(H_a - DE_L \cos \omega_L t) , \sigma_a]$$
$$- (\mathcal{T}_{rad} + \mathcal{T}_{coll}) \sigma_a \qquad (1)$$

giving the rate of variation of the atomic density matrix σ_a as a sum of
three independent rates: the free atomic evolution and interaction with the
laser (first line); radiative damping due to spontaneous emission and

collisional relaxation (second line). The relaxation matrices τ_{rad} and τ_{coll} are taken to be the same as in the absence of laser irradiation [8-9].

Another possible approach, the so-called dressed atom approach, deals with the compound system "atom plus laser photons interacting together" (dressed atom). In the absence of collisions, resonance fluorescence photons can be considered as photons spontaneously emitted by the dressed atom. All features of resonance fluorescence can be quantitatively interpreted in terms of radiative transition rates between the dressed states [18-19]. Collisional effects can be included in such a theoretical frame by adding collision induced transition rates between the dressed states [6-12].

In this paper, we present a survey of the dressed atom approach to colli-sional redistribution. We introduce the relevant parameters describing the collisional relaxation of the dressed atom (T_1 and T_2 relaxation). We show how the redistribution spectrum and absorption profile may be related to these parameters. We also discuss the validity of the Markov approximation used in writing dressed atom relaxation equations as well as optical Bloch equations. In the so called impact regime, both equations are valid and we make explicit the relation between the collisional redistribution rate w between the dressed states and the dephasing rate γ of the bare atom dipole moment. We show that the dressed atom approach remains valid outside the impact regime which is not the case for optical Bloch equations.

We will ignore in this paper other effects such as quenching, polarization redistribution and velocity changing collisions.

2. The Dressed Atom Approach

The dressed atom energy diagram is sketched on Fig.2, the left part of which gives the uncoupled states labelled by two quantum numbers (e and g for the upper and lower atomic states and n for the number of laser photons).

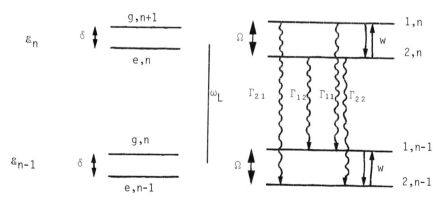

Fig. 2 Energy diagram of the dressed atom. Wavy and full arrows respec-tively describe radiative and collisional transitions between the dressed states

The two states $|e,n>$ and $|g,n+1>$ are nearly degenerate and form a two dimensional manifold $\mathcal{E}n$ (energy difference equal to the detuning δ between the laser and atomic frequencies ω_L and ω_0 and much smaller than the distance ω_L between two adjacent manifolds). The laser atom interaction has a non-zero matrix element between the two states of each manifold which is equal to $\bar{\omega}_1/2$ (ω_1, the Rabi frequency, is equal to the product of the dipole moment d by the laser field amplitude E_L). This coupling describes absorption and stimulated emission of laser photons by the atom. The dressed states $|1,n>$ and $|2,n>$ which diagonalize the total Hamiltonian are represented on the right part of Fig. 2. Their splitting is $\Omega = [\omega_1^2 + \delta^2]^{1/2}$. They can be written

$$|1,n> = \cos\Theta\,|e,n> + \sin\Theta\,|g,n+1> \tag{2a}$$
$$|2,n> = -\sin\Theta\,|e,n> + \cos\Theta\,|g,n+1> \tag{2b}$$

where $\cos\Theta = [(\Omega-\delta)/2\Omega]^{1/2}$ $\quad\sin\Theta = [(\Omega+\delta)/2\Omega]^{1/2}$. $\tag{3}$

The coupling of the dressed atom with the empty modes of the electromagnetic field is responsible for a radiative relaxation described by spontaneous transition rates between the dressed states (wavy arrows of Fig.2) and giving rise to three emission lines (fluorescence triplet) at $\omega_L+\Omega$ (transitions from $|1,n>$ to $|2,n-1>$), $\omega_L-\Omega$ ($|2,n> \to |1,n-1>$) and ω_L ($|i,n> \to |i,n-1>$ for i=1, 2). This simple description in terms of transition rates is based on a "secular approximation" [18-19] valid when the three lines are well resolved (splitting Ω large compared to the width of the lines).

In the same way, the effect of collisions can be described in this energy diagram as a relaxation mechanism, but the collision induced transitions (full arrows of Fig. 2) now occur inside each manifold (quenching neglected). Such a relaxation produces a population redistribution (T_1 type relaxation) and a coherence damping (T_2 type relaxation) respectively described by

$$\frac{d}{dt}\sigma_{1n,1n} = w\,(\sigma_{2n,2n} - \sigma_{1n,1n}) \tag{4a}$$

$$\frac{d}{dt}\sigma_{1n,2n} = -(\kappa + i\,\xi)\,\sigma_{1n,2n} \tag{4b}$$

(σ dressed atom density matrix). To summarize, there are three relevant relaxation parameters w, κ and ξ which have to be derived from the collision S matrix (see some examples of the calculation of w in [6-12]).

The dressed atom picture clearly shows that the emission spectrum has still a triplet structure in the presence of collisions. Only the positions, widths and weights of the three emission lines are modified. We want here to show how these modifications can be related to the relaxation parameters.

Solving the dressed atom relaxation equations for optical coherences gives, in the same way as for collisionless resonance fluorescence [18-19], the positions and widths of the three lines. One finds a pressure broadening equal to 2 κ for the sidebands and 4 w for the central component (full width at half maximum). The pressure shift increases the splitting Ω by an amount ξ.

The weight (integrated intensity) of a given line can be expressed as the product of the steady state population of the emitting dressed state by the radiative transition rate starting from this state. For example

$$I\,(\omega_L-\Omega) = \Gamma_{12}\,\pi_2 \qquad I\,(\omega_L+\Omega) = \Gamma_{21}\,\pi_1\,. \tag{5}$$

The steady state populations $\pi_i = \sum_n \sigma_{in,in}$ are deduced from the normalization condition $\pi_1 + \pi_2 = 1$ and from the detailed balance condition

$$(\Gamma_{12} + w) \pi_2 = (\Gamma_{21} + w) \pi_1 \qquad (6)$$

which expresses that, in the steady state, the total number of radiative and collisional transitions $|1,n> \rightarrow |2,n'>$ balances the total number of transitions $|2,n> \rightarrow |1,n'>$. In the absence of collisions (w=0), the detailed balance condition leads to a symmetric spectrum : $I(\omega_L -\Omega)= I(\omega_L +\Omega)$ according to (5). In the presence of collisions (w≠o) $\Gamma_{12} \pi_2 \neq \Gamma_{21} \pi_1$ and the spectrum becomes asymmetric [8-9].

3. Conditions of Validity

Both the optical Bloch equations (OBE) and the dressed atom relaxation equations (DARE) are first order differential equations resulting from a Markov approximation. The time derivatives appearing in (1) and (4) actually describe a "coarse grained evolution", averaged over a time interval Δt much longer than the collision time τ_c. This introduces some restrictions on the predictions which can be derived from these equations. We show in this section that these restrictions are less severe for DARE than for OBE.
The conditions of validity of OBE are well known. They can be written

$$\gamma \tau_c, \ \omega_1 \tau_c, \ |\delta| \tau_c \ll 1 \qquad (7)$$

where γ is the collisional width, ω_1 the Rabi frequency, δ the detuning. The first condition means that the collisional relaxation time $T_R = \gamma^{-1}$ is much longer than the averaging time Δt ($\tau_c \ll \Delta t \ll T_R$) and is supposed well satisfied. The two last conditions ($\omega_1 \tau_c, |\delta| \tau_c \ll 1$) which define the so-called "impact regime" express the fact that the laser atom interaction may be ignored during Δt (and in the interaction representation with respect to the atomic Hamiltonian H_A). This is why the collisional relaxation matrix $\mathcal{T}_{coll}$ is the same as in the absence of laser irradiation.

Conditions (7) may also be interpreted in the frequency domain: the coarse grained time average restricts the frequency range over which the redistribution spectrum is correctly described to an interval Δt^{-1} around the unperturbed atomic frequency ω_0 and conditions (7) just mean that the fluorescence triplet is entirely contained in this interval (Fig.3).

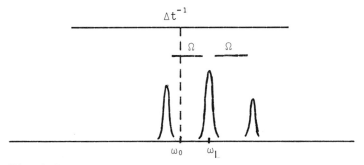

Fig. 3 Frequency range over which OBE describe correctly the redistribution. In the impact regime, the spectrum is entirely contained in this interval

In the dressed atom approach, the collisional relaxation is studied in the dressed atom interaction representation (the laser atom interaction has been first diagonalized). The spectrum derived from DARE is therefore valid in three intervals Δt^{-1} around the three Bohr frequencies ω_L, $\omega_L \pm \Omega$ of the dressed atom (Fig. 4).

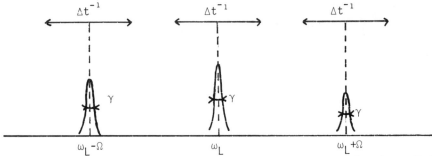

Fig. 4 Frequency intervals over which DARE describe correctly the redistribution

The only condition of validity of DARE is therefore
$$\gamma \tau_c \ll 1 \qquad (8)$$
which is much less severe than (7) : DARE remain valid outside the impact regime where OBE can no longer be used.

Non-Markovian effects, not contained in DARE (because of the coarse grained average),only appear outside the three intervals represented on Fig.4, i.e. in regions where the redistribution is negligible (the width γ of the three lines is much smaller than Δt^{-1}). It follows that DARE provides a Markovian description of the fluorescence spectrum for all values of ω_1 and δ.

4. Relations Between Optical Bloch Equations and Dressed Atom Relaxation Equations

In the impact regime, both OBE and DARE are valid. Thus, it must be possible to relate w, κ and ξ to the parameters describing the relaxation of the bare atom.

For the bare atom, the effect of a given collision (impact parameter b, relative velocity $\vec{v}$) is just to produce different phase shifts ϕ_e and ϕ_g for the upper and lower states (dephasing collisions, quenching neglected)
$$|e> \rightarrow |e> \ exp(-i\phi_e) \qquad |g> \rightarrow |g> \ exp(-i\phi_g). \qquad (9)$$
Summing over collisions (i.e. over b and $\vec{v}$) leads to a damping and a shift of the bare dipole moment described by
$$\frac{d}{dt} < e| \sigma_a | g > = -(\gamma + i\eta) < e |\sigma_a| g > \qquad (10)$$
where the two collisional parameters γ and η are given by
$$\gamma = \sum_{coll} [\ 1 - \cos (\phi_e-\phi_g)] \qquad \eta = \sum_{coll} \sin (\phi_e-\phi_g) \ . \qquad (11)$$
Suppose now that the dressed atom is before the collision in the state $|1,n> = \cos \Theta |e,n> + \sin \Theta | g,n+1>$. The impact regime conditions express that the collision time τ_c is so short that one can decouple the atom and laser photons during the collision (i.e. compute the collision S matrix as if the atom

was free). It follows that the state $|1,n>$ after the collision can be deduced from the phase shifts ϕ_e and ϕ_g :

$$|\overline{1,n}> = \cos \Theta |e,n> \exp (-i\phi_e) + \sin \Theta |g,n+1> \exp (-i\phi_g) . \tag{12}$$

Since ϕ_e and ϕ_g are not equal, $|\overline{1,n}>$ has a nonzero projection on $|2,n>$ which means that the collision has induced a population transfer from $|1,n>$ to $|2,n>$

$$|<2,n|\overline{1,n}>|^2 = 2 \cos^2\Theta \sin^2\Theta [1 - \cos (\phi_e - \phi_g)] . \tag{13}$$

Summing (13) over collisions and using (11) leads to the following expression for the redistribution rate w

$$w = 2 \gamma \cos^2 \Theta \sin^2 \Theta = \gamma \omega_1^2/[2(\omega_1^2+\delta^2)] . \tag{14}$$

Similar calculations give

$$\kappa + i \xi = (\gamma + i\eta) \cos^4\Theta + (\gamma - i\eta) \sin^4 \Theta \tag{15}$$
$$\kappa = \gamma(\omega_1^2 + 2\delta^2)/[2(\omega_1^2 + \delta^2)] \tag{16}$$
$$\xi = -\eta\delta/\Omega . \tag{17}$$

Outside the impact regime, the ω_1 and δ dependence of w and κ are no longer given by (14) and (16). They can be described by introducing two "effective parameters" $g(\omega_1,\delta)$ and $h(\omega_1,\delta)$ such that

$$w = g(\omega_1,\delta) \omega_1^2/[2(\omega_1^2+\delta^2)] \tag{18}$$
$$\kappa = h(\omega_1,\delta) (\omega_1^2+2\delta^2)/[2(\omega_1^2+\delta^2)] \tag{19}$$

the impact regime limit of g and h being equal to γ.

The failure of OBE outside the impact regime is due to an incorrect description of the collisional relaxation which is certainly no longer independent of the laser irradiation. Similar situations exist in nuclear magnetic resonance experiments performed with strong radiofrequency fields. It is well known in this case that the relaxation must be described by modified T_1 and T_2 parameters, defined in the rotating frame and depending on the direction and magnitude of the effective field [17]. This is equivalent to the study of the relaxation in the basis of the dressed states.

It must be emphasized that such modified OBE are not obtained by simply replacing in (10) γ by an effective parameter $\gamma(\omega_1,\delta)$. Such a modification would actually lead to expressions (18) and (19) with $g(\omega_1,\delta) = h(\omega_1,\delta) = \gamma(\omega_1,\delta)$ and there is no reason why g and h should be equal. For example, in the perturbative limit ($\omega_1 << \delta$), one can show that h remains equal to γ outside the impact regime which is not the case for g [20].

To summarize, the two parameters w and κ (or g and h) are independant outside the impact regime. They are not related to a single parameter as is the case in the impact regime.

5. Absorption Profile

In this section we discuss the absorption profile $A(\delta)$ which is recorded when the net absorption of laser photons is plotted versus the detuning δ. The total numbers of absorbed and reemitted photons are obviously equal so that $A(\delta)$ can be obtained by summing the weights of the three emission lines. These weights are functions of the radiative rates Γ_{ij} and of the collisional redistribution rate w and can therefore be expressed in terms of ω_1,δ,Γ (spontaneous emission rate from the upper level) and g (effective parameter appearing in the expression (18) of w). One gets in this way

$$A(\delta) = \frac{\Gamma}{2} (\Gamma + 2g) \omega_1^2 / [(\Gamma + 2g)\omega_1^2 + 2\Gamma\delta^2] . \tag{20}$$

This expression is nothing but the Karplus Schwinger formula [2] (in the secular approximation) generalized outside the impact regime [6].

In the impact regime, g is a constant (equal to γ) and the absorption profile has a Lorentzian shape. Outside the impact regime, this is no longer true because of the δ dependence of g. Thus, it clearly appears that the calculation of w outside the impact regime is closely related to the standard problem of non-Lorentzian far wing absorption.

The absorption profile $A(\delta)$ only depends on w. This explains why it may be fitted by replacing γ by $g(\omega_1,\delta)$ in OBE. Such a method would lead to a wrong result for the redistribution spectrum which also depends on κ.

6. Conclusion

We have presented in this paper a dressed atom approach to collisional effects in resonance fluorescence which provides an interpretation of collisional redistribution in terms of collision induced transition rates between the dressed states. Such an approach generalizes the perturbative picture of Fig.1 to high intensity and resonant situations.

Furthermore, when compared with the method of optical Bloch equations, the dressed atom approach not only provides a simpler physical unsight but appears to have a larger domain of validity. It gives a correct Markovian description of absorption and of spectral redistribution even outside the impact regime.

Another illustration of the advantages of the dressed atom method may be found in the following paper dealing with the interpretation of resonances between "umpopulated levels" in nonlinear optics.

References

[1] J.L. Carlsten, A. Szöke and M.G. Raymer: Phys. Rev. A15,1029 (1977)
[2] R. Karplus and J. Schwinger:Phys. Rev. 73,1020 (1948)
[3] D.L. Huber Phys. Rev. 178,93 (1969)
[4] A. Omont, E.W. Smith and J. Cooper:Astrophysical Journal 175 185 (1972)
[5] E.G. Pestov and S.G. Rautian:Sov. Phys. JETP 37,1025 (1973)
[6] V.S. Lisitsa and S.I. Yakovlenko:Sov. Phys. JETP 41,233 (1975)
[7] S.P. Andreev and V.S. Lisitsa:Sov. Phys. JETP 45,38 (1977)
[8] E. Courtens and A. Szöke:Phys. Rev. A15,1588 (1977)
[9] B.R. Mollow:Phys. Rev. A15,1023 (1977)
[10] G. Nienhuis and F. Schuller:Physica 92C,397 (1977)
[11] D. Voslamber and J.B. Yelnik:Phys. Rev. Lett. 41,1233 (1978)
[12] S. Yeh and P.R. Berman:Phys. Rev. A19,1106 (1979)
[13] J. Cooper:Astrophysical Journal 228,339 (1979)
[14] G. Nienhuis and F. Schuller:J. Phys. B12,3473 (1979)

[15] J. Fiutak and J. Van Kranendonk:J. Phys. B13,2869 (1980)

[16] Y. Rabin and A. Ben-Reuven:J. Phys. B13,2011 (1980)

[17] A. Abragam:The Principles of Nuclear Magnetism (1961, Oxford University Press, London) ch.XII.

[18] C. Cohen-Tannoudji and S. Reynaud:J. Phys. B10,345 (1977)

[19] C. Cohen-Tannoudji and S. Reynaud:in "Multiphoton Processes" ed. by J.H. Eberly and P. Lambropoulos (Wiley, New York 1978) p. 103

[20] S. Reynaud:Thèse (Paris, 1981) unpublished

Theory of Nonlinear Optical Resonances Induced by Collisions

G. Grynberg

Laboratoire d'Optique Quantique, Ecole Polytechnique
F-91128 Palaiseau, France

1. Introduction

It has been suggested by Bloembergen [1] that new resonances should appear in four-wave mixing processes in the case of large damping. These resonances occur when the difference between the frequencies of two beams is equal to a Bohr frequency for two atomic levels which are not directly populated by the light excitation. They have recently been observed by Bloembergen and his coworkers [2]. The original treatment of Bloembergen [1] was performed with the optical Bloch equations. It is well known that the validity of these equations is restricted to a small frequency range where the impact approximation can be applied [3]. We present here the theory of these resonances using the dressed-atom model [3]. The two main advantages of the present theory [4] are (i) a larger range of validity,(ii) a direct comprehension of the physical origin of these resonances.

2. Result of the optical Bloch equations

We consider atoms with a ground level g and two excited levels e and e'. The energy differences E_{eg} and $E_{e'g}$ are respectively equal to $\hbar\omega_0$ and $\hbar\omega'_0$. These atoms interact with two waves of frequencies ω and ω'. The energy detunings δ and δ' are equal to $(\omega-\omega_0)$ and $(\omega'-\omega'_0)$. We shall assume in the following that $|\delta|$ and $|\delta'|$ are much larger than (i) the relaxation rates Γ_{eg} and $\Gamma_{e'g}$ of the optical coherences (ii) the Rabi frequencies ω_1 and ω'_1 ($\omega_1 = \hbar^{-1} \langle g|\vec{D}\cdot\vec{E}|e\rangle$ and $\omega'_1 = \hbar^{-1}\langle g|\vec{D}\cdot\vec{E}'|e'\rangle$).

In order to calculate the intensity of the light generated at the frequency $(2\omega-\omega')$ one has to calculate the mean value of the electric dipole moment $\vec{D}$,

$$\langle\vec{D}\rangle = \mathrm{Tr}\ \sigma_A\ \vec{D} = \sum_{i=e,e'} (\sigma_{gi}\ \vec{D}_{ig} + \sigma_{gi}{}^*\ \vec{D}_{ig}{}^*) . \tag{1}$$

The frequency $(2\omega-\omega')$ appears at the third order of perturbation in electric field.   Knowledge of $\sigma_{gi}{}^{(3)}$ needs the calculation of $\sigma_{ee'}{}^{(2)}$. This coherence is particularly important because the resonance at $\delta'-\delta=0$ ($E_{ee'}=\hbar(\omega-\omega')$) appears when $\sigma_{ee'}$ is calculated,

$$\sigma_{ee'}{}^{(2)} = \frac{\omega_1\ \omega'_1}{4\delta\ \delta'}\ e^{i(\omega'-\omega)t}\left\{ 1 + \frac{\Gamma_{eg} + \Gamma_{e'g}^* - \Gamma_{ee'}}{\Gamma_{ee'} + i(\delta'-\delta)} \right\} . \tag{2}$$

Each relaxation rate Γ_{ij} contains a radiative part $(\frac{\Gamma}{2}i + \frac{\Gamma}{2}j)$ and a collisional part (γ_{ij}). Since $\Gamma_g = 0$, the amplitude of the resonance is proportional to $(\gamma_{eg} + \gamma_{e'g}^* - \gamma_{ee'})$ and depends only on collisional processes [1].

3. Dressed—atom approach to the Bloembergen resonances

It has been shown in [3] that the effect of collisions in the dressed-atom model is to induce transitions between states of the same manifold. The difference between [3] and the present problem is that we have two atomic excited levels and two modes of the field instead of one. Each manifold has thus a more complicated structure. Nevertheless, in the perturbational limit ($\omega_1/|\delta| \ll 1$), it is possible to restrict our attention to a manifold compounded of three states which correspond at the zero order to $|g,n,n'\rangle$, $|e,n-1,n'\rangle$ and $|e',n,n'-1\rangle$ (see fig.1).

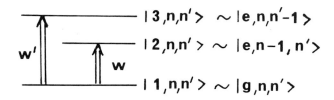

$$|3,n,n'\rangle \sim |e,n,n'-1\rangle$$

$$|2,n,n'\rangle \sim |e,n-1,n'\rangle$$

$$|1,n,n'\rangle \sim |g,n,n'\rangle$$

Fig.1 : Schematic diagram of the dressed-atom and collision-induced transitions between states of the same manifold.

Let us consider the system in the $|1,n,n'\rangle$ ($\sim |g,n,n'\rangle$) state. The collisions induced a redistribution of population inside the manifold. However the transfer of population towards the levels $|2,n,n'\rangle$ and $|3,n,n'\rangle$ are not statistically independant and some coherence is created between these two levels by the collisions [4]. As we shall see hereafter, it is the creational coherence which is at the origin of the Bloembergen resonances.

The situation encountered here is formally identical to the one usually found in level crossing spectroscopy. A pumping process creates a coherence between two levels and the resonance is observed when the energies of the two levels coherently excited become equal.

It can also be emphized that the dressed-atom model shows directly that the Bloembergen resonances are strongly connected to the collisional redistribution of radiation. It means that the experimental situation which is favorable for the observation of these resonances occurs when there is important collision—induced fluoresence.

The physics of the resonance does not, of course, depend on the value of δ. However the calculation of the rate of creation of coherence is different according to the value of δ. We discuss in the next paragraph the impact regime. Before, we recall that at the first order of perturbation in ω_1 and ω_1', the eigenvalues of the total Hamiltonian are the same as the ones calculated at the zero order and the eigenstates are :

$$|1,n,n'> = |g,n,n'> - \frac{\omega_1}{2\delta} |e,n-1,n'> - \frac{\omega_1'}{2\delta'} |e',n,n'-1>$$

$$|2,n,n'> = \frac{\omega_1}{2\delta}|g,n,n'> + |e,n-1,n'> \qquad\qquad (3)$$

$$|3,n,n'> = \frac{\omega_1'}{2\delta'} |g,n,n'> + |e',n,n'-1> \quad .$$

4. The impact regime

We assume in this paragraph that $|\delta|$ and $|\delta'|$ are much smaller than τ_c^{-1} where τ_c is a typical time for a collision. The effect of a collision on the dressed atom is particularly simple in this case [3]. For instance, if the dressed atom is in the state $|1,n,n'>$ before a collision, its state after a collision is :

$$|\overline{1,n,n'}>=|g,n,n'> - \frac{\omega_1}{2\delta} \exp{-i(\phi_e-\phi_g)}|e,n-1,n'>- \frac{\omega_1'}{2\delta'}\exp{-i(\phi_e'-\phi_g)}|e',n,n'-1> \quad (4)$$

where ϕ_g, ϕ_e and ϕ_e' are the phase shifts produced by a collision defined in [3].

The population transfer from $|1,n,n'>$ to $|2,n,n'>$ and $|3,n,n'>$ is calculated as in [3]. For instance, we obtain :

$$|<2,n,n'| \overline{1,n,n'}>|^2 = \frac{\omega_1^2}{2\delta^2} [1 - \cos (\phi_e-\phi_g)] \quad . \qquad (5)$$

In a similar way, we obtain for the creation of coherence:

$$<2,n,n'|\overline{1,n,n'}><\overline{1,n,n'}|3,n,n'> = \frac{\omega_1\omega_1'}{4\delta\delta'} \left\{1+e^{-i(\phi_e-\phi'_e)}-e^{-i(\phi_e-\phi_g)}-e^{i(\phi'_e-\phi_g)}\right\}\cdot(6)$$

Summing (5) and (6) over collisions gives the following expressions for the redistribution rate w and w' and for the rate of creation of coherence r

$$w = \frac{\omega_1^2}{2\delta^2} \operatorname{Re} \gamma_{eg} \; ; \quad w' = \frac{\omega_1'^2}{2\delta'^2} \operatorname{Re} \gamma'_{eg} \qquad (7)$$

$$r = \frac{\omega_1\omega_1'}{4\delta\delta'} \left[\gamma_{eg} + \gamma_{e'g}^* - \gamma_{ee'}\right] \quad . \qquad (8)$$

Using (8) and the dressed-atom relaxation equation [3] we find for the steady state value of the coherence between the levels $|2,n,n'>$ and $|3,n,n'>$ an expression equal to r divided by an energy denomination resonant for $\delta=\delta'$. The comparison with the expression (2) obtained using the optical Bloch equations shows that it is the same resonance which is calculated by two different methods (the time dependance comes from the coherent character of the field : the initial state should be taken as a linear superposition of the $|1,n,n'>$ states [4]).

5. Outside the impact limit

The validity of the dressed-atom relaxation equation is not restricted to the impact range [3]. It has thus been possible to calculate the values of the population transfer w and w' [5] and of the rate of creation of coherence r [4]. If we define $\alpha = r/[ww']^{1/2}$, $|\alpha|$ is a number which is equal or less than 1. In the general case, when the potential curves of levels e and e' are different, α is very small [4]. In that case, the intensity of the resonance is expected to become extremely small for large value of $|\delta|$. On the other hand, if the potential curves are nearly identical, α is closed to 1 and the resonance can be observed for large values of $|\delta|$. In particular, this is the case when e=e' (degenerate four-wave mixing). In that situation, it has been possible to observe the resonance for large values of $|\delta|$ [2].

6. Pressure-induced resonance in degenerate four wave mixing

The case of degenerate four wave mixing has been studied for isolated excited level [2]. However several new features occur when one takes into account the degeneracy of the levels [6]. For instance, for experiments performed with a single laser it is possible to observe resonances on the generated beam when a magnetic field is scanned [6]. These resonances which only exist in the presence of collisional damping correspond to level crossing resonances. Coherences between hyperfine sublevels are excited by collisions (the mechanism is identical to the one of § 3). The intensity and the shape fo these resonances can be obtained using optical Bloch equation in the impact range [6] while it is necessary to use the dressed-atom relaxation equation for large values of $|\delta|$.

7. Conclusion

This example shows that the dressed-atom theory can be very useful in non-linear optics even if at first glance the classical character of the e.m. fields do not require a quantum description. Even if the problem can be analyzed without quantification of the field, the dressed-atom picture provides an elegant and powerful way to solve it.

References

1. N. Bloembergen: In *Laser Spectroscopy IV*, ed. by H. Walther, K.W. Rothe, Springer Series in Optical Sciences, Vol. 21 (Springer Berlin, Heidelberg, New York 1979) p. 340
2. N. Bloembergen, this book and ref therein
3. S. Reynaud and C. Cohen-Tannoudji, this book and ref therein
4. G. Grynberg, J.Phys.B. Atom.Mol.Phys. 14, 2089 (1981)
5. S. Yeh and P.R. Berman, Phys.Rev. A22, 1403 (1980) and A19, 1106 (1979)
6. G. Grynberg, Opt.Commun. 1981 (to be published)

Polarization Intermodulated Excitation (POLINEX) Spectroscopy of Excited Atoms*

Ph. Dabkiewicz[1], T.W. Hänsch, D.R. Lyons, A.L. Schawlow, A. Siegel[2], Z.-Y. Wang[3], and G.-Y. Yan[4]
Department of Physics, Stanford University
Stanford, CA 94305, USA

We have demonstrated a sensitive and versatile new technique of Doppler-free saturation spectroscopy, which takes advantage of Polarization Intermodulated Excitation (POLINEX): the nonlinear interaction of two laser beams in an absorbing medium is studied by modulating the polarization of one or both beams. When the combined absorption depends on the relative polarization of the two beams, an intermodulation is observed in the total rate of excitation. In first exploratory experiments we have studied excited helium atoms in a positive column discharge [1], neon atoms in radiofrequency discharges [1,2], and copper atoms in a hollow cathode discharge [3]. Both optogalvanic detection and fluorescence detection have been used. Clean spectra free of Doppler-broadened background could be recorded despite velocity changing elastic collisions.

The POLINEX method is related to polarization spectroscopy [4] and to intermodulated fluorescence spectroscopy [5]. Although it is experimentally rather simple, POLINEX offers several unique capabilities not found in the earlier methods. For a comparison, it is worthwhile to briefly recall the essential features of the two parent techniques: like other methods of saturation spectroscopy, each provides spectra free of first order Doppler broadening by monitoring the nonlinear interaction of two counterpropagating monochromatic laser beams in a gas sample. A signal is obtained if the laser is tuned to the center of a Doppler-broadened absorption line, so that both beams are interacting with the same atoms, those with zero axial velocity.

In polarization spectroscopy [4,6], the signal is detected as a change in the polarization of one of the light beams, the probe, induced by the presence of the other beam, the pump or saturating beam. Normally, in a gas, the angular momenta of the absorbing atoms are pointing randomly in all directions, and the sample is optically isotropic. However, the absorption cross sections for the circularly or linearly polarized saturating beam are different for different atomic orientations. As this beam excites the atoms

*Work supported by the National Science Foundation under Grant PHY-80-10689, and the U.S. Office of Naval Research under Contract ONR N00014-78-C-0403.

[1] Fellow of the Deutsche Forschungsgemeinschaft.

[2] Present address: Department of Physics, University of Kaiserslautern, Kaiserlautern, Fed. Rep. of Germany

[3] Permanent address: Department of Physics, Fudan University, Shanghai, People's Republic of China.

[4] On leave from East China Normal University, Shanghai, People's Republic of China

with the largest cross sections, it leaves the remaining ones with a
preferential orientation or alignment. The gas thus becomes dichroic and
birefringent, and the probe beam, originally linearly polarized, can acquire
a polarization component which passes through a crossed linear polarizer
into a photodetector. The detection sensitivity can approach the shot-noise
limit even if the laser intensity is fluctuating. However, like all methods
directly observing some change in the laser light, polarization spectroscopy
works best if the absorption of the sample is non-negligible.

For studies of optically thin samples, higher sensitivities can be
obtained by detecting the deposition of radiation energy in the sample
indirectly, for instance by monitoring the laser induced fluorescence. Such
detection is used in intermodulated fluorescence spectroscopy [5]. To
selectively record a Doppler-free spectrum, the two counterpropagating laser
beams are chopped at two different frequencies f_1 and f_2, and the signal is
detected as a modulation of the total excitation rate at the sum or
difference frequency $f_1 \pm f_2$. Such an intermodulation occurs when the two
beams are interacting with the same atoms so that they can saturate each
other's absorption. As the number of absorbing atoms becomes smaller, both
signal and background are reduced proportionately, and a respectable signal-
to-noise ratio can be maintained down to very low concentrations. It should
even be feasible to apply the technique to a single ion in a trap. The same
intermodulation method can also be used with other indirect detection
schemes, and both optogalvanic detection [7] and optoacoustic detection [8]
have been demonstrated. However, despite its versatility and sensitivity,
the (intensity-) intermodulation method is experimentally not trivial.
Imperfections in the chopper or any nonlinear frequency mixing in detectors
or amplifiers can easily produce spurious signals, and considerable care is
necessary to avoid such artifacts.

Only some seemingly minor changes are required to convert from this older
intermodulation technique to POLINEX: the chopper is simply removed and
replaced by two polarization modulators, which modulate the polarizations of
the two beams at two different frequencies, while leaving the intensities
constant. These modulators could, for instance, be elastooptic devices
producing alternating right and left hand circular polarization. When the
laser is tuned so that both beams can interact with atoms of the same
velocity, the total rate of excitation will still be modulated at the sum or
difference frequency, because the combined absorption of the two beams will
in general depend on their relative polarization. If the two beams have
identical polarization, both lightfields will be preferentially absorbed by
atoms of the same orientation, and we expect a pronounced mutual or cross
saturation. If, on the other hand, the lightfields have different
polarizations, the two beams will tend to interact with atoms of different
orientations, and there will be less mutual saturation.

To give a simple quantitative description of the POLINEX signals for a
transition between two levels with arbitrary angular momenta J and J', we
have calculated the saturated population densities of the sublevels $|J,m\rangle$
and $|J',m'\rangle$ with the help of rate equations, expressing the relative
transition probabilities in terms of squares of Clebsch Gordan coefficients.
If multiple optical pumping cycles are ignored, the summation over the
(saturated) individual transition rates leads to analytical expressions
similar to those obtained for polarization spectroscopy [4,6], and
tabulations of the results have been published [1,3]. However, these simple
rate equation results can be valid only in the limit of low intensities and
a Doppler width large compared to the natural line width. It would be

desirable to formulate a more general semiclassical model describing the atoms with the help of a density matrix which can account for coherent superpositions between m-sublevels [9].

The POLINEX signals, as those of polarization spectroscopy, are entirely due to light-induced atomic orientation or alignment. There is one important difference, however. The older polarization technique is sensitive to both the dichroism and to the birefringence of the sample, and the combination of these two effects generally produces asymmetric line shapes. The POLINEX lines, on the other hand, remain symmetric, at least for optically thin samples, since the signal depends only on the total light absorption or the imaginary part of the nonlinear susceptibilty tensor.

At the same time, the POLINEX method preserves the advantages of indirect detection, offered by the older intermodulation techniques. But again, we find a rather important difference: in POLINEX spectroscopy, neither beam alone is capable of producing a modulated signal, because the total rate of (steady state) absorption in an isotropic medium does not depend on the sign or direction of the light polarization, even at high light intensities. Consequently, any modulation of the total excitation rate immediately gives the desired nonlinear signal, and nonlinear mixing in detector or amplifiers will no longer produce spurious signals. In fact, good selectivity for the Doppler-free signals can be maintained even if one of the polarization modulators is removed, so that $f_2 = 0$.

Such a simple setup, illustrated in Fig.1, was used in one of the first experimental tests of the POLINEX method [1]. An old He-Ne-laser discharge tube was used as the sample cell. A single-mode cw dye laser was tuned to the $2\ ^3P$ - $3\ ^3D$ line of ^3He at 587.56 nm, and the signal was detected optogalvanically, as a modulation of the discharge current. One of the two counterpropagating laser beams remained unmodulated with fixed circular polarization, the other was sent through an electrooptic modulator, which, through voltage controlled birefringence, produced alternatingly right and left hand circularly polarized light with a modulation frequency of about 800 Hz.

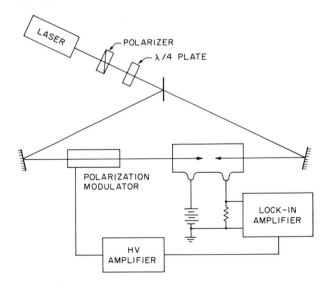

Fig.1 POLINEX spectrometer with one single electrooptic polarization modulator. The signal in the sample discharge is detected optogalvanically

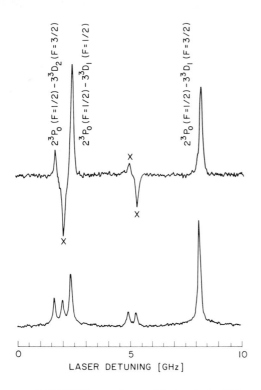

LASER DETUNING [GHz]

Fig.2 Part of the hyperfine spectrum of the 2^3P - 3^3D line of ^3He:
(a) Doppler-free POLINEX spectrum.
(b) Same spectrum recorded by the older intermodulated optogalvanic method [7]

A POLINEX spectrum obtained in this way is shown in Fig.2 (top), together with a spectrum recorded by the older intermodulated optogalvanic method [7] under otherwise identical conditions. The two methods give comparable signal-to-noise ratios. However, the POLINEX spectrum shows several line components inverted. Similar inverted signals are known from polarization spectroscopy [4,6]. They can occur only for crossover signals, which are observed when the laser is tuned halfway in between two resonances with a common lower or upper state. The signal inversion can be a welcome aid in identifying these artifacts of saturation spectroscopy.

In a second experiment [1], the POLINEX method was applied to several of the yellow 1s - 2p transitions of neon in a mild radiofrequency discharge. As shown in Fig.3, the two laser beams were sent through mechanical polarization modulators. A $\lambda/4$ retarder plate makes the light first circularly polarized, and a rotating Polaroid filter then produces linearly polarized light with the polarization axis rotating around the direction of light propagation. The signal was observed as a modulation in the fluorescent sidelight.

At first, one of the filters remained fixed, and only the other one was spun at about 150 revolutions per second. In this case, some complications arise, because the spontaneous emission of the laser-excited neon atoms is spatially anisotropic [10], and rotation of the alignment axis causes a modulated background signal, comparable to a lighthouse effect. This background signal can be supressed by selecting the proper phase at the lock-in amplifier, and the POLINEX signal can then be independently maximized by adjusting the polarization axis of the unmodulated beam.

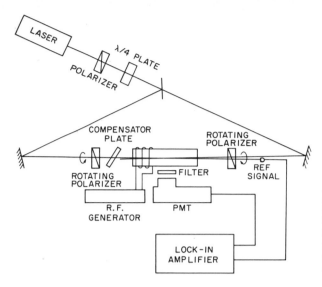

Fig.3 POLINEX
spectrometer with two
mechanical polarization
modulators. The signal
is detected as a
modulation of the laser-
induced fluorescence

The background modulation could be eliminated if the fluorescent light
could be collected over a solid angle of 4 π sterad. Similarly, the
modulation is not observed with circularly polarized beams, because of the
inherent axial symmetry. The modulation is also absent, of course, in
optogalvanic spectroscopy with its angle-independent detection sensitivity.
Such optogalvanic detection is possible even in a radiofrequency discharge,
by monitoring a modulation of the impedance of the exciting coil [2].

In order to avoid the need for careful adjustments with linearly
polarized beams and fluorescence detection, later experiments were performed
with both polarizers rotating at two different frequencies. If the
polarizers are rotating in the same sense, the signal is modulated at the
difference of the two modulation frequencies, and if the polarizers are
spinning in opposite directions, the signal appears at the sum frequency.
The proper reference is obtained by simply monitoring the intensity of a
laser beam that has passed through both polarizers.

Figure 4 shows a POLINEX spectrum of the neon $1s_5 - 2p_2$ transition at
588.19 nm, recorded in this way. The two peaks correspond to the two
naturally abundant isotopes. Shown below is a spectrum of the same line
recorded by the older intermodulated fluorescence technique. It exhibits
striking Doppler-broadened pedestals underneath the narrow Doppler-free
resonances. Similar pedestals were first observed for neon transitions
about 10 years ago [11]. They can be ascribed to elastic collisions which
redistribute the atoms over the Maxwellian velocity distribution, thus
defeating the velocity selection by spectral hole burning. The pedestals
are absent in the POLINEX spectrum, because for metastable neon atoms the
velocity changing collisions have a high probability of destroying any
atomic orientation or alignment. The disoriented atoms can then no longer
contribute to the POLINEX signal. A similar suppression of collision
broadened pedestals had earlier been observed in polarization spectroscopy
of neon discharges [12].

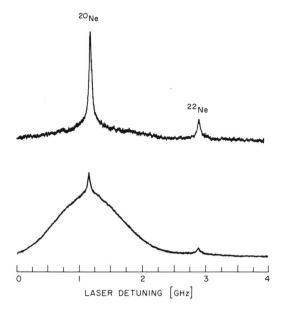

^{20}Ne

^{22}Ne

LASER DETUNING [GHz]

Fig.4 (a) Doppler-free POLINEX spectrum of the Ne $1s_5$ - $2p_2$ transition. (b) same spectrum recorded by the older intermodulated fluorescence technique

Indeed, we developed the POLINEX method originally as a means to suppress collision broadened pedestals in the saturation spectra of metastable atoms in hollow cathode discharges [13,14]. JAMES LAWLER, while at Stanford, has developed a demountable hollow cathode lamp optimized for high resolution laser spectroscopy of sputtered metal atoms [13]. A large cathode diameter (about 1 inch) allows operation at low buffer gas pressure (0.3 torr) while maintaining a central region free of large static electric fields. Initially, the tube was used for a study of the isotope shifts of visible absorption lines of metastable Mo atoms by intermodulated fluorescence and optogalvanic spectroscopy [14]. Despite the low operating pressure, strong collision broadened pedestals remained observable in all spectra. Even more pronounced pedestals were observed in the intermodulated fluorescence spectra of the Cu ($3d^{10}$ $4p$ $^2P_{1/2}$ - $3d^9$ $4s^2$ $^2D_{3/2}$) transition at 578.2 nm (the yellow Cu laser line), as shown in Fig.5a.

For POLINEX spectroscopy of the same line, an electrooptic modulator was inserted into one of the beams to produce alternating right hand and left hand circular polarization, while the circular polarization of the other beam remained fixed. The signal was observed as a modulation of the fluorescent light emitted by the upper level. Fig.5b shows a spectrum recorded under these conditions. The background is indeed suppressed, indicating that the metastable Cu atoms are easily depolarized in elastic collisions.

The last experiment indicates that the POLINEX method can yield clean Doppler-free spectra even under the non-ideal conditions of a hollow cathode discharge. Moreover, the method should permit some interesting studies of elastic collision processes. For instance, we expect that some residual Doppler broadening due to velocity changing collisions will always remain in a POLINEX spectrum, because not all elastic collisions will destroy the atomic orientation. This broadening should be substantial for certain

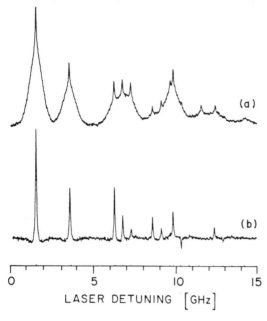

LASER DETUNING [GHz]

Fig. 5 Hyperfine spectrum of
the 578.2 nm transition of Cu.
(a) Intermodulated fluorescence
spectrum with co-rotating circular polarizations of the two
laser beams, (b) POLINEX spectrum

collision partners, such as molecules in high angular momentum states, which
behave like classical rapidly spinning tops, whose angular momentum cannot
be easily disoriented.

To conclude, we would like to point out that the general principles of
the POLINEX method are not limited to Doppler-free saturation spectroscopy
of gases. The method should offer interesting advantages in other
situations, where the nonlinear interaction of two radiation fields in a
sample is to be studied. Examples include two-photon spectroscopy, Raman
spectroscopy, or the study of picosecond relaxation phenomena in liquids or
solids with delayed pulse trains.

REFERENCES

1. T.W.Hänsch, D.R. Lyons, A.L.Schawlow, A.Siegel, Z.-Y.Wang, and
    G.-Y. Yan, Opt. Comm. 37, 87 (1981).

2. D.R.Lyons, A.L.Schawlow, and G-Y.Yan, Optics Communications,
    accepted for publication (1981).

3. Ph. Dabkiewicz and T.W.Hänsch, Optics Communications,
    accepted for publication (1981).

4. C.E.Wieman and T.W.Hänsch, Phys. Rev. Letters 36, 1170 (1976).

5. M.S.Sorem and A.L.Schawlow, Opt. Comm. 5, 148 (1972).

6. R.E.Teets, F.V.Kowalski, W.T.Hill III, N.W.Carlson, and T.W.Hänsch,
    Proc. SPIE 113, 80 (1977).

7. J.E.Lawler, A.I.Ferguson, J.E.M.Goldsmith, D.J.Jackson, and
    A.L.Schawlow, Phys. Rev. Letters 42, 1046 (1979).

8. E.E.Marinero and M.Stuke, Opt. Comm. 30, 349 (1979).

9. T.W.Hänsch and P.Toschek, Z. Physik 236, 213 (1970).

10. T.W.Hänsch, P.Toschek, Phys. Letters 22, 150 (1966).

11. P.Smith and T.W. Hänsch, Phys. Rev. Letters 26, 740 (1971).

12. C. Delsart, J.C. Keller, In *Laser Spectroscopy III*, ed. by J.L. Hall,
    J.L. Carsten, Springer Series in Optical Sciences, Vol. 7 (Springer
    Berlin, Heidelberg, New York 1977) p. 154

13. J.E.Lawler, A.Siegel, B.Couillaud, and T.W.Hänsch,
    J. Appl. Physics, to be published (1981).

14. A.Siegel, J.E.Lawler, B.Couillaud, and T.W.Hänsch,
    Phys. Rev. A23, 2457 (1981).

Laser Optical Pumping in Atomic Vapors with Velocity Changing Collisions

W.W. Quivers, Jr., R.A. Forber[1], A.P. Ghosh, D.J. Heinzen, G. Shimkaveg, M.A. Attili, C. Stubbins, P.G. Pappas[2], R.R. Dasari, and M.S. Feld

Spectroscopy Laboratory and Physics Department[3]
Massachusetts Institute of Technology
Cambridge, MA o2139, USA, and

Y. Niv[4]

Rutgers University, New Brunswick, NJ 08903, USA, and

D.E. Murnick[5]

Bell Laboratories, Murray Hill, NJo7974, USA

1.   Introduction

There is considerable interest in laser optical pumping of atomic vapors as a source of polarized nuclei.[1] Single mode dye laser radiation offers many advantages, but ordinarily the resulting efficiency in polarizing the sample is low because of incomplete Doppler coverage. Recently it has been shown that large nuclear polarizations can be efficiently achieved by means of velocity changing collisions *(vcc's)* induced by small amounts of buffer gas.[2] This paper studies the details of this process in the first resonance transitions of Yb, Li, and Na, and demonstrates that velocity thermalizing collisions can cause a Doppler broadened system subjected to monochromatic radiation to respond as if it were homogeneously broadened.

The simplest energy level scheme for optical pumping is an atomic system composed of two ground state M-levels (1 and 2), both radiatively coupled to a single excited level of energy $\hbar\omega$ and radiative lifetime τ (Fig.1). By choosing pump radiation of the appropriate polarization only $2\to0$ transitions are induced. Atoms excited from level 2 to level 0 decay to level 1 with probability (branching ratio) $\rho = \Gamma_1\tau$, Γ_1 being the radiative decay rate to level 1. Hence, atoms tend to accumulate in level 1, and are thus polarized. For laser radiation of intensity I the effective rate of pumping atoms from levels 2 to 1 is

$$r = \rho\sigma I/\hbar\omega \qquad\qquad (1a)$$

---

[1] Howard Hughes Doctoral Fellow
[2] Resident Visitor, Bell Laboratories
[3] Dr. Chaim Weizmann Fellow, supported in part by NSF
[4] Visiting Scientist, M.I.T. Regional Laser Center
[5] Work supported by NSF and U.S. Department of Energy

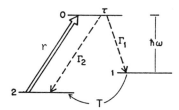

Fig.1   Three level system

with σ the 2-0 absorption cross-section. The fraction of
polarized atoms depends on the product

$$R = rT, \tag{1b}$$

with T the mean time for excess atoms in level 1 to relax back
to level 2. In the systems under study ground state M-changing
atom-buffer collisions are negligible and relaxation occurs at
the walls of the sample cell. We can also write $R=I/I_{op}$, with
optical pumping saturation intensity $I_{op}=\hbar\omega/\rho\sigma T$. Efficient
polarization requires R>1, i.e. $I>I_{op}$, for all atoms.

    In the absence of vcc processes monochromatic laser radia-
tion can optically pump only the small fraction of atoms in the
narrow range of velocities (as measured along the laser beam
axis) which are Doppler shifted into resonance. This produces
a narrow depletion in the thermal velocity distribution of
level 2, and a corresponding narrow resonant increase in that
of level 1. The occurrence of $vcc's$ causes the depleted atoms
to be replenished by atoms of other velocities as they are pumped, leading to removal of atoms over the entire velocity distribution of level 2. Similarly, $vcc's$ tend to thermalize the velocities of atoms transferred to level 1. This gives rise to two
components of the optically pumped atoms, a narrow feature
composed of atoms which are resonant with the applied field and
a broad feature indicative of complete Doppler coverage. With
increasing buffer pressure p, the broad component becomes dominant. In our experiments this occurs at p ~ 0.1-1 torr, well
below the regime of pressure broadening. Attainment of large
polarizations with a minimum of buffer atoms is important in
many applications.

    Efficient nuclear polarization requires (i) rapid excited
state decay,

$$\Gamma_1 T \gg 1, \tag{2a}$$

to insure complete optical pumping within a given atomic velocity
group; and, (ii) a large vcc rate,

$$\Gamma_v T \gg 1 \tag{2b}$$

to achieve complete Doppler coverage. In this limit (vcc regime)
the polarization Π is given by

$$\Pi = \frac{N_1 - N_2}{N_1 + N_2} = \frac{R}{1+R} , \qquad (3)$$

with N_j the velocity-integrated population of level j, and R defined by Eqs.(1). In the *vcc* regime T is determined by diffusion to the walls (and is thus proportional to p), and σ by Doppler broadening:

$$\sigma(\Delta) = \sigma_D \exp{-(\Delta/ku)^2} , \qquad (4)$$

in which the peak Doppler cross section $\sigma_D = \sigma_\Theta (1/2\tau)/(ku/\sqrt{\pi})$, with radiative cross section σ_Θ, Doppler width ku, and detuning Δ of the pump frequency from atomic line center.

When the *vcc* rates are very large $[\Gamma_v T > ku/\gamma]$ velocity changes occur so rapidly that the velocity profile is completely thermalized, and details of the *vcc* process are lost.[2] However, when

$$1 \ll \Gamma_v T < ku/\gamma \qquad (2c)$$

a residual narrow dip, which contains information about the *vcc* processes, can be observed. The following experiments were performed in this regime.

2.   Ytterbium 171

The first resonance line of Yb (λ=555.6 nm, τ = 0.875 μs [3]) is a J=0→1 transition with Doppler width ku = 464 MHz at 425°C. ^{171}Yb has nuclear spin I=1/2, giving rise to two hyperfine (hf) components split by 5.9 GHz. Hence the F=1/2→1/2 transition, when pumped by circularly polarized light, forms a non-degenerate three level system with $\sigma_0 = 4\pi\bar{\lambda}^2$ and ρ = 1/3, which is ideal for quantitative study of the effects of *vcc*'s on the laser optical pumping process.

Experiments were performed in the "Lamb-dip" configuration, using counter-propagating strong pump and weak probe laser beams. The cell was an 8.6 mm diameter x 40 cm long stainless steel tube heated to 425°C over the central ~8 cm region. Helmholtz-type coils produced a 2 G keeper field along the laser beam axis and nulled transverse fields to < 0.1 G. Optically thin samples of 95%-enriched ^{171}Yb were used, with argon buffer pressures, p, in the range 0.1-2 torr.

A single mode cw dye laser with Rhodamine 110 dye produced pump and probe intensities of 18 mW/cm^2 and 30 μW/cm^2, with beam diameters of 4 mm and 2 mm, respectively. Circularly polarized laser light was used to pump the ΔM_F=+1 transition and probe either the ΔM_F=+1 transition ($\mathrel{\hat{\leftleftarrows}}$ configuration) or the ΔM_F=-1 transition ($\mathrel{\hat{\rightleftarrows}}$ configuration). Comparison of $\mathrel{\hat{\leftleftarrows}}$ and $\mathrel{\hat{\rightleftarrows}}$ data gives information on the velocity changing collision rate of the excited state.

Figure 2 shows data taken at various values of p, at a fixed pump intensity. The change in probe absorption due to laser pumping was recorded using phase sensitive detection. Hence, a positive (negative) probe change signal indicates decreased (increased) population in the velocity group being probed. Each trace has two components, a narrow dip due to optical pumping of the resonant velocity group and a broad pedestal due to thermalizing collisions. With increasing pressure the pedestal increases in size and the dip decreases. The large pedestal area relative to that of the dip demonstrates that complete Doppler coverage can be achieved at low values of p. Also note for a given pressure that the pedestals in the two relative polarization configurations are about equal, but the dip is much smaller in the $\hat{-}$ configuration. This is due to velocity thermalization in the excited state during the long radiative lifetime.

The data was compared to a three level model[4] in which the level populations at a given velocity are described by rate equ-

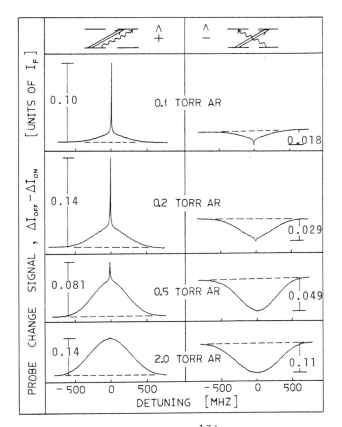

Fig.2 Optical pumping in ^{171}Yb. The figure plots the change in probe absorption for pump laser OFF compared to ON, in units of probe intensity I_p.

ations coupled by stimulated emission/absorption, radiative decay, relaxation to the walls and vcc processes. Strong collisions are assumed, in which each vcc leads to complete velocity thermaliz- ation. In the vcc limit [Eqs.(2a,c)] the probe absorption change signal is composed of two contributions, a Gaussian pedestal of width ku and a narrow Lorentzian dip of width determined by homo- geneous broadening mechanisms. The pedestal height is determined by the optical pumping saturation parameter, R [Eqs.(1)]. The dip height depends on the ordinary saturation parameter $S=I/I_S$, with

$$I_S = \frac{\hbar\omega}{\sigma_h} \Gamma_v^g , \qquad (5)$$

Γ_v^g being the ground state vcc rate and σ_h the homogeneous cross section. Figure 3 shows the peak values of pedestal and dip contributions to the probe absorption coefficient, measured at various values of p, along with a theoretical fit. The value of R, extracted from the pedestal data, gives the diffusion time T, and analysis of the dip then gives S, from which Γ_v^g is obtained. The analysis yields values for the diffusion time T which linearly increase from 55 μs at 0.1 torr to 215 μs at 2.0 torr, and a value for $\Gamma_v^g \simeq (6\pm1) \times 10^6$/sec-torr. This implies a ground state vcc cross section of 60±10 Å². The model also predicts the ratio of the dips in the $\hat{+}$ and $\hat{=}$ configurations to be $(\Gamma_v^e + \Gamma_v^g + \Gamma_1)/\Gamma_1$, with Γ_v^e the excited state vcc rate. Pre- liminary analysis of this data indicates an excited state vcc cross section several times larger than the ground state value.

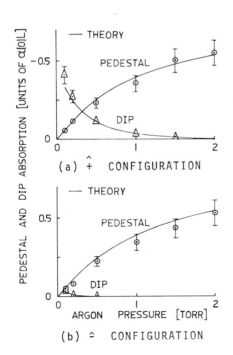

Fig.3 Pedestal and dip contributions to the probe ab- sorption coefficient at line center, extracted from the data of Fig. 2.

3. Lithium 6

Work is in progress on ^{6}Li, with the goal of producing a nuclear polarized target for study of parity non-conservation (PNC) in the nuclear reaction ^{6}Li $(\alpha,\gamma)^{10}$B*, where parity mixing between the 2^- and 2^+ levels in ^{10}B at 5.2 MeV is particularly sensitive to the isospin-changing $\Delta T=1$ component of the neutral weak current in the strong interaction.[5] The first resonance transition of Li lies at 670.8 nm, with τ = 27 ns and ku = 2.15 GHz at oven temperatures of 400-500°C. The D1 and D2 resonances are separated by only 10 GHz. The hf splittings of ground and excited states are 228 and 26 MHz, respectively, hence are negligible compared to ku.

Experiments were performed on 98.7% enriched samples in a stainless steel tube of diameter 1 cm and effective length of 2 to 3 cm, placed in a 5 G axial keeper field. The Lamb dip arrangement of the Yb experiments was used, with the pump and counter-propagating probe beams circularly polarized in the $\hat{+}$ configuration. Although the Li D1 transition is not a simple three level system, its optical pumping behavior is similar to that of Yb.   The pump field removes atoms from the 5 M_F ground states for which $\Delta M_F=+1$ transitions can be induced and transfers them to the F=3/2, M_F=3/2 state.

Studies similar to the Yb experiments showed that complete Doppler coverage can be attained at helium buffer pressures p $\geqslant$ 0.5 torr. A series of measurements was then undertaken to study the extent of nuclear polarization induced in an optically thin sample at several values of I and p. Figure 4 presents the polarization data, expressed as the fraction N_1/N_T of atoms in the F=3/2, M_F=3/2 state, taken with I tuned to the peak of the D1 Doppler profile. This quantity was extracted from the probe

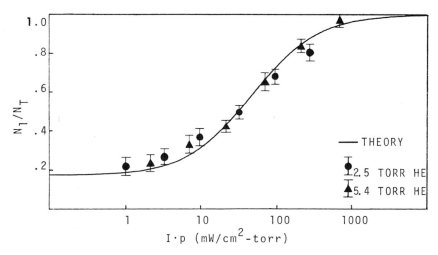

Fig.4 Polarization curve for ^{6}Li.

change signal data using an analysis similar to that of Ref.2.
The N_1/N_T data is plotted as a function of $I \cdot p$, since in the *vcc*
regime Π depends only on R[cf. Eq.(1)], regardless of the
complexity of the level structure, and so should scale accord-
ingly. The data shows that polarizations $\Pi > 0.9$ can be achieved for $I \cdot p > 500$ torr-mW/cm^2. Also shown in the figure is the
fit to an optical pumping model which includes the 6 ground and
6 excited hf states.[4] The 12 coupled rate equations were
solved under the assumption of complete *vcc* thermalization (no
narrow dip),[2] assuming the hf splitting in ground and excited
states to be negligible. The fit, which uses a value of T = 12
μs, is in good agreement with the experiments.

Since the proposed ^{6}Li PNC experiments require polarized
targets of high density ($\sim 10^{14}$/cm^2), the above studies are being
extented to optically thick media. Figure 5 shows the
probe absorption of a sample for which the absorption coefficient at line center $\alpha(o)L=9$ (1.5×10^{12} atoms/cm^2), determined
using the analysis from 6. In this experiment 6 torr of He
buffer was used, and the incident laser intensity I_0 was 140
mW/cm^2 at the entrance face of the sample ($I_0 \cdot p = 840$ mW-torr/
cm^2). The dramatic increase in transmission with the pump beam
on is an indication of the effectiveness of optical pumping.
Analysis gives $N_1/N_T \simeq 80\%$ for the pump tuned to line center.
These studies were limited by available pump laser power.
Experiments with cells of different geometries, other buffer
gases, and higher laser powers are in progress.

4. Sodium 23

The situation for optical pumping on the sodium D1 line (λ =
589.6 nm, τ = 16 nsec) is more complex than in Li because the
ground state hf structure is large, giving rise to two pairs of
Doppler broadened hf transitions split by 1772 MHz (ku = 910
MHz at 121°C). Pumping at either the F=1 or F=2 peaks leads to
transfer of atoms from that ground state to the other one (F
pumping). By tuning midway between the transitions, however,

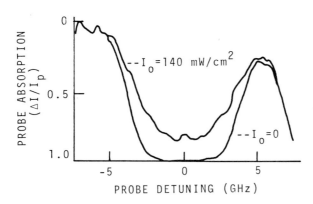

Fig.5 Optical pumping
of the D1 line in a
dense sample of ^{6}Li
vapor. Io is the inci-
dent pump intensity.
The absorption reson-
ance at the right is
due to the D2 component.

both ground state velocity distribution can be pumped simultaneously. In this case circularly polarized radiation can orient
atoms in the F=2, M_F=2 ground hf state (M_F pumping).

F and M_F-pumping were studied using two single mode cw dye
lasers, a circularly polarized pump laser fixed in frequency and
a weak probe field of the same circular polarization ($\hat{+}$ configuration), which was tuned across the Dl spectral profile. Earlier
studies in Na established that polarizations in excess of 90%
can be obtained at moderate pump intensities when velocity
thermalization is complete.[2] The present experiments were
performed at somewhat lower pressures, where (2c) is satisfied
and details of the *vcc* process can be studied.

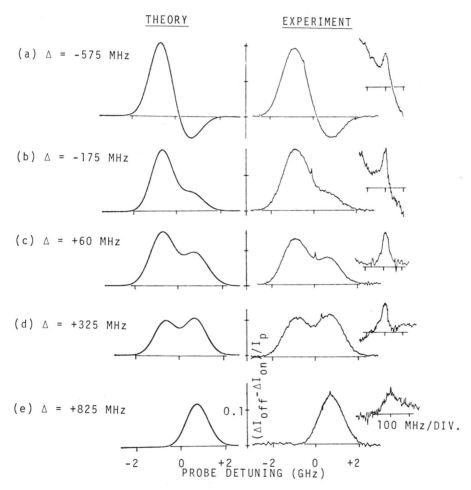

Fig.6 Probe absorption change signal, ΔI_{off}-ΔI_{on}, in units of
probe intensity I_p. The horizontal axis is the detuning of the
probe frequency from ω. Data are shown for several values of
pump frequency detuning $\Delta = \Omega - \bar{\omega}$. The vertical scale of the far
right-hand column is 5x expanded.

The pump beam (15 mm in diameter, 1-12 mW/cm^2) and the probe (2 mm in diameter, 1 μW/cm^2) were incident on a pyrex cell 19 mm diameter and 30 mm long heated to 121°C, with a 3 G axial keeper field applied. Data was taken on optically thin samples at argon buffer pressures ranging from 100 to 500 mtorr.

Figure 6 shows the probe change signals obtained in a series of measurements with I = 11 mW/cm^2 and p=265 mtorr of argon. The progression of the traces, taken by stepping the pump laser frequency through various intervals, shows the results of simultaneously pumping the F=2 and F=1 ground states to various extents, starting with F=2 only (top), progressing to approximately equal pumping of F=2 and 1 components (middle), and finally interacting primarily with the F=1 level (bottom). As can be seen, the interplay of F and M_F pumping can lead to change signal lineshapes with both positive and negative portions. The broad vcc-thermalized background predominates, accompanied by a small residual narrow feature. The far right-hand column shows this feature in an expanded scale. The left-hand column gives the results of a model in which the Dl transition is approximated by a 4-level system and complete velocity thermalization is assumed.[2] The theory was fit to the data by varying R (which) depends on the diffusion time T), and fixing the absolute frequency of the pump laser within the range of experimental uncertainty. Excellent agreement of both lineshapes and amplitudes for all the data was obtained for T = 270 μs, in reasonable agreement with the measured diffusion coefficient. [7]

Figure 7 summarizes the dependence of the observed lineshapes on the pump laser frequency. The solid curves are theoretical fits to the data; the dashed curves indicate the general behavior of the narrow feature, for which the theory is being extended.

Achieving maximum orientation in the F=2, M_F=2 state requires that both F=1 and 2 ground state hf levels be depleted at the same rate. This occurs at laser frequency

$$\Omega_o = \bar{\omega} - \frac{1}{2} \frac{(ku)^2 \ln 0.8}{\omega_1 - \omega_2} = \bar{\omega} + 52 \text{ MHz}, \quad (6)$$

where $\bar{\omega} = (\omega_1 + \omega_2)/2$, with ω_1 and ω_2 the F=1 and F=2 center frequencies, respectively. As seen in Fig.7, a relative minimun in the height of the background change signal occurs at this frequency. Since F=1 and 2 components are then depleted at the same rate they behave as a single level. Analysis shows that at 52 MHz point the narrow feature can be analyzed with the three level model used for the Yb data, except that in Eq.(5) σ_h is replaced by $\rho \sigma_h$, with effective branching ratio $\rho = 1/8$ and $\bar{\sigma}$ determined using a radiative cross section $\sigma_0 = 2.25 \ \pi \lambda^2$. The narrow resonances of Fig.7 may thus be analyzed as in the Yb case to yield S, hence Γ_v^g. Preliminary results for pressures ranging from 100 to 500 mtorr give a ground state vcc rate of $(4\pm1) \times 10^6$/sec-torr, hence a cross section of 23 ± 6 Å^2. Transient measurements to confirm this result are in progress.

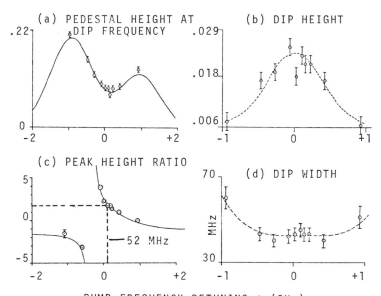

PUMP FREQUENCY DETUNING Δ (GHz)

Fig.7 Experimental parameters obtained from Figure 6 vs. pump frequency detuning, $\Delta=\Omega-\bar{\omega}$. (a) Height of background change signal (pedestal) at peak of narrow resonance (dip), in units of $(\Delta I_{off} - \Delta I_{on})$ /Ip; (b) dip height, same units as in (a); (c) ratio of F=2 and F=1 peaks in background change signal; (d) width of dip (FWHM). The peak height ratio of (c) indicates the relative importance of F-pumping with M_F-pumping. A ratio of 5/3 indicates equal pumping rates of atoms in the F=2 and F=1 components. See text for details.

5. CONCLUSION

The experiments and models presented above demonstrate the efficiency of laser optical pumping with *vcc* as a means of producing polarized atomic vapors. Such vapors are important for fundamental studies in nuclear and atomic physics, and may be useful for other applications, e.g. sensitive magnetometers, as well. The good fit of the experimental data to the models indicates that experiments of this type can be used to obtain ground state *vcc* cross sections in a straightforward way. The experiments in ^{171}Yb, ^{6}Li, and ^{23}Na with argon and helium buffers can be expected to be extended to other isotopes and other buffer gases.

REFERENCES

1. D.E. Murnick, M.S. Feld, In <u>Polarization Phenomena in Nuclear Physics 1980</u>, ed. by G.G. Ohlsen et.al. (Am. Inst. of Physics, N.Y. 1981) p.804.
2. P.G. Pappas, R.A. Forber, W.W. Quivers, Jr., R.R. Dasari, M.S. Feld and D.E. Murnick, Phys. Rev. Letters <u>47</u>, 236 (1981).

3. M. Gustavsson, H. Lundberg, L. Nilsson, and S. Svanberg, J. Opt. Soc. Am. 69, 984 (1979).
4. W.W. Quivers, Jr., et.al., to be published.
5. P.G. Bizzeti and A. Perego, Phys. Letters 64B, 298 (1976).
6. A.C.G. Mitchell and M.W. Zemansky, Resonance Radiation and Excited Atoms (Cambridge University Press, New York 1961).
7. L.C. Balling, in Advances in Quantum Electronics, vol.3, ed. by D.W. Goodwin (Academic Press, N.Y. 1975) p.1.

Dressed Atom Approach to Line Broadening

D. Pritchard and B. Walkup

Physics Department, Rm. 26-231, Massachusetts Institute of Technology
Cambridge, MA 02139, USA

I. Introduction When a two state atom is placed in an optical field whose frequency ω is near the atom's resonant frequency ω_0, the atom cycles rapidly between its ground and excited states ($|g\rangle$ and $|e\rangle$ respectively). It is often more convenient -- both mathematically and physically -- to describe this system in terms of the dressed states $|w(t)\rangle$ and $|s(t)\rangle$, i.e.

$$| \psi\rangle = a_w(t) \; |w(t)\rangle + a_s(t) \; |s(t)\rangle \tag{1}$$

where a_w and a_s are time independent in the absence of perturbations other than the oscillatory field. The fact that $|w(t)\rangle$ and $|s(t)\rangle$ are time dependent coherent superpositions of $|g\rangle$ and $|e\rangle$ is reflected in the notation.

The dressed states are eigenstates of the atom + field system in the rotating wave approximation [COT 76, COT 77] and thus form the natural basis for the discussion of properties of strongly irradiated atoms. Dressed states have been applied to the study of atomic line broadening by many authors [HAH 72, LIY 75, COS 77, MOL 77, LIS 78, NIS 79, YEB 79, CBB 80]. We shall discuss two recent observations of line broadening phenomena: dispersive corrections to the Lorentzian lineshape, and inelastic collisions involving dressed atoms.

II. Dressed Atom View of Collisions

A. Dressed Atoms

Consider a two state atom subject to an oscillating field. It is analogous to the problem of a driven classical oscillator, and we should not be surprised that the steady state solutions (defined by Eq.2) are coherent superpositions of $|g\rangle$ and $|e\rangle$ which oscillate at the driving frequency ω. Figure 1 illustrates the dressed states.

$$|w(t)\rangle = \exp(-i\omega_w t) \; [\cos\tfrac{\theta}{2}|g\rangle \; - e^{-i\omega t} \sin\tfrac{\theta}{2}|e\rangle] \tag{2}$$

$$|s(t)\rangle = \exp(-i\omega_s t) \; [\cos\tfrac{\theta}{2}|e\rangle \; + e^{i\omega t} \sin\tfrac{\theta}{2}|g\rangle]$$

with

$$\omega_0 = (E_e - E_g)/\hbar \qquad \text{resonance frequency of atom} \tag{3}$$

$$\omega_R = \langle e|\mu|g\rangle \cdot E \qquad \text{Rabi's measure of oscillating field}$$

$$\Delta = \omega - \omega_0 \qquad \text{detuning}$$

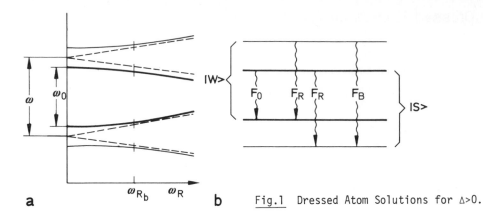

a   **b**   Fig.1 Dressed Atom Solutions for Δ>0.

$$\omega_R' = (\omega_R^2 + \Delta^2)^{1/2} \qquad \text{effective oscillation frequency in rotating frame}$$

$$\cos\frac{\theta}{2} = \left(\frac{\omega_R' + \Delta}{2\omega_R'}\right)^{1/2} \quad , \quad \sin\frac{\theta}{2} = \left(\frac{\omega_R' - \Delta}{2\omega_R'}\right)^{1/2}$$

$$\omega_W = \omega_1 - \frac{\Delta}{2} + \frac{\omega_R'}{2} \quad , \quad \omega_S = \omega_2 + \frac{\Delta}{2} - \frac{\omega_R'}{2} \qquad \text{eigenfrequencies.}$$

Part a shows the dependence of the eigenfrequencies on field strength (ac Stark effect) - note that the low field dependence is quadratic. Part b shows the two components of both dressed solutions at field ω_{Rb}, the predominate component of each dressed state is shown with a heavier line. The spontaneous decay modes of these dressed states account for all light emitted by the dressed atom: the Rayleigh component F_R is at the driving frequency, F_0 is near the atom's resonance frequency, and F_3 (the three photon peak) lies on the opposite side of ω such that the sum of the energies of a photon from F_0 and one from F_3 equals $2\hbar\omega$.

B. Collisional Transfer Rates

Now we consider the influence of collisions on the dressed atoms. We calculate the probability that a single collision causes a transition between the dressed atom states; and use classical straight line trajectories with impact parameter b. By considering only a single collision the results are valid only in the separated lines (secular) approximation [COT 77] - i.e. $\omega_R' >>$ collision rate. We suppose that the molecular potentials between the two state atom and perturber are known in the bare basis, $V_{ee}(R)$ and $V_{gg}(R)$, and that $V_{eg}(R) = 0$. The time-dependent Schrödinger's equation yields

coupled differential equations for a_W and a_S and may be further simplified by restricting to weak laser fields so that we can assume $|a_W| \approx 1$ throughout the collision. (Note $\theta \to 0$ for $\omega_R << \Delta$). The equation for a_S may then be integrated to give:

$$|a_S(b,t)|^2 = \frac{\omega_R^2}{4\Delta^2} \left| \int_{-\infty}^{t} dt_1 \; \omega_d(t_1) \; \exp[-i\Delta t + i \int_{-\infty}^{t_1} dt_2 \; \omega_d(t_2)] \right|^2 \qquad (4)$$

where $\omega_d(t_1) = V_{ee}[R(b,t_1)] - V_{gg}[R(b,t_1)]/\hbar$. The transfer rate from $|w>$ to

$|s>$ (and vice versa) is

$$K_T(\Delta) = n_p \, v \int_0^\infty 2\pi b db |a_s(b,\infty)|^2 \tag{5}$$

where v is the relative velocity and n_p is the perturber density.

C. Steady State Fluorescence Pattern

Since our treatment of the E & M field has so far been classical, we must add in spontaneous emission to predict fluorescence from the dressed states. This may be done in a semi-classical fashion; the dipole matrix elements $<s|\mu|w>$, $<s|\mu|s>$, $<w|\mu|w>$ determine the strengths of the various components of the fluorescence (indicated schematically in Fig.1b).

The steady state populations of n_s and n_w are determined from rate equations (secular approximation) which reflect the fact that the fluorescence component F_0 transfers population from $|s>$ to $|w>$ while F_3 does the opposite (F_R doesn't affect the populations). Collisions transfer population between the dressed states with a rate $K_T(\Delta)$.

For the purposes of this discussion, only the weak field limit ($\omega_R << \Delta$) need be duscussed ($F_3=0$). The fluorescence components have intensities:

$$F_R = n_a \frac{\omega_R^2}{4\Delta^2} \Gamma_N \tag{6}$$

$$F_0 = n_a K_T(\Delta) = n_a \frac{\omega_R^2}{4\Delta^2} \gamma_c(\Delta)$$

where n_a is the density of active atoms. Γ_N is the spontaneous decay rate of $|e>$, and $\gamma_c(\Delta)$ is the pressure broadening rate given by: $\quad$ (7)

$$\gamma_c(\Delta) = n_p \, v \int_0^\infty 2\pi b \, db \mid \int_{-\infty}^\infty dt_1 \, \omega_d(t_1) \, \exp[-i\Delta t_1 + i \int_{-\infty}^{t_1} dt_2 \, \omega_d(t_2)]\mid^2$$

Note that γ_c is linear in n_p, and that the total fluorescence is

$$F_\Sigma = F_R + F_0 + F_3 = n_a \frac{\omega_R^2}{4\Delta^2} [\Gamma_N + \gamma_c(\Delta)]. \tag{8}$$

III. Dispersion Correction to Lorentzian

If it is assumed that collisions between radiating two state atoms and perturbers are sudden, then the effect of the collision can be summarized by the phase shift $\eta(b,t=\infty)$ where

$$\eta(b,t) = \int_{-\infty}^t dt_1 \, \omega_d(t_1) \tag{9}$$

Mathematically this results from setting $\Delta = 0$ in Eq.7:

$$\gamma_c(0) = n_p \, v \int_0^\infty 2\pi b \, db \, 2[1 - \cos\eta(b,\infty)]. \tag{10}$$

This result is the usual expression for the pressure broadened width. In this approximation the total fluorescence lineshape is symmetric about $\Delta = 0$ and has Lorentzian wings.

Of course real collisions are not sudden; we therefore discuss the first order dependence of γ_c on Δ. This may be done by expanding the exponential in Eq.7, with the result

$$\gamma_c(\Delta) = \gamma_c(0) + \Delta\gamma_1 + \ldots \tag{11}$$

$$\gamma_1 = n_p v \int_0^\infty db \, 2\pi b \, 2 \int_{-\infty}^\infty dt \, \omega_d(t) \, (\cos[\eta(b,\infty) - \eta(b,t)] - \cos[\eta(b,t)])$$

in accord with [SUB 75].

The size of the correction γ_1 can be calculated for a power low potential, $\omega_d(R) = C_n R^{-n}$. For this potential there is a natural distance and time

$$R_0 = [\frac{C_n}{v}]^{1/n-1} \quad , \quad T_0 = R_0/v.$$

R_0 is closely related to the Weisskopf radius. γ_c can now be written

$$\gamma_c(\Delta) = n_p v \, L_n \, R_0^2 \, [1 \overset{+}{-} D_n \, \Delta T_0 + 0(\Delta^2 T_0^2)]$$

where L_n and D_n are numerical coefficients which relate the (Lorentzian) width and (dispersion) asymmetry to the natural variables, and the - sign is for attractive potentials. After averaging over a thermal distribution of relative velocities the same form holds with new coefficients $<L_n>$, $<D_n>$ (with v being the rms relative velocity). Our interest is in experimental verification of the predicted asymmetry. Consequently a short Table is helpful:

	n=4	n=6	n=12	EXPERIMENT
D_n	0.98	0.67	0.42	awfully hard
$<D_n>$	1.43	0.85	0.49	$0.85 \overset{+}{-} .05$

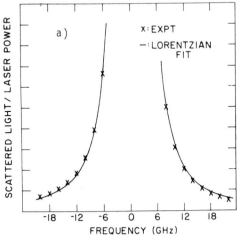

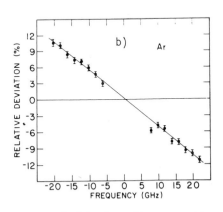

Fig.2. a. Normalized lineshape of Na $3p_{1/2}$ perturbed by Ar. The perceptive reader will note a slight deviation from the Lorentzian fit. b. Relative deviation, that is (Datum - Lorentz fit) ÷ Lorentz fit. The straight line fit shows that a dispersion correction to the Lorentzian is necessary.

Two points emerge: Soft potentials have considerably more asymmetry than hard ones, and the effect of thermal averaging is quite significant -- a result traceable to the fact that the asymmetry is proportional to the average duration of the collisions to which slow atoms contribute disproportionately.

Our observations of this effect (WSP 80) form the basis of the column marked experiment in the Table. The experimental value is the average of dispersion coefficients for Na in $3p_{3/2}$ and $3p_{1/2}$ states perturbed by Ar, Kr, and Xe -- the three targets studied for which the long range potential varies roughly as R^{-6} (He and Ne perturbers displayed considerably less asymmetry, a reflection of their harder potentials). The good agreement masks the fact that we observed a systematic 30% difference between $3p_{1/2}$ and $3p_{3/2}$, the latter having more asymmetry. A more sophisticated theory allowing for inelastic collisions and the multiplicity of excited state potentials would be required to account for this.

IV. Inelastic Collisions of Dressed Atoms

Another collisional probe of dressed atoms is inelastic collisions to a state $|e'>$ which is not radiatively coupled to $|g>$ or $|e>$. Now we discuss a system in which collisions couple $|e>$ and $|e'>$ but neither of these states to $|g>$. We restrict attention to sudden collisions.

For sudden collisions ($\Delta\tau<<1$) it is not necessary to consider the time evolution of the system explicitly as was done in Sec. III. No consideration of the effects of the oscillating field or the detuning <u>during</u> the collision is necessary, and one can assume that the effects of the collision are entirely summarized by the T (or S) matrix for the bare system. The convenience of using dressed states in this case is that they form the natural basis for describing the irradiated system and especially for determining the resulting fluorescence pattern.

A. T-Matrix Formalism

We elect to use a T-matrix approach, the general formula for any [transfer] rate being

$$K_{a \to b} = n_p \frac{2\pi}{\hbar} \rho(E) \int |<b|T|a>|^2 d\Omega \tag{12}$$

We can express all transfer rates of interest using only the elements T_{gg}, T_{ee}, and $T_{ee'}$ since $T_{e'g}$ and T_{eg} are taken to be zero. We find by straightforward calculation:

$$K_{w \to s} = n_p \frac{2\pi}{\hbar} \rho(E) \int |<s|T|w>|^2 d\Omega \tag{13}$$

$$= n_p \frac{2\pi}{\hbar} \rho(E) \sin^2(\tfrac{\theta}{2}) \cos^2(\tfrac{\theta}{2}) \int |(T_{gg} - T_{ee})|^2 \, d\Omega$$

$$\simeq \frac{\omega_R^2}{4\Delta^2} \gamma_c \text{ where } \gamma_c = n_p v \int d\Omega |f_g - f_e|^2$$

and we restrict the discussion to weak fields. γ_c is the rate for dephasing collisions and reduces to Eq.10 in classical limit for scalar potentials (f_e and f_g are elastic scattering amplitudes). Similarly

$$K_{w \to e'} = \frac{\omega_R^2}{4\Delta^2} \Gamma_{e \to e'} \tag{14}$$

where $\Gamma_{e \to e'}$ is the rate for inelastic collisions from $|e\rangle$ to $|e'\rangle$.

B. Experiment

The most novel aspect of this treatment is the non-zero transfer rate directly from the $|w\rangle$ state to $|e'\rangle$ (and back). Experimentally, this must be observed against the background process involving two successive collisions:

$|w\rangle \to |s\rangle \to |e'\rangle$. We were able to demonstrate the single collision process in the Na - Xe system by measuring the ratio of fluorescence from $|e'\rangle$ to the combined light F_Σ [fluorescence, Rayleigh scattering, and (negligible) three photon scattering from the $|w\rangle$ and $|s\rangle$ levels]. At low pressures ($\gamma_c \ll \Gamma_N$) the two collision process is negligble and the fluorescene ratio displays the linear dependence on perturber pressure characteristic of the direct $|w\rangle \to |e'\rangle$ process, as shown in Fig.3a.

At a fixed perturber pressure the dependence of the fluorescence ratio on laser frequency shows a continuous transition from off-resonance excitation ($|w\rangle \to |e'\rangle$) to resonance excitation ($|e\rangle \to |e'\rangle$) shown in Fig.3b. On resonance (within the Doppler profile) the excited atoms have a non-uniform velocity distribution, so the fluorescence ratio measures the transfer rate for Na atoms with a known velocity component along the laser beam. This has been exploited to measure the velocity dependence of the fine structure transition rate (APP 79). Off resonance (line wings) the principal contribution at low perturber pressure to the $|e'\rangle$ population comes from atoms in the $|w\rangle$ state whose velocity distribution is thermal. Hence the fluorescence ratio reflects the thermally averaged rate $\langle \Gamma_{e \to e'} \rangle$. Off resonant excitation limits the range of velocities which can be selected in experiments using the Doppler shift to control the collision velocity (SBP 80).

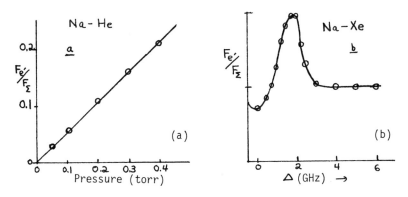

Fig.3 Pressure (a) and detuning (b) dependence of ratio $F_{e'} \div F_\Sigma$.

Problems with Virtual States

Unrefereed conference proceeding present unrivaled forums for scientific diatribes: this one is directed against the concept of "virtual levels". To begin with, the term virtual state is more appropriate, the term "level" being generally reserved for systems where the energy is specified but not the m quantum number. Our serious reservations, however, arise because the concept is fuzzy. Attempts to clarify it result in more contradictions than edification.

The lifetime of a virtual state is often regarded as $\tau = \Delta^{-1}$. This should more properly be referred to as the adiabatic following time [COS 77]. A difficulty arises in trying to discuss collisions from a virtual state. If one takes $\tau = \Delta^{-1}$ literally then collisional transfer from the virtual state is extremely unlikely since its lifetime is short compared to the time between collisions. However, dephasing collisions for example are often discussed in terms of collisional transfer from the virtual to the real state. Furthermore if forward transfer is possible then there should also be back transfer from the real to the virtual state. An attempt to describe these collisions by assigning the virtual state a population then using collision rates and lifetimes leads to contradiction.

These difficulties do not arise using dressed states; there is forward and back transfer between the dressed states with rates satisfying microreversibility. Of course, the major virtue of the dressed atom picture is the ease of discussing the physics of atoms in intense optical fields.

This work was supported by the National Science Foundation.

References

APP 79 J. Apt and D. Pritchard, J.Phys.B12 83(1979).
CBB 80 J. Cooper, R.J. Ballagh, and K. Burnett, Phys.Rev.A 22 535(1980).
COS 77 E. Courtens and A. Szoke, Phys.Rev.A 15, 1588.
COT 76 C. Cohen-Tannoudji, Frontiers in Laser Spectroscopy, Les Houches
        Summer School 1975, Session 27, ed. by R. Balian, S. Haroche, and
        S. Liberman, North Holland, Amsterdam(1976).
COT 77 C. Cohen-Tannoudji and S. Reynaud, J.Phys.B 10, 345(1977).
HAH 72 L. Hahn and I.V. Hertel, J.Phys.B 5, 1995(1972).
LIS 78 J. Light and A. Szoke, Phys.Rev.A 18, 1363(1978).
LIY 75 V.S. Lisitsa and S.I. Yakavelenko, Sov.Phys.JETP 41, 233(1975).
MOL 77 B.R. Bollow, Phys.Rev.A, 15, 1023(1977).
NIS 79 G. Nienhuis and F. Schuller, J.Phys.B. 12, 3473(1979).
SBP 80 N. Smith, T. Brunner, D. Pritchard, J.Chem.Phys. 74, 467(1981).
SUB 75 J. Szudy and W.E. Baylis, J.Q.S.R.T. 15, 641(1975).
WSP 80 R. Walkup, A. Spielfiedel, and D. Pritchard, Phys.Rev.Lett. 45,
        986(1980).
YEB 79 S. Yeh and P.R. Berman, Phys.Rev.A. 19, 1106(1979).

Collisional Processes Between Laser-Excited Potassium Atoms

M. Allegrini-Arimondo, P. Bicchi[1], S. Gozzini[2], and P. Savino

Istituto di Fisica Atomica e Moleocolare del C.N.R., Via del Giardino 3
I-56100, Pisa, Italy

[1] Also at Istituto di Fisica dell'Università - Siena, Italy
[2] Also at Istituto di Fisica dell'Università - Pisa, Italy

1. Introduction

The formation of both positive ions and highly excited atoms upon resonant
laser excitation of alkali vapors has been observed even at laser power densities considerably lower than is required for multiphoton processes to occur
[1]. Understanding the formation of these ions and excited atoms is complicated because several competitive mechanisms may occur. The relative contributions of each mechanism depend strongly on both the atomic density and
the laser power density. Experiments have been carried out at very different
values of these two parameters, namely atomic density in the range $10^{12} \div 10^{17}$
cm^{-3} and laser power density in the range $0.5 \div 10^8 W/cm^2$. The collisional mechanisms may be followed on separated time scales [2] in experiments with pulsed lasers, however it is more difficult to separate the effects of different
processes in cw experiments, where only the steady-state regime can be studied.

We may distinguish among three regimes of operation for the collisional
phenomena involving laser-excited atoms, depending upon the number of ions
and electrons created in the interaction volume. At low laser power density
($\lesssim 10^2 W/cm^2$) energy-pooling collisions [3,4] between two excited atoms lead to

$$\text{electronic energy transfer} \qquad X^* + X^* \longrightarrow X^{**} + X \pm \Delta E \qquad (1)$$

$$\text{and associative ionization} \qquad X^* + X^* \longrightarrow X_2^+ + e^- \pm \Delta E \qquad (2)$$

with cross-sections $\approx 10^{-15} \div 10^{-17} cm^2$ [1]. In sodium it has been established
that products of (2) undergo dissociative recombination or photodissociation
[4]. For high laser power density ($\gtrsim 10^6 W/cm^2$) nearly complete ionization of
the vapor has been achieved [1]. In addition to associative ionization, free
electrons may be produced by two photon ionization of X^*, and by laser-induced ionization [5]. If the atom density is large enough so that the electron mean free path is smaller than the cell dimensions, superelastic colli-

$$\text{sions of free electrons} \qquad e^- + X^* \longrightarrow e^-(\text{fast}) + X \text{ , followed·by electron}$$

impact ionization can increase the electron density N_e [2]. However, above
a critical N_e the concentration is high enough to sustain the plasma, no matter how few the number of seed electrons.

In this paper we report the results of experiments performed on laser-excited potassium vapor at intermediate laser power density ($\lesssim 10^3 W/cm^2$) and atomic density ($\approx 10^{13} \div 10^{14} cm^{-3}$). These experiments, which are the first studies of these effects in potassium, provide data that clarify the general features of these types of inelastic processes.

2. Experiment

The $4P_{1/2}$ or $4P_{3/2}$ levels of atomic potassium were excited by a low power ($\approx 10 \div 100mW$), broad-band ($\Delta\lambda \approx 0.5\text{Å}$), multimode (mode spacing $\approx 400MHz$) cw dye laser. The laser beam was focused to $\approx 0.4mm^2$, and the fluorescence volume restricted by a 3mm diaphragm. In the range $3400 \div 8200\text{Å}$ the fluorescence spectrum contained atomic potassium lines from levels as high as 12S plus the quadrupole-allowed 3D$\rightarrow$4S line at 4642Å. The levels of interest are shown in Fig.1. The group labeled c contains the 5P, 6S and 4D levels, which have energy within $\approx 4kT$ of the sum energy of two atoms in the 4P-state. The group labeled d contains all other observed levels with higher energy. The intensity of the fluorescence lines from the c and d levels is $\approx 10^{-4} \div 10^{-8}$ that of the resonance fluorescence (b$\rightarrow$a). A portion of the <u>uncorrected</u> spectrum having the strongest lines is shown in Fig.2. We have measured the intensity of each line as a function of the laser power at different temperatures and have found that the levels of groups c and d behave differently. The data in Fig.3 are typical examples of one c and one d level. The fluorescence from the c level was quadratic with N_b, the atomic density in the 4P-state, indicating that this level is populated by processes involving two b-atoms. Similar plots for the d levels show that their population depends on factors other than N_b^2. Examination of the rate equations for the system indicates that formation of the high-lying d levels depends on the electron density as well as N_b [6].

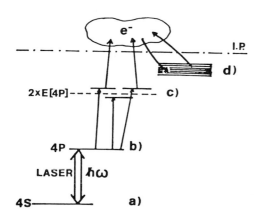

Fig.1 Schematic picture of the levels and processes involved in the experiment

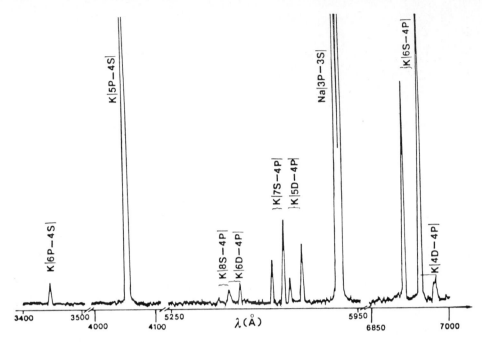

<u>Fig.2</u> Portion of a spectrum recorded in a cell containing K+5Torr of Ar at
T=170°C and $I_L \simeq 100mW$

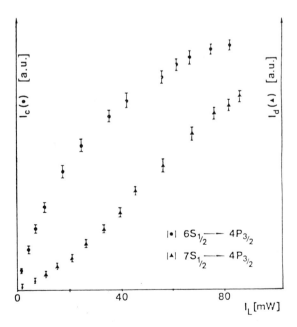

<u>Fig.3</u> Fluorescence intensities, in arbitrary units, of the $6S_{1/2}$ $4P_{3/2}$ and $7S_{1/2}$ $4P_{3/2}$ transitions vs laser intensity. The energies of the $6S_{1/2}$ and $7S_{1/2}$ levels are $\simeq 4kT$ and $\simeq 13kT$ respectively above $2 \times E[4P]$. The relative intensities of each transition are not correlated in the graph

We have also conducted experiments in a cell fitted with two parallel electrodes and containing K+5Torr of Ne. Application of a low voltage to the electrodes ($\lesssim$20V/cm) did not affect the intensity of the c levels transitions while the intensity of the d levels transitions was drastically changed. At higher electrode voltage ($\gtrsim$200V/cm) a discharge occurred; spectroscopic and optogalvanic detection was simultaneously performed (at T=200°C), by recording each fluorescence line and the change in the discharge current as the laser frequency was swept across the resonance lines. Comparison of the width of the optogalvanic signal with the width of the resonance signal showed that the current is quadratic in N_b, proving that two b-atoms were involved. We note that previous studies in a He discharge showed that He-He* associative ionization was linear with He* density [7]. Similar comparisons of optogalvanic signal with other fluorescence lines further support the quadratic dependence of c-level formation on N_b, but a more complicated dependence for the d-level formation.

The existence of a critical electron density for sustaining the plasma allows us to understand some previously unexplained experimental observations. We illuminated our cell with a second laser, tuned off the atomic resonance, with the aim of increasing the electron concentration by ionization from the highly excited states. No change in the intensities of radiation from either the c or d levels was observed. This result can be explained if the electron density, before turning on the second laser, was already larger than the critical value, but we were not able to measure this value. However, by using a microwave diagnostic technique [8], we extimate that, at the temperatures and laser intensities of our experiment, the maximum electron density is $\approx$7x10^{12} cm^{-3}, so that the critical density is certainly below this value.

References

1. T.B. Lucatorto and T.J. McIlrath: Appl. Opt. 19,3948 (1980) and references quoted therein
2. A.R. Measures and P.G. Cardinal: Phys. Rev. A23,804 (1981)
3. M. Allegrini, G. Alzetta, A. Kopystyńska, L. Moi and G. Orriols: Opt. Commun. 19,96 (1976); 22,329 (1977)
   M. Allegrini, P. Bicchi, S. Gozzini and P. Savino: Opt. Commun. 36,449 (1981)
4. G.H. Bearman and J.J. Leventhal: Phys. Rev. Lett. 41,1227 (1978)
   V.S. Kushawaha and J.J. Leventhal: Phys. Rev. A22,2468 (1980)
5. S. Geltman: J. Phys. B: Atom. Molec. Phys. 10,3057 (1977)
6. S. Gozzini: Thesis University of Pisa April 1981
7. J.E. Lawler: Phys. Rev. A22,1025 (1980)
8. M. Allegrini, P. Bicchi, S. Gozzini, I. Longo and P. Savino: to be published

Noble Gas Induced Relaxation of the Li 3S-3P Transition Spanning the Short Term Impact Regime to the Long Term Asymptotic Regime

T.W. Mossberg*, R. Kachru+, T.J. Chen, S.R. Hartmann

Columbia Radiation Laboratory, Department of Physics, Columbia University
538 W. 120th. St., New York, NY 10027, USA, and

P.R. Berman

Department of Physics, New York University, 4 Washington Place
New York, NY 10003, USA

Photon echoes have a Doppler free character which allows one to study relaxation processes which would otherwise be hidden in the inhomogeniously broadened spectral profile. It has recently been shown, for example, that contrary to expection, a radiating atom in a linear superposition of dissimilar electronic states can undergo identifiable velocity changing collisions [1]. Studies of this nature require an examination of the sub-Doppler region of the spectral line shape. The effect manifests itself, in the case of photon echoes, in a dependence of the effective relaxation cross section σ_{eff} on the excitation pulse separation τ. In this paper we report measurements in Li vapor where τ can be increased into the regime where σ_{eff} once again becomes independent of τ. In the limit $\tau = 0$ we measure σ_0 which is the phase changing cross section as calculated by Baranger [2] while in the large τ limit we measure σ_∞ the average total scattering cross section of the ground and the excited states. Our data at intermediate values of τ is used to determine the form of the scattering kernel and the average velocity change per collision. These measurements are for the 2S-2P superposition states in atomic Li perturbed by each of the noble gases. For He perturbers the scattering kernel is found to be Lorentzian, for the other perturbers it is Gaussian.

We use a N_2 laser pumped dye laser to generate a 4.5 nsec light pulse at the 6708 Å $2S-2P_{1/2}$ transition of ^7Li. The pulse which has a 6 GHz spectral width is attenuated, split, delayed an amount τ, recombined, and directed into a cell, whose effective length is 10 cm, at $525 \pm 15°K$ containing the Li vapor (at ~10^{-6} torr). For short values of τ the polarizations of the photon echo excitation pulses were orthogonal in order to reduce the effects of detector saturation which arose because of the non instantaneous response the Pockels cell shutters used for their protection.

For a superposition state relaxing at an effective rate $\Gamma_{eff} = nv\ \sigma_{eff}$ where n is the perturber density, v is the average relative velocity of the collision partners and σ_{eff} is an effective cross section, the corresponding echo intensity will decay according to

$$I = I_0 \exp(-4\Gamma_{eff}\tau) \tag{1}$$

and since Γ_{eff} varies linearly with perturber pressure P

$$I(P) = I(0)\exp(-\beta P) \tag{2}$$

---

*Present Address: Dept. of Physics, Harvard University, Cambridge, MA 02138, USA
+Present Address: Molecular Physics Lab., SRI International, Menlo Park

where the constant β, which we measure directly, is characteristic of the perturber and the collision process. We determine β at several discrete values of τ by measuring the echo intensity as a function of the perturber gas pressure. The value of σ_{eff} is obtained from

$$\sigma_{eff} = \Gamma_{eff}/nv = \beta P/4nv\tau. \tag{3}$$

In fig. (1) we summerize our work by plotting all measured values of σ_{eff} as a function of τ. A dependence on τ arises because each collision of the Li atom with a perturber gives rise to a velocity change in addition to a phase change of the Li superposition state. If only phase changing collisions occured σ_{eff} would be independent of τ. Velocity changing collisions have a delayed effect which manifests itself in a dependence of σ_{eff} on τ. Our data indicates that at the shortest values of τ σ_{eff} increases at a large and relatively constant rate while at higher τ it levels off considerably.

Echoes in the optical regime (photon echoes) are generally formed in a volume large compared to the wavelength of the optical transition. Thus any atom experiencing a velocity change sufficient to displace it an appreciable fraction of a wavelength from the position it would otherwise have taken in the phased array which radiates the echo will not necessarily reinforce the echo signal. As τ is increased the resulting displacement increases and the effect of a particular velocity change is enhanced. This proceeds up to a point that being when τ is so large that all atoms experi-

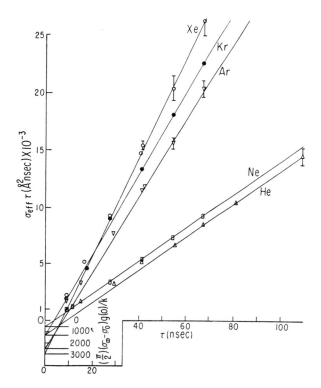

Fig. 1 Plot of $\sigma_{eff}(\tau)$ versus τ. Error bars represent statistical uncertainty.

encing a velocity change are effectively eliminated from the echo formation processes. The data of fig. (1) at large τ shows this effect clearly in the weakening dependence of σ_{eff} on τ.

In what may be called the collision kernel approximation Flusberg [3] has shown that σ_{eff} may be expressed as

$$\sigma_{eff} = \sigma_0 + \sigma_v \left[1-(1/\tau) \int_0^\tau dt \; \tilde{g}(kt)\right] \tag{4}$$

where $\sigma_0(\sigma_v)$ is the phase changing (velocity changing) cross section and

$$g(kt) = \int_{-\infty}^{\infty} \exp(ikt\Delta v) \; g(\Delta v) \; d(\Delta v). \tag{5}$$

The collision kernel $g(\Delta v)$ gives the probability of a particular change Δv in the component of the velocity along the laser pulse direction. For $k\tau \ll 1$

$$\sigma_{eff} = \sigma_0 + \sigma_v \frac{1}{6} k^2 \tau^2 <\Delta v^2> \tag{6}$$

where $<\Delta v^2>$ is the second moment of the collision kernel. For $k\tau \gg 1$

$$\sigma_{eff} = \sigma_0 + \sigma_v \left[1 - \pi g(0)/2k\tau\right] \tag{7}$$

where $g(0)$ is the amplitude of the collision kernel at $\Delta v = 0$.

Our data at short τ does not fit (6) well, shorter excitation pulses would have been required to enter the regime where this approximation is valid. Our data does suffice however to use (6) to estimate σ_0 and we find that except for He we agree to within a few percent with measurements of σ_0 made from line broadening experiments [4]. Our estimate of σ_0 for He runs ~10% high.

The solid line curves of fig. (1) were obtained using an explicit form of the collision kernel. For all perturbers except He we have used a Gaussian kernel

$$g(\Delta v) = (1/\sqrt{\pi} \; u_0) \exp(-\Delta v^2/u_0^2) \tag{8}$$

while for He we have used the Lorentzian kernel

$$g(\Delta v) = (u_0/\pi)/(u_0^2 + \Delta v^2). \tag{9}$$

We vary u_0 and σ_v to obtain the best fit. All relevant parameters are tabulated in table I.

Table 1

Perturber	σ_0	σ_v	$\sigma_\infty = \sigma_0 + \sigma_v$	u_0	$\sigma_\infty = \sigma_0 + \sigma_v$ (from fig.2)
He	99 Å^2	49 Å^2	148 Å^2	247 cm/sec	146
Ne	101	47	148	1140	146
Ar	181	145	326	1400	338
Kr	206	170	376	1320	356
Xe	233	200	434	1320	434

An alternative proceedure for presenting our data is to plot $\sigma_{eff}\tau$ as a function of τ, see fig. (2), in which case we expect that from (7)

$$\sigma_{eff} = (\sigma_0 + \sigma_v)\tau - \sigma_v \pi g(0)/2k \qquad (10)$$

and we should obtain an assumptotic fit to a straight line whose slope yields $\sigma_0 + \sigma_v = \sigma_\infty$ and whose negative intercept yields the product of σ_v with $g(0)$. The values of $\sigma_0 + \sigma_v = \sigma_\infty$ so obtained are compared with that calculated from the data of table I.

Supported financially by the U.S. Office of Naval Research (Contracts N00014-78-C-517 and N00014-77-C-0553) and by the U.S. Joint Services Electronics Program (Contract DAAG29-79-C-0079).

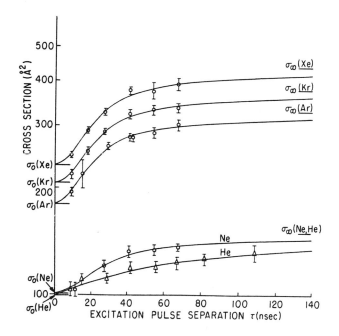

Fig. 2 $\sigma_{eff}\tau$ plotted versus τ.

REFERENCE

1. T.W. Mossberg, R. Kachru, and S.R. Hartmann, Phys. Rev. Lett. 44, 73 (1980).
2. Michel Baranger, Physics Rev. 111, 481 (1958).
3. A. Flusberg, Opt. Commun. 29, 123 (1979).
4. N. Lwin, Thesis (University of Newcastle upon Tyne) 1976, as reported by E. L. Lewis, Phys. Rep. 58, 1 (1980).

Double Resonance Studies of Collision-Induced Transitions in a Molecular Beam

T. Shimizu, F. Matsushima, and Y. Honguh

Department of Physics, University of Tokyo
7-3-1 Hongo, Bunkyo-ku, Tokyo 113, Japan

1. INTRODUCTION

Various types of collision-induced transitions occur among molecular levels. Relevant molecular energy levels and transitions among them are schematically shown in Fig.1 and 2. The α- and β-transitions, named by Oka on his micro-wave-microwave double resonance experiments [1], are the collision-induced transitions between rotational levels and inversion doubling levels, respectively. Transitions between vibrational levels mostly occur as a V-V energy transfer, because an energy discrepancy at the collision is too large to be absorbed by a translational degree of freedom [2]. Since the rotational

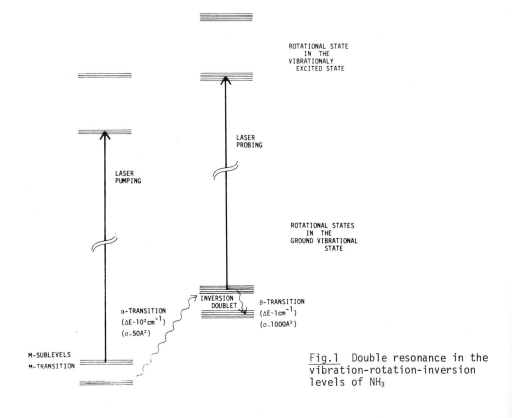

ROTATIONAL STATE
IN THE
VIBRATIONALY
EXCITED STATE

LASER
PROBING

LASER
PUMPING

ROTATIONAL STATES
IN THE
GROUND VIBRATIONAL
STATE

INVERSION
DOUBLET

α-TRANSITION
($\Delta E \sim 10^2 cm^{-1}$)
($\sigma \sim 50 A^2$)

β-TRANSITION
($\Delta E \sim 1 cm^{-1}$)
($\sigma \sim 1000 A^2$)

M-SUBLEVELS
M-TRANSITION

Fig.1 Double resonance in the vibration-rotation-inversion levels of NH_3

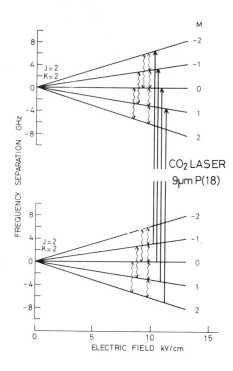

FREQUENCY SEPARATION GHz

M

J = 2
K = 2

CO₂ LASER
9μm P(18)

ELECTRIC FIELD kV/cm

Fig.2 Double resonance detection
of collision-induced transitions
between M-sublevels of CH₃F

state has a dual parity in a case of symmetric top molecule, direct transitions between closely spaced M-sublevels occur.

The double resonance in a four-level system is a powerful method to obtain
so-called state-to-state cross sections of the collision-induced transitions.
It was originated by Oka using two microwaves. An application of the infrared lasers has improved the sensitivity because of much larger changes in the
populations produced by a efficient pumping [3]. Several advantages may be
found in the experiment with laser-excited molecular beam [4]. "The collision" can be controlled by adjusting intensity, direction, and collimation
of the beam, strength and direction of the external field, and frequency and
intensity of the laser radiation. A "spectroscopic selection" of molecular
velocity is also possible. Since two independent Stark tuning fields are
applicable to the molecular beam in the pump and probe sections, the infrared-
infrared double resonance with untunable lasers can be performed.

In the present experiments on NH_3, the cross section of the α-transition ($\Delta J=$
± 1, $\Delta K=0$, $+\leftrightarrow-$) was obtained isolatedly from that of the β-transition ($\Delta J=0$,
$\Delta K=0$, $+\leftrightarrow-$). A difference in the cross sections of the upward and downward
transitions was detected. The collision—induced transitions among closely
spaced M-sublevels of CH_3F were investigated.

2. α-TRANSITION

The $\nu_2 asQ(3,2,3)$ transition of NH_3 is pumped by the N_2O P(8) laser radiation
in the upper stream of the molecular beam and the $\nu_2 asQ(2,2,2)$ transition is

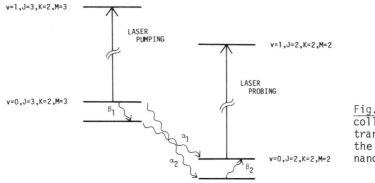

Fig.3 Sequential collision-induced transitions give the double resonance signal

probed by the N_2O P(9) laser downstream. The double resonance signal $\Delta I/I$ is brought by the $\beta_1 \to \alpha_1$ or the $\alpha_2 \to \beta_2$ successive collision-induced transitions as shown in Fig.3. Since the cross section of the β-transitions is much larger than that of the α-transition, the latter is the rate-determining step of these processes.

The signal intensity may be given by

$$\frac{\Delta I}{I} = \frac{\Delta n_s(t)}{n_s^0} = 4\pi k_\alpha (t-t_0)\eta P_B \frac{\Delta n_p(t_0)}{n_p^0} \tag{1}$$

where k_α is the rate of α-transition, η the collision efficiency in the beam, P_B the effective pressure in the beam, and the $\Delta n_p/n_p^0$ the relative number of molecules pumped upstream. Putting observed values of $\Delta I/I = 5.6 \times 10^{-3}$, $\Delta n_p/n_p^0 = 0.2$, $(t-t_0) = 1.2 \times 10^{-4} s$, and $P_B = 10^{-4}$ Torr and the estimated value of $\eta = 0.47$, into (1), we obtain

$$k_\alpha = \frac{1}{2\pi T_1^\alpha P_B} = 4.0 \times 10^5 \text{ /s.Torr} \tag{2}$$

which coresponds to the line-broadening parameter of 0.4 MHz/Torr or the cross section of 25 A^2.

3. β-TRANSITION

In an observation of the line-broadening parameter, the β-transition dominates among other processes. The rate of β-transition for the J=7, K=6 inversion doublet has been obtained as $[2\pi T_1^\beta p]^{-1} = 32$ MHz/Torr. At the beam pressure of 8.3×10^{-6} Torr, the time constant of collisional transfer is $T_1^\beta = 6 \times 10^{-4} s$, while the mean transit time of the molecule between the pump and the probe sections is $1.2 \times 10^{-4} s$. In this condition the β-transition may not occur frequently so that the highly sensitive detection of the collision-induced signals clarifies some detail of the collisional processes. As shown in Fig. 4 when one pumps the $\nu_2 as^q Q(7,6,7)$ transition by the P(36) CO_2 laser at the

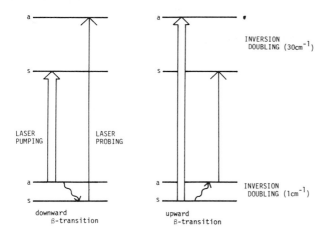

INVERSION DOUBLING (30cm^{-1})

LASER PUMPING   LASER PROBING

INVERSION DOUBLING (1cm^{-1})

downward β-transition   upward β-transition

Fig.4 The upward and the downward β-transitions studied by the laser double resonance

tuning field of 64.6 esu and probes the $\nu_2$sa^{q}Q(7,6,7) transition by the R(4) CO_2 laser at 71.9 esu, the rate of downward β-transition is obtained. By interchanging the roles of the pump and the probe lasers, we can obtain the rate of upward β-transition.

A change in the excess population in the probed and pumped levels may be given by

$$\Delta\dot{n}_s = 2\pi\eta k_\beta P_B \Delta n_p - 2\pi\eta k'_\beta P_B \Delta n_s - 2\pi\eta k'_\alpha P_B \Delta n_s , \qquad (3)$$

$$\Delta\dot{n}_p = 2\pi\eta k'_\beta P_B \Delta n_s - 2\pi\eta k_\beta P_B \Delta n_p - 2\pi\eta k'_\alpha P_B \Delta n_p \qquad (4)$$

where k'_α represents the coupling to a thermal bath. In the case of $k_\beta, k'_\beta \gg k'_\alpha$, the double resonance signal is written as

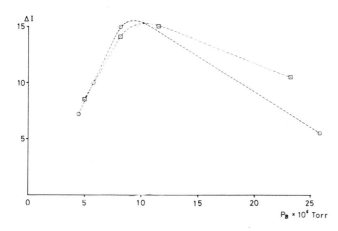

Fig.5 Double resonance signals due to the upward (▣) and the downward (◉) β-transitions

$$\Delta I = [k_\beta'/(k_\alpha'-k_\beta-k_\beta')]$$
$$\times \{[\exp[-2\pi(k_\beta+k_\beta')nP_B(t-t_0)]-\exp[-2\pi n k_\alpha' P_B(t-t_0)]\}\Delta n_p^0 . \quad (5)$$

The observed pressure dependence of the double resonance signal is shown in Fig.5. A difference in the initial gradients of the curves gives an excitation temperature of the beam.

4. M-TRANSITION

The cross section of collision-induced transition between closely spaced M-sublevels was determined on the J=2, K=2 rotational levels of CH_3F by using the 9μm-P(18) CO_2 laser (Fig.2)[5]. The experiment was carried out at several configurational arrangements of the quantization axes. In every case the quantum number M was well defined and the adiabatic condition for a transition between the M-sublevels holds well. The double resonance signal was proportional to the beam intensity, and its pressure dependence is well reproduced by a rate equation analysis. Tentative values of the cross section are 1250 A^2 for the M=±1 ← 0 transition and 1490 A^2 for the M=±2 ← ±1 transition.

References

[1]T.Oka, Adv. At. Mol. Phys., 9 127 (1973)
[2]S.Kano, T.Amano, and T.Shimizu, J.Chem. Phys., 64 4711 (1976)
[3]T.Shimizu and T.Oka, Phys. Rev., A2 1177 (1970)
[4]F.Matsushima, N.Morita, S.Kano, and T.Shimizu, J. Chem. Phys., 70 4225 (1979), Appl. Phys., 24 219 (1981)
[5]R.L.Shoemaker, S.Stenholm, and R.G.Brewer, Phys. Rev., A10 2037 (1974)

Part V
Nonlinear Processes

New Phenomena in Coherent Optical Transients

R.G. Brewer, R.G. Devoe, S.C. Rand, A. Schenzle, N.C. Wong, S.S. Kano, and
A. Wokaun

IBM Research Laboratory, 5600 Cottle Road
San Jose, CA 95193, USA

It is often stated, and with justification, that the concepts underlying
coherent optical transient phenomena can be found in the literature of spin
transients. For example, the Bloch vector [1] description of spin echoes
[2] and free precession [3] applies equally well to photon echoes [4] and
optical free induction decay [5]. While this conclusion remains largely
true, it is not equally valid to say that the techniques are the same or to
prophesy that the applications of optical and rf transients are identical
also. In fact, the decay channels of optically excited atoms, molecules
and solids usually involve a variety of interactions which contrast with
the magnetic interactions encountered in spin resonance. This is a major
distinction, one which has permitted new studies at optical frequencies.
Moreover, it is also true that some of the practioners of magnetic
resonance are beginning to learn a few lessons from quantum optics as well.
In this paper, we have selected some of the new developments which are
representative of the field of coherent optical transients.

PHENOMENA

Scientific discoveries sometimes proceed in the following stages [6]:

Prehistoric age	Might have been discovered, but wasn't
Stone age	Discovered, but not noticed
Ice age	Noticed, but not believed
Bronze age	Believed, but not interesting
Iron age	Wow!

I am not sure where the new phenomena I wish to discuss stand, perhaps
between the bronze and iron age. They are: (1) oscillatory free induction
decay; (2) stimulated photon echo effect; and (3) molecular spin-rotation
effect.

Oscillatory Free Induction Decay [7]

Imagine that a square pulse of coherent electromagnetic radiation of
duration T resonantly excites an inhomogeneously broadened two-level
quantum system. We ask what is the nature of the coherent emission, free
induction decay, following the pulse. It is remarkable that anything new
is to be discovered here since all coherent transient effects involve some
form of pulse preparation. However, analytic and numerical solutions of

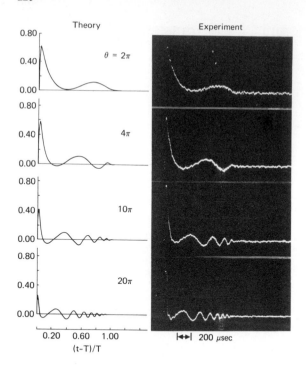

Figure 1. Oscillatory free induction decay. Left: theory of Schenzle et al.(Ref.7); right: FID of protons in water by Hashi et al.(Ref.10).

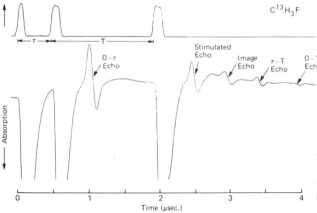

Figure 2. Stimulated and two-pulse infrared photon echoes in $C^{13}H_3F$ (lower trace) resulting from Stark switching pulses (upper trace).

the Bloch equations reveal new features which are shown in Fig. 1. In particular, large amplitude oscillations appear as the pulse area χT increases, χ being the Rabi frequency. Physically, Rabi sidebands introduced in the preparative period are preserved in the polarization following the pulse and give rise to the predicted oscillations. The oscillation frequency is not constant, but increases with time due to interference among the packets of the inhomogeneously broadened lineshape. As the packets get out of phase, due to differences in their frequency, an interference results and causes the transient to vanish precisely at time 2T, i.e., one pulse width following the pulse [8]. The initial rapid rise near time t=T arises from the first order FID, discussed elsewhere [9]. Hence, there are three time scales: the initial rapid response (T_2^*), the period of oscillation ($\sim 1/\chi$), and the duration of the emission ($T \leq t \leq 2T$). The uniqueness of the phenomenon is due to a combination of nonlinear behavior, which becomes evident at elevated intensities, and a transition that is strongly inhomogeneously broadened.

These predictions have been fully confirmed in the radio frequency region recently (Fig. 1) [10] and in the future should be readily detectable at optical frequencies as well.

Stimulated Echo Effect

Hahn's discovery of the spin echo or two-pulse echo of a two-level spin system also lead to the discovery of the stimulated or three-pulse echo [2] (Fig. 2). In the latter, two rf pulses at times t=0 and t=τ produce the normal spin echo at t=2τ, while a third pulse at t=T+τ creates a stimulated echo at t=T+2τ (and subsequent two-pulse echoes which can be ignored here). The first pulse prepares two-level spin packets in quantum mechanical superposition which thereafter dephase. The second pulse reverses their phase relationship so that they come back into phase to produce the spin-echo pulse. In addition, a modulated population distribution among the upper and lower spin states results with a modulation frequency $1/\tau$, the inverse of the time between the two initial pulses. The modulated population possesses a precise phase in the frequency domain, within the inhomogeneous lineshape, a property that can persist long after the temporal coherence of the superposition state has vanished. Finally, the third pulse reads out the precise phase information contained in the modulated population of both upper and lower spin states by again introducing a two-level coherent superposition that interferes in such a way as to produce an echo, the stimulated echo. The new aspect of this subject is that the frequency of the third pulse can differ from the first two pulses, allowing either the lower or upper initial spin states to participate in a second transition which similarly produces a stimulated echo [11,12]. Thus, population relaxation, which washes out the modulation pattern, can be examined selectively in either initial spin state - a point not realized in the NMR literature. The recent optical studies in gases [11] and solids [12] support these conclusions. In some solids at liquid helium temperature, for example, the ground state population modulation persists for 30 minutes or more [12].

Spin-Rotation Effect [13]

The optical Stark switching technique [14] has permitted observing many of the optical analogs of spin transients and also some new effects as well. Consider one of these, the two-pulse photon echo of $^{13}CH_3F$ gas which shows a pronounced interference or modulation behavior in the echo amplitude as the pulse delay time is advanced [13]. The modulation arises from the molecular spin-rotation interaction $\mathcal{H} = C\mathbf{I} \cdot \mathbf{J}$, where I and J are the spin and molecular rotation quantum numbers. The smallness of the interaction constant $C \lesssim 20$ kHz suggests that a significant modulation is unlikely, but rather surprisingly, a depth of modulation of 35% or more is readily detectable. To understand the phenomenon, we imagine that two Stark pulses switch $^{13}CH_3F$ molecules twice into resonance with a fixed frequency (9.66 μm) CO_2 laser so as to produce an infrared photon echo. The Stark pulse amplitudes are sufficiently large that the laser drives only one pair of levels in a given molecule. When the pulses are switched off, the formerly driven level pair (still in superposition) mixes with the neighboring undriven levels through the spin-rotation interaction. It can be shown that four levels are now in superposition in place of the original two, and because of the small spin-rotation splitting, transitions of slightly different frequency interfere. Even though the splitting is small and normally difficult to detect, a large modulation in the echo envelope function results. Spin-rotation modulation in $^{13}CH_3F$ has been monitored for the ^{13}C nucleus and the ^{19}F nucleus in both ground and excited vibrational states. Precise determinations of the interaction parameter C should be possible in future measurements.

ELASTIC COLLISIONS

Any process that introduces random phase changes in an ensemble of coherently prepared atomic dipoles will produce damping. One example is the subtle effect of elastic collisions of atoms in a coherent two-level superposition state. It is the optical superposition state that distinguishes this elastic scattering problem from previous work. Is it possible to predict in general whether the two states which are in superposition behave in the same way or not during elastic scattering ? We think that the answer is no, and we must, at least for the present, be guided by observations.

Initial photon echo studies [15] in the infrared region involving the superposition of a pair of vibration-rotation states of $^{13}CH_3F$ show that the elastic collisions are of the velocity changing type. Here, the two states are described by the same scattering trajectory. Experiments can be interpreted by the following simplified agrument. The dipolar phase change corresponding to a change in velocity Δv along the optic axis direction is given by $\Delta\phi = k\Delta vt$, where $k = 2\pi/\lambda$, λ is the optical wavelength and t is the time. In a two-pulse echo experiment with pulse delay time τ, such random phase changes will cause the molecular dipoles to get out of step, and the echo amplitude will shrink by the phase factor

$$S \sim \langle e^{ik\Delta v\tau}\rangle_{collision} . \tag{1}$$

Eq. (1) takes limiting forms in the short- and long-time regimes,

$$S \sim \exp[-k^2\Delta u^2\Gamma\tau^3/4] , \quad k\Delta v\tau \ll 1$$

$$S \sim \exp[-\Gamma\tau] , \qquad k\Delta v\tau \gg 1 \tag{2}$$

where Γ is an average rate of a binary collision and Δu is a characteristic change in velocity per collision. The nonexponential decay at short times followed by an exponential decay is the signature for the velocity changing collision mechanism (Fig. 3). In addition to the $^{13}CH_3F$ results, recent infrared echo measurements [16] in SF_6 obey (2) also, in support of the velocity changing collision mechanism. It may be that all molecules prepared in coherent superposition in the infrared exhibit elastic velocity changing collisions in this way.

In the visible region, however, the superposition involves two electronic states which may behave quite differently during an elastic

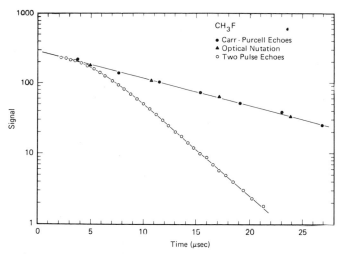

Figure 3. Population decay (upper curve) and dipole dephasing (lower curve) in CH₃F, detected by infrared coherent transients, where the difference in the two results from velocity changing collisions.

collision. State-dependent elastic collisions can produce phase interruptions of the form $\Delta\phi=\Delta\omega t_c$, where $\Delta\omega$ is the angular shift in the transition frequency and t_c is the collision duration. The decay law for this model is exponential (Fig. 4). Optical free induction decay of the visible transition of I_2 vapor does indeed reveal an exponential decay [17], and recently measurements [18] have been extended to times as short as 10 nsec with no evidence for an $\exp(-Kt^3)$ decay law. Photon echo studies in helium, neon and argon plasmas also obey exponential behavior [19]. In contrast to the above, it is claimed [20] that photon echoes of the sodium D lines display both velocity changing and phase interrupting character in different time regimes. This interpretation, if correct, raises the issue of a need for a more general treatment of the effect of elastic scattering of atoms in superposition states, rather than the limiting cases of velocity changing and phase interrupting collisions considered here.

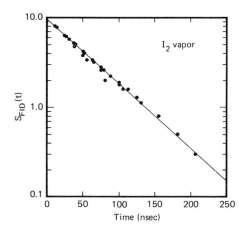

Figure 4. Optical FID in a visible transition of I_2. This shows exponential decay down to ~10 nsec due to phase interrupting collisions in contrast to Fig. 3.

SOLIDS

Remarkably narrow optical homogeneous linewidths of the order of 1 kHz have now been observed in the low temperature zero-phonon transition $^3H_4 \leftrightarrow ^1D_2$ of the impurity ion Pr^{3+} in LaF_3 and other host crystals using optical free induction decay [21,22]. A cw ring dye laser that is frequency-locked to a stable external reference cavity coherently prepares a single Pr^{3+} packet within its homogeneous width, producing a narrow hole within the strain-broadened inhomogeneous lineshape (~5 GHz width). When the exciting laser beam is frequency switched a few megaHertz, using an external Bragg modulator [21], the free precession commences and the usual FID heterodyne beat signal results as seen with a P-I-N photodiode in transmission (Fig. 5). In the absence of power-broadening, the decay of this signal is one-half the dipole dephasing time ($T_2/2$).

For Pr^{3+}:LaF_3, the homogeneous linewidth can be no narrower than 160 Hz, due to the 0.5 ms radiative lifetime of the $^1D_2$ state. At 2 K, phonon processes contribute at most a few Hertz. Laser power broadening is estimated to be less than 100 Hz. The dominant dynamic process arises from the magnetic-dipolar spin interactions, due first to the fluctuating resonant flip-flops among pairs of ^{19}F spins and second to the resulting nonresonant heteronuclear $^{141}Pr-^{19}F$ interaction which reflects the time dependence of the homonuclear events. That such a mechanism determines the optical homogeneous width follows first from the magnitude of the broadening which is comparable to that encountered in NMR. Secondly, the linewidth or dephasing time depends on the strength of the applied external magnetic field. For example, in the Earth's field, T_2=3.6 μsec, but when the external field exceeds the local field of ~20G, T_2=15.8 μsec corresponding to a 10 kHz HWHM linewidth. Obviously, when the external field exceeds the local field, the induced precession rate is fast enough to time average the slower chaotic motions associated with the field fluctuations along the external field direction. Magnetic fluctuations in the transverse directions are unaffected, however, and thus the linewidth narrows but does not vanish.

An even more dramatic and unambiguous demonstration of the dipolar line-broadening mechanism is found in the application of the magic angle line-narrowing technique of NMR to optical spectroscopy [22]. The optical FID experiment is repeated, but now in the presence of an rf and a static magnetic field (Fig. 6). By transforming the Bloch formalism and the homonuclear (F-F) and the heteronuclear (Pr-F) dipolar interactions into a suitable rotating frame, it can be shown [22] that the optical linewidth in

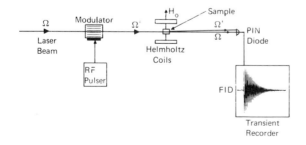

Figure 5. Laser frequency switching technique for observing coherent optical transients such as FID.

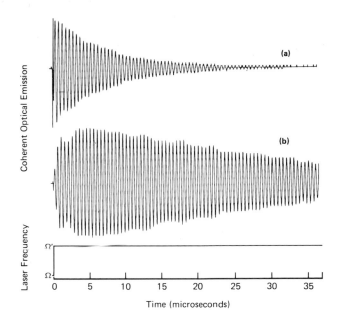

Coherent Optical Emission

Laser Frecuency

(a)

(b)

Ω'

Ω

| | | | | | | | |
0   5   10   15   20   25   30   35

Time (microseconds)

Figure 6. Optical FID in Pr^{3+}:LaF$_3$ with (a) no rf and (b) with rf at the magic angle condition.

the presence of an rf field is given by

$$\Delta\nu(\beta) = \Delta\nu(0)|\cos\beta \cdot 1/2(3\cos^2\beta-1)| \quad . \tag{3}$$

Here, β is the angle the effective field makes with the applied static magnetic field and $\Delta\nu(0)$ is the optical linewidth in the absence of the rf field. Note that in the rotating frame the $1/2(3\cos^2\beta-1)$ term takes the form of a secular dipolar interaction and we see that the linewidth vanishes when $\beta=\pi/2$, the fluorine resonance condition, and $\beta=54.7°$, the magic angle condition. Physically, the induced precession resulting from the applied static and oscillating magnetic fields provides a time averaging mechanism which tends to eliminate the chaotic motions arising from the local ^{19}F field fluctuations. The argument therefore is similar to the time averaging achieved with a static field, as described above, but in the magic angle case, the averaging is more effective.

The form of (3) has been verified in detail by FID studies of Pr^{3+}:LaF$_3$, and furthermore shows that the linewidth narrows from ~10 to ~2 kHz [22] (Fig. 7). The source of the residual broadening is still being examined.

Finally, a Monte Carlo theory [23] of optical dephasing in Pr^{3+}:LaF$_3$ has been developed to explain the current FID observations. A numerical approach was attempted since the heteronuclear dipolar interaction is neither weak enough to fit a Gaussian theory ($\delta\omega\tau\ll 1$) nor strong enough to fit the Klauder-Anderson theory [24] ($\delta\omega\tau\gg 1$). Instead, $\delta\omega_{max}\tau\sim 1$, where $\delta\omega_{max}$ is the maximum Pr^{3+} frequency jump, due to a fluorine spin flip, and τ is the observed Pr^{3+} dephasing time. A Monte Carlo computer routine was utilized that assumed: (1) the LaF$_3$ crystal structure and (2) a sudden fluorine spin-flip model, thus avoiding many of the approximations of past analytic theories in magnetic resonance. The Pr^{3+} decay behavior is obtained by sampling statistically the Pr^{3+} phase history as subgroups of ^{19}F spins flip randomly in space and time. Only a few ^{19}F spins contribute

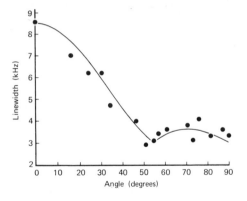

Figure 7. Optical linewidth of $Pr^{3+}LaF_3$ versus the angle β expressed in degrees. Solid circles: experimental points. Solid line: Eq. (3).

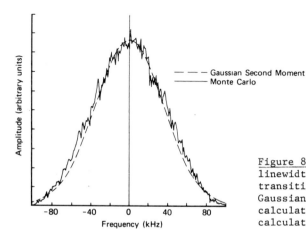

Figure 8. Magnetic inhomogeneous linewidth of a ^{141}Pr quadrupole transition of $Pr^{3+}:LaF_3$ showing the Gaussian lineshape of a Monte Carlo calculation and also a second moment calculation.

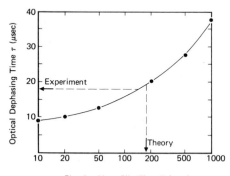

Figure 9. The Pr^{3+} optical dephasing time $1/\gamma_\phi$ versus the fluorine mean flip time T. The optical FID result is $1/\gamma_\phi$=15.8 μsec and the theoretical fluorine spin-flip time T=170 μsec.

significantly to the homogeneous width, a result which shows for the first time that spin-flip correlations are not significant. Furthermore, a Pr^{3+} ion polarizes and detunes the nearest fluorines, forming a frozen core that is incapable of spin flipping with the bulk fluorines. Although the core grows radially as the ^{141}Pr (I=5/2) magnetic moment increases with I_z, the

optical linewidth changes little since the fluorines which participate in
spin flipping become more distant. Using no free parameters, a Lorentzian
lineshape of 8.4 kHz HWHM is calculated which compares very well to the
optical FID observation of a 10.1 kHz Lorentzian. (See Figs. 8 and 9.)

CONCLUSION

There exist many other cases of coherent optical transients and their
applications which have not been mentioned due to lack of space. For
example, long optical dephasing times have been observed for the first time
in stoichiometric EuP_5O_{14} which appears to exhibit a weak delocalized
excitation among the Eu^{3+} ions [25]. Dephasing studies in the picosecond
range [26] are also attracting considerable attention for exploring very
fast processes in condensed matter. All of these investigations make use
of the selectivity and precision inherent in the coherence techniques which
lasers now provide.

REFERENCES

1. F. Bloch, Phys. Rev. 70, 460 (1946).
2. E. L. Hahn, Phys. Rev. 80, 580 (1950).
3. E. L. Hahn, Phys. Rev. 77, 297 (1950).
4. N. A. Kurnit, I. D. Abella and S. R. Hartmann, Phys. Rev. Lett. 13, 567 (1964).
5. R. G. Brewer and R. L. Shoemaker, Phys. Rev. A 6, 2001 (1972).
6. Variations on a theme of H. J. Lipkin in The Mossbauer Effect, edited by H. Frauenfelder (W. A. Benjamin Inc., New York, 1962), p. 13.
7. A. Schenzle, N. C. Wong and R. G. Brewer, Phys. Rev. A 21, 887 (1980).
8. See the theorem A. Schenzle, N. C. Wong and R. G. Brewer, Phys. Rev. 22, 635 (1980).
9. R. G. DeVoe and R. G. Brewer, Phys. Rev. A. 20, 2449 (1979).
10. M. Kunitomo, T. Endo, S. Nakanishi and T. Hashi, Phys. Lett. 80A, 84 (1980).
11. T. Mossberg, A. Flusberg, R. Kachru and S. R. Hartmann, Phys. Rev. Lett. 42, 1665 (1979).
12. J.B. Morsink, O.A. Wiersma, In Laser Spectroscopy IV, ed. by H. Walther, K.W. Rothe, Springer Series in Optical Sciences, Vol. 21 (Springer Berlin, Heidelberg, New York 1979) p. 404
13. J. A. Kash and E. L. Hahn (to be published).
14. R. G. Brewer and R. L. Shoemaker, Phys. Rev. Lett. 27, 631 (1971).
15. J. Schmidt, P. R. Berman and R. G. Brewer, Phys. Rev. Lett. 31, 1103 (1973); P. R. Berman, J. M. Levy and R. G. Brewer, Phys. Rev. A 11, 1668 (1975).
16. B. Comaskey, R. E. Scotti and R. L. Shoemaker, Optics Letters 6, 45 (1981).
17. R. G. Brewer and A. Z. Genack, Phys. Rev. Lett. 36, 959 (1976).
18. R. G. Brewer and S. S. Kano in Nonlinear Behavior of Molecules and Ions in Electric, Magnetic or Electromagnetic Fields (Elsevier, Amsterdam, 1979), p. 45.
19. M. Woodworth and I. D. Abella, J. Opt. Soc. Am. 70, 1567 (1980).
20. T. W. Mossberg, R. Kachru and S. R. Hartmann, Phys. Rev. Lett. 44, 73 (1980).
21. R. G. DeVoe, A. Szabo, S. C. Rand and R. G. Brewer, Phys. Rev. Lett. 42, 1560 (1979).

22. S. Č. Rand, A. Wokaun, R. G. DeVoe and R. G. Brewer, Phys. Rev. Lett. 43, 1868 (1979).
23. R. G. DeVoe, A. Wokaun, S. C. Rand and R. G. Brewer, Phys. Rev. B 23, 3125 (1981).
24. J. R. Klauder and P. W. Anderson, Phys. Rev. 125, 912 (1962).
25. R. M. Shelby and R. M. Macfarlane, Phys. Rev. Lett. 45, 1098 (1980).
26. D. A. Wiersma, Advances in Chemical Physics (to be published in 1981).

Stimulated Raman Studies of CO_2-Laser-Excited SF_6

P. Esherick, A.J. Grimley, and A. Owyoung

Sandia National Laboratories
Albuquerque, NM 87185, USA

1. Introduction

Advances in high-resolution stimulated Raman spectroscopy (SRS) have provided a means for performing gas phase Raman studies with unprecedented spectral resolution ($\sim .002$ cm^{-1}) and high sensitivity [1-3]. Studies in static cells at low (< 4 Torr) pressures have revealed a wealth of spectral detail inaccessible to conventional Raman techniques. Investigations in flames [4] and supersonic free-expansion jets [5] have illustrated the spatial resolving power offered by SRS. In this paper we report a preliminary investigation into the use of SRS for high-resolution studies of transient species. Specifically, we are using SF_6 molecules excited by a CO_2 laser as a proto-typical system for examining these capabilities.

In recent years the discovery of infrared multiphoton dissociation processes in SF_6 has created considerable interest in the vibrational spectroscopy of this molecule. Direct infrared absorption [6] and infrared-infrared double resonance studies [7-8] have contributed to a better understanding of the infrared excitation processes. Recently, lower spectral resolution time-resolved spontaneous Raman studies of CO_2 laser excited SF_6 have shown the potential utility of Raman probing as a means of studying the excitation dynamics [9]. Since stimulated Raman techniques offer prospects for accessing any desired spectral region with very high temporal and spectral resolving power, they appear well suited to the extension of such studies.

2. Experimental Observations in CO_2 Laser Pumped SF_6

In Fig. 1 we show a Raman spectrum of dynamically cooled SF_6 in the region of its ν_1 fundamental near 774 cm^{-1}. The very regular $J(J + 1)$ functional form of the spectrum makes this Q-branch spectrum ideal for monitoring the redistribution of population among the rotational states. Moreover the pulsed supersonic free expansion jet [10] provides an excellent environment, characterized by high densities ($\sim 5 \times 10^{17}$ cm^{-3}) and reduced collision broadening. In addition, the spectrum is simplified by the cooling of a large portion of the population into the ground vibrational manifold.

The energy diagram of Fig. 2a illustrates the use of the ν_1 band as a means of monitoring the rotationally specific removal of population from

---

*This work supported by the United States Department of Energy.

the ground state via CO_2 laser excitation in the ν_3 band. Experimentally this is achieved by operating the SRS system at 10 Hz and synchronizing the CO_2 laser to excite the sample at 5 Hz. Two alternately triggered boxcar averagers are then used to obtain both the unperturbed, "normal", spectrum and the "perturbed", CO_2-laser-pumped spectrum.

The spectrum shown in Fig. 3 illustrates the selective depletion of the ν_1 fundamental when pumped by the P(14) line of the grating tuned CO_2 TEA laser. The spectrum was recorded at the peak of the 100 nsec CO_2 laser probe ($\pm$ 25 nsec) at a fluence level of ~ 200 kW/cm^2. Comparison with the "normal" spectrum shows significant depletion in the

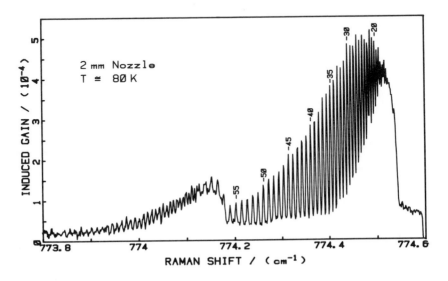

Fig. I Raman gain spectrum of the ν_1 band of SF_6 obtained in a pulsed free-expansion jet by probing 7 mm downstream from a 2 mm nozzle. Backing pressure = 86 psia. Instrumental resolution is ~ 70 MHz.

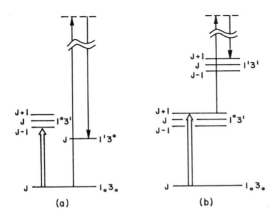

(a)   (b)

Fig. 2 Energy level schematic illustrating the use of SRS to probe SF_6 molecules excited by a CO_2 laser in the ν_3 band. (a) Monitoring of the ν_1 fundamental to measure ground state depletion. (b) Monitoring of the $\nu_1 + \nu_3 \leftarrow \nu_3$ hot band populated by the CO_2 laser.

J = 28 region where the P(14) line of CO_2 at 947.479 cm^{-1} overlaps the ν_3 R branch [6]. Additional scans taken at longer delay times (up to 1 μsec) show a thermalization of the rotational states, which results in a depletion of the entire band and a filling of the "hole" near J = 28. At higher fluence levels (~600 kW/cm²), more extensive depletion is seen over the entire band, even at short delay times. The lines around the region of strongest depletion are also found to be strongly pulled in frequency toward the most heavily saturated line. This frequency shift arises from the near resonant interaction of the CO_2 laser intensity with the ν_3 R-branch transitions. The resultant up- and down-shifting of the ν_3 levels are consequently reflected in the ν_1 transitions through the Stark shift in their common ground states. Careful examination of Fig. 3 reveals that such a shift is also present in this spectrum, although not nearly as evident as it would be at higher fluences.

We have also investigated the effects of using the P(18) line of CO_2 to pump the ν_3 band in the vicinity of P(33). As expected, the sign of the Stark shift is reversed, resulting in a pushing of the lines away from the "hole" near J = 33. A stronger overall depletion of the band is also observed, consistent with more efficient multiphoton excitation using CO_2 laser lines that are red shifted with respect to the ν_3 band maximum [11]. Entirely unexpected, however, is the appearance of a strong selective depletion of the band, in the vicinity of J = 16, which can not be explained with our current understanding of one- and two-photon processes in SF_6. We are continuing our investigation of this unexplained depletion in an attempt to understand its origin.

In addition to detecting depletion of the ground state populations, SRS techniques can also yield both spectroscopic and dynamic information on the excited species. This is illustrated in Fig. 2b where SRS is employed

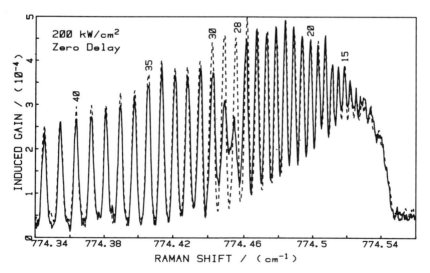

Fig. 3 Raman gain spectra of the ν_1 fundamental showing the effect of excitation of the ν_3 band near R(28) by the P(14) line of the CO_2 laser. The CO_2 laser "perturbed" spectrum (solid line) is compared to the "normal" spectrum (dashed line).

to monitor the CO_2 laser populated $\nu_1 + \nu_3 \leftarrow \nu_3$ hotband. The free-expansion jet becomes an essential part of this experiment because of the necessity to eliminate thermally populated "hotbands" which, at room temperature, would obscure the presence of the CO_2 pumped band. We have unambiguously identified the population of this band via pumping by the P(14) line of CO_2. However, at this time, our data is insufficient to report fully on the character of this band.

3. Conclusions

The preliminary results reported in this work demonstrate the applicability of SRS techniques to studying the dynamics of multi-photon excitation processes in molecular systems. These experiments illustrate the potential capability for obtaining time resolved spectral information with SRS, and may lead the way to eventual Raman probing of molecular photofragments or other transient species.

We gratefully acknowledge the contributions of J. J. Valentini in developing the pulsed free expansion jet apparatus used in these studies. We also wish to acknowledge the expert technical assistance of R. E. Asbill and R. W. Willey and thank G. A. Fisk for helpful discussions in the preparation of this manuscript.

References

1. A. Owyoung, In Laser Spectroscopy IV, ed. by H. Walther, K.W. Rothe, Springer Series in Optical Sciences, Vol. 21 (Springer Berlin, Heidelberg, New York 1979) p. 175
2. A. Owyoung, P. Esherick, in Lasers and Applications, ed. by W.O.N. Guimaraes, C.T. Lin, A. Mooradian, Springer Series in Optical Sciences, Vol. 26 (Springer Berlin, Heidelberg, New York 1981) p. 67
3. P. Esherick and A. Owyoung, in Advances in Infrared and Raman Spectroscopy Vol. 10, eds., R. J. H. Clark and R. E. Hestor, to be published by Heyden and Son Ltd.
4. A. Owyoung and P. Esherick in Proc. VII Int. Conf. on Raman Spectroscopy, Ottawa (1980), ed., W. F. Murphy, North-Holland, NY, 1980, p. 656.
5. J. J. Valentini, P. Esherick and A. Owyoung, Chem. Phys. Lett. 75, 590 (1980).
6. R. S. McDowell, H. W. Galbraith, B. J.Krohn, C. D. Cantrell and E. D. Hinkley, Optics Comm. 17, 178 (1976).
7. P. F. Moulton, D. M. Larsen, J. N. Walpole and A. Mooradian, Opt. Lett. 1, 51 (1977).
8. C. C. Jensen, T. G. Anderson, C. Reiser, and J. I. Steinfeld, J. Chem. Phys. 71, 3648 (1979).
9. V. N. Bagratashvili, Yu. G. Vainer, V. S. Doljikov, V. S. Letokhov, A. A. Makorov, L. P. Malyavkin, E. A. Ryabov and E. G. Silkis, Opt. Lett. 6, 148 (1981).
10. J. B. Cross and J. J. Valentini, submitted to Rev. Sci. Instr.
11. R.V. Ambartzumian, In Tunable Lasers and Applications, ed. by A. Mooradian, T. Jaeger, P. Stokseth, Springer Series in Optical Sciences, Vol. 3 (Springer Berlin, Heidelberg, New York 1976) p. 150

Pulsed and CW Molecular Beam CARS Spectroscopy

R.L. Byer, M. Duncan[1], E. Gustafson, P. Oesterlin[2], and F. Konig[3]

Edward L. Ginzton Laboratory, Stanford University
Stanford, CA 94305, USA

I. Introduction

The advantages of molecular beam coherent anti-stokes Raman spectroscopy include increased spectral resolution from sub-doppler linewidths, improved signal-to-noise resulting from spectral simplification due to rotational cooling upon expansion, and the possibility of studying molecular complexes generated by the expansion process. In this paper we report recent high resolution CARS studies using pulsed laser sources combined with a reliable pulsed molecular beam source and the first cw CARS measurements in a steady state high Mach number supersonic expansion.

High resolution pulsed molecular beam CARS measurements of the Q-branch of acetylene were used to characterize the properties of the pulsed molecular beam expansion. The onset of saturation broadening in a Raman spectrum was observed for the first time in the resolved Q-branch spectrum of expansion cooled acetylene. The previously unresolved ν_2 Q-branch of ethylene was resolved in the expansion cooled spectrum.

We have obtained the first high resolution cw CARS spectra of the CH_4 Q-branch in a steady state supersonic expansion. The supersonic jet expansion is a very useful spectroscopic tool that provides a convenient method of generating a range of molecular temperatures and densities. We observed, and have included in the CARS lineshape theory, the effects of transit time broadening evident in the tightly focused geometry used to obtain the supersonic jet CARS spectra of CH_4.

2. Pulsed Molecular Beam CARS

The possibility of high resolution CARS spectroscopy in a supersonic molecular beam was theoretically analyzed by DUNCAN and BYER [1] in 1979. Initial estimates of the required laser power and resulting signal-to-noise ratios showed that conventional effusive molecular beam sources could not be used in the proposed measurements because of their inherently low molecular density. The molecular beam CARS experiment was, therefore, designed to utilize the pulsed nozzle source introduced by GENTRY and

---

[1]Present address: Naval Research Laboratory, Code 6540, Washington DC 20375, USA
USA
[2]Present address: Freiburg University, D-7800 Freiburg, Fed. Rep. of Germany
[3]Present address: Heidelberg University, D-6900 Heidelberg, Fed. Rep. of
Germany

GIESE [2] . Simultaneously with the development of a 10 Hz, reliable, pulsed nozzle [3] a high peak power, single axial mode, unstable resonator Nd:YAG source was successfully demonstrated using injection locking techniques [4] . A Nd:YAG pumped dye amplifier chain was also implemented to provide high peak power amplified output from a cw dye laser source. The doubled Nd:YAG source and the dye laser source both operated at their Fourier transform linewidth limits near 100 MHz. The laser sources are described in more detail in [6] . Figure 1 shows a schematic of pulsed molecular beam CARS apparatus.

During the construction of our pulsed molecular beam CARS apparatus shown in Fig. 1, low resolution CARS of a continuous nitrogen supersonic flow was demonstrated by HUBER-WALCHLI, GUTHALS and NIBLER [7], early results of cw CARS of CH_4 in a supersonic jet were reported [8] and a series of high resolution Raman Gain Spectroscopy measurements of CH_4 followed by other molecules were obtained by VALENTINI, ESHERICK and OWYOUNG [9].

The principal advantage of the supersonic molecular beam in these experiments is the spectral simplification provided by the reduced rotational temperature. The temperature for a continuum expansion in a supersonic jet is given by [10].

$$T = T_0/[1 + 1/2(\gamma - 1)M^2], \tag{1}$$

where
$$M = A\left(\frac{X - X_0}{D}\right)^{\gamma-1} - \frac{1}{2}\left(\frac{\gamma + 1}{\gamma - 1}\right)\left[A\left(\frac{X - X_0}{D}\right)^{\gamma-1}\right]^{-1} \tag{2}$$

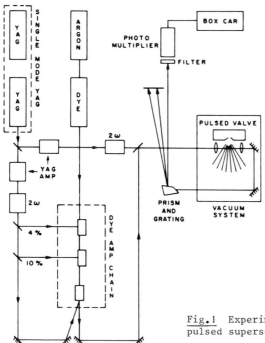

<u>Fig.1</u> Experimental arrangement for recording pulsed supersonic molecular-beam CARS spectra

is the Mach number, γ is the ratio of specific heats, A is a constant determined by γ, X is the distance from the nozzle and D is the nozzle diameter.

We have used CARS spectra of the ν_2 Q-branch of acetylene where $\gamma = 7/5$ to verify the expansion cooling provided by the pulsed nozzle. The measurements showed that during the steady state expansion phase of the 100 μsec long gas pulse emitted by the nozzle (1) is indeed obeyed when dimer formation is not significant at low gas densities. At higher gas densities dimer formation, with the resultant heat release [11] , prevents cooling to the theoretically predicted temperatures. Figure 2 shows the Q-branch spectra of acetylene obtained in a static cell, at 80°K and at 20°K expansion temperatures. Figure 3 shows the rotational temperature in the supersonic expansion vs X/D. Dimer formation clearly prevents acetylene from reaching predicted temperatures except at low gas density or when seeded in helium.

The spectral simplification provided by the supersonic expansion is illustrated by the ν_2 Q-branch of ethylene. Figure 4 shows the spectrum at room temperature in a static cell at 60 torr pressure. Figure 5 shows the same spectrum at successively colder locations along the jet expansion. The temperatures indicated in the figure are estimates based on the known rotational constant for ethylene and a best fit to the envelope of the rotational component amplitudes. Efforts are under way to fit the spectrum and to determine more precise ethylene spectral constants.

The above molecular beam CARS spectra were generated with input pulse energies of less than 1 mJ. Higher input energies near 5 mJ lead to saturation broadening of the Raman spectrum. The population changes induced by

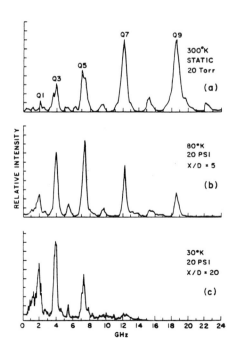

Fig.2 Spectra of the ν_2 Q-branch of acetylene taken with the CARS technique in (a) static gas, (b) in the molecular beam at X/D = 5, and (c) in the molecular beam at X/D = 20.

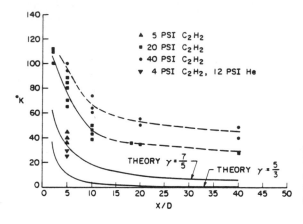

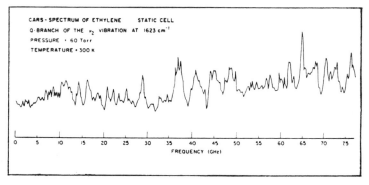

Fig.3 Plot or rotational temperature in the supersonic expansion vs X/D. Theoretical curves for gases with $\gamma = 7/5$ and $\gamma = 5/3$ are shown.

Fig. 4 CARS spectrum of the ν_2 Q-branch of ethylene at 60 torr in a static cell.

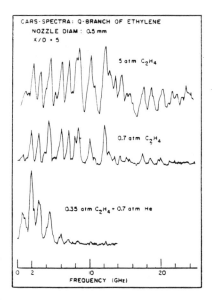

Fig. 5 CARS spectra of the ν_2 Q-branch of ethylene at $\dot{X}/D = 5$ at various input pressures and seeded in helium.

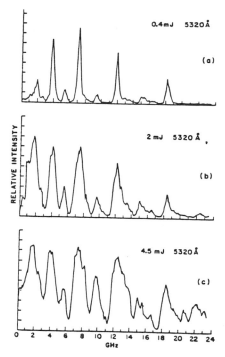

Fig. 6 CARS spectra of the ν_2 Q-branch of acetylene; a) unperturbed spectrum at 0.4 mJ 5320 Å energy; b) at 2.5 mJ; c) same spectrum with 4.5 mJ 5320 Å energy. In all spectra the Stokes energy at 5944 Å was kept at 0.1 mJ.

stimulated Raman scattering lead in turn to significant spectral broadening. We have observed saturation broadening in the expansion cooled acetylene spectra as shown in Fig. 6 [12]. Saturation broadening sets a limit to the sensitivity of the CARS process in high resolution studies. Improvements in sensitivity can be gained by decreasing the incident laser power in the inter-action focal volume by either increasing the focal spot size or by increasing the pump and probe laser pulse widths. The latter option is desirable as it also leads to improved Fourier transform limited laser linewidths. However, longer pulse lengths are not easily achieved under normal Q-switched laser operation.

The early onset of saturation in the molecular beam Raman spectrum of acetylene suggests that Raman saturation spectroscopy, in the form first demonstrated by OWYOUNG and ESHERICK [13], is a practical approach for sub-Doppler Raman spectroscopy. The shift of substantial population into the Raman level also suggests the possibility of state elective collisional trans-fer studies. Stepwise ionization from the Raman level followed by detection of the ion or Raman ionization spectroscopy, is also possible.

There is interest in obtaining the Raman spectra of van der Waals complexes for the determination of the structure of these systems. The formation of complexes in the pulsed molecular beam expansion led us to search for a Raman signal due to ehtylene complexes. Early results were encouraging in that a broad peak shifted 4.8 cm^{-1} from the previously resolved ν_2 Q-branch peak of ethylene was observed. The magnitude of the peak varied with input pressure squared indicating either an ethylene complex or

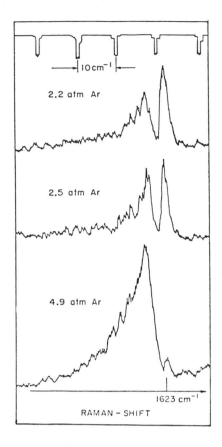

2.2 atm Ar

2.5 atm Ar

4.9 atm Ar

1623 cm^{-1}

RAMAN – SHIFT

Fig. 7 CARS spectra of a supersonic molecular beam of an ethylene/argon mixture. The ethylene pressure was constant for all measurements, (0.27 atm.). The peak at 1623 cm^{-1} is the unresolved Q-branch of the ν_2 vibration of C_2H_4. The peak at a smaller Raman shift is due to the ethylene molecular complex.

ethylene/argon complex was present. Figure 7 shows the CARS spectrum taken at 1 cm^{-1} resolution. The peak at 1623 cm^{-1} is the previously resolved (see Fig. 5), ν_2 Q-branch of ethylene. Measurements vs position from the nozzle showed that the complex began forming at X/D = 5, the signal strength increased to X/D = 10 and then decreased due to the decrease in density with further expansion.

Spectral scans attempted at higher resolution did not resolve structure in the spectrum but instead led to a significant decrease in signal strength that was suggestive of pre-dissociation of the complex. These results are preliminary and must be confirmed by further measurements.

Our pulsed molecular beam CARS system has now operated reliably for more than one year. Contrary to early expectations, measurements at good signal-to-noise with less than 1 mJ of input energy have been readily achieved. The combination of the single axial mode pulsed laser sources and the reliable pulsed nozzle have allowed significant advances to be made in high resolution Raman spectroscopy.

3. cw Supersonic Jet CARS

Supersonic cooling in steady state flows can be achieved in the laboratory using available vacuum pukps if the nozzle diameter is less than

approximately 1 mm. We have constructed a ½ mm diameter supersonic jet
backed by an old nitrogen laser vacuum pump. The jet is enclosed in a
1 cm x 1 cm dye laser cuvette for easy optical access. Control of the
back pressure and of the input puressure allows a wide range of Mach flows
up to Mach 8 to be generated at minimum gas consumption. The supersonic
jet was used in cw CARS studies of the flow density, temperature and
velocity. The study showed that the supersonic jet is an inexpensive,
straightforward method of generating molecular pressures from 1 atm to
less than 1 torr with corresponding temperatures from room temperature to
less than $20^{\circ}K$ under very well controlled conditions.

A schematic of the supersonic jet with spectra of the methane Q-branch
taken at three positions along the expansion is shown in Fig. 8. The
spectra were taken using the cw CARS apparatus described previously [14].
Spatial location in the supersonic jet was assured by using a tight
focusing geometry with 3.7 cm focal length lenses providing a 2 μm spot

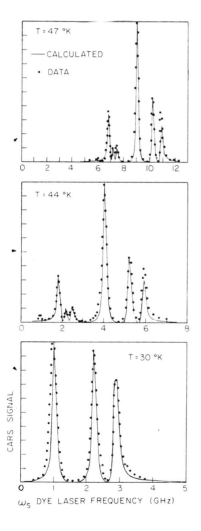

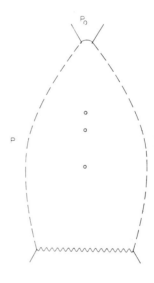

Fig.8 Schematic of the supersonic jet expansion showing the location of the CARS
measurements along the jet axis. The dashed
line indicates the barrel shock and the
wavy line the location of the Mach disk.
cs CARS spectra of CH_4 Q-branch at the temperatures indicated. The dots are the data,
the solid line the calculated spectra.

size with 90% of the CARS signal generated in less than 100 μm length.
It is interesting to note that less than 10^8 molecules were in the
interaction volume of 6 x 10^{-10} cm^3 at 10 torr pressure.

The spectral simplification of the methane Q-branch to the ground
quantum state is illustrated in the T = $30^{\circ}K$ spectrum of Fig. 8. Here
the J = 0, 1 and 2 rotational components are fully resolved. The
spectral resolution was set by residual Doppler broadening of 100 MHz,
transit time broadening of 100 MHz, and laser linewidth jitter of 30 MHz.
The theoretical fit to the data points was made using a CARS lineshape
theory that included transit time broadening. The cw CARS spectrum was
obtained using a 3W argon ion laser and 100 mW of cw dye laser input.
Typical signal count rates were 1000 cps at pressures of a few torr. The
signal levels were high due to the spectral simplification that resulted
from the expansion cooling. Signal levels were high enough to allow an
accurage measurement jet flow velocity and temperature.

These results illustrate the spectroscopic utility of a simple steady
state supersonic jet. The independent control of input pressure and
backing pressure coupled with position along the expansion axis of the
jet allows ready access to a wide range of temperatures and pressures.
The small diameter jet provides adequate interaction length for nonlinear
spectroscopy and yet minimizes gas consuption at even high Mach number flows.
For example, the present jet consumed approximately 1 bottle of methane
for eight hours of operation.

4. Conclusion

We have demonstrated that high resolution Raman spectra of complex
molecules can be obtained using a combination of pulsed high peak power
single axial mode laser sources and a pulsed supersonic beam source. The
pulsed nozzle reduces the vacuum pumping requirements by a factor of 10^4
so that even a small 6" diameter diffusion pump provides adequate pumping.
Spectral simplification due to expansion cooling allows the Raman spectra
of even complex molecules to be resolved, improves the signal-to-noise
level and introduces the possibility of Raman studies of molecular complexes
formed by the expansion process. Saturation effects were observed.
Improvements in both the pulsed laser sources and the pulsed nozzle should
greatly enhance the ease of pulsed supersonic expansion spectroscopy.

Our investigation of the fluid flow properties of a steady state
supersonic jet has shown that cw CARS is readily performed in this
simplest of flows. The ease of construction and operation of the steady
state supersonic jet should lead to its use in a much wider range of
spectroscopic studies.

Acknowledgements

This work was supported by the National Science Foundation under
Grant #CHE79-12673 and by the National Aeronautics and Space Administration
under Grant #NCC2-50.

References

1. M.D. Duncan and R.L. Byer, IEEE J. Quant. Electron. QE-15, 63 (1979).

2. W.R. Gentry and C.F. Giese, Rev. Sci. Instrum. 49, 595 (1978).

3. R.L. Byer and M.D. Duncan. J. Chem. Phys. 74, 2174 (1981).

4. Y.K. Park, G. Giuliani and R.L. Byer, Opt. Letts. 5, 96 (1980).

5. P. Drell and S. Chu, Opt. Commun. 28, 343 (1979).

6. M.D. Duncan, P. Oesterlin and R.L. Byer, Optics Letters, 6, 90 (1981).

7. P. Huber-Walchi, D.M. Guthals and J.W. Nibler, Chem. Phys. Letts. 67, 233 (1979).

8. M.A. Henesian and R.L. Byer, "CARS Spectroscopy : Theory and Experiment", presented at the XIth International Quantum Electronics Conference, Boston, Mass. June 1980.

9. J.J. Valentini, P. Esherick and A. Owyoung, Chem. Phys. Letts. November 1980.

10. H. Ashkenas and F.S. Sherman in IVth International Symposium on Rarefied Gas Dynamics, J.H. deLeeuw ed. (Academic Press, New York 1966) vol. 2, p.84.

11. T.A. Milne and F.T. Greene, Adv. High Temp. Chem. 2, 107 (1969).

12. M.D. Duncan, P. Oesterlin, F. Konig and R.L. Byer, "Observation of Saturation Broadening of the CARS Spectrum of Acetylene in a Pulsed Molecular Beam", (to be published in Chem. Phys. Letts.).

13. A. Owyoung and P. Esherick, Optics Letters, 5, 421 (1980).

14. M.A. Henesian, M.D. Duncan, R.L. Byer and A.D. May, Optics Letters, 1, 149 (1977).

Spectroscopy of Higher Order CARS and CSRS

L. Songhao, W. Fugui, Y. Bingkun, L. Min, C. Yisheng, and Z. Fuxin

Shanghai Institute of Optics and Fine Mechanics, Academia Sinica, P.O.Box 8211 Shanghai, China

We report here the experimental results of higher order coherent Raman scattering (CRS) in the calcite with a YAG laser CARS system. The 1st to 6th order CARS and 1st to 2nd order CSRS were observed. The cascade four-wave mixing is the dominant process.

In the study of higher order CARS of liquids and solids, observation of up to the 4th order CARS spectra has been reported and some theoretical analysis have been made [1,2].

Our experimental arrangement is shown in the schematic of Fig 1. The light source is a YAG laser of 30 mj, 10 ns with a repetition rate of 5–10 Hz. The dye laser was tuned over 5600 Å –6200 Å range with an output of 12.6 kW.

In our experiment the 6th order CARS have been recorded. The near filed patterns of scattered light from calcite are shown in Fig 2 and Fig 3. Higher order CARS and CSRS distribute almost symmetrically around ω_1 and ω_2 with CARS on one side and CSRS on the other side, forming a series of crescents. As the order increases, each crescent is stretched and becomes narrower and narrower. The intensity of the scattered light decreases rapidly with the increasing order. The crescent

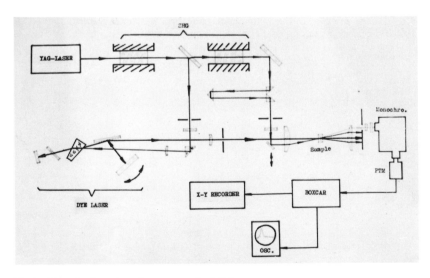

Fig. 1 Schematic of experimental arrangement for CARS

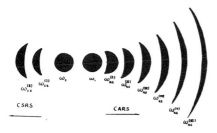

Fig. 2 Schematic of CRS near field pattern

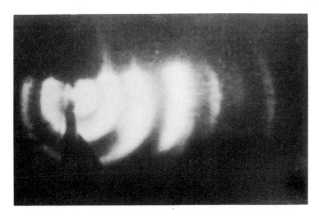

Fig. 3 CRS near field pattern

series of scattered beams span over the whole visible spectrum: red, orange (CSRS), yellow (ω_2), green (ω_1), indigo, blue, violet (CARS) region.

Below the stimulated scattering threshold of calcite, only the 1st to 5th order CARS and the 1st to 2nd order CSRS are observable. The stimulated scattering ring appears as the pumping power increases, CARS and CSRS show up simultaneously by the two sides of the ring. At this time, CARS intensity saturates, especially in the 1st order and the 2nd order, and meanwhile the 6th order CARS appears in this spectra. (Fig. 3).

Whenever the tunable beam ω_2 is blocked, the coherent scattering of all orders will disappear at once, whereas the stimulated scattering ring remains. In our experiment, the stokes back-scattering of ω_1 is used to monitor the ω_1 power to keep it below the sample stimulated scattering threshold.

Our experimental results show that higher order CARS and CSRS and their modes are directly related to the pumping light [3] [4]. CARS and CSRS can easily be produced with relatively high conversion efficiency when a single mode pumping light is used.

With multimode pumping light, the 1st order CARS and CSRS appear in multimode structure corresponding to the pumping light. Higher order CARS break into multiline structures (with CARS of each order changing into several crescent lines). The higher the order, the narrower the lines become, and further apart they get.

Fig. 4 shows a near field pattern of the 1st and 2nd order CARS and the 1st order CSRS produced

Fig. 4
Near field pattern of the 1st to 2nd
order CRS by multimode pumping
light

by multimode pumping light. The multimode structure of CARS and CSRS and the splitting of the 2nd order CARS crescent is observed.

Taking into consideration the spatial characteristics of higher order CARS and CSRS, using a photo-electronic system, the resonance curves of $I_{as}^{(N)}$ (intensity of CARS) against $\Delta\omega$ ($\omega_1-\omega_2$) for 1st to 3rd order CARS and 1st to 2nd CSRS taken with a x–y recorder are shown in Fig 5.

From the profile it can be seen that the resonance peak of the calcite appears to be at 1086 cm⁻¹ of the Raman mode, the characteristic asymmetry is observed.

Conservation of energy and momentum must be satisfied for CARS and CSRS of all orders. For this, accurate adjustment of phase matching is required, ω_2 must be carefully tuned into resonance and the two light beams are synchronized both temporally and spatially. Meanwhile the light intensity ratio is adjusted to 3:1. At that condition, optimum results are observed.

The F–P interferogram is shown in Fig 6 using F–P etalon to measure the 1st order CARS linewidth. Its value (0.37 Å) is of the same order of magnitude as the pumping light.

In our investigatin on higher order CARS and CSRS, the existence of backward CSRS has been also noticed, which is now under study.

Different processes can be produced in CARS and CSRS. The relations for corresponding frequencies, wave vectors and intensities of CARS spectra are as follows:
tensities of CARS spectra are as follows:

$$\omega_{as}^{(N)} = (N+1)\,\omega_1 - N\,\omega_2$$

$$\vec{K}_{as}^{(N)} = (N+1)\,\vec{K_1} - N\,\vec{K_2}$$

$$I_{as}^{(N)} \propto |\,X^{(3)}\,|^{2N}\,I_1^{(N+1)}\,I_2^{N}\ \text{(Cascade)}$$

$$I_{as}^{(N)} \propto |\,X^{(2N+1)}\,|^2\,I_1^{(N+1)}\,I_2^{N}\ \text{(Multi-Raman or Multi-photon)}$$

The expressions for CSRS is similar to above CARS equations.

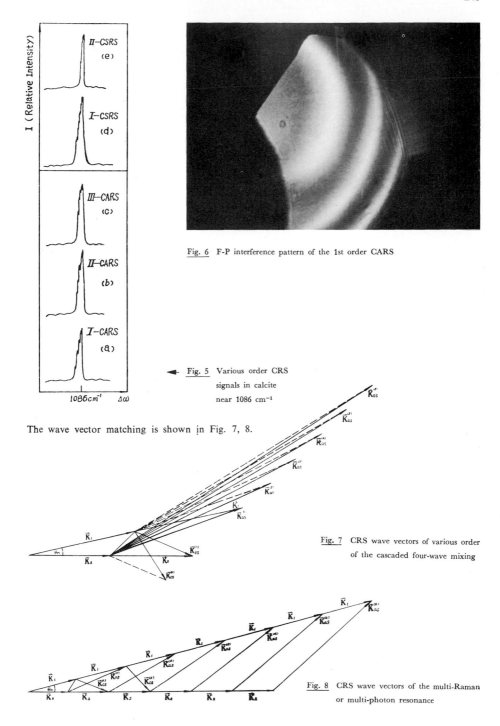

I (Relative Intensity)

II–CSRS (e)

I–CSRS (d)

III–CARS (c)

II–CARS (b)

I–CARS (a)

1086 cm⁻¹   Δω

Fig. 6   F-P interference pattern of the 1st order CARS

Fig. 5   Various order CRS signals in calcite near 1086 cm⁻¹

The wave vector matching is shown in Fig. 7, 8.

Fig. 7   CRS wave vectors of various order of the cascaded four-wave mixing

Fig. 8   CRS wave vectors of the multi-Raman or multi-photon resonance

All our results led us to explain the production of higher order CARS and CSRS by cascade four-wave mixing. Our experiments have verified the consistency of such explanation with others' work.

Reference

[1] I. Chabay, G. K. Klauminzer, B. S. Hudson, *Appl. Phys. Lett.* **28**, 27 (1976)

[2] A. Compaan, E. W. Avnear, S. Clandra, *Phys. Rev.* (*A*), **17**, 1083 (1978)

[3] M. A. Yuratich, *Mol. Phys.* **38**, 625 (1979)

[4] R. L. St. Peters, *Optics Lett.* **4**, 403 (1979)

Multiphoton Free-Free Transitions

A. Weingartshofer, J.K. Holmes, and J. Sabbagh*

Department of Physics, Saint Francis Xavier University
Antigonish, Nova Scotia, B26 1CO, Canada

In an electron-atom (molecule) scattering process the electron changes its direction; it is accelerated and so it can emit electromagnetic radiation, so-called Bremsstrahlung. For electrons with energies in the order of 10eV the spontaneous emission of a photon with a measurable energy has such a low probability that it is totally negligible. However, in a sufficiently intense external laser field we can have induced emission and absorption; and if the intensity is moderately high, very distinct non-linear processes can be observed in which the scattered electrons change their energy by integer multiples of the photon energy. Because in such transitions the electron proceeds from one continuum state to another, these processes are also called free-free (F-F) transitions [1,2] and can be represented as

$$e^-(E_i) + \text{Atom} + \text{Laser} \quad \rightarrow \quad e^-(E_i \pm N\hbar\omega) + \text{Atom} + \text{Laser} . \qquad (1)$$

It is of interest to know the dependence of the F-F cross-section on the incoming electron energy E_i, the electron scattering angle θ, the laser frequency ω, the laser flux density F, the direction of the laser polarization vector $\vec{\varepsilon}$ relative to the electron momenta $\vec{p}_i$ and $\vec{p}_f$ and also the properties of the target, i.e., how does a possible structure like a resonance of the electron target scattering influence the F-F cross-section.

To answer these questions we examine an expression that has been derived in the low frequency limit [3,4] and that permits a spatial separation of the *electron-target* and the *electron-laser* interactions giving a simple physical insight into the problem which is illustrated in Fig. 1. A free electron with momentum $\vec{p}_i$ and energy E_i enters the laser interaction region where it changes its energy by k photons with the amplitude equal to the Bessel function of order k and argument $-\alpha$. It then reaches the region of potential V (it is assumed that the wavelength of the laser is long in comparison to atomic dimensions) and scatters on the target, changes its angle θ and may also excite the target (we assume no interaction between the laser and the atom). The electron on leaving the potential region V can interact again with the laser field and change its energy by N-k photons with an amplitude equal to the Bessel function of order N-k and argument β. In the final state we have a free electron with momentum $\vec{p}_f$ and energy E_f. To obtain the complete scattering amplitude we have to sum over all possible intermediate states obtaining the following general expression

$$f^N(E_i,\theta) = \sum_{k=-\infty}^{+\infty} J_{N-k}(\beta)\, f^{el}(E_i+k\hbar\omega,\theta)\, J_k(-\alpha) . \qquad (2)$$

*Also at: Laboratoire de Recherche en Optique et Laser, Département de Physique, Université Laval, Québec, Québec, Canada

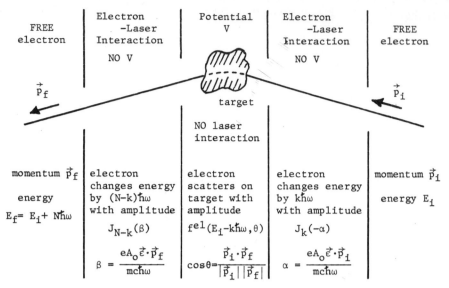

Fig.1 Diagram of low frequency approximation for free-free transitions

In our experiments we have been investigating until now the special case of elastic scattering where f^{el} is independent of the energy. Under these circumstances the amplitude in (2) reduces to a simpler form which after squaring gives the well-known formula for the differential F-F cross-section [4]

$$\frac{d\sigma^N}{d\Omega} \text{ F-F}(E_i,\theta) = \frac{P_f}{P_i} J_N^2(\beta-\alpha) \frac{d\sigma^{el}}{d\Omega}(E_i,\theta) . \tag{3}$$

To get a feeling for these expressions it is essential to consider numerical values for the argument of the Bessel function (disregarding the geometrical dependence). Non-linear effects become only important when the argument is greater than 1. In Table 1 we present some typical values of interest in our experiments with $E_i = 10$ eV. Three different laser photons are considered in column one. Calculated arguments of the Bessel function, corresponding to the laser flux densities shown in column two, are given in column three. We notice, in general, that the argument is proportional to $\sqrt{F}$, to $1/\omega^2$, and to $\sqrt{E_i}$ ($E_i = p^2/2m$). Our choice to use a TEA CO_2 laser was based on these considerations. The peak flux density of 10^8 Watts/cm^2 (after focusing) of our "Lumonics" laser is adequate.

The differential F-F cross-section in (3) leads to the differential sum rule

$$\Sigma_N d\sigma_{F-F}^N/d\Omega = d\sigma^{el}/d\Omega . \tag{4}$$

It expresses the law of conservation of scattering probability and has been the object of detailed theoretical analysis [5-7]. We illustrate the practical application of the sum rule in Fig.2 which represents a typical multi-photon F-F spectrum obtained in our laboratory. The laser was polarized

Table 1 Values of the argument of the Bessel function for three types of laser photons and varying flux densities, $E_i = 10$ eV

$\hbar\omega$	$\dfrac{A_o^2\omega^2}{8\pi c} = F[W/cm^2]$	$\dfrac{e\,A_o\,\lvert\vec{p}\rvert}{mc\hbar\omega} = \dfrac{e\sqrt{8\pi cF}\,\lvert\vec{p}\rvert}{mc\hbar\omega}$
117 meV	10^6	0.25
CO_2 laser	10^8	2.5
	10^{10}	25.0
1.2 eV	10^8	0.025
Neodymium laser	10^{10}	0.25
	10^{12}	2.5
1.8 eV	10^{10}	0.1
Ruby laser	10^{12}	1.0
	10^{14}	10.0

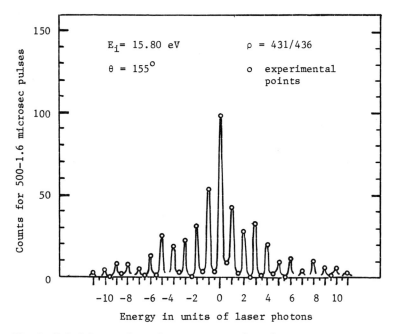

Fig. 2 Multiphoton free-free spectrum for electron-argon scattering. ρ expresses the sum rule in (4): sum of counts in multiphoton peaks/counts in elastic peak

linearly in the scattering plane and operated in multimode optical configuration. The energy resolution of the electron spectrometer (less than 40 meV) is sufficient to show distinct F-F transition quantum states. This allows a direct test of the validity of the sum rule by summing up the peak counts over all detected electron energies, $E_f = E_i \pm N\hbar\omega$ and comparing this sum to the counts in the elastic peak measured without laser. It is remarkable to find that in Fig.2 the sum of all the peaks from -11 to +11 photons agrees within experimental error with the counts in the elastic peak: the ratio of these two numbers is ρ as shown in Fig.2.

We have now turned our attention to the problems outlined in the second paragraph of this text. In particular, the case where f^{el} has an isolated resonance and, therefore, is strongly dependent on the energy. Theory predicts that the sum rule does not apply in this case.

We gratefully acknowledge the continuous support of the Natural Sciences and Engineering Research Council of Canada, and thank the Saint Francis Xavier Council for Research for its support. One of us (J. S.) expresses appreciation to the Laboratoire de Recherche en Optique et Laser for financial assistance and to Professor S. L. Chin for advice and encouragement.

References

1 A. Weingartshofer, J. K. Holmes, G. Caudle, E. M. Clarke, and H. Krüger, Phys. Rev. Lett. 39, 269 (1977)
2 A. Weingartshofer, E. M. Clarke, J. K. Holmes and C. Jung, Phys. Rev. A19, 2371 (1979)
3 H. Krüger, and M. Schultz, J. Phys. B 9, 1899 (1976)
4 H. Krüger, and C. Jung, Phys. Rev. A 17, 1706 (1978)
5 C. Jung, Phys. Rev. A 20, 1585 (1979)
6 C. Jung, Phys. Rev. A 21, 408 (1980)
7 P. Zoller, J. Phys. B 13, L249 (1980)

Review articles:

a S. Geltman, J. of Research NBS 82, 173 (1977)
b M. Gavrila, and M. Van der Wiel, Comments Atom. Mol. Phys. 8, 1 (1978)
c A. Weingartshofer and C. Jung, Phys. in Canada 35, 119 (1979)
d D. Andrick, Electronic and Atomic Collisions, N. Oda and K. Takayanagi, editors (North-Holland, Amsterdam, 1980) pp. 697-704

Precision Studies in 3-Level Systems

B.W. Peuse, R.E. Tench, P.R. Hemmer, J.E. Thomas, and S. Ezekiel

Research Laboratory of Electronics, Massachusetts Institute of Technology
Cambridge, MA 02139, USA

We have performed precision high resolution studies of the interaction
of two monochromatic laser fields with a cascade 3-level system in an
atomic beam and with a folded 3-level system in a vapor. Such experimental studies provide important and fundamental information about the nature
of atom-field interaction and can be compared directly with theoretical
predictions [1-3].

Study of 3-level system in an atomic beam

Our atomic beam studies consist of a highly collimated atomic beam of Na
and two frequency stabilized dye lasers. The sodium atoms are prepared in
the $3^2S_{1/2}(F=2, m_F=2)$ sublevel, i.e., level #1 in Fig. 1(a), by optical
pumping in a separate interaction region. These atoms then interact with
two counterpropagating, circularly
polarized laser fields perpendicular to
the atomic beam. One laser, at 5890 Å,
provides the driving (or pumping) field
ω_1 between the $3^2S_{1/2}(F=2, m_F=2)$ and
$3\ ^2P_{3/2}(F=3, m_F=3)$ levels corresponding
to levels 1 and 2 in Fig. 1(a). The
second laser, at 5688 Å, provides the
probing field ω_2 between the $3^2P_{3/2}(F=3,
m_F=3)$ and $4^2D_{5/2}(F=4, m_F=4)$ levels corre-
sponding to levels 2 and 3 in Fig. 1(a).
Both laser beams are right circularly
polarized, insuring that only the three
levels mentioned are involved in the
interaction. Absorption of the probe is
measured by monitoring fluorescence from
the 4D level.

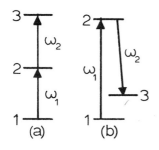

Fig.1 Cascade and folded
3-level systems

The dye lasers used are first short term frequency stabilized by locking
them to passive Fabry-Perot reference cavities. The laser providing the
pump field, which is also used for optical pumping, is long term stabilized
by locking it to the $1-2$ transition in Na. The long term frequency stability of the pump laser is transferred to the probe laser via a reference
cavity. In this way the probe laser can be tuned very precisely by means
of an acousto-optic modulator. The details of the experimental setup have
been described elsewhere [2].

Density matrix calculations predict [1-3], for low pump and probe intensities, a Lorentzian lineshape for probe absorption whose FWHM is equal to

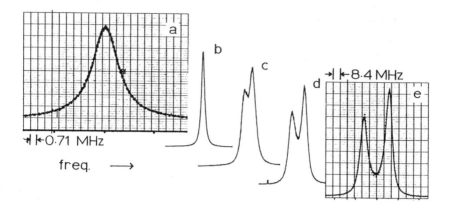

Fig. 2   Probe lineshape data for cascade system. Pump intensity:
         (a) 130 μW/cm^2, (b) 2 mW/cm^2, (c) 25 mW/cm^2, (d) 46 mW/cm^2,
         (e) 71 mW/cm^2

the sum of the widths of the first and third levels, but completely independent of the width of the intermediate level. For our system, level 1 is
a ground state with essentially zero width, level 2 has a width of 10.0 MHz,
and level 3 a width of 3.15 ± 0.03 MHz [4]. A typical low-power lineshape
is displayed in Fig. 2(a), where the FWHM linewidth is measured to be 3.18
MHz. The superimposed dotted curve in the figure is the corresponding theoretical lineshape calculated with the measured pump and probe intensities
and is in good agreement with the data.

    For high pump intensities, calculations predict a symmetrical splitting
of the absorption lineshape into two peaks separated by the Rabi frequency.
The observed probe lineshapes are shown in Fig. 2(b-e) for various pump intensities. The observed splitting is clearly asymmetric, in contrast with
the data of Ref. 3. The pecularities of the observed strong field lineshapes may be explained by including in the calculations the effects of the
many momentum conserving atomic recoils that must take place as the atom
traverses the pump field [5]. Figure 2(e) shows the calculated recoil modified lineshape (dotted curve) which is clearly in good agreement with the
data. It should be noted that the recoil from both the pump beam and the
optical pumping beam was included in this calculation.

    Our study also included [6] the case in which the pump laser was held at
a fixed detuning from the 1 — 2 transition. In order to correctly interpret
the observed lineshape, atomic recoil from both the pump beam and the optical pumping beam had to be again included in the calculations. However, as
the detuning is made larger the recoil contribution from the pump field becomes less significant.

Study of 3-level system in a vapor

We have also studied the interaction of two monochromatic fields with a
folded, Doppler-broadened [2] three-level system shown in Fig. 1(b). These
experiments were performed with copropagating pump and probe fields in a
150 cm long cell containing I$_2$ vapor. Here, the pump laser is an argon

laser at 5145 A which has spectral width of 3 kHz and is long term stabilized to an I_2 hyperfine resonance in another cell. The probe laser is a frequency-stabilized dye laser at 5828 Å. The long term stability of the pump at ω_1 is transferred to the probe at ω_2 in the manner described in the previous section. Probe absorption is measured by modulating the intensity of the pump field at 2kHz and synchronously detecting the probe field.

When the pump is weak, the probe lineshape for a folded Doppler-broadened system is calculated [2] to be a Lorentzian with a width (FWHM) Γ_p given by

$$\Gamma_p = \Gamma_1 + \Gamma_2(1 - \frac{\omega_2}{\omega_1}) + \Gamma_3(\frac{\omega_2}{\omega_1}) \quad .$$

Note that in this case there is a contribution from Γ_2. In the I_2 molecule under study, levels 1 and 3 are hyperfine components of the $v'' = 0$, $J'' = 15$ and $v'' = 11$, $J'' = 15$ rovibrational levels in the ground electronic state. These levels have essentially zero natural width, so Γ_1 and Γ_3 are limited by transit time. Level 2 is one of the hyperfine components of the $v' = 43$, $J' = 16$ level in the first excited electronic state with $\Gamma_2 \approx 100$ kHz. With beams of ≈ 1 cm dia. the calculated probe linewidth Γ_p is approximately 17kHz. Figure 3(a) shows the experimental weak field lineshape with a FWHM of 65 kHz which at present is determined primarily by the spectral width of the dye laser.

For intense pump fields, the calculated probe lineshape is no longer a Lorentzian but splits into two peaks due to the velocity averaged ac Stark effect. The solid line in Fig. 3(b) shows the experimental lineshape for a pump intensity of 8.4 W/cm^2. In order to apply the steady state theory to our experiment, we must sum over contributions from all rotational M-levels and integrate over the Gaussian intensity distribution of the pump [7]. The dotted curve in Fig. 3(b) shows a least-squares fit to the data. As can be seen, the fit is reasonably good except for a deviation in the low frequency tail due to the influence of a nearby I_2 hyperfine component not considered in the fit.

The above data was taken by detecting the full probe beam. However, when the detector was apertured so that only the central, uniform intensity region of the probe beam was demodulated, a strongly asymmetric lineshape was obtained as shown in Fig. 4(a). This asymmetry was found to vary as a function

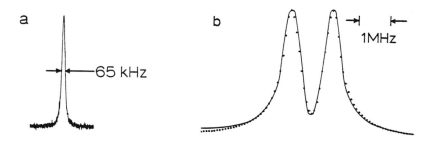

Fig. 3 Probe lineshape data for folded Doppler-broadened system. Pump intensity: (a) 4 mW/cm^2, (b) 8.4 W/cm^2

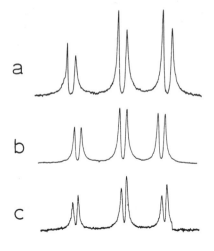

a

b

c

Fig. 4

Probe lineshapes as a function of apertured detector position Δ from beam center. (a) $\Delta = 0$, (b) $\Delta = 0.5 \, w_0$, (c) $\Delta = 0.9 \, w_0$. w_0 is beam radius.

of detector position (Fig. 4(b,c)). We attribute this effect to self-focusing of the pump beam and we are presently studying it in more detail.

This work was supported by the National Science Foundation and the Joint Services Electronics Program.

References

[1] S.H. Autler and C.H. Townes, Phys. Rev. 100, 703 (1955); M.S. Feld and A. Javan, Phys. Rev. 177, 540 (1969); T. Hansch and P. Toschek, IEEE J. Quant. Elect. 4, 467 (1968); B.J. Feldman and M.S. Feld, Phys. Rev. A5, 899 (1972); I.M. Beterov and V.P. Chebotayev, Progress in Quantum Electronics, edited by J.H. Sanders et al, Pergamon Press 1973; C. Delsart and J.C. Keller, J. Phys. (Paris) 39, 350 (1978); B.R. Mollow, Phys. Rev. A5, 1522, (1972); R.M. Whiteley and C.R. Stroud, Phys. Rev. A14, 1498 (1976); C. Cohen-Tannoudji and S. Reynaud, J. Phys. B10, 345 (1977).

[2] R.P. Hackel and S. Ezekiel, Phys. Rev. Lett. 42, 1736 (1979) and references therein.

[3] H.R. Gray and C.R. Stroud, Opt. Commun. 25, 359 (1978).

[4] W.L. Wiese, M.W. Smith, and B.M. Miles, Atomic Transition Probabilities, vol. 2, NSRDS, NBS (1969).

[5] P.R. Hemmer, F.Y. Wu, and S. Ezekiel, Opt. Commun., to be published.

[6] P.R. Hemmer, B.W. Peuse, F.Y. Wu, J.E. Thomas, and S. Ezekiel, to be published.

[7] R.E. Tench, J.E. Thomas, and S. Ezekiel, to be published.

Non-Linear Hanle Effect: Saturation Parameters and Power Enhancements in Lasers

F. Strumia, M. Inguscio, and A. Moretti

Istituto di Fisica dell'Università di Pisa, Piazza Torricelli 2
I-56100 Pisa, Italy

The use of intense laser sources has opened the possibility of observing the non-linear version of well known effects of the linear spectroscopy. The non-linear Hanle effect in the past years has been observed in the visible and infrared both on atoms and molecules. Recently it was also introduced to explain the power enhancements observed on optically pumped three-level lasers [1].

The non-linear Hanle effect is caused by the different saturation of the M components of an inhomogeneous line in presence or not of an external field. Differently from the conventional linear case, the effect can be observed directly as a change in the absorption of the saturating laser radiation. As is well known, in presence of saturation the absorption coefficient of a Doppler broadened line can be written

$$I(\omega) = \frac{\alpha |\mu|^2}{\sqrt{1 + \sigma |\mu|^2}} e^{-\left[(\omega-\omega_0)/\Delta_D\right]^2} = \frac{\alpha |\mu|^2}{\sqrt{1 + I/I_s}} e^{-\left[(\omega-\omega_0)/\Delta_D\right]^2} \tag{1}$$

where μ is the matrix element of the transition, ω_0 its center frequency, Δ_D the Doppler linewidth and σ a coefficient proportional to the intensity of the saturating beam. $I/I_s = S$ is defined as the saturation parameter. When the degeneracy of the M sublevels is removed by an external field each transition saturates independently. As an example for a transition $J = 0 \rightarrow 1$ and a laser beam linearly polarized orthogonal to the static field (selection rule $\Delta M = \pm 1$) the absorption is given by eq. (1) when the static field intensity is negligible and by

$$I^F(\omega) = \frac{\alpha |\mu^+|^2}{\sqrt{1 + \sigma |\mu^+|^2}} e^{-\left[(\omega-\omega_0^+)/\Delta_D\right]^2} + \frac{\alpha |\mu^-|^2}{\sqrt{1 + \sigma |\mu^-|^2}} e^{-\left[(\omega-\omega_0^-)/\Delta_D\right]^2} \tag{2}$$

when the degeneracy of the levels is removed. Since $|\mu|^2 = |\mu^+|^2 + |\mu^-|^2$, we have an increase in the absorption given by

$$R(S) = \frac{I^F}{I} = \sqrt{\frac{1 + 2S}{1 + S}} \qquad (3)\ddot{u}$$

assuming Δ_D much larger than the homogeneous linewidth Δ_H. R ranges from 1 to $\sqrt{2}$ for $S = 0$ and $S \to \infty$, respectively. As a consequence the non-linear Hanle effect is as large and easily detectable effect as a change in the absorption.

For intermediate static field intensities the coherence between the M sublevels must be taken in account. A density matrix formalism treatment was performed by Feld et al. [2]. It is worth noting that in the case of molecular transitions large J values are generally involved and the density matrix method introduces lengthly and often meaningless calculations. In contrast, a simple calculation of R defined by eq. (3) can be performed as a function of J and ΔJ selection rule, using the Clebsh-Gordon coefficients formalism for the two limiting cases. In absence of the static field the quantization axis is defined by the polarization and direction of the laser beam. For a linearly polarized beam and an electric dipole transition the direction is parallel to the oscillating electric field. The saturated absorption can then be computed by summing over all the $\Delta M = 0$ transitions with the corresponding $|\mu_M|^2$ values.

When the applied static field is large enough to resolve the M sublevels and is orthogonal to the beam direction the saturated absorption is computed summing over all the $\Delta M = \pm 1$ transitions respectively for a parallel or orthogonal polarization. It is worth noting that in the case of monochromatic beam and of an inhomogeneous line the $\Delta M = -1$ and the $\Delta M = +1$ transitions starting from a given M sublevel involve molecules with a different velocity component along the beam direction, i.e., different molecules. We assume also $\Delta_H << \Delta_D$, which for infrared and visible lines and low densities is true. The exponential terms in eq. (1) and eq. (2) can be neglected and a simple expression for R can be obtained for the two limiting cases. For a $J \to J$ transition (Q branch) we find for $\Delta M = \pm 1$ selection rules:

$$R_Q(S) = \frac{\displaystyle\sum_{-J}^{+J} \left(\frac{J(J + 1) - M(M - 1)}{\sqrt{1 + \frac{3}{4} S \frac{J(J + 1) - M(M - 1)}{J(J + 1)}}} + \frac{J(J + 1) - M(M + 1)}{\sqrt{1 + \frac{3}{4} S \frac{J(J + 1) - M(M + 1)}{J(J + 1)}}} \right)}{4 \displaystyle\sum_{-J}^{+J} \frac{M^2}{\sqrt{1 + S \frac{3M^2}{J(J + 1)}}}} \qquad (4)$$

where a proper normalization has been introduced in order to have for S the same expression as in eq. (1). Similar equations can be written also for P and R branches' lines, in particular yielding R_p $(J + 1; S) = R_R$ $(J;S)$. Obviously $R(S) = 1$ for the selection rules $\Delta M = 0$.

In Figure 1 the computed results are shown for several J values. It has been assumed that the intensity of the saturating beam is constant over the whole section. We have the same maximum absorption increase ($R = \sqrt{2}$) for Q_1, R_0, and P_1 lines. By increasing the J value the Q_J and the P_J or R_J lines show a rapid convergence to two limiting curves (marked ∞ in Figure 1).

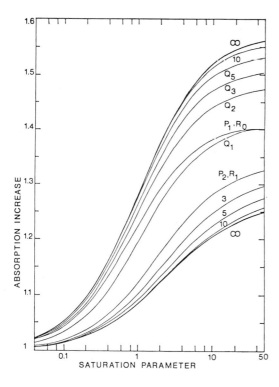

Figure 1

$R_0(J,S)$ increases with J up to a maximum value of 1.57 while $R_p(J + 1,S)$ = $R_R(J,S)$ decrease to an asymptotic value of 1.27. In both cases the effect is quite large and can be easily detected.

We measured the changes in the absorption for different molecular species and selection rules using a Stark-optoacoustic technique [1]. The linearly polarized radiation from a CO_2 laser was fed into a cell equipped with an electret microphone and parallel plate electrodes for applying a static electric field. The laser polarization could be rotated in order to induce

ΔM = 0 or ΔM = ± 1 transitions. The reduced dimensions of the cell allowed
assuming the optoacoustic signal to the absorption on a small optical length
(~ 1 cm). The sensitivity of the technique allowed work at low pressures
(a few tens of millitors) of the absorbing gas (CH_3OH, CH_3F). In all the
measurements the laser frequency was tuned at the center of the unperturbed
molecular line. A typical dependence of the absorption versus the electric
field (ΔM = ± 1 pump selection rule) is shown in Figure 2 for two different
CH_3OH molecular transitions. The upper part measurements refer to molecules
irradiated with the 9P(36) CO_2 laser line, inducing a transition of the Q
branch. Results are shown for two different CH_3OH pressures (47 and 90 mTorr
respectively for the upper lower curve). At low electric field (up to 0.6
KV/cm) a large absorption increase due to the non-linear Hanle effect, is
detected. The relative increase is lower for the higher pressure, as expected
from the saturation theory. The signal decrease at higher electric field
shows the Doppler profile of the absorption line. In the lower part of
Figure 2 measurements are reported by using the 9R(10) CO_2 laser radiation,
which is absorbed by CH_3OH via a ΔJ = 1 (R branch) transition. The gas pressure was 70 mTorr, the laser intensity was a factor 1.6 higher than in the
upper curves, nevertheless the Hanle effect absorption increase was lower,
as expected from the theoretical results shown in Figure 1. The S value can
be increased by lowering the gas pressure or increasing the laser power
density. Consistent with the different asymptotic behaviour, we have observed with our experimental apparatus a maximum value R = 1.48 for the
Q_{16} transition and R = 1.5 for the R_{26} transition. Measurements of R as a
function of S were carried out by changing the laser intensity I. The results
for the Q_{16} are shown in Figure 3 for two different values of the pressure
(92 mTorr the dot points and 133 mTorr the square points). The agreement
with the theoretical behaviour (continuous line) is good.

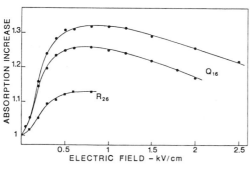

Figure 2

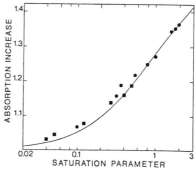

Figure 3

In conclusion we have shown that this new non-linear technique can be very useful for an easy and straightforward measurement of saturation parameters. It is also worth noting that in molecular spectroscopy, where in general high J values are involved, the technique can yield quantitative results even on unassigned transitions. In fact the theoretical results of Figure 1 show a fast convergence of the curves by increasing J. Moreover it is straightforward to assign the transition to Q or P or R branches, simply from the relative enhancement magnitude. A further consequence of the non-linear Hanle effect is an increase of the number of molecules excited in the upper state. That explains the power enhancement observed in FIR molecular laser optically pumped in the presence of small electric fields (1).

References

1. M. Inguscio, A. Moretti, F. Strumia: IEEE J. Quant. Electron. *QE-16*, 955 (1980) and references therein
2. M.S. Feld, A. Sanchez, A. Javan, B.J. Feldman: Proceedings of Aussois Symposium "Methodes de Spectroscopie sans Largeur Doppler de Niveaux Excités de Systemes Moleculaires Simples", May 1973 (Publ. N.217 du CNRS, Paris 1974) pp.87-104

Atomic Coherence Effects in Resonant Four-Wave Mixing Spectroscopy of Sodium

D.G. Steel, J.F. Lam, and R.A. McFarlane

Hughes Research Laboratories, 3011 Malibu Canyon Road
Malibu, CA 90265, USA

This paper presents preliminary experimental and theoretical results on two new aspects of sub-Doppler spectroscopy using resonant four-wave mixing and cw tunable dye lasers. The first technique uses degenerate four-wave mixing (DFWM) to study optically induced Zeeman coherences while the second technique uses nondegenerate FWM to study excited atomic states.

The starting point in the analysis of four-wave mixing processes is the quantum mechanical transport equation for the density operator ρ [1]. The perturbation calculation is carried out to include terms in the series expansion of ρ up to third order. In the calculations we assume counterpropagating pumps (E_f and E_b) and a nearly collinear probe (E_p). Phase-matching results in the signal (E_s) counterpropagating to E_p. The analysis assumes a Maxwellian velocity distribution yielding after velocity integration a third order polarization in terms of Plasma Dispersion functions.

For a two-level system with degeneracies, we calculate the contribution to the DFWM signal arising from optically induced spatial modulation of atomic populations and atomic coherences (Zeeman coherence) between degenerate levels. Damping is included as a phenomenological decay from the two levels at rates γ_1 and γ_2. The frequency dependence of the signal (in the extreme Doppler limit and for small $\alpha_o\ell$) is given by $E_s = C\, E_f E_b E_p^* / (\gamma_{12} + i\Delta)$ where $\Delta = \omega - \omega_o$, $2\gamma_{12} = (\gamma_1 + \gamma_2)$ is the linewidth, and $\alpha_o\ell$ is the absorption length product. The Doppler-free property of the signal arises because only the $v=0$ velocity group can interact simultaneously with all four waves.

When all four beams are copolarized, the dipole selection rules imply that all four waves interact with the same transition ($\Delta m = 0, \pm 1$) resulting in a population dependent interaction. However, with the appropriate choice of pump and probe field polarizations, the presence of magnetic degeneracies allows the possibility of generating optically induced Zeeman coherences between degenerate states. Consider the choice of electric field polarization vectors given by $\hat{e}_f = \hat{\varepsilon}_+$, $\hat{e}_b = \hat{\varepsilon}_-$ and $\hat{e}_p = \hat{\varepsilon}_-$. The action of the forward pump E_f and input probe E_p, generates a spatial modulation of the Zeeman coherence between states $|\alpha J_\alpha M_\alpha\rangle$ and $|\alpha J_\alpha M_\alpha + 2\rangle$, which are connected to a common state $|\beta J_\beta M_\alpha + 1\rangle$. The four-wave mixing signal is generated via a scattering of the backward pump field E_b from the spatial interference, and it has an electric field polarization given by $\hat{\varepsilon}_+$. Physically, the Zeeman coherence induced by E_f and E_p can be viewed as an electric quadrupole. The Zeeman coherence is susceptible to depolarizing collisions in contrast to the population which experiences velocity changing collisions.

In the case $\hat{e}_f = \hat{e}_b = \hat{x}$ and $\hat{e}_p = \hat{y}$, the DFWM signal is polarized in the y-direction. The strength of the signal arising from both ground and excited state coherences depends on the respective life-times. The relative strengths are given by I_g and I_e and $C \propto (I_g + I_e)$. As an example for this polarization case, $I_g = (8\gamma_D)^{-1} \sum_{M2} |<J_2M_2|\mu|J_1M_2-1>|^2 |<J_2M_2|\mu|J_1M_2+1>|^2 -$

$(16\gamma_1)^{-1} \sum_{M2} [|<J_2M_2|\mu|J_1M_2+1>|^4 + |<J_2M_2|\mu|J_1M_2-1>|^4 -$

$|<J_2M_2|\mu|J_1M_2+1>|^2 |<J_2M_2+2|\mu|J_1M_2+1>|^2 - |<J_2M_2|\mu|J_1M_2-1>|^2|<J_2M_2-2|\mu|J_1M_2-1>|^2]$.

The first term is proportional to $\gamma_D^{-1} = (\gamma_1 + \gamma_{ph})^{-1}$ while the second term is proportional to γ_1^{-1}. γ_{ph} is determined by depolarization collisions. In general $I_e \neq I_g \neq 0$ but in the case $J=1 \to J=0$ if $\gamma_{ph}=0$ then $I_g=I_e=0$ and no DFWM signal is expected. In the presence of depolarizing collisions, $\gamma_{ph} \neq 0$, and therefore $\gamma_D \neq \gamma_1$ leading to a DFWM signal even in this special case. Similar results are obtained by Bloch and Ducloy [2].

For the second technique of nondegenerate FWM spectroscopy via a non-degenerate 3-level cascade-up system, we consider as a general example the case for which the backward pump oscillates at Ω_1 (which is resonant with transition frequency ω_{21}) and the forward pump and input probe oscillate at frequency Ω_2 (which is resonant with transition frequency ω_{32}). We assume that the level energies in the cascade-up 3-level system are given by $E_3 > E_2 > E_1$. The physical contribution to the FWM signal arises from the generation of a 2-photon coherence between levels $|1>$ and $|3>$ via the simultaneous action of E_f and E_b. In collinear geometry the resonance condition that must be satisfied in order for all four waves to interact with the same velocity group is given by $\Omega_2 - \omega_{32} = -(k_2/k_1)(\Omega_1 - \omega_{21})$. The signal E_s is proportional to

$$E_s \propto E_f E_b E_p^* \, e^{i(k_1 z + \Omega_1 t)} \left[\gamma_{13} + \left(\frac{k_2}{k_1} - 1\right)\gamma_{12} - i\left(\Delta_{32} + \frac{k_2}{k_1}\Delta_{21}\right)\right]^{-1}$$

$$\left\{ (ik_1 u_o)^{-1} Z' \left(-\frac{\Delta_{21} + i\gamma_{12}}{k_i u_o}\right) \right.$$

$$+ \left(\frac{k_2}{k_1} - 1\right)\left[\gamma_{13} + \left(\frac{k_2}{k_1} - 1\right)\gamma_{12} - i\left(\Delta_{32} + \frac{k_2}{k_1}\Delta_{21}\right)\right]^{-1} \left[Z\left(\frac{\Delta_{21} + \Delta_{32} + i\gamma_{13}}{\left(\frac{k_2}{k_1} - 1\right)k_1 u_o}\right) - Z\left(-\frac{\Delta_{21} + i\gamma_{12}}{k_1 u_o}\right)\right]\right\}$$

where $\Delta_{21} = \Omega_1 - \omega_{12}$, $\Delta_{32} = \Omega_2 - \omega_{32}$, $\gamma_{ij} = (1/2)(\gamma_i + \gamma_j)$ and Z is the plasma dispersion function. In the Doppler limit, the frequency response of E_s is a Lorentzian whose linewidth is given by the 2-photon linewidth γ_{13}. We are presently extending the theory to include the effects due to AC Stark shift [3].

The experimental studies of these effects were made in atomic sodium using a stabilized Coherent Inc. ring dye laser. Using an experimental configuration discussed earlier [1] we studied the optical field induced Zeeman coherence using counterpropagating pumps arranged to be linear and copolarized. With the probe beam linearly polarized and the axis of quantization assumed parallel to the pump electric field, the population dependent physics could then be isolated from the Zeeman coherence by setting the probe polarization either parallel or orthogonal to the pumps. Fig. 1a is a DFWM spectrum of the D_2 line ($\gamma=0.5890$ μm) with the pumps and probe copolarized showing the signal resulting from population physics. Due to hyperfine (hf) optical pumping, only two components appear: the $3^2S_{1/2}(F=1)-3^2P_{3/2}(F=3)$ transition and the $3^2S_{1/2}(F=1)-3^2P_{3/2}(F=0)$ transition. The latter transition shows a splitting due to saturation. The insets show spectra at pump intensities well below 6 mW/cm^2, the saturation intensity for the first transition. Rotating the probe polarization 90° we obtain the signal resulting from

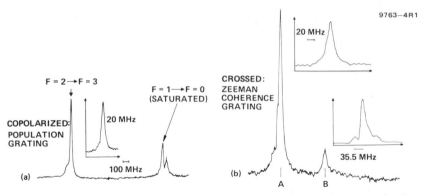

Fig.1   Frequency dependence of DFWM signal arising from population grating and Zeeman coherence

Zeeman coherence shown in Fig. 1b (arbitrary vertical scale). The low frequency transition (structure A) was observed as expected from the analysis and the high frequency transition was missing as predicted above. The central component (structure B) is a crossover resonance.

A more detailed analysis shows that the Zeeman signal should be reduced when upper or lower level coherence is removed. Due to the transit time and the finite laser linewidth, the ground state coherence is destroyed by application of a small magnetic field parallel to the pump electric field. Since the lower energy level splitting is of order 700 kHz/gauss, we expect a significant signal reduction with a field of a few hundred milligauss. The effect of the sudden drop-off is shown in Fig.2a which is an oscillogram of the Zeeman signal versus field. Fig.2b shows the ordinary population signal dependence on magnetic field. The rise in both signals is presently not understood. Note at higher fields near 10 gauss, corresponding to a line splitting on the order of the natural linewidth, both signals begin falling.

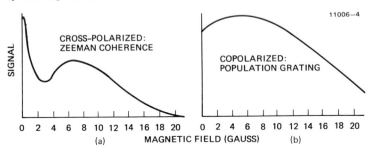

Fig.2   Dependence of DFWM signal on magnetic field

Demonstration of sub-Doppler nondegenerate four-wave mixing spectroscopy of excited states was achieved by coupling the $3^2S_{1/2}-3^2P_{3/2}$ transition to the $3^2P_{3/2}-4^2D_{5/2}$ and $3^2P_{3/2}-4^2D_{3/2}$ transitions. As an example of the potential usefulness of this coupling we show in Fig.3 an example of the resulting spectrum. Using two stabilized Coherent Inc. ring dye lasers, the forward pump and probe were tuned to the upper transition near 0.5688 μm while the backward pump was tuned to the lower transition at 0.5890 μm. A signal

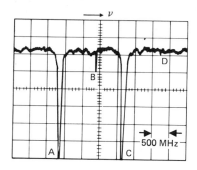

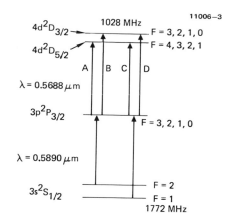

Fig.3 Frequency spectrum of optically induced two-photon coherence as observed by non-degenerate FWM. Level diagram shows observed transitions

was observed at 0.5890 µm as the 0.5688 µm laser was scanned. This interaction is perfectly phase matched for nearly collinear geometry. The four structures showing in Fig.3 are from the following transitions:
A, $3^2S_{1/2}(F=2)-3^2P_{3/2}(F=3)-4^2D_{5/2}(F=4)$; B, $3^2_{1/2}(F=2)-3^2P_{3/2}(F=3)-4^2D_{3/2}(F=3)$; C, $3^2S_{1/2}(F=1)-3^2P_{3/2}(F=0)-4^2D_{5/2}(F=1)$; D, $3^2S_{1/2}(F=1)-3P_{3/2}(F=0)-4^2D_{3/2}(F=0)$. The other hf transitions were not observed because of optical pumping. The linewidth of the intense transitions was less than 10 MHz comparable to the expected two-photon linewidth.

It is a pleasure to acknowledge useful discussions of these results with Professor T.W. Hänsch of Stanford University. This work was supported in part by the Army Research Office under Contract No. DAAG29-81-C-0008.

References

1. J.F. Lam, D.G. Steel, R.A. McFarlane, and R.C. Lind, Appl. Phys. Lett. **38** June 15, 1981.
2. D. Bloch and M. Ducloy (private communication).
3. T.W. Hänsch and P. Toschek, Z. Physik **236**, 213 (1970).

On the Validity of the Judd-Ofelt Theory for Two-Photon Absorption in the Rare-Earths

M. Dagenais

Advance Technology Laboratory, GTE Laboratories, 40 Sylvan Road
Waltham, MA 02254, USA, and
G. McKay Laboratory, Harvard University, 9 Oxford St.
Cambridge, MA 02138, USA, and

M. Downer, R. Neumann*, N. Bloembergen

G.McKay Laboratory, Harvard University, 9 Oxford St.
Cambridge, MA 2138, USA

The one-photon absorption theory of JUDD and OFELT[1] has been rather succes-
ful in describing the oscillator strength of many forced electric-dipole
f-f transitions in the trivalent rare-earths.[2,3] Such a success has very
often been taken as a vindication of all the assumptions behind the Judd-
Ofelt theory when really it shows only that the formalism of representing
the intensity of an f-f transition by the sum of phenomenological parameters multiplied by the squares of tabulated reduced matrix elements
$U^{(\lambda)}$ (λ = 2, 4, 6) is valid. There are cases where the Judd-Ofelt theory of
one-photon absorption does not lead to accurate predictions. Some well-
known examples are given by the hypersensitive transitions.[2] These transitions are characterized by large values of $U^{(2)}$ matrix elements and are in
general rather sensitive to the environment of the lanthanide ions. Since
the Judd-Ofelt theory has also its weak points, it is very important to
design new experiments to pinpoint its limitations. We would like to report
here on a completely new test of the Judd-Ofelt theory. It is based on a
theoretical calculation of AXE[4] on two-photon absorption in the rare-earths.
His work is a direct extension of the Judd-Ofelt theory for one-photon
absorption to two-photon absorption. The theoretical predictions on relative two-photon absorption cross-sections are even simpler and involve only
the $U^{(2)}$ reduced matrix elements. It should be noted here that both the Axe
and the Judd-Ofelt theories ignore the band character of the 5d states. In
our work, we have observed direct two-photon f-f transitions in the Gd^{3+}

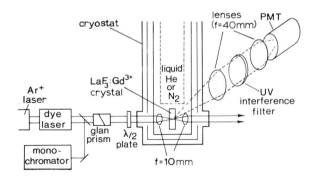

Fig. 1   Schematic
diagram of the experi-
mental arrangement

*On leave from Physikalisches Institut, D-6900 Heidelberg, Fed. Rep. of
Germany

ions in a LaF$_3$ host, using a CW linearly polarized laser beam of about 100 mW power.

Figure 1 shows the experimental arrangement. The LaF$_3$ crystal, which contained 0.5% Gd^{3+}, was mounted with its c axis either parallel or penpendicular to the direction of beam propagation. Two-photon transitions were induced between the $^8S_{7/2}$ ground state of Gd^{3+} and the lowest lying excited multiplets $^6P_{7/2}$ at 32230 cm^{-1}, $^6P_{5/2}$ at 32790 cm^{-1}, and $^6P_{3/2}$ at 33360 cm^{-1} above the ground state. The relevant intermediate states, which belong mostly to the $4f^6$ 5d configuration, lie about 100,000 cm^{-1} above the ground state. UV fluorescence from one-photon decay back to the ground state was observed at right angles to the laser beam with a suitably filtered EMI 9635 photomultiplier tube. A rejection ratio against scattered light larger than 10^{12} was obtained. The observed fluorescence is emitted almost entirely from the lowest-energy crystal-field component of the $^6P_{7/2}$, since higher-lying ^6P levels decay to this level through rapid non-radiative processes. Fluorescence yield of the lowest level, on the other hand, approaches unity, since, because of the large energy gap separating it from the ground state, radiative decay dominates. Nevertheless, to check that fluorescence following excitation of different excited states was not selectively quenched, which would lead to erroneous measurements of relative transition intensities, single-photon absorption and excitation spectra were taken with a Cary 14 spectrophotometer and found to yield identical relative intensities to all of the ^6P_J and ^6I_J excited 4f states (except for an inconclusive result with the $^6P_{3/2}$, which was too weak to be observed by direct one-photon absorption).

Two-photon intensity measurements were made at both liquid-nitrogen and liquid-helium temperatures. As suspected, no temperature dependence was observed, since the ground-state splitting is less than 0.3 cm^{-1},[5] so that all Stark components are essentially equally populated even at 4K. Level positions for the two-photon transitions were checked at liquid-helium temperature, and found to agree within the resolution of our monochromator (± 1 cm^{-1}) with those previously reported in the literature.[6] Crystal-field splittings of the ^6P multiplets were fully resolved, as shown by the experimental curves in Fig. 2. Measurements at 4K on the $^6P_{7/2}$ state with a single-mode dye laser (~ 1 MHz width) yielded transition linewidths of 7 GHz for the three lowest crystal field components, and a linewidth of 21 GHz for the highest. The last component is homogeneously broadened because of spontaneous emission of phonons.

For photons of the same frequency and polarization, AXE'S theory[4] gives the line strength S for a two-photon transition from a ground state with components $|f^n \psi J M>$ to an excited state with components $|f^n \psi'J'M'>$ in the form

$$S = C \Omega_2 < f^n \psi J \| U^{(2)} \| f^n \psi'J' >^2 . \tag{1}$$

C is a polarization dependent factor equal to 1 for linearly polarized light and 3/2 for circularly polarized light. Ω_2 contains integrals over the radial parts of the $4f^n$ and $4f^{n-1}$ 5d wavefunctions, and is analogous to the phenomenological T_2 parameter in Judd's theory of one-photon intensities, in which the role of the second photon is replaced by the non-centrosymmetric part of the crystal field. Eq. (1) assumes that total angular momentum J is a good quantum number. The J-mixing is believed to be small in our crystal. Effective spherical symmetry is then obtained by summing over all components of such J multiplet. If we assume, as is

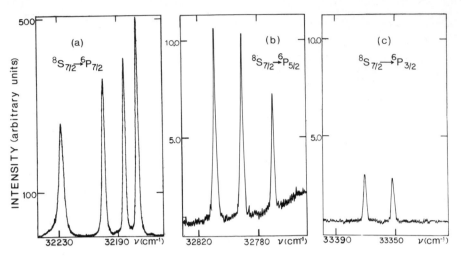

Fig. 2 Observed uv fluorescence intensity at 77K following two-photon absorption with incident light linearly polarized perpendicular to c axis from the $^8S_{7/2}$ ground state to the (a) $^6P_{7/2}$, (b) $^6P_{5/2}$, and $^6P_{3/2}$ excited states. Note that the vertical scale in (a) is different. The indicated scales measure the relative intensities. For these measurements, the laser linewidth was 0.3 cm^{-1}.

customary, that Ω_2 has the same value for all excited states, then Eq. (1) shows that, for a given polarization, the relative integrated intensities of two-photon transitions to state $|f^n \psi'J'>$ should be given by the ratio of the quantities $< g.s. \| U^{(2)} \| f^n \psi'J' >^2$. These quantities have been calculated for most lanthanides by using atomic intermediate-coupling wave functions,[7,8] and the appropriate ratios for the $^8S_{7/2} \to {}^6P_{7/2}$, $^6P_{5/2}$, $^6P_{3/2}$

TABLE I Calculated and experimental ratios of integrated intensities for three two-photon transitions in Gd^{3+} with incident light linearly polarized perpendicular and parallel to crystal c axis.

Intensity ratios	Calculated[a]	Experimental Polarization Perpendicular to c axis	Experimental Polarization Parallel to c axis
$S(^6P_{7/2})/S(^6P_{5/2})$	2.3	58 ± 5	83 ± 5
$S(^6P_{5/2})/S(^6P_{3/2})$	29	5.4 ± 0.4	>15
$S(^6P_{7/2})/S(^6P_{3/2})$	69	320 ± 10	>900

[a]Ratios of the quantities $(g.s. \| U^{(2)} \| f^7 \psi' J')^2$ as calculated in Refs. 6 and 7 for the states indicated.

transitions in Gd^{3+} are shown in Table I. This table also illustrates, however, the serious discrepancies between this theory and the experimental results, considerably larger than discrepancies normally encountered in single-photon intensity data. Contrary to our experimental findings, AXE's theory predicts the wrong relative two-photon absorption cross-sections and furthermore predicts no polarization dependence for transitions between two J multiplets. Recently, we have performed two-photon absorption measurements on the I and D states with a pulsed nitrogen pumped dye laser. It is found that the line strength of the two-photon transitions to the I and D states are on the average respectively smaller and almost equal to the P state's transition line strength, as would be expected from AXE's theory. But still, serious discrepancies are observed.

Our results show that a free-ion calculation, as the one performed by AXE, does not account at all for the measured two-photon absorption cross-sections. A more exact treatment, probably including many-body interactions, should be worked out. The ligand fields might be playing a very important role. Two-photon absorption experiments uniquely test a prediction of the Judd-Ofelt method which does not involve phenomenological parameters and, for this reason, is much more sensitive to the physical assumptions on which the method rests.

Acknowledgments

We want to acknowledge the participation of Dr. A. Bivas in obtaining the most recent results connected with this experiment. We also want to thank Dr. W. T. Carnall for having kindly provided us with the latest value of the $U^{(2)}$ matrix element for the $P_{5/2}$ and $P_{3/2}$ states.

References

1. B. R. Judd, Phys. Rev. 127, 750 (1962); G. S. Ofelt, J. Chem. Phys. 37, 511 (1962).

2. R.D. Peacock, "The Intensities of Lanthanide f-f Transitions", In Rare Earths, Structure and Bonding, Vol. 22 (Springer Berlin, Heidelberg, New York 1975) p. 82-121.

3. S. Hüfner, Optical Spectra of Transparent Rare Earth Compounds (Academic Press, New York, 1978).

4. J. D. Axe, Jr., Phys. Rev. 136, A42 (1964).

5. D. A. Jones, J. M. Baker, and D. F. D. Pope, Proc. Phys. Soc. London, 74, 249 (1959).

6. R. L. Schwiesow and H. M. Crosswhite, J. Opt. Soc. Am. 59, 602 (1969).

7. W. T. Carnall, H. Crosswhite, and H. M. Crosswhite, "Energy Level Structure and Transition Probabilities of Trivalent Lanthanides in LaF$_3$" (Argonne National Laboratory Report, 1977).

8. B. G. Wybourne, Phys. Rev. 148, 317 (1966).

Molecular Spectroscopy Using Diode and Combined Diode-CO_2 Lasers

G. Winnewisser, K. Yamada, R. Schieder, R. Guccione-Gush*, and H.P. Gush**

I. Physikalisches Institut, Universität Köln
D-5000 Köln, Fed. Rep. of Germany

1. Introduction

An increasing number of molecules have been detected in interstellar space
and envelopes of stars using radio and infrared techniques in astronomy [1].
It is apparent that extensive laboratory studies of these molecules, and
others thought susceptible of being in space, are warranted as an aid in
the identification and understanding of newly discovered features, and to
predict promising spectral regions for astronomical research. In this context we report briefly on the results of two types of investigation using
an infrared diode laser spectrometer. The first consists in simple
absorption measurements on the long chain carbon molecules HC_3N and HC_5N,
and a recently synthesized molecule, C_3OS, in the 5μ region. The second
consists in the use of combined CO_2-diode lasers for obtaining two-photon
spectra in the 10μ region.

2. Absorption of Long Chain Molecules

A program is underway to study the cyanopolyynes HC_nN, n = 3-9 (recently
identified in space using radio-telescopes) in the infrared region using
a diode laser spectrometer (Laser Analytics, Inc., Model LS-3). Diodes were
obtained which work in the spectral regions of the stretching vibrations of
the carbon triple bond, and the carbon-nitrogen bond, near 2080 and 2270
cm^{-1} respectively. Some spectra of HC_3N have recently been published [2,3],
and we present in Fig. 1 part of the spectrum of the isotopically substituted species. Superimposed on this band are hot-band transitions associated
with the ν_7 vibration. It has been possible to identify the major features
of these spectra and deduce improved molecular constants [2,3]. In Fig. 2
is part of a very recently obtained spectrum of the HC_5N molecule, belonging
also to the ν_2 band. Because of the larger moment of inertia the rotational
spacing is considerably less than in the spectra of HC_3N, giving rise to a
serious crowding of the spectral lines. The line width is due to combined
Doppler and pressure broadening. The main band is overlaid by a hot band,
tentatively assigned to a ν_{11} transition. In Fig. 3 is shown a spectrum of
the ν_1 vibration of the C_3OS molecule, with tentative rotational assignments;
further studies will be necessary to unravel the structure.

---

*Permanent address, 804-4639 West 10th, Vancouver, Canada.
**Permanent address, Dept. of Physics, University of British Columbia,
  Vancouver, B.C. V6T 1W5, Canada.

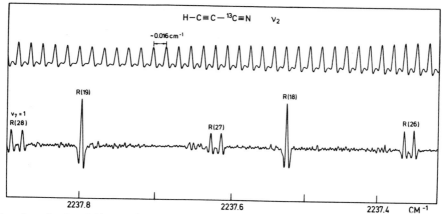

Fig. 1 Part of the ν_2 band of isotopically substituted cyanoacetylene, HC$_2$^{13}CN, and a superimposed hot band from ν_7

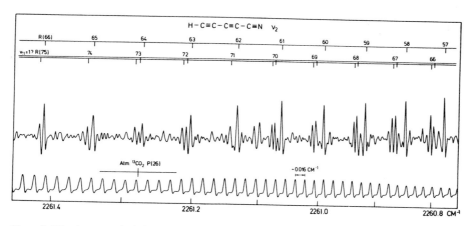

Fig. 2 First recorded infrared of spectrum of HC$_5$N. The hot band assignment ν_{11} is tentative

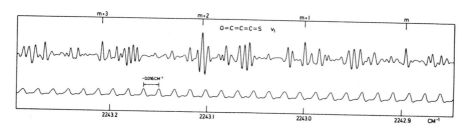

Fig. 3 IR spectrum of C$_3$OS with tentative rotational assignment

3. Two-Photon Absorption in NH₃

The assignment of lines in complex molecular spectra is sometimes ambiguous, and additional information to that given in a simple infrared spectrum is usually essential in the analysis. We are evaluating the feasibility of using combined CO_2 and diode lasers to obtain two-photon spectra, which, since different selection rules pertain, yield such information [4]. A sketch of the experimental apparatus is shown in Fig. 4. Beams from a CW, CO_2 laser (ca. 15 watts, frequency modulated) and a diode laser (ca. 1 mw) propagate in opposite directions through the sample cell, overlapping for a distance of about 5 cm. The transmitted intensity of the diode beam is measured with a cooled detector, and a phase sensitive detector locked to the CO_2 modulation frequency. The first molecule chosen for study was NH_3 because of its large electric dipole moment, and since the spectrum

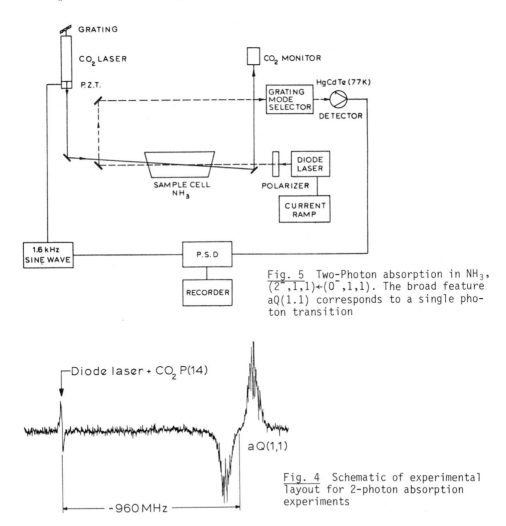

Fig. 5 Two-Photon absorption in NH_3, $(2^-,1,1) \leftarrow (0^-,1,1)$. The broad feature aQ(1.1) corresponds to a single photon transition

Fig. 4 Schematic of experimental layout for 2-photon absorption experiments

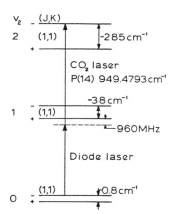

Fig. 6   Energy level diagram pertaining to Fig. 5

is relatively well known. Features corresponding to two-photon transitions have been obtained and are illustrated in Fig. 5. The intense broad feature lies at the already known one-photon absorption frequency, the two-photon absorption being the weaker, much narrower feature. The former does not arise from two-photon processes, but is believed to be largely due to scattered CO_2 radiation affecting slightly the frequency of oscillation of the diode - a signal thence results when scanning through a very strong absorption line. The transitions involved in the spectrum of Fig. 5 are shown in Fig. 6. Because the two laser beams travel in opposite directions through the cell, the Doppler effect practically vanishes, and sub-Doppler spectra are obtained. This means that it is possible to measure transition frequencies very precisely, and also to study pressure broadening at low pressure. Several other expected transitions of NH_3 were searched for without success; the reason for this is not understood, but we are currently modifying the apparatus to improve its sensitivity for a repeat experiment.

Acknowledgements: H.P.G. gratefully acknowledges partial financial support by the Heinrich-Hertz Stiftung during his sabbatical stay at Cologne University.

References

1 For a recent example see P. Thaddeus, M. Guelin, and R.A. Linke, Astrophys. J. 246,L41 (1981).
2 K. Yamada, R. Schieder, G. Winnewisser, and A.W. Mantz, Z. Naturforsch. 35a, 690 (1980).
3 K. Yamada and G. Winnewisser, Z. Naturforsch. 36a, 23 (1981).
4 R. Guccione-Gush, H.P. Gush, R. Schieder, K. Yamada, and G. Winnewisser, Phys. Rev. A23, 2740 (1981).

Two-Photon Light Shift in Optically Pumped FIR Lasers

W.J. Firth, C.R. Pidgeon, B.W. Davis, and A. Vass
Department of Physics, Heriot-Watt University
Riccarton, Edinburgh, U.K.

1. INTRODUCTION

We have made the first observation of a far-infrared two-photon light shift (2PLS). The 2PLS - an excess splitting of the gain peaks due to bidirectional pumping - is monitored by beating the output of two $^{15}NH_3$ OPFIRL's [1]. The effect is related to, but distinct from that observed by BIALAS, FIRTH and TOSCHEK [2], and is described by the three-level system model of FELDMAN and FELD [3].

Light shifts occur when atomic or molecular systems interact with non-resonant light fields [4]: we define a two-photon light shift as the shift of a two-photon resonance - the OPFIRL transition - by a third, non-resonant field - in this case the second pump component. Such shifts clearly have implications for FIR laser spectroscopy and frequency synthesis.

2. THEORETICAL ANALYSIS

FELDMAN and FELD (FF) [3] gave an essentially complete analysis of a three-level system under bidirectional pumping. They show the probe response contains terms arising from all spatial harmonics of the pump field. In the molecule frame, the frequencies of the pump fields split, and the response includes all beat harmonics. The full FF response contains infinite products of continued fractions, and subsequent authors have examined special cases in which physical interpretation is easier. KYROLA and SALOMAA [5] analyse the "Raman Limit" case[1] $k_2 \to k_1$ while FIRTH, TOSCHEK and SALOMAA [6] describe 2PLS effects in the "pump transparency" case, where the "dark" population difference n_{21}^0 is zero.

We consider the opposite extreme: $k_2/k_1 \to 0$ and $n_{32}^0 = 0$. We concentrate on the case of a detuned pump: FF assert that this is equivalent to the superposition of two opposite travelling-wave pumps, but we find that this is not quite true; the 2PLS effects represent an interaction between the pump waves. We begin with a qualitative description of the detuned limit: we then outline a detailed calculation which yields the spectral widths of the 2PLS contributions.

---

[1] We adopt the notation (1,2,3) for (initial, intermediate, final) levels, and (1,2) for (pump,probe) fields. Thus $\Delta_1 = \omega_1 - \omega_{21}$ is the pump detuning, and k_1 is the pump wavevector.

An outline level scheme is shown in Fig.1,where the pump field is tuned above resonance.   One pump component, E_+, is redshifted to resonance in the frame of a molecule with velocity $v \simeq \Delta_1/k_1$.   The opposite component E_-, is blue-shifted above resonance.   It light-shifts levels 1 and 2 together (Fig.1b) by $2\omega_s$;

$$\omega_s \simeq \beta^2/2\Delta_1 \; ; \; \beta = |\underline{\mu} \cdot \underline{E}_{+}|/2\hbar \; .$$

As a consequence, the resonant velocity is actually given by $v=(\Delta_1+2\omega_s)/k_1$. The resonant molecules emit a single frequency in their rest frame, which transformed to the lab.frame yields the distinct forward and backward peaks. These are split by $|2k_2v|$, which we see obeys

$$2k_2v= 2(k_2/k_1)(\Delta_1 + 2\omega_s) = 2(k_2/k_1) \; \Delta_1 \; (1 + \beta^2/\Delta_1{}^2) \; . \tag{1}$$

The natural frequency of the $2 \rightarrow 3$ transition is itself reduced due to the non-resonant pump beam, so that the mean frequency of the two FIR peaks obeys

$$\omega_{32}(\beta) = \omega_{32}(0) - \omega_s/2 = \omega_{32} \; -\beta^2/2\Delta_1 \; . \tag{2}$$

Since typically k_2 is an order of magnitude smaller than k_1 we see that (2) is a larger 2PLS than (1).   It is harder to observe, however, since it requires absolute, rather than relative, frequency measurement.

Exactly the same arguments apply to the second velocity group, resonant with E_-, and (1) and (2) follow as above.   Tuning below the transition ($\Delta_1 < 0$) leaves (1) unchanged (peaks still pushed apart), while the mean frequency is increased, since (2) is an odd function of Δ_1.   This is consistent with the effect predicted and observed by TOSCHEK and co-workers [2,6] in the visible/near-infrared.

The above considerations establish the existence of the 2PLS effects, but are only valid for $|\Delta_1|$ large.   To predict the spectrum, i.e. the widths, of the effects, it is necessary to undertake a more detailed analysis, which we can only outline here [7].   The approach is based on the FF formalism, with notational changes.   The probe susceptibility may be written in symbolic form

$$\chi_{probe} = \frac{n_{32}^{0} + (Pump + Raman) \; n_{21}^{0}}{L_{32} + Shift} \quad , \tag{3}$$

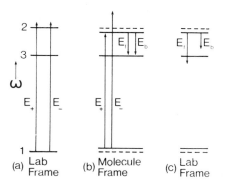

(a) Lab Frame   (b) Molecule Frame   (c) Lab Frame

Fig.1   The pump fields, and forward and backward FIR emissions in laboratory and resonant-molecule frames, showing level shifts due to E_-.

L_{32} is the FIR emission Lorentzian, while "Shift" is a pair of continued fractions which represent the effects of virtual interactions with the pump field [6]. "Pump" is basically the spatially-uniform component of level 2 population due to bidirectional pumping, and is of prime interest here. "Raman" describes coherent $1 \to 3$ transitions, and has a complex structure, but is mainly important for $k_2 \simeq k_1$ [5]. In the FIR we have $k_2/k_1 \to 0$, and on velocity integration "Raman" reduces to a small correction which we ignore. (We are concerned with pump intensities smaller than required for Autler-Townes splitting).

We can write the continued fraction characterising "Pump" in the form [3]

$$Z_0 = 1/(F_1 + 1/(F_2 + 1/(F_3 + \ldots \qquad (4)$$

In the rate equation approximation, only the first term contributes, and

$$Z_0 = 1/F_1 = \beta/(\gamma_{21} + i(k_1 v + \Delta_1)) + \beta/(\gamma_{21} + i(k_1 v - \Delta_1)) . \qquad (5)$$

We are interested in $k_1 v \simeq \Delta_1$, with $|\Delta_1| \gg \gamma_{12}$. The first term in (5) is small, and gives no significant shift [7]. The higher terms in (4) obey

$$|1/F_n| \sim \beta/|n k_1 v|$$

so that at most F_2 is significant. With these approximations, to lowest order,

$$"Pump" = - (\beta^2/\gamma_2) [(\gamma_{21} + i(k_1 v - \Delta_1 - \beta^2/k_1 v)^{-1} + c.c] . \qquad (6)$$

We can perform a similar analysis on the denominator of (3). "Shift" includes a resonant term which affects the width but not the frequency of the double resonance; the other term, due to interaction with the counter-propagating pump wave, leads to an approximate denominator of the form

$$L_{32} + Shift \simeq \Delta_2 - k_2 v + i\gamma_{32} + \beta^2/(\Delta_1 + k_1 v) .$$

Performing a velocity average in the Doppler limit, we find

$$<\chi_{probe}> \sim \beta^2 / (\delta + i\gamma)$$

where

$$\delta = \Delta_2 - (k_2/k_1) \Delta_1 [1 + \beta^2/(\Delta_1^2 + k_1^2 \gamma_{32}^2 /k_2^2)] - \beta^2 \Delta_1/(2\Delta_1^2 + \gamma_{21}^2/2) \qquad (7)$$

and

$$\gamma = \gamma_{32} + k_2 \gamma_{21}/k_1 + O(\beta^2) .$$

We see that (7) agrees with (1) and (2) for large Δ_1, but that we have obtained the additional information that the "splitting" 2PLS is broad, with spectral width $\sim k_1 \gamma_{32}/k_2 \gg \gamma_{32}$, whereas the "mean" 2PLS is narrow, with width $\sim \gamma_{21}/2$. The latter result is outside the validity range of our present calculation, but is consistent with calculations of the TOSCHEK group.

3. EXPERIMENTAL RESULTS

The apparatus has been largely described elsewhere [8], with the addition of second OPFIRL chain for the FIR beat detection experiments. The c.w. CO_2 pump lasers (Edinburgh Instruments), PZT tunable over $\pm$ 40 MHz, were frequency stabilised by offset locking to an opto-galvanically-locked reference laser. The primary FIR laser was an invar-stabilised dielectric waveguide system (Edinburgh Instruments 195), pumped on the $\nu_2$asQ(5,4) $^{15}NH_3$ line, emitting at 152.9 μm.

The FIR cavity was continuously tunable across the doubled-peaked gain profile. In our most recent experiments, simultaneous frequency measurements were obtained by beating the FIR laser output with that of a second fixed-frequency FIR laser in a Putley detector. This system gave adequate short-term stability for determination of the frequency differences of the gain peaks, but reference FIR laser drift prevented absolute frequency measurements.

Figure 2 shows the separation of the peaks, where resolved, as a function of CO_2 laser tuning, as deduced from cavity scan data. Line centre is marked by the transferred Lamb dip (solid circle). A clear and substantial excess splitting is apparent. (Similar, but smaller, effects have been observed in CH_3OH at 119μm [8]). Recent experiments with direct frequency measurements confirm a 2PLS spectrum of this form, but demonstrate that the cavity tuning itself is nonlinear, indicating significant dispersion. We estimate that dispersion exceeds 2PLS by about 2:1 in Fig.2. This dispersion arises from the fact that each excited velocity group is off resonance on one pass of the FIR cavity. The magnitudes of the effects are in accord with expectations - the dispersion is slaved to the gain, which must balance the losses, while the size of the observed 2PLS is reasonable for operation below the threshold for Autler-Townes splitting.

In conclusion we have identified a strong coherent two-photon light shift, absent from rate equation models, and demonstrated that it is ob-

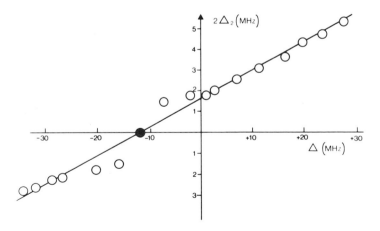

Fig.2 Gain peak splitting in $^{15}NH_3$ at 152.9 μm as a function of pump offset Δ deduced from cavity scans (0).   Line is $\Delta_2 = (k_2/k_1)\Delta_1$

servable at low powers in c.w. OPFIRL's.   We predict also an even larger
2PLS in the mean frequency of the Doppler-split gain peaks.   From a
practical point of view, these observations demonstrate that linear tuning
of the CO_2 pump laser does not normally lead to linear tuning of the OPFIRL.

We are grateful to Professor S.D. Smith for his support of this programme
and we acknowledge helpful conversations with M.F. Kimmitt and R.J. Temkin.
One of us (A.V.) has benefited from a Science Research Council studentship.
We gratefully acknowledge partial support from the European Research Office
of the US Army.   The views, opinions and/or findings contained are those
of the authors, and should not be construed as an official Department of
the US Army position.

REFERENCES

1. R.A. Wood, B.W. Davis, A. Vass, C.R. Pidgeon, Optics Lett.5, 153 (1980)

2. J. Bialas, W.J. Firth, P.E. Toschek, Opt. Commun.36, 317 (1981)

3. B.J. Feldman, M.S. Feld: Phys. Rev. A5, 899 (1972)

4. C. Cohen-Tannoudji, F. Hoffbeck, S. Reynaud, Opt. Commun.27, 71 (1978)

5. E. Kyrola, R. Salomaa: Phys. Rev. A23, 1874 (1981)

6. W.J. Firth, P.E. Toschek, R. Salomaa: to be published

7. W.J. Firth, C.R. Pidgeon, E. Abraham: to be published

8. R.A. Wood, A. Vass, C.R. Pidgeon, W.J. Firth: Opt. Commun.35,105 (1980)

Optical Stark Splitting of Rotational Raman Transitions*

R.L. Farrow and L.A. Rahn

Applied Physics Division 8342, Sandia National Laboratories
Livermore, CA 94550, USA

We report the observation of rotational state splitting in molecular nitrogen induced by a nonresonant optical field. The splittings are observed in high resolution inverse Raman spectra obtained during irradiation of the molecules by a pulsed 1.06 μm laser field. The magnitudes and signs of the Stark shifts are in agreement with predictions from polarizability theory for perturbing optical frequencies far from resonance with the molecular transitions.

Optical Stark effects on levels coupled by a one or two photon transition have been widely studied using perturbing optical frequencies near resonance with the energy level difference [1]. Recently, Stark shifts of rotational and vibrational transitions were reported for molecules interacting with nonresonant light via the derivative of the isotropic polarizability with respect to internuclear coordinate [2]. We report here the observation of J-dependent Stark splitting of rotational transitions induced by nonresonant light interacting with the anisotropic polarizability. An elementary polarizability theory appropriate for static fields is extended to nonresonant optical fields and is used to describe the results.

Measurements were performed on the S(2) pure rotational transition of nitrogen at 27.852 cm^{-1} with a crossed-beam, high resolution, "quasi-cw" inverse Raman system (IRS) [3]. A single frequency, cw krypton laser at 568.2 nm was employed as the probe laser, while a pulse-amplified, single frequency, cw ring dye laser provided the tunable pump laser. The dye amplifier was pumped by a Molectron MY34-20 frequency-doubled Nd:YAG laser equipped with optics for single axial mode operation. A wavemeter with accuracy better than 0.1 cm^{-1} was used to monitor both probe and pump laser frequencies (see Fig. 1).

The perturbing optical field was provided by residual 1.06 μm output from the Nd:YAG laser (up to 125 mJ was available at the sample). The perturbing infrared (IR) pulse shape was nearly Gaussian in time with a width of ∿20 nsec (FWHM). The optical path lengths were adjusted so that the 10 nsec (FWHM) dye laser pulse arrived at the sample during the peak of the IR intensity, providing approximate temporal uniformity of the perturbing field during the measurement. Spatial uniformity was obtained

---

*Research supported by the U.S. Department of Energy

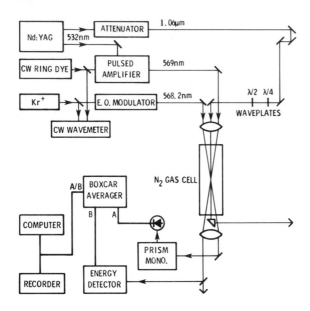

Fig.1 Schematic for inverse Raman system (IRS) and 1.06 μm optical path. The IRS probe volume is overlapped spatially and temporally by the perturbing 1.06 μm optical field to measure induced Stark splittings in rotational spectra of nitrogen

by crossing the pump and IR beams at an angle less than 1.5° and focusing the crossed pump and probe beams to a spot size smaller than that of the IR beam. Based on pinhole measurements, the visible beams were ∿50 μm diameter at the focal waists and the IR beam had a ∿100 μm waist. A 45 cm lens was used to focus all three beams. The IR polarization was aligned parallel to that of the pump and probe lasers.

Spectra of nitrogen in a room temperature cell at a pressure of 25 torr were initially measured with the IR beam blocked to determine the degree of Stark splitting induced by the IRS pump laser alone. As seen in Fig. 2, use of the full ∿15 mJ output caused substantial asymmetric broadening of the rotational lineshape. This effect results from unresolved Stark splitting of the transition by the spatially and temporally varying pump intensity. (Calculations of stimulated Raman pumping rates show saturation effects to be negligible.) Data taken with lower pump energies (1.3 mJ in Fig. 2) displayed a greatly reduced linewidth with some residual Stark-induced asymmetry.

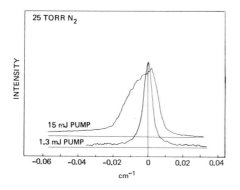

Fig.2 Comparison of measured lineshapes for the 27.852 cm^{-1} S(2) line of nitrogen using different IRS pump laser energies (1.06 μm beam blocked). Intensity at the focus for the 1.5 mJ pulse energy was approximately 7 GW/cm^2. Resolution is estimated to be ∿.003 cm^{-1}

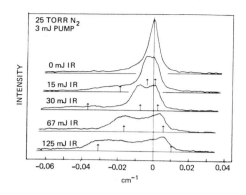

Fig.3 Spectra of the S(2) line for nitrogen molecules irradiated by a pulsed 1.06 μm optical field. Intensity at the focus for a pulse energy of 125 mJ was approximately 80 GW/cm². Arrows indicate Stark-split peak positions calculated from (2)

Keeping the pump fixed at low energy, the IR input to the sample was varied from 0 to 125 mJ to produce the spectra of Fig. 3. Due to residual nonuniformities of the IR field in the measurement volume, the Stark-shifted peaks are expected to be inhomogeneously broadened approximately in proportion to their shift; this effect appears to be observed in the data. Nevertheless, a Stark splitting is clearly revealed in this series: the S(2) line divides into two components which separate linearly with IR intensity. A third, weaker component is believed to contribute the unresolved low frequency shoulder in the 15 mJ scan. The lower frequency peak of the dominant pair is shifted more rapidly with IR intensity than the other, while the weak component is evidently shifted off scale in the two highest energy scans.

These results can be analyzed by a well-known polarizability theory which describes magnetic quantum number splitting in the rotational states of a homonuclear diatomic molecule in a static electric field [4]. The energy of the molecule in a time-dependent field is given by:

$$H = -\frac{1}{2}\alpha E^2(t) \, ,\tag{1}$$

where α is the polarizability. The correspondence to static field equations is found by writing the electric field as $1/2E_0^2(1+\cos 2\omega t)$ and considering the time-independent term. The time-dependent term is assumed negligible when 2ω is much larger than the rotational transition frequency. Using a spherical harmonic basis set for the rotational wavefunctions, first order perturbation theory yields the following expression for energy level shifts:

$$\Delta E_{J,M} = \left(\frac{1}{6}\right) E_0^2 \left[\frac{3M^2 - J(J+1)}{(2J-1)(2J+3)}\right] (\alpha_{aa} - \alpha_{bb}) \, .\tag{2}$$

The J=2 state splits into three levels, and, for transitions to J=4 with the selection rule $\Delta M=0$ (appropriate for parallel polarizations), three lines should be observed (see Fig. 4). The line positions of the triplet were calculated using Stark shift coefficients obtained from (2) and estimating the peak IR field intensity from the measured energy, pulsewidth, and focal waist. Results are indicated for different IR

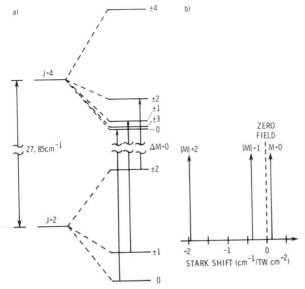

Fig.4 Energy level diagram (a) for J=2 and J=4 calculated from (2) for an applied field of 1 terwatt (TW) per cm^2. Transitions for $\Delta M=0$ are indicated by vertical lines. The predicted transition shifts are shown in (b) in units of cm^{-1} per TW/cm^2

intensities by the vertical arrows in Fig. 3, and are seen to be consistent with the approximate observed line positions. Quantitative comparisons were not attempted here due to uncertainties in the IR intensity (±30%) and Raman shift (±.005 cm^{-1}) for each scan.

In conclusion, the J-dependent Stark splittings reported here are in good agreement with static field polarizability theory. Implications of this study are of fundamental importance for high resolution rotational Raman spectroscopy. Results indicate that rotational lineshape perturbations induced by measurement lasers may be difficult to avoid, particularly when collisional linewidths are small (e.g. in flames or low pressure environments). Further work is in progress to investigate J-dependence, polarization and M-dependent intensity variations, and possible pressure-narrowing effects for the Stark-split multiplets.

The authors are grateful to P. L. Mattern for helpful discussions and to J. Brandt and A. R. Van Hook for technical assistance of the highest caliber. We are also indebted to Molectron Corporation for their cooperation and assistance in the installation of the single mode unit.

References

1. See, for example, P. F. Liao and J. E. Bjorkholm, Phys. Rev. Lett. **34**, 1 (1975), and A. M. Bonch-Bruevich, N. N. Kostin, V. A. Khodovoi, and V. V. Khromov, Sov. Phys. JETP **29**, 82 (1969).

2. L. A. Rahn, R. L. Farrow, M. L. Koszykowski, and P. L. Mattern, Phys. Rev. Lett. **45**, 620 (1980).

3. A. Owyoung, in Laser Spectroscopy IV, ed. by H. Walther, K.W. Rothe, Springer Series in Optical Sciences, Vol. 21 (Springer Berlin, Heidelberg, New York 1979)

4. See, for example, W. H. Flygare, <u>Molecular Structure and Dynamics</u> (Prentice-Hall, New Jersey, 1978).

Part VI
Rydberg States
(Panel Discussion)

Resonant Collisions Between Two Rydberg Atoms

T.F. Gallagher, K.A. Safinya, F. Gounand, J.F. Delpech, and W. Sandner
SRI International Molecular Physics Laboratory, PS091, 333 Ravenswood Ave.
Menlo Park, CA 94025, USA

Resonant collisions, that is, those in which one of the two colliding atoms (or molecules) loses as much internal energy as the other gains, are expected to have far larger cross sections than nonresonant processes [1]. However, systematic studies of the effect of resonance have been difficult for several reasons. First, it is necessary to rely upon chance coincidence or the same atom for a resonance, and there is thus no good way to tune through a resonance. In addition, usual gas kinetic cross sections lead to collision times so short that the resonance is very broad and thus not pronounced.

Atomic Rydberg states offer a unique possibility to study resonant collisions because their energy separations may be easily changed in discrete steps by changing n or ℓ quantum numbers or continuously by electric or magnetic field tuning. Changing n has already been used to observe resonant electronic to rotational and vibrational energy transfer in Rydberg atom—molecule collisions [2,3]. Here we report the use of electric field tuning to study sharply resonant energy transfer in collisions between two Rydberg atoms.

We have observed that when an electric field is used to tune Na levels so that the ns level ($16 \leqslant n \leqslant 27$) lies midway between the np and $n - 1$ p levels that the population in the s states is very rapidly transferred to the p states. As an example we show in Fig. 1 the population in the 20p state 2 μs after population of the 20s state as a function of the tuning field for two powers of the exciting laser. As shown in Fig. 1 there are four collisional resonances due to the existence of $m = 0$ and 1 p states. They are labelled by the m value of the upper and lower p states. Note that the black body radiation induced population transfer at, 200 V/cm and 250 V/cm, is linear in the laser power while the resonant collisional signal is quadratic in the laser power. By comparing the resonant collision signal with the total number of excited atoms we find that cross section for the process scales as $n^{*(4.2\pm0.3)}$ with a magnitude of 1.2×10^9 Å^2 at $n^* = 20$. Here n^* is the effective quantum number defined by $W = -1/2n^{*2}$. When we convert the observed widths, in V/cm, to frequency, we find that the widths scale as $n^{*(-1.9\pm0.2)}$ with a value of 0.5 GHz at $n^* = 20$. Both the magnitudes and widths of the observed cross sections are in agreement with calculations based on the long range dipole-dipole interaction of the two atoms. This work was supported by the ONR Physics Division under contract N00014-79-C-0202.

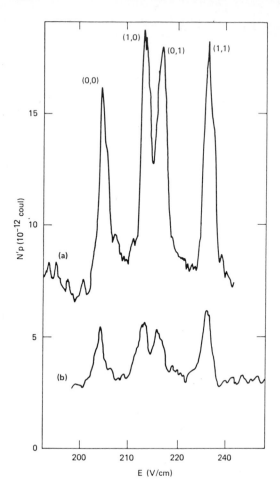

Fig. 1 The 20p state population vs. electric field after populating the 20s state.

REFERENCES

1. N. F. Mott and H.S.W. Massey, The Theory of Atomic Collisions (Clarendon Press, Oxford, 1950).

2. T. F. Gallagher, G. A. Ruff, and K. A. Safinya, Phys. Rev. A 22, 843 (1980).

3. K. A. Smith, F. G. Kellert, R. D. Rundel, F. B. Dunning, and R. F. Stebbings, Phys. Rev. Lett. 40, 1362 (1978).

Rydberg Atoms in Magnetic Fields – The Landau Limit of the Spectrum

J.C. Gay and D. Delande

Laboratoire de Spectroscopie Hertzienne de l'E.N.S.
Tour 12 - EQ1, 4 place Jussieu, F-75230 Paris, Cédex 05, France

The understanding of the physics of atoms —especially Rydberg atoms —in magnetic fields is far from being complete. Experiments and numerical simul-
ations [1-3] have recently shown that several series of lines, presumably
weakly interacting —one of them having dominant character in optical absorption —are participating in the building up of the atomic quasi-Landau
spectrum. Such series are also existing above threshold where they are associated with resonances embedded in the continua. Although the basic features of the quasi-Landau phenomena (the $3/2\hbar\omega_c$ spacing and the $n_r^3.B = $ Cste
quantization law at threshold) are well understood from WKB theories, there
is not yet any quantum approach at the spectrum making clear the connections
between the Rydberg and Landau limiting cases.

 We here present some of the more striking features of a high resolution
optical study of the Landau regime above threshold, performed on highly
hydrogenic states of Caesium. The optical source is a 10 MHz width, 800 mW
power, c.w. single mode ring dye laser. Detection of Rydberg states as high
as 160 and Landau resonances up to 120 cm^{-1} above threshold is achieved by
the use of a shielded thermoionic detector placed in the clear bore of a 8 T
superconducting solenoïd. Owing to the well controlled character of the laser
lineshape and frequency, we have been able to deduce some information from
the intensities of the lines.

 Figure 1 shows the aspect of the Landau-Rydberg spectrum in field for an
electronic energy 10 cm^{-1} above threshold. Similar data has been collected
for energies up to 120 cm^{-1} above threshold. Figure 2 plots the radial quantum number N_r as a function of $1/B$ for energies of the electron between 0
and 120 cm^{-1}. The plot characterizes the spectrum of the electron weakly
bound to the ionic core and experiencing a magnetic field. It is far different from that of the free electron in a magnetic field as expected. Near
zero energy, a $N_r^3.B = $ Cste quantization law associated with the quasi-Landau
regime is fulfilled [2]. Far above threshold, the quantization law is

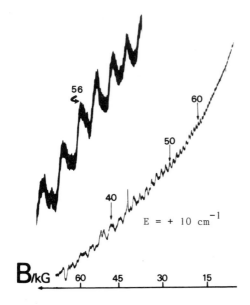

B/kG   60   45   30   15

$E = + 10 \text{ cm}^{-1}$

Fig.1. Aspects of the spectrum above threshold

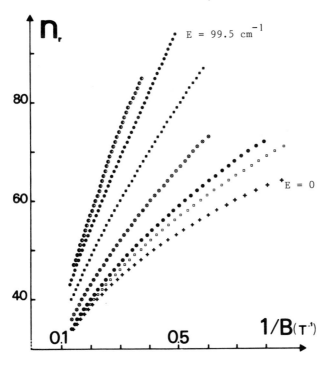

$E = 99.5 \text{ cm}^{-1}$

$E = 0$

$1/B \text{(T}^{-1}\text{)}$

Fig.2. Plot of the radial quantum number N_r versus $1/B$ for various energies of t of the electron ($E = 0, 11.5, 17.2, 37.3, 65.7, 99.5, 121.5 \text{ cm}^{-1}$)

$N_r.B = Cste$ associated with the almost pure Landau regime. For $E = 100 \ cm^{-1}$, the spacing of resonances is $1.1 \ \hbar\omega_c$ for $B \sim 20$ kG while for higher field values departure from the pure Landau regime occurs.

The main intensity features of the spectrum are displayed on Figure 1. While a steep edge structure is increasing in resolution, the average signal is decreasing with B. The heights of the edges are varying as $1/N_r^\alpha$ with $2 \leq \alpha \leq 3$. α is slowly increasing with the energy. The shape of the signal could be explained in terms of a band structured model of the quasi-Landau spectrum in which the edges are associated with the bottom of a radial band — that is with states confined near the $Z = 0$ plane. Indications of the existence of other resonances between the main ones exist, but due to their autoionizing characters and motional Stark effects, they are not clearly re-solved. Consequently the band looks like an almost continuous one except at the bottom.

Aside from its fundamental interest, such a situation, providing almost complete discretization of the atomic and electronic continua, seems of interest for many studies involving electrons of subthermal energies.

References

1. J.C. Castro et al.: Phys. Rev. Lett. *45*, 1780 (1980)
2. J.C. Gay et al.: J. Phys. B *13*, L729 (1980)
3. C.W. Clark, K.T. Taylor: J. Phys. B *13*, L737 (1980)

High Precision Energy Level Measurements in Helium

E. Giacobino, A. Brillet, and F. Biraben

Laboratoire de Spectroscopie Hertzienne de l'ENS, Université de Paris VI
4, Place Jussieu, F-75230 Paris Cedex 05, France, and

Laboratoire de l'Horloge Atomique, Université de Paris Sud
F-91405 Orsay Cédex, France

The wavefunctions and the atomic parameters of helium have formed the subject of many important theoretical studies and we now have reliable values
for a certain number of levels.

Using Doppler-free two-photon spectroscopy, we have shown in previous works
that we could measure with a high precision the fine and hyperfine structures [1] in the two isotopes of helium, ^{3}He and ^{4}He, for the nS (n = 4-6)
and nD (n = 3-6) levels. We have also determined the isotopic shift ^{3}He -
^{4}He for the two-photon transitions between the 2^3S level and these levels
[2]. These experimental values have been found to be in full agreement with
the values calculated from the theory [3].

In this paper, we are presenting further results about the same levels,
now concerning the energies of the two-photon transitions. Using a high
precision scanning Michelson interferometer ("lambda meter"), we have compared the laser wevelengths resonant for the two-photon transitions with
an iodin stabilized He - Ne laser. We have determined the energies of these
transitions with an accuracy of about ± 15 MHz.

Let us briefly recall the experimental method. The helium atoms are excited
to the 2^3S metastable level by a pulsed discharge. Then they are excited to
other S and D levels by absorbing two photons from a c.w. single mode ring
dye laser. The cell containing the helium atoms is placed inside a near-
concentric Perot-Fabry cavity whose length is servo-locked to the laser
wavelength. The fluorescence light emitted from the level under investigation is detected sidewards only during the afterglow (using a gated amplifier). With this set-up, the needed laser powers range between 100 mW and
several hundreds of mW for the highest levels. The laser wavelength is set
to the maximum of the two-photon transition and a small part of the laser
beam is sent into the lambda-meter and carefully aligned on the helium-neon
laser beam. The apparatus is not evacuated so we had to correct the results
for the dispersion of air.

We have verified that the light shift due to the laser irradiation was negligible as expected from the theoretical estimation and that there was
no Stark shift due to the presence of charged particles (as we detect the
signal only in the afterglow, the number of charged particles is very small).

On the other hand, we have found a significant pressure shift of the two-
photon lines especially for the S levels (up to more than 100 MHz/torr for
the 6^3S level) that had to be corrected for.

The values are the following :

Transition	Energy (cm $^{-1}$)
$2^3S - 3^3D_1$	26 245.572 2 (± 5)
$2^3S - 4^3D_1$	31 588.526 2 (± 5)
$2^3S - 5^3D_1$	34 061.187 6 (± 5)
$2^3S - 6^3D_3$	35 404.097 7 (± 5)
$2^3S - 4^3S$	30 442.139 8 (± 5)
$2^3S - 5^3S$	33 491.017 6 (± 6)
$2^3S - 6^3S$	35 080.145 4 (± 8)

These values are in agreement with those of W.C. MARTIN [4] to within a few 10^{-3} cm^{-1}.

Most calculations and especially that of [3] for the S levels do not take into account the Q.E.D. correction. So the difference between the calculated energies and the experimental ones leads to the Lamb shifts ; as the theory is usually expected to have an accuracy of 10^{-3} cm^{-1}, we can then obtain precise values of the Lamb shift which can in their turn be compared with theory, in particular to check the two-electron effects. Up to now there have been rather few precise calculations of these effects and hopefully such experimental results will stimulate new theoretical developments.

References

1. F. Biraben, E. de Clercq, E. Giacobino and G. Grynberg - J. Phys. B Atom. Molec. Phys. 13 (1980) L685.

2. E. de Clercq, F. Biraben, E. Giacobino, G. Grynberg and J. Bauche - J. Phys. B Atom. Molec. Phys. 14 (1981) L183.

3. Y. Accad, C.L. Pekeris and B. Schiff - Phys. Rev. A 4 (1971) 516.

4. W.C. Martin - J. Phys. Chem. Ref. Data 2 (1973) 257.

Rydberg-Rydberg Interactions in Dense Systems of Highly Excited Atoms

S. Haroche, G. Vitrant, and J.-M. Raimond

Laboratoire de Physique de l'Ecole Normale Superieure
24, rue Lhomond, F-75231 Paris Cédex 05, France

Electric dipoles of Rydberg atoms prepared in a small volume interact very strongly with each other. At large interatomic distances r, they are coupled via the radiative dipole-dipole interaction (decreasing as $1/r$), responsible for superradiance. At short distances (r of the order of a few Rydberg diameters), the $1/r^3$ Van der Waals interaction becomes dominant. It quenches the superradiant emission and induces in the optical and microwave domains strong spectral perturbations of the Rydberg transitions.

We have observed these effects on the $6P\frac{3}{2} \rightarrow nS$ and nD lines of Cesium atoms excited in a dense atomic beam by a tightly focused powerful pulsed dye laser. Figure 1a shows the spectrum of the Rydberg transition for n values around 40, corresponding to a "moderate" density of Rydberg atoms excited at the center of each line ($N \sim 10^{10}$ cm^{-3}). Figure 1b exhibits the same spectrum when N is increased to 2.10^{11} cm^{-3}. The average interatomic distance is then about seven times the Rydberg atom diameter ($2a_0 n^2 \sim 0.15\mu$m where a_0 is the Bohr radius). Strong line-broadenings are observed and the resonances corresponding to successive Rydberg levels start to overlap each other. Careful tests have allowed us to rule out radiative effects and collisions with low-excited state atoms as the origin of these perturbations (these effects are orders of magnitude smaller than the observed broadenings). Stark broadenings due to ions and electrons produced in the Rydberg medium have also been ruled out by measuring the number of free charges in the system. The observed perturbations are thus produced by Van der Waals dipole-dipole interactions between Rydberg atoms. Clusters of atoms passing close to each other are simultaneously excited by two-photon (or multiphoton) processes when the difference between the laser and the unperturbed Rydberg frequency equals the instantaneous Rydberg-Rydberg Van der Waals shift. Simple order of magnitude estimates of the coupling and transition rates sustain well this interpretation.

Spectral line broadenings and shifts have also been observed at somewhat lower Rydberg densities ($N \sim 5.10^{10}$ cm^{-3}) on mm-wave transitions between neighbour Rydberg states. All these effects reveal an unusual situation where the electric coupling between different atoms is no longer negligible compared to the unperturbed Rydberg atom energy splittings. Instead of ordinary Rydberg levels, one should rather consider "excitations" of the dense Rydberg gas made of a collection of strongly interacting Rydberg atoms. The field-ionization characteristics of this gas are also different from these of isolated Rydberg states : ionization current versus electric field strength measurements show that the onset of ionization occurs for fields much smaller than the free atom threshold field F_0, revealing transfers to higher Rydberg levels due to the Van der Waals interaction. At the same time, some atoms resist ionization in fields larger than F_0, due to transfers to less excited levels. These field ionization studies have shown that, above a well defined threshold Rydberg density, the Rydberg gas becomes unstable and decays into a plasma by an avalanche mechanism: seed electrons produced by Rydberg-Rydberg collisions break the Rydberg atoms at an exponentially increasing rate. The existence of a threshold for this effect results from the competition between the avalanche and the expansion of the system at thermal velocities. All these studies -whose results will be published in detail elsewhere- will certainly stimulate theoretical studies of these new interesting systems.

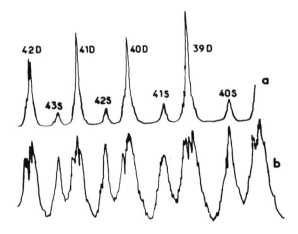

Figure 1 : Optical spectra of the $6P\frac{3}{2} \to nS$, nD transitions in Cs
a) small Rydberg atom density
b) large Rydberg atom density
(frequency scale : 136 GHz between 40S and 41S lines)

Turning Off the Vacuum

D. Kleppner

Research Laboratory of Electronics and Department of Physics
Massachusetts Institute of Technology
Cambridge, MA 02139, USA

Spontaneous emission is usually regarded as a fundamental property of matter. An excited atom in empty space inevitably decays to its ground state, the decay rate depending on the nature of the particular radiative transition. The transition rate is given by the Einstein A coefficient, and can be expressed in terms of the "Golden Rule"

$$A = \hbar^{-2} \; |<f|H|i>|^2 \rho(\nu)$$

where $|i>$ is the initial state (excited atom, no photon), $|f>$ is the final state (ground state atom, 1 photon), H is the interaction Hamiltonian, and $\rho(\nu)$ is the frequency density of final photon states. In free space, $\rho_0(\nu) = 4\pi\nu^2/c^3$.

It was pointed out many years ago by Purcell [1] that the spontaneous emission rate in a tuned cavity is larger than in free space by a factor of Q, where Q is the quality factor of the cavity. Essentially, the mode density in the cavity is $\rho_c = 1$ mode$/(\Delta\nu_c V)$ where $\Delta\nu_c = \nu/Q$ is the cavity frequency width, and $V \simeq (\lambda/2)^3$ is the cavity volume. From this it follows that $\rho_c \sim Q\rho_0$. Within the cavity an atom undergoes enhanced spontaneous emission.

The converse effect, inhibited spontaneous emission, can also be realized [2]. In a non-resonant cavity the mode density can be lower than in free space. If an atom's radiation frequency lies below the fundamental cavity frequency, or lies somewhere between the lowest modes, the cavity response can be so small that the atom cannot radiate. In effect, the cavity isolates the atom from the zero point fluctuations which induce spontaneous emission: the vacuum is "turned off".

The experimental problem in observing enhanced and inhibited spontaneous emission is to locate a transition with long enough wavelength to permit the construction of a low mode, high Q, cavity, but a short enough lifetime for the atom to radiate in less than its time of flight. These conditions can be satisfied by Rydberg atoms. The mode density can be altered by a tuned cavity, parallel conducting plates, or by a simple waveguide. By employing states with $\ell = n - 1$ all dipole decay channels save one, $n \rightarrow n - 1$, are disallowed. Such states are well suited for studies of highly inhibited spontaneous emission. We have begun experiments on inhibited radiation and have observed the "turn off" of blackbody absorption. Space limitations prevent detailed description here, but the work will be described in forthcoming publications.

There are many ramifications to the study of inhibited and enhanced emission. Radiative interactions such as the Lamb shift are altered by a cavity. Various atom-cavity interactions must be considered. In the case of enhanced emission in a high Q cavity, one can achieve a situation where the atom-cavity system undergoes periodic energy transfer rather than exponential decay. The condition for this is that the atom decay rate, QA, must be large compared to the cavity decay rate, ω/Q. Using $A \simeq \alpha^3\omega^3r^2$ (atomic units), and $\omega \sim n^{-3}$, $r \sim n^2$ (where n is the principle quantum number), one obtains $Q \geq n/\alpha^{3/2}$. For n = 50, $Q \geq 8 \times 10^4$, a value easily achieved.

The primary motivation for studying inhibited and enhanced spontaneous emmission is to observe fundamental phenomena under novel conditions. Beyond this, one may look to practical applications in spectroscopic studies or in radiative devices. Any technique for substantially lengthening the natural radiative lifetime lends itself to high precision spectroscopy. It should be pointed out, however, that the atom-cavity interaction can lead to measureable frequency shifts. From the point of view of high precision spectroscopy such shifts are a nuisance. From the point of view of studying the atom-cavity system, of course, the shifts are interesting in their own right.

This work was supported by the Joint Service Electronic Program, the National Science Foundation, and the Office of Naval Research.

References

1. E.M. Purcell, Phys. Rev. 69, 681(1946).
2. D. Kleppner, Atomic Physics and Astrophysics, (Gordon & Breach, 1971), p.5, M. Chretien and E. Lipworth, ed.

Influence of the Earth Magnetic Field of the Photoionization Stark Spectrum of Excited Sodium Atoms

S. Feneuille, S. Liberman, E. Luc-Koenig, J. Pinard, and A. Taleb

Laboratoire Aimé Cotton, Centre National de la Recherche Scientifique 11, Campus d'Orsay, Bâtiment 505, F-91405 Orsay, France

Although most of the characteristics of photoionization Stark spectra of alkali atoms seem to be qualitatively understood, a few problems remain concerning the strong field mixing especially in the vicinity of the field-free ionisation limit. In order to study the properties of such structures and namely their dependence versus light polarization we have performed an experiment on sodium atoms in a well controlled m_ℓ state. More precisely the experiment has been done on the sodium atoms of an atomic beam which are submitted to two laser excitations in the presence of a d.c. electric field. The first laser excitation is provided by a cw single mode dye laser right-hand polarized, tuned and servolocked on the transition $3\ ^2S_{1/2}$, $F=2 \rightarrow 3\ ^2P_{3/2}$, $F=3$; it thus permits exciting the sodium atoms in the pure state : $3\ ^2P_{3/2}$, $F=3$, $m_F=3$ which corresponds to the well defined magnetic orbital quantum number $m_\ell=1$. The second laser excitation is provided by a nitrogen pumped dye laser continuously tunable over the range 408 - 413 nm , and polarized in either one of the σ^+ , π or σ^- polarizations so that it permits to study the three different m_ℓ projections of the photoionization spectrum : $\left| m_\ell \right| = 0,\ 1,\ 2$ (the experimental configuration is set in such a manner that $m_\ell = 0$ and $m_\ell = 2$ photoionization spectra are obtained with the two laser beams counterpropagating collinearly, whereas $m_\ell = 1$ photoionization spectrum is obtained with the pulsed laser beam propagating perpendicularly to the cw one).

According to the three polarizations, typical recordings of the photoion current versus the pulsed laser frequency are displayed on the figure : trace a corresponds to $m_\ell = 0$, trace b to $m_\ell = 1$ and trace c to $m_\ell = 2$. As the latter is quite different from the two others, one can be convinced that the intermediate state $m_F=3$ $(m_\ell = 1)$ is a pure state (otherwise it would have been polluted by similar resonances as those observed in the $m_\ell = 0$ and $m_\ell = 1$ spectra). Moreover, there is a big difference between the resonance shapes in the three spectra : Fano profiles can be identified in the $m_\ell = 0$ spectrum, whereas less dissymmetric resonances mainly appear in the $m_\ell = 1$ spectrum and almost symmetric and narrow ones characterize the $m_\ell = 2$ spectrum. All these observations are interpreted through the more or less hydrogenic character of the spectra. Under that viewpoint one can try to label the observed resonances using n and n_1 quantum numbers, and perturbation theory to fourth order, approximately one resonance out of two can be labelled, as it is shown on

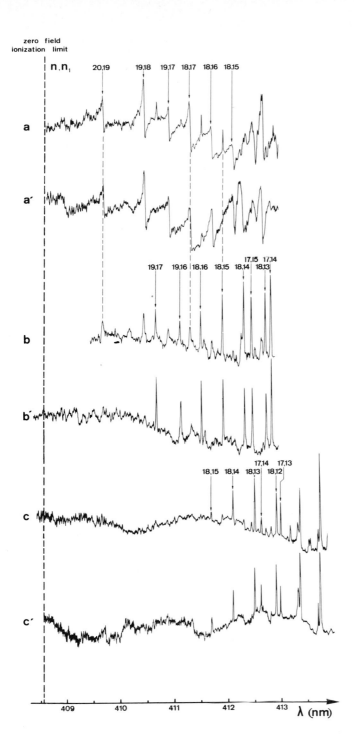

the figure. Actually some resonances appear in the three spectra at the same frequency, which corresponds to an apparent degeneracy in m_ℓ. It could be seen in some way as a breaking of the symmetry around the d.c. field direction, the origin of which should be external. This leads the earth magnetic field to play some essential role in such an assumption. In order to check the validity of the explanation, the experiment has been done again while cancelling the earth magnetic field in the interaction region. The resulting recordings are displayed on traces a', b' and c': all the resonances not identified according to the hydrogenic model have completely vanished. This result is experimental evidence of a strong perturbation of the photoionization Stark spectra due to interactions with very weak fields.

Rydberg Superradiance

L. Moi*, C. Fabre, P. Goy, M. Gross, S. Haroche, and J.-M. Raimond

Laboratoire de Physique de l'Ecole Normale Supérieure, 24 rue Lhomond
F-75231 Paris Cêdex 05, France

Rydberg atom masers and superradiant systems, which are now being developed for far infrared and mm-wave oscillators and amplifiers [1], provide a link between macroscopic eletrodynamics and quantum optics. They offer a very interesting testing ground for various aspects of the superradiance theory. We summarize here some of these tests, after recalling the main features of these systems.

Preparing the Rydberg atoms by pulsed laser excitation of an atomic beam, we detect the superradiant emission of transition to lower lying states either by monitoring the level populations as a function of time, or by direct detection of the mm-wave radiation. The Rydberg medium radiates either in free space (pure superradiance) or in an etalon-like resonator which enhances the self-radiated field acting on each atom by a factor equal to the cavity finesse (superradiant maser). Emission frequencies range from the 100 GHz to the terahertz domain, depending on the transitions. Inversion thresholds and emission delays are in good agreement with theory. The orders of magnitude for thresholds ($\sim 5.10^3$ inverted atoms for $\Delta n = 1$ transition around $n \simeq 30$) are quite unusual and reflect the huge size of the atom-field coupling.

Taking advantage of the large emission wavelengths of these systems, we have studied the dependence of superradiance on the size of the emitting medium. Intersecting the atomic beam with a laser beam of variable focus, we have prepared needle shaped samples of small Fresnel numbers, with a fixed length $L \sim 2$ cm and a diameter w varying from 1.5 mm to 0.1 mm. The emission wavelengths λ ranged from 0.4 mm to 1 mm. We have observed that, for a fixed number of Rydberg atoms, superradiance disappears as soon as w becomes smaller than 0.6 λ. If then the atom number is increased, the superradiant emission is not restored, but instead strong transfers to a distribution of neighbour levels, more and less excited than the initial one, are observed. This theoretically predicted [2] quenching of superradiance is due to the competition with short range Van der Waals Rydberg-Rydberg interaction which randomly dephases the dipoles before they can lock in phase and radiate the superradiant pulse.

In the original Dicke paper [3] on superradiance, this dephasing was neglected and a symmetric coupling of all the atoms to the fields in the small size medium was assumed. Under this assumption, a complete deexcitation towards the lower state of the transition was predicted (total superradiance corresponding to a π pulse emission). Our experiments have verified that the Van der Waals dephasing prevents the observation of Dicke superra-

---

*Permanent address: Laboratoire di Fisica Atomica e Moleocolare, C.N.R., Via del Giardino 3, I-56100 Pisa, Italy

diance for small samples radiating in free space. Large samples on the other hand introduce other dephasings due to diffraction and propagation, resulting in an incomplete decay, a large fraction remaining in the initially prepared level, as also verified in our experiments. We have shown that the total superradiance of the Dicke theory can however be observed if a small size Rydberg medium radiates in an open Fabry-Perot cavity. The symmetry conditions for the atom-field coupling are then satisfied. On the other hand, the field enhancement due to the coupling with the resonator allows us to reduce the medium density below the value where Van der Waals coupling becomes damaging to superradiance. A complete deexcitation of the upper to the lower state of the transition is then observed provided the small medium is located at an antinode position of the cavity mode.

In another set of experiments, we have studied the initiation of the mm-wave Rydberg emission. In this wavelength domain, superradiance is triggered by blackbody radiation and not by spontaneous emission as in optical experiments. We have checked this point by measuring the size of the small injection signals sufficient to alter the characteristics of the "spontaneous" superradiant emission. We have shown that the maser emission delays, polarization and phases were modified as soon as the injected signal was of the order of the room temperature backbody flux in the bandpass of the emitting system. Implications of this sensitivity to detection technology are obvious.

Another interesting feature of Rydberg superradiance lies in the properties of the energy diagrams of Rydberg states. The existence of nearly degenerate ns → np and np → (n-1)s transitions makes it possible to study peculiar cascading superradiance effects. In a maser whose cavity was simultaneously tuned to both transitions, we have observed a direct collective stepwise deexcitation bringing the atoms from the ns to (n-1)s level. An improvement of our cavity quality factor should allow us to observe true degenerate two-photon maser emission in the mm-wave domain. This is only an example among many other non-linear phenomena which could certainly be studied in the microwave domain on Rydberg atoms.

References

[1] L. Moi, C. Fabre, P. Goy, M. Gross, S. Haroche, P. Encrenaz, G. Beaudin and B. Lazareff, Opt. Comm. 33, 47 (1980)
[2] R. Friedberg and S.R. Hartmann, Phys. Rev. A 10, 1728 (1974)
[3] R.H. Dicke, Phys. Rev. 93, 99 (1954)

Collision Effects with High-Rydberg State Atoms

B.P. Stoicheff

Department of Physics, University of Toronto
Toronto, Ontario M5S 1A7, Canada

It is well-known that Rydberg atoms increase in radius as
$r \propto n^2$(n being the principal quantum number) so that at high
n, they offer a huge cross-sectional area for collisions with
particles. Moreover, as n increases beyond about 40, the
electron may be considered as essentially free, and therefore
an interesting projectile of low kinetic energy for scattering
experiments. Thus highly-excited Rydberg atoms are extremely
delicate sensors for studies of collision effects.

Recently, several investigations using laser spectroscopy
have been reported, concerning collision effects of Rydberg
atoms(A**) and ground state atoms A or B. Amongst these, mention may be made of 1-mixing[1] and broadening[2] of Na levels
by rare gas perturbers; the self-broadening of Cs levels[3];
the collisional quenching[4] of Rydberg states of alkalis by
ground state atoms; and the frequency shifts, line broadenings
and phase-interference effects in Rb**+Rb collisions[5].

Here, I would like to present our most recent data on K**+K
collision effects. In the figures are shown the linewidths
and shifts for nS←4S transitions versus n at several pressures.

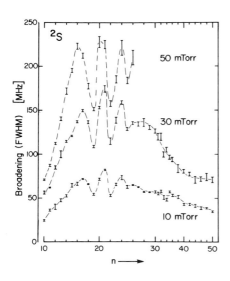 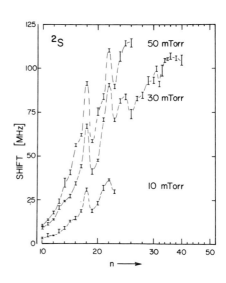

The main features are the prominent oscillations in both line-width and shift in the region n ~15 to 30. At different pressures the oscillations show different amplitudes (proportinal to pressure), but the maxima and minima remain at the same quantum numbers, and the oscillations of widths and shifts are out of phase.

The explanation of these oscillations is not yet clear. A theory by MATSUZAWA[6] based on resonance scattering of a highly-excited Rydberg electron and a colliding partner, predicts just such oscillations in width and shift, but requires essentially free electrons ($n \gtrsim 40$). PRESNYAKOV[7] has treated the resonance as an electron attachment process, and recently has suggested scattering by autoionized levels of negative ions. We have applied Matsuzawa's theory to the K**+K (and to our earlier Rb**+Rb) data, and obtain values of E = 1.97 and Γ = 1.5 meV for K^- (and E = 1.5 and Γ = 1.5 meV for Rb^-) where E and Γ may be interpreted as the resonance energy and width of the autoionized levels of the negative ions. These values are in reasonable agreement with theoretical values[8].

While further experimental and theoretical work is required to confirm this explanation, it is clear that laser spectroscopy is an extremely sensitive tool for probing interactions of highly-excited Rydberg atoms with other atoms and molecules.

References

1. T. F. Gallagher, S. A. Edelstein, and R. M. Hill, Phys. Rev. A15, 1945(1977).
2. R. Kachru, T. W. Mossberg, and S. R. Hartmann, Phys. Rev. A21, 1124(1980).
3. K.-H. Weber and K. Niemax, Opt. Commun. 28, 317(1979).
4. M. Hugon, F. Gounand, P. R. Fournier, and J. Berlande, J. Phys. B12, 2707(1979); B13, L109(1980).
5. B. P. Stoicheff and E. Weinberger, Phys. Rev. Lett. 44, 733(1980).
6. M. Matsuzawa, J. Phys. B 8, L382(1975); B10, 1543(1977).
7. L. P. Presnyakov, Phys. Rev. A2, 1720(1970): JETP Lett. 30, 53(1980).
8. A.-L. Sinfailam and R. K. Nesbet, Phys. Rev. A7, 1987 (1973).

Perturbations in Rydberg Sequences Probed by Lifetime, Zeeman-Effect and Hyperfine-Structure Measurements

S. Svanberg

Department of Physics, Lund Institute of Technology, P.O.Box 725
S-220 07 Lund, Sweden

In alkaline-earth atoms strong perturbations in sequences of highly excited (Rydberg) states occur due to interaction with low-lying valence states of series converging to a higher ionization limit. Multi-channel quantum-defect theory (MQDT) has been used to describe the energy-level structure. Configuration mixing not only affects the level positions but is also reflected in the radiative properties, in the Zeeman and Stark effects and in the hyperfine structure, and measurements of these phenomena can serve for sensitive probing of the mixing of wavefunctions.

We have performed extensive measurements of natural lifetimes for the $6sns$ $^1S_0$ (n=11-21) and $6snd$ $^1D_2$ (n=12-30) sequences of Ba, which are perturbed by states belonging to the $5d7d$ configuration. We have also determined the lifetimes of the $6snd$ $^3D_2$ states in the regions of perturbation and of the shortlived $5d7d$ $^3F_2$, $^1D_2$ and $^3P_0$ perturbers [1,2]. Pulse modulation of a CW dye-laser beam in conjunction with delayed-coincidence electronics was used [3]. The even-parity states were reached via the short-lived $6s6p$ $^1P_1$ state using step-wise excitation in atomic-beam experiments. Influences of black-body radiation-induced transitions were also investigated. In Fig. 1 part of our lifetime results are plotted. The strong decrease in lifetime value at about n=14 and n=27 of the $^1D_2$ sequence is very spectacular as is the corresponding effect at n=18 for the $^1S_0$ sequence. The pulling of the corresponding short-lived perturbers is clearly seen in the figure. The perturbation at n=27 was also observed by GALLAGHER et al. [4] and by AYMAR et al. [5]. These authors have also performed a MQDT analysis of their results and a similar analysis for the other perturbations is in progress [2].

Measurements of Landé g_J factors have for a long time served to determine coupling conditions in atoms. WYNNE et al. have performed optical Zeeman-effect measurements by multi-photon ionization spectroscopy for interacting Rydberg sequences in Sr [6]. We have performed extensive measurements of g_J factors for the $6snd$ states shown in Fig. 1. Most of the states were measured by observing Zeeman quantum beats but optical double-resonance and collimated atomic-beam fluorescence spectroscopy was also used [7].

Using Doppler-free two-photon spectroscopy, BEIGANG et al. have measured the hyperfine structure in the above-mentioned interacting Rydberg series in Sr [8]. We have performed hyperfine-structure measurements and determinations of isotope shifts for the n=11-17 members of the $6snd$ $^{1,3}D_2$ sequences of Ba, i.e. around the perturbation at n=14 [9]. In the $^1D_2$ as well as the $^3D_2$ sequence the hyperfine structure increases in the perturbed region. As in the Sr case this indicates strong polarization effects. Experiments on other alkaline-earth atoms are in preparation.

302

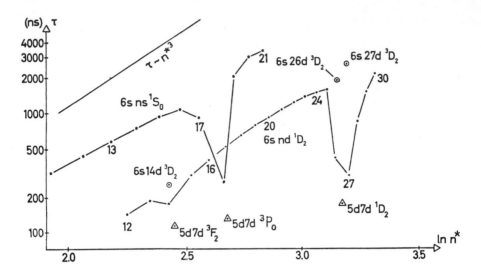

Fig.1 Experimental lifetime values for even-parity states of Ba showing
strong configuration interaction effects

References

1. K. Bhatia, P. Grafström, C. Levinson, H. Lundberg, L. Nilsson and
   S. Svanberg, submitted to Z. Physik.
2. P. Grafström, C. Levinson, H. Lundberg, S. Svanberg and M. Aymar, to be
   published.
3. H. Lundberg and S. Svanberg, Z. Physik A 290, 127 (1979).
4. T.F. Gallagher, W. Sandner and K.A. Safinya, to be published.
5. M. Aymar, R.-J. Champeau, C. Delsart and J.C. Keller, to be published.
6. J.J. Wynne, J.A. Armstrong and P. Esherick, Phys. Rev. Lett. 39, 1520
   (1979).
7. P. Grafström, P. Grundevik, C. Levinson, H. Lundberg, L. Nilsson and
   S. Svanberg, to be published.
8. R. Beigang, A. Timmermann and E. Matthias, to be published.
9. P. Grafström, Jiang Zan-Kui, G. Jönsson, S. Kröll, C. Levinson, H. Lundberg and S. Svanberg, to be published.

Part VII
Methods of Studying
Unstable Species

Optogalvanic Spectroscopy in Recombination-Limited Plasmas with Color Center Lasers*

R.J. Saykally, M.H. Begemann, and J. Pfaff
Department of Chemistry, University of California
Berkeley, CA 94720, USA

Abstract

Strong infrared optogalvanic signals have been observed in the region from
3600-4100 cm^{-1} for H, He, Li, Ne, and Ar excited in hollow cathode discharges
as a result of transitions induced among low Rydberg states of the atoms by
a cw F-center laser. On the order of fifty transitions have been assigned
in both neon and argon, eleven in helium, and one each in lithium and hydrogen. Studies of the quenching of helium and neon optogalvanic signals by
the addition of hydrogen, deuterium and nitrogen to the inert gas plasmas
indicate that a quasi-resonant process is occurring for the quenching of
helium $n = 4 \rightarrow 6$ transitions by H_2. The high signal-to-noise ratios observed for many of these transitions with only a few milliwatts of laser
power illustrates the potential of this technique for studying excited states
of atoms and molecules in plasma environments and suggests the use of atomic
Rydberg optogalvanic spectra for frequency calibration in the infrared.

The optogalvanic effect (OGE), in which intense monochromatic light is
used to cause an impedance change in a plasma medium through the excitation
of bound-bound transitions, has been rather broadly exploited at visible
wavelengths with the use of tunable dye lasers. By contrast, only a few OGE
experiments have been reported in the infrared region. The first of these
involved the observation [1] of large changes in the voltage across a CO_2
plasma tube when lasing occurred and the exploitation [2] of this effect
for stabilizing the laser. Recently Jackson et al. [3] have used a cw
F-center laser to study the $n = 4 \rightarrow 6$ infrared transitions of helium and
hydrogen in a positive column plasma. We report here the use of a cw F-center

*This work was supported by the National Science Foundation Grant # CHE-80-
07042.

laser to detect infrared optogalvanic spectra of H, He, Li, Ne, and Ar atoms excited in recombination-limited hollow cathode plasmas.

The experimental arrangement used in our earliest experiments is shown in Figure 1. A Kr^+ laser (Spectra Physics 171) operating with a 1-W output power on the 647 nm line was used to pump the F-center laser (Burleigh FCL-20), which in turn produced powers near 5 mW with the Type $F_A(II)$ (KCl:Na) crystal used for all of the measurements reported here. The infrared radiation, chopped at 500 Hz was directed through a photoacoustic cell (Burleigh) filled with dry nitrogen and into one of several commercial atomic absorption hollow cathode lamps. The photoacoustic spectrum of residual water in the nitrogen, used for wavelength calibration, and the laser-induced voltage change across the discharge ballast resistor were simultaneously monitored with separate lock-in amplifiers (PAR HR-8, PAR 124A) and displayed on a two-channel XY recorder (HP 7046A). The optogalvanic spectra from 3600-4100 cm^{-1} obtained in a sodium hollow cathode lamp with argon fill gas and in a lithium lamp buffered with neon are shown in Figures 2A and 2B, respectively. Many lines appear as doublets due to the existence of a spatial hole-burning mode 0.03 cm^{-1} from the fundamental of the F-center laser. Pressures in the lamps were in the range 5-10 Torr and the current was near 10 mA. The laser-induced voltage changes were very small, on the order of a few ppm, but the discharge noise levels were also very low, resulting in about the same signal-to-noise ratios observed by Jackson et al. in a positive column. In all about fifty transitions were found for neon, and assigned to the 4s → 4p, 4p → 4d, 4d → 4f, and 5s → 6p arrays. For argon approximately fifty lines

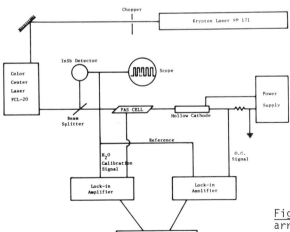

Fig.1. Diagram of experimental arrangement used to obtain infrared optogalvanic spectra

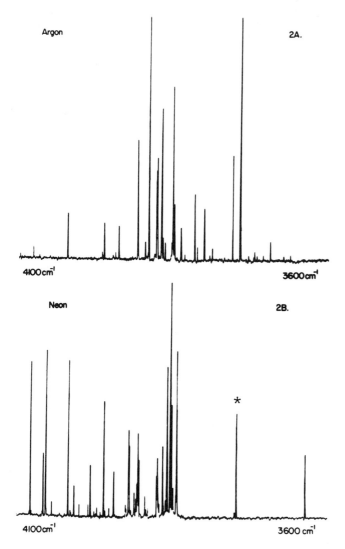

<u>Fig.2.</u> Optogalvanic spectra of argon and neon from 4100-3600 cm^{-1} obtained with an F-center laser. a) argon spectrum obtained in an argon-buffered sodium hollow cathode lamp. Pressure ~5 Torr. Operating current = 8 mamps. b) neon spectrum obtained in a neon-buffered lithium hollow cathode lamp. Pressure ~ 5 Torr. Operating current = 12 mamps. The line marked with an astrisk is the lithium 3s-3p transition

were assigned as 3d → 5p, 5s → 5p, and 5p → 5d. No lines were found in the argon-buffered sodium lamps that could be assigned to sodium atom transitions, but several broad features werc observed, the origin of which has not yet been established. A copper and a multielement cathode with neon buffer gas were found to produce all of the same lines observed with the lithium cathode, except for a single extra lithium line, which was assigned as 3s → 3p. Most transitions observed in these commercial hollow cathodes exhibited the same polarity, corresponding to a decrease in the discharge impedance during the laser irradiation. This is consistent with recent detailed modeling [4] of discharge processes which indicates that as populations in higher excited states is increased the ionization rates and electron densities increase correspondingly. The large signal-to-noise ratios observed for both neon and argon on many lines throughout the frequency region studied here and the high probability for extending these measurements to other regions (e.g., 10 μ) suggests that the use of optogalvanic atomic spectra for frequency calibration in the infrared might prove valuable.

Our goal in pursuing these infrared optogalvanic experiments is to determine whether this technique can be developed for systematic molecular spectroscopy studies. Because the need to vary pressures, gas compositions and currents requires a specially designed discharge system, we have constructed a hollow cathode discharge cell similar to those used in the metal vapor laser systems developed by Warner et al. [5]. The discharge consists of the negative glow confined between two water-cooled copper cathodes and an aluminum block anode. The cathode consists of two X-band waveguide pieces (6" × 1" × 0.5" outer dimensions) in a parallel arrangement separated by 0.4", thus forming a discharge channel with a cross section of 0.4" × 1". It can be operated at pressures in the range .05-10 Torr at currents up to 5 amps. The dynamics of a helium plasma in these devices has been worked out in detail [5]. The electron temperature is very low (1000°K), the neutral gas temperature remains nearambient, and the plasma density is high (10^{13} cm^{-3}), making the plasma suitable for molecular spectroscopy experiments.

The helium n = 4 → 6 infrared transitions studied by Jackson et al. [3] were the first to be observed in this system. Our signals were considerably larger than theirs, which were detected in a very different type•of plasma with current roughly ten times lower than those used here (0.2 amps). As in their experiments, the hydrogen n = 4 → 6 line appeared with no macroscopic source of hydrogen. An exponential decrease in signal intensity is observed upon an increase in the pressure.

Initial attempts to produce a stable hydrogen plasma were met with difficulty. Therefore hydrogen was added in small amounts to a helium discharge. The helium transitions were observed to quench more rapidly with the addition of a few percent of hydrogen than with deuterium or N_2. It was realized that the dissociation energy of H_2 (D_0 = 36,119 cm^{-1}) is about 700-1000 cm^{-1} greater than the energy gap between n = 6 of He and the He 2^3S state while that of D_2 is 1350-1670 cm^{-1} greater. Hence, it seems possible that we are observing dissociative collisions of H_2 molecules with He^* (n = 6) atoms, which is a resonant process requiring additional translational energy of about 2-3 times kT. In the case of D_2 an additional translational energy of 5-6 times kT is needed, which makes the dissociation much more unlikely. This dissociation process must have a very large cross section to compete effectively with ionization from n = 6 of helium and thereby decrease the size of the optogalvanic signals. These signals exist in the first place because the n = 6 level is more easily ionized than n = 4, both because it is 0.5 eV closer to the ionization limit and therefore can be ionized by a larger fraction of the thermal electrons in the Boltzmann tail and because its radiative lifetime is about 3 times longer than for n = 4. When H_2 is added to the plasma the laser possibly effects a "trading" of one n = 4 He for two ground state H atoms and a 2^3S He atom, which cannot be ionized by the thermal (0.1 V) electrons. No difference in the effects of adding D_2, H_2, or N_2 on neon OGE signals was detected. Clearly some interesting resonance effects are operating in the He/H_2 system.

Very recently we have found that the strongest helium OGE signals actually change sign when about 10% H_2 is added, and in addition exhibit a large phase shift of 35° relative to the pure helium case at modulation frequencies from 3-5 kHz. Although no sign reversal of the signals was observed when D_2 was added, it too caused a large phase shift. Changes in the kinetic processes involved in producing the optogalvanic signals on a millisecond time scale is evidenced by these phase shifts, probably resulting from diffusion of heavy particles in the plasma.

Acknowledgments. RJS thanks the Dreyfus Foundation, Research Corporation and the American Chemical Society for Grants. JP thanks the Chemical Manufacturer's Association for a postdoctoral fellowship. We thank Dr. Bruce Warner for his advice in constructing the discharge cell, and Dr. Tony O'Keefe for interesting discussions and suggestions.

References

1. A.I. Carswell, J.I. Wood: J. Appl. Phys. *38*, 3028 (1967)
2. M.L. Skolnick: IEEE J. Quantum Electron. *QE-6*, 139 (1970)
3. D.J. Jackson, E. Arimondo, J.E. Lawler, T.W. Hänsch: Optics Comm. *33*, 51 (1980)
4. R. Shuker, A. Gallagher, A.V. Phelps: J. Appl. Phys. *51*, 1306 (1980)
5. B.E. Warner, D.C. Gerstenberger, R.D. Reid, J.R. McNeil, R. Solanki, K.B. Persson, G.J. Collins: IEEE J. Quantum Electron. *QE-14*, 569 (1978)

LIF Studies of Fragment Ions: CH+ and CD+

A. O'Keefe, F.J. Grieman*, and B.H. Mahan

Department of Chemistry, University of California and
Materials and Molecular Research Division, Lawrence Berkeley Laboratory
Berkeley, CA 94720, USA

1. Introduction

We have combined the techniques of laser induced fluorescence (LIF) and three dimensional trapping of ions [1] to obtain the excitation spectra of several positive molecular ions. The radio frequency quadrupole ion trap we use allows mass spectrometric isolation and identification of the ions which greatly aids in the assignment of spectra. In addition, the ions are confined in a region easily probed by a laser and at a low enough pressure to produce collisionless conditions. The ions whose spectra we've obtained thus far include N_2^+ [2], CO^+ [3], H_2S^+, 1,3,5-trifluorobenzene$^+$, $ClCN^+$, and $BrCN^+$ [4]. However, because we can store ions, one real advantage of our method is in the study of fragment ions. To this end, we have recently observed the A-X bands of CH^+ and CD^+ [5]. Another advantage of ion storage in a collision-free environment is that it allows the measurement of reliable molecular ion radiative lifetimes.

2. Experimental

The experimental apparatus, a schematic of which is given in Fig.1, has been described in detail previously [4,5,6]. Briefly, $CH^+(CD^+)$ ions are created from $CH_4(CD_4)$ by a 2 msec electron beam pulse inside an r.f. quadrupole trap of cylindrical geometry [7]. At a mass resolution of 1 amu (m/Δm=40) a concentration of 10^7 ions·cm^{-3} can be maintained for several msec at a neutral gas pressure of 10^{-5} torr. After a 200 usec relaxation period, the ions are interrogated by a 10 nsec laser pulse from a N_2 laser pumped dye laser. Any subsequent fluorescence is then monitored by a "naked" photomultiplier tube and a gated photon counting system. The laser power and ion density are also measured with each laser pulse and are used to normalize the fluorescence signal. The experiment is controlled by a PDP-8 computer which scans the laser by 0.1 A increments and averages the signal over 500 laser pulses at each wavelength.

When radiative lifetimes are measured, the N_2 pump laser is replaced by a Nd:YAG laser because of its pulse-to-pulse power stability. The laser wavelength is fixed on a resonance and the fluorescence decay is monitored continuously by a signal averager, time base system. The system contains 1024 channels with individual channel widths of 10 nsec. Fluorescence and background signals are collected over several hundred thousand laser

*Present address: University of Oregon, Physics Dept., Eugene, OR 97403, USA

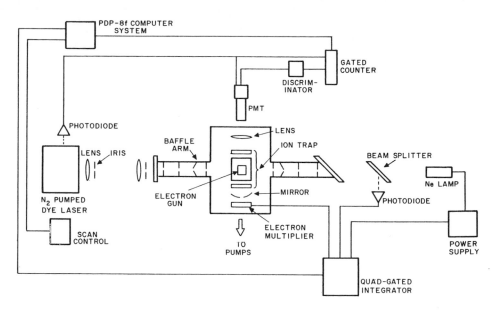

<u>Fig.1</u> Experimental arrangement used in the frequency scanning experiments

pulses. This system has been tested [3] with several ions having lifetimes ranging from 60 nsec to 10 μsec and good agreement has been found
with accepted values.

3. Results and Discussion

3.1 The $A^1\Pi$-$X^1\Sigma^+$ System of CH^+ and CD^+

We have observed the (0,0) and (2,1) band of the $A^1\Pi$-$X^1\Sigma^+$ transition of
CH^+ and CD^+. A representative portion of the CH^+ spectrum is presented in
Fig.2. Examination of the spectrum reveals a line width noticeably larger
than that of the bandwidth of the laser (1 cm^{-1}). The source of this
broadening is due to the large Doppler profile typical of ions stored in
an r.f. quadrupole trap [8]. The translational temperature of the CH^+ and
CD^+ ions is estimated to be ~4000 K under our conditions. A result of
this large translational energy is that the ion rotational population
distribution may be thermalized by collisions to a temperature much higher
than room temperature [2]. The rate at which this occurs depends on the
pressure of the background gas. At 10^{-5} torr, the pressure at which these
spectra were taken, we find that each ion will experience only one collision per msec using the Langevin estimate (usually considered an upper
bound) for the ion neutral collision rate. Thus, the CH^+ and CD^+ rotational and vibrational distribution we observe in our spectra is the
initial product state distribution of the $X^1\Sigma^+$ state created by electron
impact of CH_4 and CD_4 after fast radiative decay.

In our spectrum we observe a rotational temperature of ~3000 K which is
considerably higher than that found in the emission spectra [9,10]. Be-

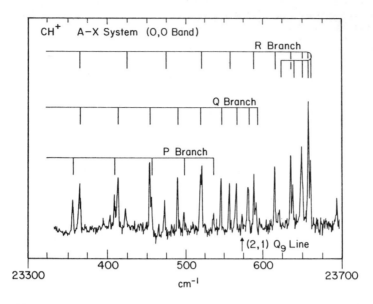

Fig.2 Representative portion of the (0,0) band of the CH$^+$ A-X system. The Q$_9$ line of the (2,1) band is identified by an arrow.

cause of this higher temperature, we were able to extend the (0,0) band of CH$^+$ and the (0,0) and (2,1) bands of CD$^+$ to higher J. Recently, quasibound levels corresponding to very high J of the A and X states of CH$^+$ have been observed by photofragment spectroscopy [11]. Our extension of the bound level spectroscopy to higher J has aided in the assignment of these very high J quasibound levels. The assignment of our spectra was based on the rotational constants of [9] for CH$^+$ and [10] for CD$^+$ both of which accurately reproduced lines corresponding to lower J, but increasingly overestimated transition frequencies at higher J. We first tried to fit our spectra by calculating higher order terms in the Dunham expansion formula (i.e. H$_y$) using the previously determined molecular constants, but could not obtain adequate agreement between the calculated and observed frequencies. Therefore, an unweighted least squares fit of all observed line frequencies was used to produce a new set of molecular constants. The molecular constants obtained for the (0,0) band of CH$^+$ and the (0,0) and (2,1) bands of CD$^+$ are reported in [5]. Because no new transitions in the (2,1) band of CH$^+$ were observed, no molecular constants were determined for this band.

3.2 Radiative Lifetimes of the A$^1\Pi$ State of CH$^+$ and CD$^+$

The measurement of the radiative lifetime of CH$^+$ has been the subject of numerous studies [12] where lifetimes ranging from 70 nsec to 630 nsec have been found for the (0,0) band. Because of this discrepancy and because our ion trap eliminates the problem of rapid spatial dissipation of ions due to the strongly repulsive ion-ion forces, we have made direct measurements of the A$^1\Pi$ radiative lifetime for the v'=0 level for both CH$^+$ and CD$^+$. The line used to pump this transition was the strong bandhead which consists of the overlapping R_3 and R_5 lines. The result for CH^+ is shown in Fig.3 where a radiative lifetime of 815±25 nsec was determined. The lifetime for CD^+ was found to be 820±50 nsec, the larger uncertainty being due to a weaker signal level.

Our measured lifetime for the A state of CH^+ is larger than any other previous experimental determination. However, our value agrees well with the theoretical results of [13] which predict a radiative lifetime of 660-800 nsec. Using the ratio of our measured lifetime to that of the calculated lifetime, the oscillator strength for the (0,0) band of CH^+ can be estimated from the calculated value of $f_{oo} = 6.45 \times 10^{-3}$. The result is $f_{oo} = (5.8 \pm 0.2) \times 10^{-3}$.

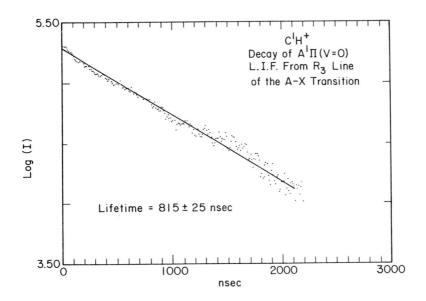

Fig.3 Least squares fit to the data collected from the decay of the $A^1\Pi(v'=0)$ of CH^+

References

1. H.G.Dehmelt: Adv. At. Mol. Phys. 3, 53 (1967); 5, 109 (1969)
2. B.H.Mahan, A.O'Keefe: J. Chem. Phys. 74, 5606 (1981)
3. B.H.Mahan, A.O'Keefe: to be published in Astrophys. J. 248 (1981)
4. F.J.Grieman, B.H.Mahan, A.O'Keefe: J. Chem. Phys. 74, 857 (1981)
5. F.J.Grieman, B.H.Mahan, A.O'Keefe, J.Winn: Disc.Faraday Soc. 71 (1981)
6. F.J.Grieman: LBL Report LBL-10021 (1979) (Ph.D. Thesis)
7. M.Benilan, C.Audoin: Int. J. Mass Spec. Ion Phys. 11,421 (1973)
8. R.D.Knight, M.H.Prior: J. Appl. Phus. 50, 3044 (1979)
9. A.E.Douglas, J.R.Morton: Astrophys. J. 131, 1 (1960)
10. A.Antic-Jovanovic, V.Bojovic, D.S.Pesic, S.Weniger: J. Mol. Spec. 75, 197 (1979)

11. H.Helm, P.C.Cosby, M.M.Graff, J.T.Moseley: Bull. Am. Phys. Soc. <u>25</u>, 1138 (1980); submitted to Phys. Rev. A
12. W.H.Smith: J. Chem. Phys. <u>54</u>, 1384 (1971); R.Anderson, D.Wilcox, R.Sutherland: Nuc. Inst. Meth. <u>110</u>, 167 (1973); J.Brzosowski, N.Elander, P.Erman, M.Lyyra: Astrophys. J. <u>193</u>, 741 (1974); N.H.Brooks, W.H.Smith: Astrophys. J. <u>196</u>, 307 (1974); P.Erman: Astrophys. J. <u>213</u>, L89 (1977)
13. M.Yoshimine, S.Green, P.Thaddeus: Astrophys. J. <u>183</u>, 899 (1973)

Sub-Doppler Spectroscopy of Some Gaseous Metal Oxide Molecules

A.J. Merer

Department of Chemistry, University of British Columbia
Vancouver, B.C. V6T 1Y6, Canada

The diatomic oxides of four of the first row transition metals, Sc, Ti, V and Fe have electronic spectra that are of interest in the astrophysics of cool star atmospheres. ScO and TiO have comparatively simple spectra which have been analysed at Doppler-limited resolution by grating methods [1,2]. The bands of VO are complicated by extensive hyperfine structure since ^{51}V has a nuclear spin I = 7/2 and a large nuclear magnetic moment, g_I = 5.15. FeO has an unusually complex band system in the orange region where large spin-orbit effects produce an almost random distribution of upper state sub-levels best described by case (c) coupling.

Controversy has surrounded the nature of the ground state of FeO, since the only bands that have been analysed from grating spectra have surprisingly simple rotational structures [3,4] which have been assigned as Ω' = Ω'' = 0 sub-bands. Arguments over whether the lower state is actually the ground state [5,6] have now been settled by the matrix IR work of GREEN, REEDY and KAY [7]; the lower state of the orange bands is the ground state. Theoretical predictions of the nature of the ground state give variously $^5\Sigma^+$ [8] and $^5\Delta$ [9].

We have investigated one of the simple bands, at 5820 Å, by laser-induced fluorescence at sub-Doppler resolution, using the technique of intermodulated fluorescence developed by SOREM and SCHAWLOW [10]. FeO was made by a microwave discharge in a flowing mixture of ferrocene, argon and oxygen, at a total pressure of about 1 torr. The results show that the simple bands consist of doubled lines, unresolved at Doppler-limited resolution, which must be interpreted as Λ-doubling in a transition between orbitally degenerate states. Further experiments have shown that the bands are actually Ω' = Ω'' = 4 sub-bands, from the first lines of the branches. This gives unambiguous evidence that the ground state of FeO is $^5\Delta$, because the other possible candidate, $^5\Sigma^+$, cannot produce Ω = 4 spin-orbit components.

Doppler-limited laser-induced fluorescence has since been used to map and identify bands involving the other four spin-orbit components of the ground state. The ground state, $X^5\Delta_i$, can be fitted by a simple model giving r_0 (Fe-O) = 1.619 Å. The upper states on the other hand are totally chaotic, with massive rotational perturbations in all observed levels.

The (0,0) band of the $C^4\Sigma^-$ - $X^4\Sigma^-$ electronic transition of VO has been recorded by intermodulated fluorescence over the range 17300-17427 cm^{-1}. VO was prepared by a microwave discharge in flowing

$VOCl_3$ and argon, and the C-X system gives bright yellow-orange fluorescence on laser excitation. Our line widths are limited by the pressure of the gas mixture to about 100 MHz, but this is sufficient for the hyperfine structure to be almost completely resolved, except in the crowded band head region.

The most interesting aspect of the band system is that internal hyperfine perturbations between the F_2 (N = J−1/2) and F_3 (N = J + 1/2) electron spin components occur in both electronic states. These perturbations are caused by matrix elements of the hyperfine Hamiltonian diagonal in N and F, but off diagonal in J [11]. In the absence of hyperfine effects the F_2 and F_3 electron spin components would cross (because of the particular values of the rotational and electron spin parameters) at N = 15 in the $X^4\Sigma^-$ state and at N = 5 in the $C^4\Sigma^-$ state. The result of the perturbations is that levels with the same F value in the two electron spin components mix with each other, and give rise to extra lines with the selection rules $\Delta N = \Delta F = \pm 1$; J = 0, ± 2.

The hyperfine patterns are illustrated in Figs. 1-3. Figure 1 shows the "normal" hyperfine patterns taken from the high-N P lines. The spread of the hyperfine structure is governed by the difference of the isotropic hyperfine paramters for the two electronic states, following the diagonal matrix elements of $b\mathbf{I}\cdot\mathbf{S}$,

$$
\begin{aligned}
W(F_1) &= 3\ b\ \mathbf{I}\cdot\mathbf{J}\ /(2N + 3) \\
W(F_2) &= \quad b\ \mathbf{I}\cdot\mathbf{J}\ (2N + 9)/[(2N + 1)(2N + 3)] \\
W(F_3) &= -\ b\ \mathbf{I}\cdot\mathbf{J}\ (2N - 7)/[(2N - 1)(2N + 1)] \\
W(F_4) &= -3\ b\ \mathbf{I}\cdot\mathbf{J}\ /(2N - 1)
\end{aligned}
\tag{1}
$$

where $\mathbf{I}\cdot\mathbf{J} = [F(F + 1) - I(I + 1) - J(J + 1)]/2$. Note the 3:1:-1:-3 pattern for the four electron spin components, which is very characteristic and allows a line to be assigned at once to the appropriate spin component. Figure 2 shows the hyperfine energy level structure for the

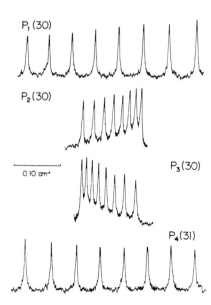

$P_1(30)$

$P_2(30)$

0.10 cm⁻¹

$P_3(30)$

$P_4(31)$

Fig. 1 Hyperfine structures of P lines from the four electron spin components of the C-X (0,0) band of VO.

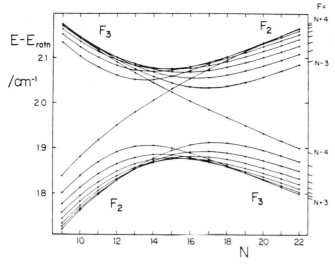

Fig. 2 Pattern of hyperfine levels at the $X^4\Sigma^-$ $v = 0$ internal hyperfine perturbation in the VO molecule.

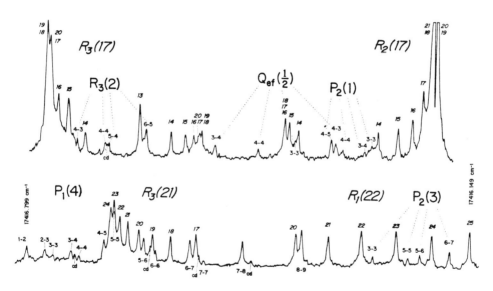

Fig. 3 Two regions from the centre of the VO $C^4\Sigma^- - X^4\Sigma^-$ (0,0) band. Small numbers are the F hyperfine quantum numbers. High-N R branch assignments are given in italics.

ground state internal hyperfine perturbation near N = 15. Seven of the eight hyperfine levels in the two electron spin components undergo an avoided crossing, but the levels with F = N + 4 from the F_2 component and F = N - 4 from the F_3 component have no levels in the other component to interact with and pass through the perturbation unaffected. Figure 3 shows two regions from the centre of the band. The upper tracing shows the $R_2(17)$ and $R_3(17)$ lines together with various low N lines. The extra lines induced by the internal hyperfine perturbation are clearly seen as the two clumps of lines near the centre of the tracing. Even though the transition is $^4\Sigma^-$ - $^4\Sigma^-$ weak Q branch lines occur as a result of the uncoupling to case (a) at low rotational quantum numbers: the Q_{ef} (1/2) line is a good example, where all four of the possible hyperfine components have been seen. Hyperfine lines with $\Delta F \neq \Delta J$ occur at low N values, and cross-over signals (labelled c.d., for centre dip) can be seen on the lower tracing. These additional lines give direct hyperfine combination differences, and have proved very useful in breaking the least squares correlation between the upper and lower state constants imposed by the parallel selection rules of the electronic transition.

The $C^4\Sigma^-$ state of VO is unfortunately affected by many small electronic perturbations, so that it has been difficult to achieve a least squares fit which does justice to the precision of the data. In the end only the lines up to N' = 24 and the ground state $\Delta_2 F$ combination differences were used; the standard deviation was 23 MHz. The results are given in Table 1.

Table 1 Rotational, spin and hyperfine constants for the v = 0 levels of the $C^4\Sigma^-$ and $X^4\Sigma^-$ states of VO, in cm^{-1}.

		$C^4\Sigma^-$		$X^4\Sigma^-$	
	T_0	17420.1025_7	$\pm\ 0.0001_7$	0.0	
	B	0.4937896_6	$\pm\ 0.0000003_3$	0.5463833_3	$\pm\ 0.0000002_9$
10^7	D	6.44	$\pm\ 0.03$	6.50_9	$\pm\ 0.01_4$
	γ	-0.018444	$\pm\ 0.000069$	0.022516	$\pm\ 0.000066$
	λ	0.74697	$\pm\ 0.0003$	2.03087	$\pm\ 0.0002_4$
10^7	γ_D	5.43	$\pm\ 0.50$	0.56	$\pm\ 0.32$
10^6	λ_D	-4.3	$\pm\ 0.5$	0.0	fixed
10^5	γ_S	-23.1	$\pm\ 1.4$	-1.0	$\pm\ 1.5$
	b	-0.00881	$\pm\ 0.00003$	0.02731	$\pm\ 0.00004$
	c	-0.00114	$\pm\ 0.00009$	-0.00413	$\pm\ 0.00008$
	e^2Qq	0.00139	$\pm\ 0.00023$	0.00091	$\pm\ 0.00088$
10^6	C_I	-3.9	fixed	4.7	fixed
10^5	b_S	4.5	$\pm\ 1.8$	0.0	fixed

The capable experimental work of Dr. A.M. Lyyra and Mr. A.S-C. Cheung is gratefully acknowledged.

References

1. R. Stringat, Ch. Athénour and J-L. Féménias, Can. J. Phys. <u>50</u>, 395 (1972).
2. W.H. Hocking, M.C.L. Gerry and A.J. Merer, Can. J. Phys. <u>57</u>, 54 (1979).
3. R.K. Dhumwad and N.A. Narasimham, Proc. Ind. Acad. Sci. <u>A64</u>, 283 (1966).
4. R.F. Barrow and M. Senior, Nature, <u>223</u>, 1359 (1969); S.M. Harris and R.F. Barrow, J. Mol. Spectrosc. <u>84</u>, 334 (1980).
5. P.C. Engelking and W.C. Lineberger, J. Chem. Phys. <u>66</u>, 5054 (1977).
6. T.C. De Vore and T.W. Gallaher, J. Chem. Phys. <u>70</u>, 4429 (1979).
7. D.W. Green, G.T. Reedy and J.G. Kay, J. Mol. Spectrosc. <u>78</u>, 257 (1979).
8. H.H. Michels, unpublished work quoted in ref. 4.
9. P.S. Bagus and H.J.T. Preston, J. Chem. Phys. <u>59</u>, 2986 (1973); see also ref. 5.
10. M.S. Sorem and A.L. Schawlow, Optics Comm. <u>5</u>, 148 (1972).
11. D. Richards and R.F. Barrow, Nature, <u>219</u>, 1244 (1968).

The Infrared Spectrum of H_3^+

T. Oka

Herzberg Institute of Astrophysics, National Research Council of Canada
Ottawa, Ontario K1A OR6, Canada

1. Introduction

Just as the most basic task of an astronomer is to find a new star, the most basic task of a spectroscopist is to find a new spectrum. But just as an astronomer is happier if an extraordinary object rather than another ordinary star is found, a spectroscopist is happier if a spectrum of some unusual species is found. And just as the new method of radioastronomy has led to the discoveries of such exotic objects as quasars, pulsars, 3 K blackbody radiation and molecular clouds, we would hope that laser spectroscopy will enable us to discover new spectra of exotic species.

The H_3^+ molecular ion is the simplest polyatomic system, with two electrons keeping three protons in an equilateral triangle configuration. I consider H_3^+ as a system in which a hydrogen molecule has swallowed a proton; and it swallows with a tremendous appetite. The formation energy of the reaction $(H_2 + p \rightarrow H_3^+ + 4.8 \text{ eV})$ is even greater than the formation energy (4.5 eV) of H_2 from two hydrogen atoms. The proton is well bound to H_2.

Because of this stability H_3^+ is the most abundant hydrogenic ion in laboratory plasma and in dark molecular clouds. However there has previously been no spectroscopic observation of this species in any range. This is probably because H_3^+ is predissociated in electronic excited states and does not have a discrete optical spectrum. The vibrational spectrum in the infrared region seems to be the only way to study this ion spectroscopically. This is a beautiful jewel of nature left for the laser spectroscopist.

2. Method

A direct infrared absorption method combining a multiple-reflection discharge cell and a frequency-tunable infrared source was chosen as the detection method [1]. A block diagram of the apparatus is shown in Figure 1.

A difference-frequency laser system developed by Alan PINE [2] was constructed and used as the frequency tunable infrared source. Mixing radiation from an Ar laser (ν_A) and that from a dye laser (ν_D) in a temperature-controlled $LiNbO_3$ crystal, we obtained an infrared radiation source with a power of a few microwatts whose frequency ($\nu_A - \nu_D$) was tunable over the range 4400 - 2400 cm^{-1}. This continuous coverage over a wide infrared region was essential in the detection and identification of H_3^+ because the spectrum extends over a region of ~ 600 cm^{-1} and the estimated accuracy of the theoretical prediction [3] was ~ 50 cm^{-1}.

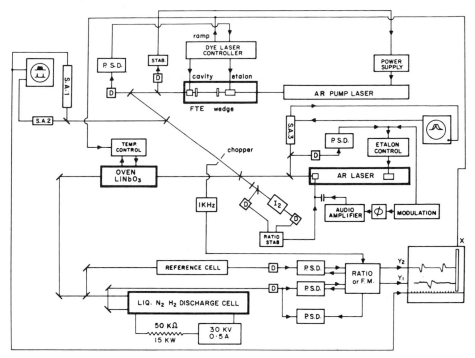

Fig.1 Apparatus

A Spectra Physics 580 dye laser pumped by a 171 Ar laser generated red to yellow radiation with a power of 20 ∿ 60 mWatts and spectral purity of 5 ∿ 30 MHz. The scanning of the etalon was locked to that of the cavity to prevent mode hops and to minimize amplitude fluctuation. The latter was further reduced to the level of less than 1% by feedback to the pump Ar laser. The dye laser output was mixed with either 5145 Å or 4880 Å radiation (∿ 80 mWatts) from a 165 Ar laser on a dichroic mirror. The frequency of this Ar laser was modulated with an amplitude of ∿ 400 MHz and a frequency of 2.5 kHz for frequency modulation of the infrared radiation. In order to do this an etalon mounted on a piezoelectric element was inserted in the Ar laser cavity and modulated synchronously with the cavity. Here also the etalon was locked to the cavity to minimize amplitude modulation.

The multiple-reflection discharge cell was 2 m long with a diameter of 2 cm and was cooled with liquid nitrogen as was done by WOODS and his collaborators for their microwave detection of molecular ions [4]. The discharge tube was sealed with two CaF_2 Brewster angle windows and placed inside a multiple reflection mirror system. This arrangement allowed easy optical alignment and a wide variation of discharge conditions. The radiation loss due to Brewster windows was less than that due to the reflecting mirrors. An absorption path of 32 meters was normally used. The discharge current density of the plasma and the pressure of H_2 varied between 40 ∿ 120 mA/cm^2 and 0.3 ∿ 10 Torr, respectively. For observing transitions starting from higher J levels, water was used for cooling instead of liquid nitrogen.

The overall sensitivity for the search was ∿ 1.5×10^{-2} and the speed of scanning was ∿ 8 cm^{-1}/hr. This is by no means a high sensitivity spectroscopy but a reading of previous papers on hydrogen discharge [5][6] convinced me that this sensitivity should be sufficient.

3. Observed Spectrum

After two years' assembling and adjusting the spectrometer and another two years' search and continuous improving of the apparatus, the first absorption line of H_3^+ was observed on April 25, 1980. It took another 8 months to reveal the whole spectral pattern between 2450 cm^{-1} – 3050 cm^{-1}. It is shown in Fig. 2 for liquid nitrogen cooling (temperature of discharge $\sim$ 200 K) and ice water cooling ($\sim$ 450 K).

Altogether 30 absorption lines have been observed and assigned. For liquid nitrogen cooling no other lines were observed except for the Bracket line (n = 5 ← 4) of the hydrogen atom at 2467.746 cm^{-1} which happened to appear with an intensity similar to that of the H_3^+ lines. For ice water cooling, many spurious lines were observed; the H_3^+ lines were identified from their intensity dependence on discharge conditions.

The spectrum was analyzed by Jim WATSON [1][7] and least squares fitted to 13 vibration-rotation parameters. The lack of any obvious regularity or symmetry in the spectral pattern in Fig. 2 is due to a large ℓ-type resonance with ℓ-doubling constant q = -5.383 cm^{-1}. The lower order parameters derived by Watson agree well with those predicted by CARNEY and PORTER [8] from ab initio calculations.

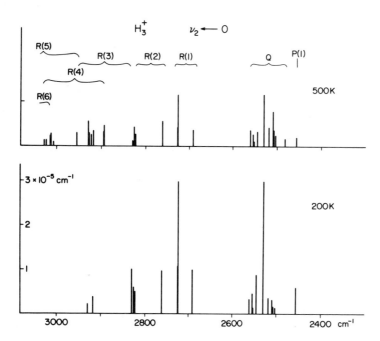

Fig.2 Observed spectrum

Shortly after my submission of the manuscript to Physical Review Letters
[1], I received a telephone call from Bill WING and learned that his group
has observed [9] the infrared spectrum of D_3^+.

4. Discussion

The success of our observation of the infrared spectrum of H_3^+ seems to open
the following paths for future research in various fields: (1) systematic
infrared spectroscopy of molecular ions in discharges, (2) in situ monitoring of molecular ions for ion mobility measurements and ion-molecule
reactions, (3) detection of H_3^+ in interstellar space.

An observation of the infrared spectrum of H_3^+ was attempted on March 23-
25, 1981 with the high resolution Condé Fourier transform spectrometer on
the 4 m Mayall telescope of the Kitt Peak National Observatory [10]. The
result was negative. More observations will be attempted for this important
and fundamental molecular ion.

References

1. T. Oka: Phys. Rev. Lett. 45, 531 (1980)
2. A.S. Pine: J. Opt. Soc. Amer. 64, 1683 (1974); 66, 97 (1976).
3. G.D. Carney, R.N. Porter: J. Chem. Phys. 65, 3547 (1976).
4. T.A. Dixon, R.C. Woods: Phys. Rev. Lett. 34, 61 (1975).
5. M. Saporoschenko: Phys. Rev. 139A, 349 (1965).
6. D.L. Albritton, T.M. Miller, D.W. Martin, E.W. McDaniel: Phys. Rev.
    174, 94 (1968).
7. J.K.G. Watson: to be published.
8. R.N. Porter: private communication.
9. J.T. Shy, J.W. Farley, W.E. Lamb, Jr., W.H. Wing: Phys. Rev. Lett. 45,
    535 (1980).
10. T. Oka: "Molecules in Interstellar Space", the Royal Society Meeting
    for Discussion, London, May (1981).

Infrared Vibrational Predissociation Spectroscopy of Small Molecular Clusters

J.M. Lisy, M.F. Vernon, A. Tramer, H.-S. Kwok, D.J. Krajnovich, Y.R. Shen, and Y.T. Lee

Materials and Molecular Research Division, Lawrence Berkeley Laboratory, University of California, Berkeley, CA 94720, USA

1.   Introduction

The structure and interaction of small molecular clusters have been the subject of extensive studies. Molecular beam electric resonance spectroscopy [1], Fourier-transform microwave spectroscopy [2] and pressure induced infrared absorption spectra [3] have provided information concerning the ground state structure and the intermolecular potential energy surface for a large number of binary systems. Recently, vibrational predissociation experiments using molecular beam techniques and tunable infrared lasers have measured the infrared absorption spectra and dynamical properties of many van der Waals and hydrogen bonded clusters.

SCOLES et al. [4] first observed vibrational predissociation of $(N_2O)_2$ by exciting near the ν_3 transition of N_2O using a tunable diode laser. The vibrational predissociation lifetime, τ, was found to be within the range 10^{-12} secs $< \tau < 10^{-4}$ secs. The lower limit was determined by attributing the total width of the band to predissociation induced homogeneous broadening. The upper limit was fixed by the time of flight from the irradiation zone to the bolometer detector. Van der Waals complexes of C_2H_4 with Ne, Ar, Kr, C_2H_4, C_2F_4 and larger clusters have been predissociated using a CW CO_2 laser near the ν_7 mode of C_2H_4 by JANDA and coworkers [5]. The bandwidths and lineshapes were shown to be consistent with a homogeneous broadening mechanism. The corresponding predissociation lifetimes ranged from 0.3 to 1 psec for the various ethylene complexes.

For weak van der Waals bonding, there is little or no frequency shift of the vibrational predissociation spectrum from the infrared absorption spectrum of the monomer. However, for complexes which involve hydrogen bonds, substantial frequency shifts are observed for the pertinent X-H stretching motions. In our study of small water clusters [6], the dimer frequencies displayed red shifts of approximately 50 cm^{-1}, while the bands for the trimer through hexamer extend from 3100 to 3720 cm^{-1}. The vibrational predissociation spectra of these larger clusters bear a remarkable resemblance to the infrared spectra of liquid H_2O.

In this paper, we take $(C_6H_6)_{2,3}$ and $(HF)_{2-6}$ as examples to discuss both the spectroscopic and dynamic aspects of infrared vibrational predissociation processes.

2. Experimental

Molecular beams containing a small fraction of molecular clusters are produced by supersonic expansion through a nozzle. The infrared laser is a Nd-YAG pumped $LiNbO_3$ optical parametric oscillator based on the design of BYER [7]. In the frequency range 3000-4000 cm^{-1}, the pulse energy is typically 1 to 4 mJ with a pump energy fluence of 1.0 J/cm^2. The OPO linewidth (FWHM) varies from 4 to 10 cm^{-1}.

The vibrational predissociation process of the molecular clusters is monitored using a mass spectrometer in two different arrangements described below. Using mass spectrometric detection in these arrangements avoids two problems which occur in gas-phase and matrix-isolation absorption work, the interference from monomer absorptions and the assignment of spectral features to specific clusters.

2.1 Perpendicular Laser-Molecular Beam Arrangement

An in-plane view of this configuration is shown in Fig.1. In this arrangement, the vibrational predissociation process is monitored by detection of the cluster <u>fragments</u>. The electron impact mass spectrometer detector, which rotates about the laser-molecular beam interaction zone, observes only those molecules emanating from the interaction zone (~8 mm^3). The predissociation fragment products are detected by rotating the mass spectrometer by an angle θ relative to the molecular beam. The clusters which dissociate within 2 μsec after the laser pulse (before traveling beyond the detector's viewing range) and whose products recoil along the angle θ, can be detected.

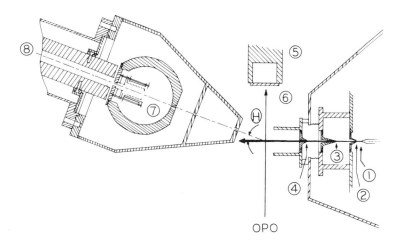

OPO

Fig.1 In plane view of perpendicular laser molecular beam apparatus. Labeled components are: 1. 0.178 mm quartz nozzle heated to 70°C, 2. First skimmer, 3. Second skimmer, 4. Third skimmer, 5. Power meter, 6. Germanium filter, 7. Ionizer assembly, 8. Quadrupole mass spectrometer. θ measures the angle of rotation of the detector from the molecular beam.

The infrared laser enters the vacuum chamber through a BaF$_2$ lens mounted on a movable tube in order to focus the laser at the interaction zone. The laser is linearly polarized perpendicular to the plane defined by the laser and molecular beam. The energy is monitored by a power meter in the vacuum chamber just beyond the interaction zone and determines the power dependence of the absorption.

The mass spectrometer signal is collected by a 255 channel, variable channel width, multichannel detector (MCS) triggered by the laser pulse. The time interval between the laser pulse and the arrival of molecules at the detector, corrected for the detector ion flight times, gives the fragment time of flight from the interaction zone. The signal is averaged over 2,000 to 10,000 laser pulses. The angular and velocity distributions of the predissociation products can be used to determine the translational energy distribution of the product molecules.

2.2 Coaxial Laser-Molecular Beam Arrengement

A cross-sectional view of this arrangement is shown in Fig.2. The ion optics and quadrupole mass spectrometer are perpendicular to the molecular beam, permitting the laser to travel along the molecular beam flight path from the nozzle source to the ionizer while the mass spectrometer continuously monitors the molecular beam. Vibrational predissociation is observed by the depletion of the mass spectrometer signal from the predissociating parent cluster. The signal is again detected by the 255 channel MCS. The molecular beam flight time from the nozzle to the ionizer is approximately 1 msec. This determines the upper limit to the vibrational predissociation lifetimes observable with this apparatus. The lower limit is determined by the minimum MCS channel width of 1 μsec. The OPO energy is monitored by measuring a fraction of the output reflected off a BaF$_2$ beam splitter. By directly measuring the depletion of the parent clusters, this arrangement is not sensitive to the angular or translational energy distributions of the fragments. The advantage of this configuration is a large increase in sensitivity which arises from two factors: a 65 cm length of the molecular beam is probed as opposed to

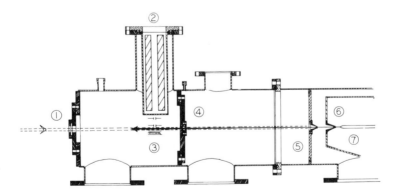

Fig.2 Side view of the coaxial laser-molecular beam apparatus. Labeled components are: 1. BaF$_2$ entrance window for the OPO beam, 2. Quadrupole mass spectrometer, 3. Ionizer assembly, 4. Final molecular beam defining aperture, 5. Second skimmer, 6. First skimmer, 7. Nozzle.

a 2 mm length in the perpendicular experiment, and all of the dissociation events are detected as opposed to the small solid angle sampling of the rotating detector.

3. Results

The vibrational predissociation spectra taken at m/e = 78 ($C_6H_6^+$) and 390 (($C_6H_6)_5^+$) for benzene clusters produced in two different expansion conditions using the perpendicular laser–molecular beam arrangement are shown in Fig.3. These fragments must have originated from clusters containing at least two or six benzene molecules respectively. The signals represent the total laser induced signal corrected for background and photon number and are thus proportional to the absorption cross section. The frequency shifts between the two spectra are small with respect to the laser linewidth of 3–4 cm^{-1}.

The angular distributions of the m/e = 78 and 156 cluster fragments, shown in Fig.4, are obtained in a separate experiment with expansion conditions chosen to avoid the formation of larger clusters. The extent of cluster formation is checked by rotating the detector to 0° along the molecular beam axis. The ratios of m/e 156:234:312 in the molecular beam were found to be 100:3:0.3. Under these conditions, the off–axis signal at m/e = 78 is assigned to the monomer fragment from reaction 1 and the m/e = 156 signal, based on the observed intensity, is assigned to the dimer fragment of reaction 2:

$$(C_6H_6)_2 \xrightarrow{\ h\nu\ } 2\ C_6H_6 \tag{1}$$

$$(C_6H_6)_3 \xrightarrow{\ h\nu\ } C_6H_6 + (C_6H_6)_2. \tag{2}$$

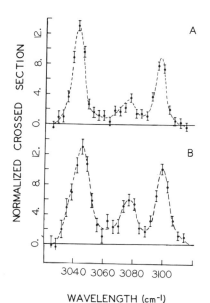

Fig.3 Wavelength dependence of the predissociation cross sections. A) Mass 390, ($C_6H_6)_5^+$, at 400 torr Ar with T_{nozzle} = 25°C, B) Mass 78, $C_6H_6^+$, at 250 torr Ar with T_{nozzle} = 25°C.

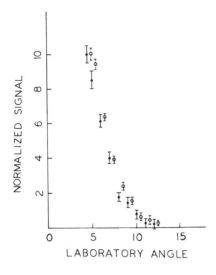

Fig.4 Wavelength dependence of the mass 78 angular distributions. Open circles ν_{OPO} = 3040 cm^{-1}, solid circles, ν_{OPO} = 3095 cm^{-1}. The 3040 cm^{-1} data have been displaced +0.5° relative to the 3095 cm^{-1} data to allow easier comparison.

The laboratory velocity distributions of all fragments are found to be almost identical with the initial molecular beam velocity distribution. The similar fragment angular distributions and narrow angular range ($\theta <$ 10°), together with the time of flight data indicate qualitatively that only a small amount of translational energy is imparted to the fragments when the cluster dissociates. This was found to be true for all the clusters investigated including hydrogen bonded $(HF)_n$ and $(H_2O)_n$ clusters. The predissociation signal is observed to be independent of the OPO polarization.

The vibrational predissociation spectra of $(HF)_n$ were obtained by the coaxial laser–molecular beam arrangement in order to increase the rate of data collection. As $(HF)_n$ cracks extensively under electron impact, $(HF)_n$ predissociation was observed by detecting the depletion of the predominant ionization fragment $(H_nF_{n-1})^+$ after the laser pulse. Fig.5 contains the $(HF)_2$ spectrum and the vibrationally averaged ground state structure as determined by molecular beam electric resonance spectroscopy [8]. The bandwidths are laser linewidth limited (~8 cm^{-1}). The spectra of $(HF)_{n=3-6}$ displayed in Fig.6, are broader than the laser linewidth (4–5 cm^{-1}), with a pronounced red shift from the HF monomer frequency of 3960 cm^{-1}. No other features were observed in the 3000–4000 cm^{-1} region. The signals have a linear dependence on photon number. Depletion of the molecular beam signal was found to occur directly after the laser pulse, indicating predissociation lifetimes of less than 1 μsec for all $(HF)_n$.

The assignment of the spectra to specific clusters was done through a series of experiments by varying the source pressure, carrier gas and seeding ratios. The mass spectrometer was essential in confirming assignments by comparing spectra taken at different m/e settings. For example, $(HF)_3$ bands are observed only at m/e = 41 while the $(HF)_4$ bands can be observed at both m/e = 41 and 61.

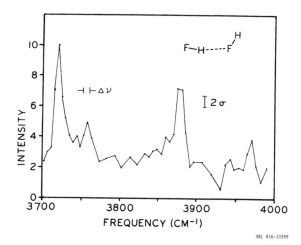

Fig.5 Vibrational predissociation spectrum of $(HF)_2$ corrected and normalized for photon number, with the structure of $(HF)_2$ as determined by molecular beam electric resonance spectroscopy [8].

XBL 816-10399

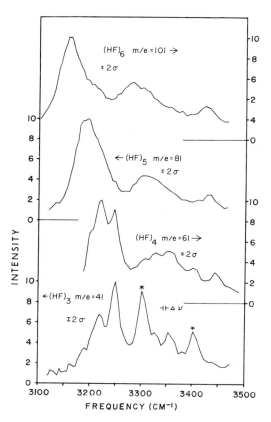

Fig.6 Vibrational predissociation spectra of $(HF)_n$, n = 3-6, corrected and normalized for photon number. The peaks marked with asterisks in the m/e = 41 spectrum are from $(HF)_3$, the other peaks are due to larger clusters (see text).

4. Discussion

From the perpendicular laser molecular beam experiments, the main product channels for the predissociation of benzene dimer and trimer are found to be given by reactions (1) and (2).

The predissociation lifetimes are less than 2 μsec, and a lower limit of 1 psec may be deduced from assuming that all of the observed linewidths (5 cm^{-1}) arise from homogeneous broadening caused by the predissociation. This is consistent with the lack of polarization dependence of the predissociation cross-sections, since after photon absorption, the cluster would rotate on this timescale prior to dissociation. Predissociation of $(C_6H_6)_2$ near 1040 cm^{-1} has been observed [9,10] with a linewidth [10] of 3 cm^{-1} using a CW CO_2 laser with a linewidth <0.01 cm^{-1}.

The translational energy distribution obtained from the laboratory angular and velocity distributions indicates that the average translational energy is small compared to the excess energy available for breaking the van der Waals bond between benzene molecules. Since the initial internal excitation in the benzene clusters is small due to adiabatic cooling in the supersonic expansion, the excess energy of approximately 2300 cm^{-1}, which corresponds to the difference between the excitation energy (3050 cm^{-1}) and the bond dissociaion energy (~750 cm^{-1}) [11,12], is mainly retained as internal excitation of the fragments.

An important question is how this energy is distributed between the monomer and dimer fragments from the trimer predissociation. Since the dimer fragment would itself predissociate should it contain more than 750 cm^{-1} of energy, the observation of dimer fragments places a lower limit to the internal energy of the monomer fragment of 1550 cm^{-1}. Vibrational energy transfer between the benzene molecules in the cluster must be slow with respect to predissociation. It seems likely that the photon excitation is isolated to a single molecule, and it becomes the internally excited monomer fragment. This mechanism would predict that the dimer predissociation will produce one monomer fragment vibrationally "hot" with the other vibrationally "cold". The observation of small average translational energy of the fragments with the translational energy distribution peaking at zero is in agreement with the momentum gap law suggested by EWING [13] or the energy gap laws of BESWICK [14].

The $(HF)_n$ spectra obtained from the colinear arrrangement permit a number of observations. The $(HF)_2$ bands have been assigned as follows: the 3720 cm^{-1} band to the HF stretch of the hydrogen bonded proton, the 3878 cm^{-1} band to the HF stretch of the "free" proton and the 3970 cm^{-1} to a combination band involving an intra- and intermolecular mode. The latter two bands have been observed in the gas-phase [15,16] and our results confirm their identification as dimer bands. The observed rotational structure [16] of the 3970 cm^{-1} band with linewidths of approximately 0.15 cm^{-1}, combined with our time of flight results enables the predissociation lifetime, τ, to be limited to 30 psec < τ < 1 μsec. The predicted lifetime of $(HF)_2$ based on an empirical analysis [17] of vibration to translation energy transfer is ~200 psec, which falls within the above range.

The $(HF)_{n=3-6}$ clusters exhibit a large frequency shift from the HF fundamental. The lack of absorption above 3500 cm^{-1} indicates the absence of a terminal ----H-F or ---F-H groups, indicating that $(HF)_{n=3-6}$ are cyclic structures with each HF monomer both donating and accepting a proton. This is in agreement with molecular beam electric deflection results [8] which indicate negligible permanent dipole moments in these species. The band structure of the $(HF)_{n=3-6}$ spectra can be fit by the frequency equation $\nu_n = \nu_n' + m\nu_n''$ where $m = 0,1,2,$ ν_n' is the frequency of the intramolecular H-F mode decreasing with the cluster size n, and ν_n'' is the frequency of the intermolecular F---HF mode increasing with n. In analogy with other hydrogen bonded systems [18] these spectral bands are assigned as a combination band series involving ν_n' and ν_n'' modes. The frequency variations with cluster size, n, are in accord with the strengthening of the intermolecular interaction and weakening of the intramolecular bands in progressively larger clusters.

5. Conclusion

The combination of tunable infrared radiation and molecular beam techniques has enabled spectroscopic and energy disposal information on the vibrational predissociation dynamics of molecular clusters to be measured. Examples of weak, (C_6H_6), and strong, (HF), intermolecular bonding have been presented. Increased laser resolution and isotopic studies will extend the preliminary information gained here and, it is hoped, stimulate the continued development of theoretical models for these processes.

References

1. T. A. Dixon, C. H. Joyner, F. A. Baiocchi, W. Klemperer, J. Chem. Phys. 74, 6539, 6544, 6550 (1981)
2. A. C. Legon, P. D. Soper, W. H. Flygare, J. Chem. Phys. 74, 4936 (1981), ibid 74, 4944 (1981).
3. A. R. W. McKellar, H. L. Welsh, J. Chem. Phys. 61, 4636 (1974).
4. T. E. Gough, R. E. Miller, G. Scoles, J. Chem. Phy. 69, 1588 (1978).
5. M. P. Casassa, D. S. Bomse, K. C. Janda, J. Chem. Phys. 74, 5044 (1981).
6. M. F. Vernon, D. J. Krajnovich, H. S. Kwok, J. M. Lisy, Y. R. Shen, Y. T. Lee, to be published.
7. S. Brosnan, R. L. Byer, IEEE Jour. Quant. Elect. QE-15, 415 (1979).
8. T. R. Dyke, B. J. Howard, W. Klemperer, J. Chem. Phys. 56, 2442 (1972).
9. W. R. Gentry, unpublished results.
10. K. C. Janda, private communication.
11. D. J. Evans, R. O. Watts, Mol. Phys. 29, 777 (1975), ibid 31, 83 (1976), ibid 32, 93 (1976).
12. T. B. MacRury, W. A. Steele, B. J. Berne, J. Chem. Phys. 64, 1288 (1976).
13. G. E. Ewing, J. Chem. Phys. 72, 2096 (1980).
14. J. A. Beswick, J. Jortner, Adv. Chem. Phys. XLVII, Part 2, p.363-506 (1981).
15. D. F. Smith, J. Chem. Phys. 28, 1040 (1958).

16. J. L. Himes, T. A. Wiggins, J. Mol. Spectrosc. 40, 418 (1971).
17. W. Klemperer, Ber. Bunsenges. Phys. Chem. 78, 1281 (1974); D. A. Dixon, D. R. Herschbach, W. Klemperer, Faraday Discuss. Chem. Soc. 62, 341 (1977).
18. a) Y. Marechal, A. Witkowski, J. Chem. Phys. 48, 3697 (1968); b) M. Wojcik, Mol. Phys. 36, 1757 (1978); c) G. N. Robertson, Phil. Trans. R. Soc. Lond., A286, 25 (1977).

The Direct Photodissociation of Van Der Waals Molecules

E. Carrasquillo M., P.R.R. Langridge-Smith, and D.H. Levy

The James Franck Institute and The Department of Chemistry
University of Chicago, 5640 S. Ellis Ave., Chicago, IL 60637, USA

The laser induced photodissociation of a number of van der Waals molecules has now been studied. Prior to this work, all of these van der Waals photo-dissociation reactions have been predissociations [1] involving laser excitation to a quasi-bound state. In this case the energy that will eventually be used to break the van der Waals bond is initially stored in some other vibrational mode, usually one of the vibrational modes of the chemically bound part of the molecule. After a finite period of time energy is transferred from the storage mode to the dissociating van der Waals mode, and the bond is broken.

We have now observed the direct dissociation reaction of a number of nitric oxide-rare gas van der Waals molecules. These van der Waals molecules are produced in free jet expansions of nitric oxide seeded into various mixtures of rare gases. Fluorescence excitation and dispersed fluorescence spectra were excited using a Nd:YAG pumped rhodamine 6G dye laser which was frequency doubled and hydrogen Raman shifted to excite the A$\leftarrow$X electronic transition of NO near 44200 cm^{-1}. In these molecules the van der Waals potential curve of electronically excited NO is shifted outward from that of ground state NO, and the Franck-Condon allowed transitions are from the zero-point level of the ground electronic state to the repulsive wall of the excited electronic state as shown in Fig. 1.

In the case of predissociation the laser-induced fluorescence excitation spectrum of the van der Waals complex is fairly sharp and only slightly shifted in frequency from the spectrum of the uncomplexed molecule [2]. As seen in Fig. 2, the fluorescence excitation spectra of the directly dissociating van der Waals molecules NOAr, NONe, and NOHe are broad bands several hundred cm^{-1} to the high frequency side of the sharp features produced by uncomplexed NO.

Because the direct dissociation reaction is fast compared to the fluorescence lifetime, all of the observed fluorescence is due to the product NO* produced in the reaction NOX* $\rightarrow$ NO* + X. In Fig. 3 we show the dispersed fluorescence spectrum that is produced when (top) uncomplexed NO is excited or (bottom) the NOAr complex is excited near the peak of its fluorescence excitation spectrum. At the low resolution used in this figure, the dispersed fluorescence spectra look very similar as would be expected considering that the same species is responsible for the fluorescence in both cases. The one difference between the top and bottom spectra is the fact that the top spectrum is better resolved. This may be seen by noting that the minimum between the fine structure doublets reaches the baseline in the case of the NO excited spectrum while it does not in the case of the NOAr excited spectrum.

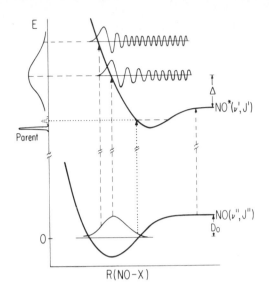

E

Parent

$NO^*(\nu',J')$

Δ

$NO(\nu'',J'')$

D_0

0

R(NO-X)

Fig. 1 Potential energy curves of the van der Waals molecule NOX (X = He, Ne, Ar) in the ground and first excited electronic state. The strong Franck-Condon transitions are from the bound zero-point level of the ground state to the repulsive wall of the excited state and give rise to a broad spectral feature as shown. The sharp feature which would result from a bound-bound transition (dotted line) was too weak to observe.

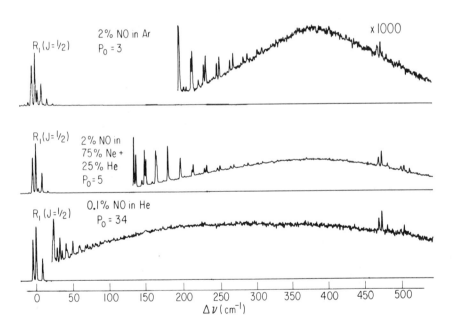

$R_1(J=1/2)$

2% NO in Ar
$P_0 = 3$

x 1000

$R_1(J=1/2)$

2% NO in
75% Ne +
25% He
$P_0 = 5$

$R_1(J=1/2)$

0.1% NO in He
$P_0 = 34$

0   50   100   150   200   250   300   350   400   450   500

$\Delta \nu \, (cm^{-1})$

Fig. 2 The fluorescence excitation spectra of NOAr (top), NONe (middle), and NOHe (bottom). The gas mixture used to prepare these species are indicated. The total lacking pressure, P_0, is given in atmospheres. The abscissa is displacement from the $A^2\Sigma^+(v'=0) \leftarrow X^2\Pi_{1/2}(v''=0)$ transition of uncomplexed NO. The sharp structure at small $\Delta \nu$ is due to rotational structure of the uncomplexed NO band. The sharp structure near $\Delta \nu = 470$ is a sequence band of uncomplexed NO.

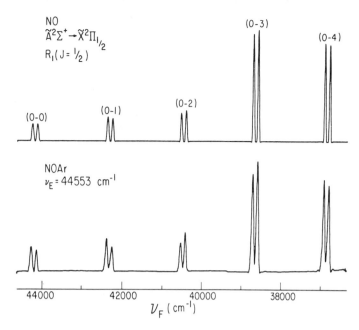

Fig. 3 The dispersed fluorescence spectrum produced when (top) NO is excited on the R_1(J=1/2) line of the A←X transition and (bottom) NOAr is excited in the middle of its broad band. The features spaced by ∿2000 cm^{-1}
are various vibrational features of the type A(v'=0) → X(v''=0-4). The doublet structure is due to transitions to the two ($^2\Pi_{3/2}$, $^2\Pi_{1/2}$) fine structure
components of the ground state. The apparent decrease in intensity at high
frequency is due to the instrumental response. The spectral resolution is
32 cm^{-1}.

The instrumental resolution was identical in both cases, and the reason
for the decreased resolution in the van der Waals spectrum is the broader
rotational state distribution of the emitting NO* molecules. Under the conditions in our supersonic free jet, both the NO and NOAr are expected to be
prepared largely in the ground (J=1/2) rotational level of the ground vibronic state. Optical excitation can add, at most, one unit of angular momentum, and therefore the initial excited state population of both NO* and
NOAr* must consist of only a couple of rotational levels. In the case of
NO*, this is the rotational state population of the emitting species, and
therefore the emission spectrum has a narrow rotational profile. In the
case of NOAr*, however, the dissociation reaction can add several units of
angular momentum to the product NO*, angular momentum being conserved by the
orbital angular momentum of the recoiling argon atom. Therefore the rotational state distribution of the emitting NO* can be broader leading to a
broader spectral envelope.

The effect of this rotational excitation of the fragment NO* produced in
the photodissociation reaction can be most easily seen in Fig. 4 which shows
one band of the dispersed fluorescence spectrum taken at higher resolution.
The bands result from excitation at various points of the broad NOAr feature.

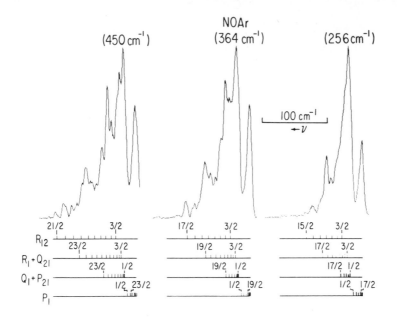

Fig. 4 The dispersed fluorescence spectrum produced when NOAr is excited at three different positions in the broad band. Frequency shifts noted above each spectrum are displacements of the exciting frequency from the band origin of uncomplexed NO. A stick diagram of the six rotational branches of the NO* emission spectrum is drawn below each trace. The spectral resolution is 6 cm⁻¹.

In all cases the rotational envelope contains contributions from many rotational levels, and the degree of rotational excitation may be seen to increase as more excess energy is added to the complex.

This material is based upon work supported by the National Science Foundation under Grant CHE-7825555.

References

1. D. H. Levy, in Photoselective Chemistry, Advances in Chemical Physics, Vol. 47, J. Jortner, R. D. Levine, and S. A. Rice, Eds. (Wiley-Interscience, New York, 1981), Part I, pp. 323-362; D. H. Levy, Ann. Rev. Phys. Chem. 31, 197 (1980); and references cited therein.

2. R. E. Smalley, D. H. Levy, and L. Wharton, J. Chem. Phys. 64, 3266 (1976); R. E. Smalley, L. Wharton, and D. H. Levy, J. Chem. Phys. 66, 2750 (1977).

Optothermal Infrared Spectroscopy

T.E. Gough and G. Scoles

The Guelph-Waterloo Centre for Graduate Work in Chemistry
Department of Chemistry, University of Waterloo
Waterloo, Ontario N2L 3G1, Canada

Introduction

To measure the absorption of radiation by optically thin samples one frequently employs indirect methods of detection, thereby escaping the constraints imposed by amplitude instabilities of the radiation source. In optothermal spectroscopy one detects the absorption of radiation by transferring the excitation energy to a thermal detector. The transfer is accomplished by allowing the molecules excited by the radiation to impinge upon a thermal detector, whose surface is able to accommodate the internal energy of the excited molecules. In gaseous samples, the technique is best suited to the collision-free regime in which the mean free path is greater than the distance between the excitation region and the thermal detector. The technique therefore is different from and complements optoacoustic spectroscopy which requires the excitation energy to be converted by collisions into translational energy before the excited molecules reach the walls of the experimental apparatus.

Optothermal infrared spectroscopy was developed at the University of Waterloo using molecular beams as the collision-free medium and low temperature silicon bolometers as the thermal detector [1]. The principal goal of these experiments is the determination of the population of molecular states following supersonic expansions and after scattering events, both inelastic and reactive. However, the use of a supersonic molecular beam confers definite spectroscopic advantages. Such beams are collision-free, and are composed of internally cold molecules moving in common direction with only a small spread in velocity. The low internal temperature of the molecules can lead to dramatic simplification of the rotational structure of a spectrum while the directionality of the molecular beam allows the Doppler-width of the spectrum to be reduced to negligible proportions. Below we present illustrative examples from our experiments and review the progress made by other laboratories in pursuing similar experiments.

More recently, some optothermal spectroscopic experiments have been performed on gaseous samples using pyroelectric elements as thermal detectors [2,3]. In addition Richards has performed important experiments in which the optothermal effect in condensed phases is coupled with F.T.I.R. spectroscopy achieving sub-monolayer sensitivity in the spectroscopy of gases adsorbed on the surface of metals [4].

Experimental

Figure 1 shows a schematic diagram of the molecular beam machine. The supersonic source is of the conventional differentially pumped design and consists of a nozzle, 35 μ diameter, and skimmer, 300 μ diameter, the nozzle-skimmer distance being approximately 1 cm. The nozzle operates at stagnation pressures up to 10 atm. under which conditions the primary and secondary chamber pressures are 10^{-4} and 10^{-6} torr respectively. The mass spectrometer operates in a pressure 10^{-9} torr and is used to monitor the beam for condensed species following the expansion.

The laser beam crosses the molecular beam orthogonally 20 cm downstream from the nozzle in a region where the beam undergoes collisionless free molecular flow. After being excited by the laser radiation the molecules of the beam travel a further 50 cm to the bolometer, which is maintained below the λ - point of helium. The molecules condense on the bolometer, releasing their energy thereby increasing the temperature of the bolometer and decreasing its resistance. Chopping the laser radiation will periodically change the molecules' energy contents and therefore produce synchronous fluctuations in the resistance of the bolometer. These fluctuations are detected by a lock-in amplifier and displayed on a chart recorder. The bolometer is 1 x 4 mm with the larger dimension oriented perpendicular to both laser and molecular beams and is located 70 cm from the source so that the maximum Doppler shift produced by a terminally accelerated beam moving at 1.7×10^5 cm sec^{-1} and absorbing at 4000 cm^{-1} is ±0.5 MHz. The detailed shape of the broadening depends upon the Mach number of the beam but is certainly less than 1 MHz F.W.H.M. Note also that the width is directly proportional to the absorption frequency.

Normally laser instabilities are the dominating source of broadening. However, there are further line broadening effects due to the characteristics of the laser beam. If the laser beam employed is a diffraction limited Gaussian beam it may be characterised by the position and size of its beam waist. The existence of the beam waist implies a transverse constraint of the photon flux and hence a transverse component of momentum. When the laser beam is Gaussian this transverse momentum generates a Gaussian broadening of the transition such that the half width at 1/e points is equal to $2v/\omega_0$, v being the beam velocity and ω_0 the 1/e radius of the laser beam at the beam waist. (1/e half width may be converted to F.W.H.M. by multiplying by 1.66) This width is independent of the position of the beam waist with respect to the molecular beam.

Results

In Table 1 we list the molecules and transitions studied so far by optothermal spectroscopy of molecular beams. The range of molecules studied has been extended from linear molecules to spherical and symmetric tops. Transitions have been observed from 1,000 to 4,000 cm^{-1}, and with the continued development of tunable lasers there is every reason to believe this range will be extended. The range will be limited at the high frequency end by decreasing fluorescence lifetimes, unless intramolecular quenching is present and at the low frequency end by the lower energy of the exciting photons.

The experimental intensities of the laser-induced, bolometrically detected signals for hydrogen fluoride is consistent with intensities calculated from the N.E.P. of the detector and the anticipated degree of excitation of the molecular beam. The measurement and simulation of Rabi oscillations in sulfur hexafluoride entail quantitative understanding of the effects of laser and molecular beam divergence, of transit-time broadening, and strong field limit excitation of the beam. Thus we consider the excitation and detection

Table 1

Laser	Molecule	Experiment	University	Reference
Diode	CO	Rot. Relaxation	Trento	[7]
Diode	N_2O	Rot. Relaxation	Waterloo	[1c]
Diode	$(N_2O)_2$ ν_3	Photodissociation	Waterloo	[1c]
Diode	NO	Hyperfine splitting	Waterloo	[1b]
Colour Centre	HF, CO_2	Rot. Relaxation	Waterloo	[8]
"	HF	μ_1	Waterloo	[5]
"	HCN	μ_{001}	Waterloo	[5]
"	$(CO_2)_2$ and $(CO_2)_n$	Photodissociation	Waterloo	[6]
"	NO	Hyperfine splitting	Waterloo	[8]
N_2O	SF_6 P3 F_2 and P3 A_2	Rabi oscillations	Paris-Nord	[9]
CO_2 c.w.	SF_6 P4 A_1 F_1 and E	Rabi oscillations	Paris-Nord	[10]
" c.w.	CH_3F	Rot. Relaxation and Stark tuning	Waterloo	[11]
" (pulsed)	SF_6	Multiphoton spectrosc.	Trento	[12]
" (pulsed)	$(SF_6)_n$	Photodissociation	Trento	[12]
" c.w.	SF_6	Collisional up-pumping	Columbia	[13]
HF	HF	Surface accommodation	Toronto	[14]

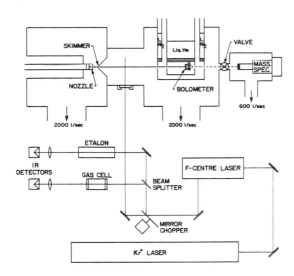

Fig.1 Molecular beam infrared optothermal spectrometer: general layout.

processes to be well understood, and capable of discussion in quantitative terms. As an analytical probe of state populations, we consider our latest studies on saturated transitions of methyl fluoride to be reproducible to 1 or 2 per cent.

The narrowest linewidth achieved in our experiments has been ~1.5 MHz F.W.H.M. This linewidth was obtained using a Burleigh FCL-20 colour centre laser and corresponds to a resolution of 1 part in 10^8. To the present time no experiments have been performed in which the laser frequency has been measured with an absolute precision comparable to the resolution of the ex-

periment. Thus the full potential of the technique as a source of spectroscopic transition energies has yet to be exploited. One can of course choose
to measure frequency differences, rather than absolute frequencies, calibrating
the laser scan by the transmission curve of a Fabry-Perot etalon. This approach was adopted in experiments designed to measure dipole moments of vibrationally excited states. A Stark field was applied at the intersection between
laser and molecular beams and Stark splittings were analysed to yield the
ratio of dipole moments in ground and excited states [5]. The precision to
which dipole moments can be measured from the Stark-split infrared spectra of
molecular beams is comparable to the precision obtainable from the Stark-split
microwave spectra of bulk gases [5].

One of the main applications of molecular beam spectroscopy is in the direction of the study of species which are unstable under normal gas kinetic
conditions. Indeed with techniques very similar to the one described above
we have obtained photodissociation spectra for N_2O dimers [1c] and, more
recently, CO_2 dimers and higher clusters [6]. The study by means of infrared
spectroscopy of these molecular aggregates opens up interesting new possibilities to increase our understanding of the size effects in condensed systems
and the transition between the diluted and the condensed state of matter.

Two directions in future developments seem to hold particular promise. The
first is the study of coherence effects in the spectra and their perturbation
by long range collisions [9]. The second is the transfer of the know-how
originated in the molecular relaxation studies to the field of gas-surface
interactions. The first experiments of this type have been successfully
carried out recently in Toronto [14].

References

1. T.E. Gough, R.E. Miller and G. Scoles, (a) Appl. Phys. Lett. 30, 338(1977);
   (b) J. Mol. Spectrosc. 72, 124(1978); (c) J. Chem. Phys. 69, 1588(1978).
2. (a) C. Hartung, R. Jurgeit, and H.-H. Ritze, Appl. Physics 23, 407(1980);
   (b) C. Hartung and R. Jurgeit, Sov. J. Kvant. Elektron. 5, 1825(1978) and
   Sov. J. Opt. i Spektrosc. 45, 1169(1979).
3. R.V. Ambartzumian, L.M. Dorozhkin, G.N. Makarov, A.A. Puretzky, and
   B.A. Chayanov, Appl. Phys. 22, 409(1980).
4. R.B. Bailey, I. Iri, and P.L. Richards, Surf. Science 100, 626(1980).
5. T.E. Gough, R.E. Miller, and G. Scoles, Royal Soc. Chem., Faraday Disc.
   No. 71, paper 6, 1981.
6. T.E. Gough, R.E. Miller, and G. Scoles, to be published.
7. D. Bassi, A. Boschetti, S. Marchetti, G. Scoles, and M. Zen, J. Chem.
   Phys. 74, 2221(1981).
8. R.E. Miller, Ph.D. Thesis, University of Waterloo (1980).
9. S. Avrillier, J.-M. Raimond, Ch.J. Bordé, D. Bassi, and G. Scoles, Optics
   Communications, to be published.
10. S. Avrillier, Ch.J. Bordé, D. Bassi, and G. Scoles, unpublished work.
11. C. Douketis, T.E. Gough, and G. Scoles, unpublished work.
12. D. Bassi, A. Boschetti, G. Scoles, M. Scotoni, and M. Zen, 8th Int. Mol.
    Beams Symp., Book of abstracts, Cannes 1981, page 36.
13. R. Coulter, F.R. Grabiner, L.M. Casson, G.W. Flynn, and R.B. Bernstein,
    J. Chem. Phys. 73, 281(1980). M.I. Lester, D.R. Coulter, L.M. Casson,
    G.W. Flynn, and R.B. Bernstein, J. Phys. Chem. 85, 751(1981).
14. J.C. Polanyi, private communication.

Laser Magnetic Double Resonance Spectra at 5 and 10μm

R.S. Lowe and A.R.W. McKellar

Herzberg Institute of Astrophysics, National Research Council of Canada
Ottawa, Ontario K1A OR6, Canada

1. Introduction

Laser magnetic resonance (LMR) spectroscopy, in which molecular transitions
are Zeeman-shifted into coincidence with fixed laser frequencies, is now a
well-established technique for studying unstable molecules. It was first
developed in the far-infrared to study pure rotational spectra; progress in
this region has been reviewed by EVENSON et al. [1]. More recently, LMR
has been used in the mid-infrared region of CO (5-8 μm) and CO_2 (9-11 μm)
lasers to study vibration-rotation spectra; this has been reviewed by
MCKELLAR [2]. Infrared-radiofrequency or -microwave double resonance [3]
is another fruitful spectroscopic technique that utilizes mid-infrared
lasers. This paper presents some results combining the double resonance
and LMR techniques. Such a combination yields precise optically-detected
EPR measurements within excited and ground vibrational levels, and can also
serve as a useful aid for the assignment of difficult LMR spectra. Our
technique is analogous to the combination of infrared-microwave double
resonance and CO_2 laser Stark spectroscopy used by TANAKA et al. [4], and
of infrared-optical double resonance and CO_2 LMR used by AMANO et al. [5].

2. Experimental Details

The experiment utilized a previously described [6] intracavity LMR apparatus
based on a 30 cm electromagnet with a 6 cm gap. A radiofrequency (rf)
double resonance cell was mounted in the gap, occupying the space usually
taken by a pyrex free radical flow cell. The rf cell was a modification of
one used [7] to study zero-field double resonance spectra of NO. It is
constructed [8] as a 50 ohm transmission line with tapered transition
sections between the cell body and the input and output cables; this careful
impedance matching gives reasonably flat transmission to beyond 8 GHz. The
main cell body is a stainless steel tube with sufficient electrical skin
depth to pass the 15 kHz LMR Zeeman modulation field while still containing
the rf radiation. Though Zeeman modulation is not required for double
resonance spectra, it is useful for monitoring LMR signals during an
experiment. The cell has a diameter of 4.5 cm and a length of $\sim$ 20 cm, but
for unstable molecules the active length is defined by the 12 cm spacing
between the gas inlet and outlet connections. The cell is oriented such
that the Zeeman field is parallel to the electric vector of the laser
radiation ($\Delta M = 0$ infrared transitions) and perpendicular to that of the
radiofrequency radiation ($\Delta M = \pm 1$ rf transitions).

Double resonance spectra were recorded by sweeping the rf, to which a small frequency modulation (∼ 30 kHz) was applied, and processing the laser output signal in a phase sensitive amplifier, with resulting first derivative lineshapes. There are other detection possibilities that were not explored: for example, the field may be swept while the rf is held fixed (as long as the LMR transition remains sufficiently in coincidence), or Zeeman modulation may be used in the swept rf mode.

3. Results on NO at 5.3 μm

Since NO is among the few stable paramagnetic molecules, and gives strong LMR spectra with a CO laser [9], it was an obvious choice to test the laser magnetic double resonance technique. Spectra were obtained using near coincidences between CO laser lines and the v = 1←0 band R(1.5) transitions of both the $^2\Pi_{3/2}$ and $^2\Pi_{1/2}$ substates of NO. The rf spectra of the two substates are rather different, and will be described separately.

Figure 1 shows a double resonance spectrum obtained with the $^{12}C^{16}O$ 9-8 P(13) laser line. At a field of 825 Gauss, the M_J = -1.5 ← -1.5 component of $^2\Pi_{3/2}$ R(1.5) is coincident with the laser frequency, and the possible rf transitions are M_J = -0.5 ↔ -1.5 within the v = 0, J = 1.5 rotational level, and M_J = -0.5 ↔ -1.5 or -1.5 ↔ -2.5 within v = 1, J = 2.5. Each M_J level is tripled by ^{14}N hyperfine structure and further doubled by Λ-doubling. The resulting 6 transitions within v = 0, J = 1.5 are those in Fig. 1(b); the observed doubling (∼ 2 MHz) of each of the 3 hyperfine transitions is just twice the Λ-doubling of the J = 1.5 energy level. A total of 12 transitions are expected for v = 1, J = 2.5 in Fig. 1(a); in this case, the hyperfine tripling and Λ-doubling are fully resolved, but the M_J splitting is small and is only partially resolved in the first three lines.

In contrast to $^2\Pi_{3/2}$, the $^2\Pi_{1/2}$ state in NO exhibits much greater Λ-doubling but only a weak Zeeman effect for low J. The $^{13}C^{16}O$ 8-7 P(10)

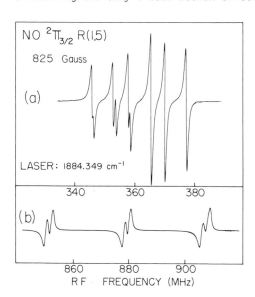

NO $^2\Pi_{3/2}$ R(1.5)

825 Gauss

(a)

LASER: 1884.349 cm⁻¹

340   360   380

(b)

860   880   900
R F   FREQUENCY (MHz)

Fig.1 Laser magnetic double resonance spectrum of NO. The laser ($^{12}C^{16}O$ 9-8 P(13)) coincides with one M_J component of the $^2\Pi_{3/2}$ R(1.5) transition of NO at 825 G, and the observed rf lines are hyperfine Λ-doubling transitions within v=1, J=2.5 (trace a) and v=0, J=1.5 (trace b). The change of phase between (a) and (b) has no significance

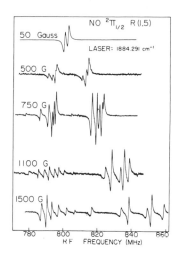

Fig.2 Laser magnetic double resonance spectrum of NO. The laser ($^{13}C^{16}O$ 8-7 P(10)) coincides with the $^2\Pi_{1/2}$ R(1.5) NO transition over a wide range of field; the Zeeman splitting of one hyperfine Λ-doubling component is shown here

laser line occurs within the Doppler width of $^2\Pi_{1/2}$ R(1.5), and zero-field double resonance measurements utilizing this coincidence were reported by LOWE et al. [7]. Figure 2 shows the effect of a magnetic field on one hyperfine component (v = 0, J = 1.5, F = 2.5 ↔ 2.5) of this Λ-doubling rf spectrum. Spectra may be obtained over a wide range of fields because the Zeeman splitting of the infrared transition is small and the laser remains within the Doppler width of all M_J components. The rf spectra appear complicated because the hyperfine and Zeeman splittings are of the same magnitude, but all observed lines in Fig. 2 may be assigned. Their frequencies give information on parameters in the Zeeman Hamiltonian, but these are already rather well known for NO [10].

4. Results on HO_2 at 9.2 μm

Although NO provided a useful test, the present technique is most valuable for transient free radicals. To demonstrate this application, we turn to HO_2, whose ν_3 band LMR spectrum was studied by JOHNS et al. [11]. HO_2 was produced in a flow system by mixing the products of an rf discharge in O_2 with allyl alcohol at the inlet to the double resonance cell. Figure 3 shows a spectrum obtained at 1812 G using the $^{12}C^{18}O_2$ 9R(6) laser line. In the lower trace, with the discharge turned off, two unassigned double resonance transitions of allyl alcohol are observed. When the discharge is turned on (upper trace), a number of new lines due to HO_2 appear. The laser coincides with the M_J = 2.5 ← 2.5 component of the HO_2 ν_3 $^qP_2(4)$ transition at this field [11], and rf transitions occur between M_J components of asymmetry doublets: $(\nu_1 \nu_2 \nu_3)$ = (000), J = 4.5, and $N_{K_aK_c} = 4_{22} \leftrightarrow 4_{23}$ for the lines < 700 MHz; (001), 3.5, $3_{21} \leftrightarrow 3_{22}$ for those > 700 MHz. Each HO_2 line in Fig. 3 is broadened by unresolved hyperfine structure, as well as by pressure and (rf) power broadening. By minimizing the latter two effects, the proton hyperfine structure can be resolved as shown in Fig. 4. This spectrum was recorded at 2510 G, where M_J = 1.5 ← 1.5 of the same HO_2 infrared transition is resonant with the laser; the particular rf component shown is: (001) J = 3.5, 3_{21} ← 3_{22}, M_J = 2.5 ← 1.5.

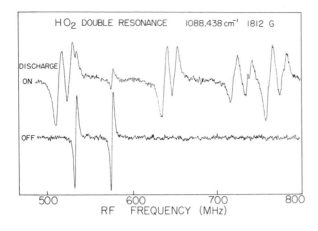

Fig.3 Laser magnetic double resonance spectrum of HO_2. With the discharge off, two double resonance lines of the parent allyl alcohol molecule remain, while the HO_2 lines disappear

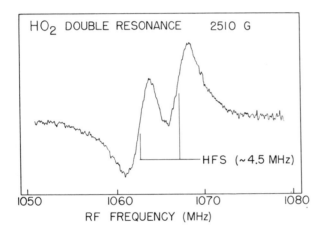

Fig.4 One M_J component of the HO_2 spectrum at high resolution, showing resolved proton hyperfine structure

5. Conclusions

A new laser magnetic double resonance technique has been demonstrated on the NO and HO_2 molecules using CO and CO_2 lasers; the latter constitutes the first detection of an unstable free radical by infrared-rf or -microwave double resonance. The technique enables precise measurements of low frequency transitions to be made in excited vibrational states of transient molecules, and should also be a useful aid in the assignment of difficult unknown LMR spectra.

References

1. K.M. Evenson, R.J. Saykally, D.A. Jennings, R.F. Curl, J.M. Brown: in *Chemical and Biochemical Applications of Lasers*, ed. by C.B. Moore (Academic Press, New York, 1980), 5, 95.

2. A.R.W. McKellar, Faraday Discussions Chem. Soc. 71, to be published (1981).
3. E. Arimondo, J.G. Baker, P. Glorieux, T. Oka, J. Sakai: J. Mol. Spectrosc. 82, 54 (1980).
4. T. Tanaka, C. Yamada, E. Hirota: J. Mol. Spectrosc. 63, 142 (1976).
5. T. Amano, K. Kawaguchi, M. Kakimoto, S. Saito, E. Hirota: to be published.
6. R.S. Lowe, A.R.W. McKellar: J. Mol. Spectrosc. 79, 424 (1980).
7. R.S. Lowe, A.R.W. McKellar, P. Veillette, W.L. Meerts: J. Mol. Spectrosc., to be published.
8. A. Karabonik, T. Oka: unpublished.
9. R.M. Dale, J.W.C. Johns, A.R.W. McKellar, M. Riggin: J. Mol. Spectrosc. 67, 440 (1977).
10. W.L. Meerts, L. Veseth: J. Mol. Spectrosc. 82, 202 (1980).
11. J.W.C. Johns, A.R.W. McKellar, M. Riggin: J. Chem. Phys. 68, 3957 (1978).

High Sensitivity Color Center Laser Spectroscopy

R.F. Curl, Jr., J.V.V. Kasper, P.G. Carrick, E. Koester, and F.K. Tittel
Departments of Chemistry and Electrical Engineering and Rice Quantum
Institute, Rice University, Houston, TX 77001, USA

1. Introduction

The development of cw color center lasers [1] has inevitably led to
their application to spectroscopy. The commercial availability of a color
center laser provides ready access to this technology to any laboratory.
In this paper, we wish to examine the prospects of color center lasers as
tunable sources for high sensitivity infrared absorption spectroscopy, describe the development of a computer-controlled color center laser spectrometer based upon the commercial Burleigh FCL-20 laser, and report on several
spectroscopic investigations using the computer-controlled color center
laser.

2. Sensitivity of Absorption Spectroscopy

The sensitivity of any absorption spectroscopy experiment in terms of
absorbers/unit volume is meaningless without some specification of the effective pathlength. In a laboratory experiment, the physical pathlength is
generally limited by the available space, the reflectivity of mirrors in
multipass optics or in some cases by the overlap of multipass beams with
the consequent generation of unwanted standing waves. For some laser systems, the effective pathlength can be made very much greater than the sample length by placing the sample inside the laser cavity [2]. This length
enhancement increases as the laser cavity loss is reduced and the sensitivity of laser gain to circulating power level decreases. The rate of change
of gain with power level is smallest when laser action is not the dominant
mechanism for the loss of population inversion. Color center lasers do not
give great enhancement of effective pathlength over physical pathlength for
intra-cavity absorption both because the crystals are generally fairly
lossy (~5%/pass) and because they have a high quantum efficiency with laser
action the dominant mechanism for loss of population inversion. Thus we
will limit our sensitivity considerations to extra-cavity absorption experiments and bear in mind that there may be trade-offs between various approaches to absorption spectroscopy when the possibility of increasing the
pathlength is introduced.

There are three sources of noise which could limit the minimum detectable change in laser power in absorption spectroscopy: source noise, detector noise, and quantum noise. (For further discussion of these noise
sources, see SHIMODA [3]). In contrast with some other laser absorption
spectroscopy experiments such as far-infrared LMR [2], color center laser

absorption spectroscopy is dominated by source noise considerations. Color center lasers are noisy, primarily because the ion lasers which pump them are affected by acoustic vibrations, AC ripple, and noise from the instabilities in the electric discharge. Acoustic vibrations and AC ripple are dominant at low frequencies, but appear to decrease to a negligible level at a few kHz. Above a few kHz, however, noise remains essentially constant to above 100 kHz although it should fall off above 3 MHz because the ion laser cavity acts as a tank circuit. (Here we summarize our own observations using Stark modulation of methanol and acetonitrile and an Ar+ laser pump. In contrast, DELEON, JONES and MUENTER [4] report noise decreasing as 1/f to above 100 kHz using a Kr+ pump).

The sensitivity limitations imposed by source noise may be reduced or removed by several different approaches:
1. Reduce source noise by active amplitude stabilization (noise eaters).
2. Balance it out with two detectors (double beam compensation).
3. Balance it out by interference (balanced bridge or magnetic rotation).
4. Operate at a modulation frequency (above 3 MHz?) where source noise is reduced.
5. Increase the pathlength at the expense of power (multiple reflection cells).

Let us comment briefly about each of these options. First, noise eaters are highly effective [5] in improving sensitivity for simple absorption spectroscopy without modulation because the laser exhibits excess noise at low frequencies and regular modulation because of power line ripple. On the other hand, active feedback loops, especially those which sample the infrared beam rather than the ion laser pump, seem to be difficult to build and adjust. The second approach of using two beams and two detectors is limited by any instability in the beam direction and by drift in either detector, and ultimately by the total detector noise of the detector pair. If sophisticated signal processing, is employed slow relative drifting of the two detectors may be at least partially compensated. The third approach of interference, as realized by magnetic rotation spectroscopy, can be quite effective in reducing source noise [6]. It is limited by the quality of available polarizers and eventually by detector noise. Because S/N is increasing as $1/\sqrt{P}$ when better quality polarizers are introduced, the gain in S/N when detector noise limit is reached is far less than if the source noise were simply not present. The fourth approach of moving the modulation frequency to a value sufficiently high that noise is falling off is not easily implemented with Stark or Zeeman modulation although it remains an interesting possibility with frequency modulation. The final approach of increasing the pathlength by multiple reflection can be highly effective when sample dimensions permit especially if done in conjunction with Stark or Zeeman modulation. The S/N improves linearly with pathlength until the laser beam is attenuated to the extent that detector noise becomes a factor. This approach has been used with great success in diode laser spectroscopy by the group of HIROTA [7].

Our recent efforts have been in the direction of increased pathlength, because the signal increases linearly with pathlength. Since S/N is independent of infrared power when source noise limited, more is to be gained by increasing the pathlength at the expense of power at the detector until detector noise limited than by employing higher quality polarizers in magnetic rotation spectroscopy until detector noise limited. Recently construction of a 2 meter White Cell similar to the design of OKA [8] with

Brewster windows protecting the mirrors has been completed. Because of the strong atmospheric water absorptions in this region, the mirrors are placed in evacuable compartments. The cell is wound with a solenoid for Zeeman modulation and an electric discharge can be maintained between the cell ends. With Zeeman modulation at 1800 Hz and detection at 3600 Hz, the minimum detectable absorption is 10(-6) /cm (pathlength: 80 meters). The sensitivity obtained is illustrated by Fig.1 which displays the R(3) transition of OH obtained in both simple absorption and Zeeman modulation.

3. Computer Controlled Color Center Laser Spectrometer

The Burleigh FCL-20 has been converted into a computer controlled color center laser spectrometer capable of continuous single mode scans of more than 10 cm-1 [9]. The spectrometer is capable of acquiring extremely high resolution spectra which can be precisely measured. The system consists of the laser, diagnostics used to monitor the laser output, hardware control units for the wavelength selective elements, detectors and amplifiers for data acquisition, a CAMAC system which controls the tuning ele-

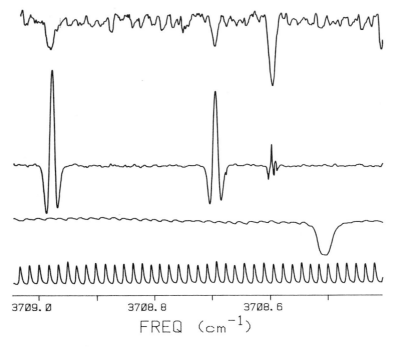

FREQ (cm^{-1})

Fig.1 The R(3) transition of OH in both simple absorption (top trace) and Zeeman modulation (next trace) observed through a neon sign transformer discharge through water. The peak absorption is about 10% through an 80 m pathlength which corresponds to an OH pressure of 0.04 mTorr. Also shown: NO overtone reference spectrum and reference marker cavity. TC(Zeeman): 10 msec.

ments (grating angle, etalon length, and cavity length) and acquires the data, and DEC PDP-11/V03 microcomputer.

The method of absolute frequency calibration can be seen by examining Fig.1. The line frequencies of the NO overtone reference spectrum are accurately known from Fourier transform spectroscopy. The temperature compensated Invar reference cavity has been observed to be stable to 0.001 cm-1 for several hours. A feature can thus be measured with a precision of ~0.0003 cm-1 and an accuracy of ~0.001 cm-1. A detailed discription of the spectrometer can be found elsewhere [9].

4. Spectroscopic Studies

Several spectroscopic investigations have been carried out using computer controlled color center lasers. The fundamental vibration of OH has been observed by magnetic rotation spectroscopy [6] in the products of a microwave discharge in water. The magnetic dipole allowed transition between the fine structure levels of the Br atom has been observed using magnetic rotation [10]. This transition had been observed in emission with very high resolution by Fourier transform spectroscopy [11]. However, the resolution obtainable by color center laser spectroscopy was sufficient to resolve completely the isotopic structure which was incompletely resolved in the Fourier transform study. The isotopic mass shift in the fine structure transition frequency was determined by color center laser spectroscopy to be 14 MHz.

The spectra of two stable molecules, nitrogen dioxide and methanol, have been investigated with the color center laser spectrometer. In the case of nitrogen dioxide the transition observed is the combination band $(1,1,1)<-(0,0,0)$, while for methanol the OH stretch fundamental has been studied. These investigations will be reported elsewhere.

Acknowledgements

Contributions to the work described here were made by C. R. Pollock, G. Litfin, and W. Dillenschneider. This work was supported in part by the Robert A. Welch Foundation and the Department of Energy. J. Kasper is a member of the Chemistry Department, University of California at Los Angeles, Los Angeles, CA 90024, USA. E. Koester is on leave from Institut fur Angewandte Physik, Universitat Hannover, D3000 Hannover, Germany.

References

1. L. F. Mollenauer and D. H. Olson: J. Appl. Phys. 46, 3109 (1975)
2. K. M. Evenson, R. J. Saykally, D. A. Jennings, R. F. Curl, Jr., and J. M. Brown: in Chemical and Biochemical Applications of Lasers, ed. C. B. Moore, (Academic Press, New York, 1980) Vol. 5, p.95
3. K. Shimoda: Appl. Phys. 1, 77 (1973)
4. R. L. DeLeon, P. H. Jones, and J. S. Muenter: Appl. Optics 20, 525 (1981)
5. J. L. Hall and S. A. Lee: in Tunable Lasers and Applications, eds. A. Mooradian, T. Jaeger, and P. Stokseth (Springer Verlag, Berlin, 1976) p 361

6. G. Litfin, C. R. Pollock, R. F. Curl, and F. K. Tittel: J. Chem. Phys. 72, 6602 (1980)
7. E. Hirota: in Chemical and Biochemical Applications of Lasers, ed. C. B. Moore, (Academic Press, New York, 1980) Vol. 5, p.39
8. T. Oka: Phys. Rev. Letters, 45, 531 (1980)
9. J. V. V. Kasper, C. R. Pollock, R. F. Curl, and F. K. Tittel: to be submitted
10. J. V. V. Kasper, C. R. Pollock, R. F. Curl, and F. K. Tittel: Chem. Phys. Letters 77, 211 (1981)
11. E. Luc-Koenig, C. Morrillon, and J. Verges: Physica, 70, 175 (1973)

Part VIII
Cooling, Trapping, and Control
of Ions, Atoms, and Molecules

Mono-Ion Oscillator for Ultimate Resolution Laser Spectroscopy

H.G. Dehmelt

Department of Physics, University of Washington, 1600 43rd Ave. E.
Seattle, WA 98195, USA

1. Introduction

The 1965 experiment [1] of FORTSON, MAJOR and DEHMELT in which the 0-0 hyperfine transition in the ground state of ^{3}He$^+$ near 8 GHz was observed with a
linewidth of $\sim$ 10 Hz or a resolution of about 1 part in 10^9 marked the advent
of high resolution microwave spectroscopy of stored ions proposed earlier by
the author [2]. Samples of $\sim 10^7$ ions were used in these experiments. Already then in our 1966 publication we pointed out the potential usefulness
of ion storage for laser spectroscopy. In 1973 MAJOR and WERTH [3], now with
NASA, reported the record resolution of 2 parts in 10^{10} in an optical pumping experiment [4] on the hyperfine structure transitions near 40 GHz on $\sim$
10^5 stored ^{199}Hg$^+$ ions. Also in 1973 I proposed laser fluorescence spectroscopy on a highly refrigerated, individual Tl$^+$ ion stored in an rf quadrupole
trap, "The Tl$^+$ Mono-Ion Oscillator," [4] with a projected resolution of 1
part in 10^{14}. In 1975 I added a laser double resonance scheme for effectively amplifying about million-fold the fluorescence intensity used for the detection of the forbidden transition of interest [5]. This scheme also made
the study of the similar ions In$^+$, Ga$^+$, Al$^+$ attractive. Incorporation into
the Tl$^+$ experiment of the side band cooling mechanism [6], searched for by
WINELAND and DEHMELT in 1974 in experiments on stored electron clouds and
found in 1976 by VAN DYCK, EKSTROM and DEHMELT in their geonium experiment
[7], was proposed by WINELAND and DEHMELT in 1975. Also in 1974 the award of
a HUMBOLDT prize enabled me to initiate a Ba$^+$ mono-ion oscillator [8] experiment at the Universitaet Heidelberg in collaboration with P. TOSCHEK and his
group to prepare the ground for the more difficult Tl$^+$ work. At the same time,
I proposed a Sr$^+$ mono-ion oscillator experiment [9] together with H. WALTHER,
then at the Universitaet Köln.

In 1978 the Heidelberg experiment [10] led to the first isolation and uniform confinement, initially to < 5 μm and eventually to $\sim$ 2000 Å via laser
side band cooling to $\sim$ 6 K and eventually [11] to $\sim$ 10 mK as well as continuous visual and photoelectric observation of an individual Ba$^+$ ion in a small
PAUL rf quadrupole trap. Recently WINELAND and ITANO have also succeeded in
a Mg$^+$ mono-ion oscillator experiment [12] and they have established a localization of the ion to $\leq$ 15 μm in their PENNING trap.

2. Concept of Experiment

The goal of our current preparatory experiment is the observation of copious
laser scattering at a very narrow electronic transition at ω_0 on an indivi-

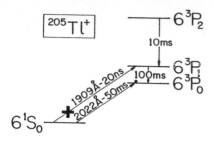

Fig.1 The 4 lowest electronic levels of the $^{205}Tl^+$ ion. In the proposed double resonance scheme the forbidden ω_0 transition at 2022 Å of interest to the $^3P_0$ level of 50 ms natural life time is observed indirectly: during the temporary shelving of the electron in the metastable $^3P_0$ level it is obviously impossible to observe laser scattering at $\lambda_2 = 1909$ Å, from [5].

dual ion stored at the bottom of the potential well formed by a small PAUL rf quadrupole trap. The ions of the Group IIIA elements, namely Tl^+, In^+, Ga^+, Al^+, B^+ are excellent candidates for such work. Their metastable lowest $^3P_0$ levels have extraordinary long life times; e.g., the corresponding natural line width of the $3\,^3P_0$ level of Al^+ is estimated as ~ 10 µHz. However in a mono-ion oscillator such a narrow width can only be exploited when a powerful amplification mechanism for the correspondingly low scattering is devised. Monitoring of the absence of the optical electron from the $^1S_0$ ground state via laser fluorescence at the strong $^1S_0 - {}^3P_1$ intercombination line of frequency ω_2 provides such an <u>atomic</u> amplification mechanism. For Tl^+, see Fig. 1, where the estimated lifetime of the $^3P_0$ level is ~ 50 ms, absorption of a single photon at the forbidden $^1S_0 - {}^3P_0$ ω_0-transition at 2022 Å suppresses ω_2-laser fluorescence at 1909 Å for about 50 ms, during which interval otherwise $\sim.10^6$ photons might have been scattered. Frequency shifts of the sharp 2022 Å line due to the 1909 Å excitation are obviated by appropriate pulsing schemes. Centering of the ion at the bottom of the trap is achieved via side band cooling by tuning the 1909 Å laser a few megahertz below the $^1S_0- {}^3P_1$ resonance at ω_2 thus that its Doppler side band at $\omega_2 - \omega_v$ is preferentially excited. The vibration frequency in the parabolic trap is denoted by ω_v. Three-dimensional cooling to < 1 mK is achieved by means of a single strategically directed laser beam.

3. Apparatus

The heart of the apparatus is the small Paul rf trap containing the ion. The design used in the Heidelberg Ba^+ experiments [10] is well suited for the future work on Tl^+ etc. and will therefore be sketched here. The trap,

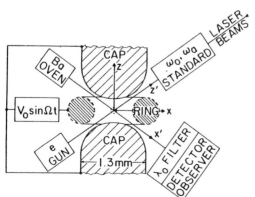

Fig.2 Apparatus for trapping in ultra-high vacuum, cooling, and visually observing an individual Ba^+ ion (schematic). Ba oven and electron gun were actually not in the xz plane. Vacuum envelope, pump etc. are not shown, from [10].

see Fig. 2, is formed by spherically ground wire stubs, the "caps" and a wire
"ring". Application of an rf voltage $V_0 \simeq 200$ Volt at $\Omega \simeq 2\pi \cdot 18$ MHz created
a 3-dimensional well of axial depth ~ 10 volt (for Ba^+) in which the ion os-
cillated at an axial frequency $\omega_{vz} \simeq 2\pi \cdot 2.4$ MHz and perpendicular frequen-
cies $\omega_{vx*}, \omega_{vy*} \simeq 2\pi \cdot 1.2$ MHz with $|\omega_{vx*} - \omega_{vy*}| \simeq 2\pi \cdot 10$ KHz. Ideally the
cooling laser beam is directed along the body diagonal [6] of the $x*, y*, z* \simeq$
z coordinate system formed by the principal axes of the effective trapping potential. It suffices to approximate these conditions by slightly deforming
the ring electrode elliptically and appropriately directing the laser beam
through the gaps between ring and cap electrodes. The resonance fluorescence
is likewise observed through the ring-cap gap. A vacuum of $\sim 10^{-11}$ Torr is
maintained by an ion getter pump in the sealed off, baked glass envelope.
The trap was filled by crossing and ionizing a very weak Ba atomic beam with
a very weak electron beam in the center of the trap.

4. Cooling

The following description [13,14] brings out the essential features of optical cooling. Due to the Doppler effect, when moving at velocity v opposite to the propagation of a plane light wave of frequency ω, an atom in its
rest frame sees an excitation at the frequency $\omega' = \omega(1 + v/c)$ near an atomic
resonance at ω_0. Then, again in its rest frame, the atom re-emits photons at
ω' in all directions in accordance with its characteristic radiation pattern.
Now to an observer in the laboratory due to the Doppler effect these re-
emitted photons will exhibit various frequency shifts depending on the angle
between the directions of re-emission and motion. Nevertheless the average
energy of the re-emitted photons will be $\hbar\omega' = \hbar\omega + \hbar\omega(v/c)$. The average
energy excess in the energy of the re-emitted over that of the absorbed pho-
tons has to come from the kinetic energy in the translational motion which
is thereby cooled.

This picture is easily adapted to the case of one-dimensional periodic motion at frequency ω_v parallel to the light beam. Now due to the modulated
Doppler shift the atom of resonance frequency ω_0, in its rest frame, when
irradiated by a laboratory light source at the sharp frequency ω sees a
whole excitation spectrum at $\omega'_n = \omega \pm n\omega_v$, $n = 0,1,2. \ldots$ For a sharp spectral line of width $\Delta\omega_0 \ll \omega_v$ and $\omega'_m = \omega + m\omega_v \simeq \omega_0$, $m > 0$, the effect of the
m-th side band predominates. Again the average energy of the photons re-
emitted in all directions is $\hbar\omega + m\hbar\omega_v$ and the energy excess $m\hbar\omega_v$ has to be
provided by the oscillatory motion. For the oscillation amplitude $z_0 \ll$
$\lambda_0/2\pi$ (in the LAMB-DICKE regime) the spectrum seen by the atom consists
only of the carrier at ω of power E plus two weak symmetric side bands at
$\omega \pm \omega_v$ of Power $E_\pm \ll E(z_0/2\lambda_0)^2$ and the problem remains manageable even
when the requirement $\Delta\omega_0 \ll \omega_v$ is dropped, see Fig. 3. The top section of
Fig. 3 shows the absorption profile of the stored ion when the laser frequency ω is tuned through the electronic resonance at ω_0. The multiple excitations of the ion in its rest frame at $\omega - \omega_v$, ω, $\omega + \omega_v$ according to the
position of these spectral components with respect to the Lorentz profile
centered about ω_0, see middle section of Fig. 3, result in the re-emission
of these same three components. When viewed from the lab frame each of the
3 re-emitted components will develop its own two Doppler side bands, see
bottom section of Fig. 3, without changing the average photon energy. Comparing top and bottom parts of Fig. 3 showing absorption and re-emission as
viewed from the lab frame illustrates the cooling process: excitation by
an upper/lower or cooling/heating side band photon extracts/adds the energy

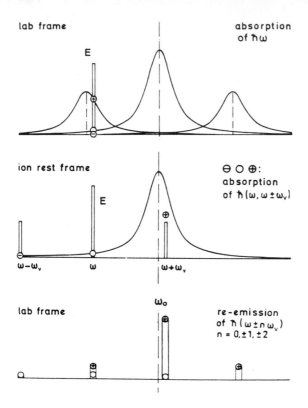

lab frame

absorption
of ℏω

E

ion rest frame

⊖ ◯ ⊕:
absorption
of ℏ(ω, ω ± ω_v)

E

ω−ω_v   ω   ω+ω_v

lab frame

ω_o

re-emission
of ℏ(ω ± nω_v)
n = 0, ± 1, ± 2

Ω

Fig.3 Spectral components effective in side band cooling (schematic). As shown in middle section, the ion vibrating at ω_v in the parabolic trap parallel to the propagation direction of the exciting electromagnetic wave at ω sees spectral components (of power E_-, E, E_+) at $\omega - \omega_v, \omega, \omega + \omega_v$. The case of oscillation amplitude $\ll c/\omega$ and $E_\pm \ll E$, the LAMB-DICKE regime, is assumed here. To an observer in the lab-frame the ion appears to have the response spectrum (e.g., for power absorption) shown in top section. In the ion rest frame each of the three exciting components is re-emitted with unshifted frequency and a strength determined by the Lorentzian response profile. When seen from the lab frame (compare bottom section) each re-emitted component develops its own two Doppler side bands, after [13].

$\hbar\omega_v$ from/to the vibrational motion. For the case depicted while photons of energy ℏω are absorbed by the ion those re-emitted have a <u>larger</u> average energy $\sim h(\omega + \omega_v)$ for the tuning $\omega \simeq \omega_o - \omega_v$ shown.

This semiclassical picture indicates strong cooling and an exponential approach in time of absolute zero temperature. Quantum effects establish a finite minimum temperature, however. Obviously, when the ion is in the lowest vibrational state v = 0, it cannot be cooled any further and the power in the cooling side band must vanish while that in the heating side band remains finite. This asymmetry in side band strength does not begin with v = 0 but is always present, becoming more and more pronounced as v decreases. A similar asymmetry is present in the re-emission side bands. The attainable minimum vibrational temperature and the corresponding vibrational quantum number $\langle v \rangle_{min}$ are determined by the requirement that the heating effect of the stronger but more off-resonant spectral components at $\omega - \omega_v$ and ω is balanced by the cooling effect due to the weaker, more resonant side band $\omega + \omega_v$. Using only familiar elementary quantum mechanics of atomic and molecular spectra I analysed the problem in 1976 and applied it to the Tl⁺ mono-ion oscillator obtaining a minimum temperature [13,14] of $\sim$ 0.1 mK for cooling with a $\lambda \leq 1909$ Å laser beam. In this chapter ω_o refers to the allowed transition used for cooling.

The discussion is easily extended to the realistic 3-dimensional case of an individual stored ion. This requires irradiation along $\pm \hat{i}^*, \hat{j}^*, \hat{k}^*$ where i*,j*,k* refer to the principal axes of the ellipsoidal trapping potential [6]. For the minimal vibrational quantum number we obtained [11]

$$\langle v \rangle_{min} \simeq (g_\omega + g_-)/(g_+ - g_-) \qquad (1)$$

which applied to vibration along all three principal axes. Here g_ω, g_+, g_- denote the values of the Lorentz profile at ω, $\omega + \omega_v$ and $\omega - \omega_v$. Our results were later confirmed by more elaborate treatments [15,16].

5. Shifts and Broadening of the Resonance Line

Numerical estimates of such shifts have been made for Tl^+ in a small rf trap [4]. For an estimated temperature ~ 0.1 mK one finds for δ_D, the second order Doppler shift, $\delta_D/\nu_o = 3kT/Mc^2 \leq 10^{-18}$. The estimated quadratic Stark shift $\delta_S \simeq 2 \, [mHz/(V/cm)^2] \cdot E^2$ amounts to 10^{-4} Hz or $\delta_S/\nu_o \leq 10^{-19}$ at 0.1 mK. For small, deep traps it is also necessary to discuss an electric "trap shift" proportional to the applied trap voltage which should occur when ground or excited states have an appreciable <u>electronic</u> electric quadrupole moment eQ. For heavy atoms in a P state with HFS quantum number F ≥ 1 eQ may be as large as 3 e$\AA^2$. The order of magnitude of the corresponding m_F-dependent shift δ_Q is $eQ\phi_{zz}$, ϕ_{zz} referring to the electric trapping field. For the Ba^+ rf trap one finds $|eQ \, \phi_{zz}| \leq 30$ KHz. However since it is an AC shift, it averages out to zero. A neighboring ion at the equilibrium distance in the well [14] of ~ 5 μm would cause an analogous quadrupole shift of ~ 200 Hz which would not be averaged out by the motion of the ions. For the Penning trap of [12] one finds a DC shift $\delta_Q \simeq 2$ Hz. Our Group IIIA mono-ion oscillator experiments are immune to such shifts.

One might think the $\Delta m_F = 0$ transitions for our ω_o line would show no (nuclear) Zeeman effect. However for the isoelectronic ^{199}Hg isotope a "nuclear" g-factor for the ^3P_o state about twice as large as for the ^1S ground has been measured. This is explained by a small admixture of $^3P_1^o$ to the ^3P_o state via the strong HFS interaction. Estimated Zeeman splittings for Tl^+ amount to ~ 2 KHz/Gauss requiring fairly elaborate shielding to bring this δ_z shift down to the level of δ_S and δ_D. Outside the Lamb-Dicke dominant carrier regime the amplitude of the carrier is strongly dependent on the vibration amplitude. Thus variations in the vibration amplitude can broaden the carrier [4]. So far it has been tacitly assumed that laser sources of arbitrarily high sharpness and short-term stability are available. However the record in minimal laser spectral width still appears to be a few Hertz. This is still far above fundamental limitations which lie around $\sim 10^{-4}$ Hz.

6. Results of Preparatory Experiments

The most important result so far obtained in our mono-ion oscillator work [11] is the 3-dimensional localization in free space to a region of ~ 2000 Å diameter of an individual Ba^+ ion, see Fig. 4. The photographic image of the ion has a diffraction limited diameter of 2 μm. It is possible to deconvolute the much smaller localization range by comparing the 1-ion image with a 3-ion image which, obtained under otherwise identical conditions, has a ~ 20 times larger area and a ~ 30 times lower peak brightness. This implies $z_o \simeq \lambda_o/2\pi$ and that the boundary of the LAMB-DICKE regime of the dominant carrier has been reached. The best spectral resolution obtained so far by us [17] is a line-width of ~ 40 MHz for the $6^2S_{1/2} - 6^2P_{1/2} - 5^2P_{3/2}$ two-photon transition [8] to the highly metastable $5^2D_{3/2}$ state of

358

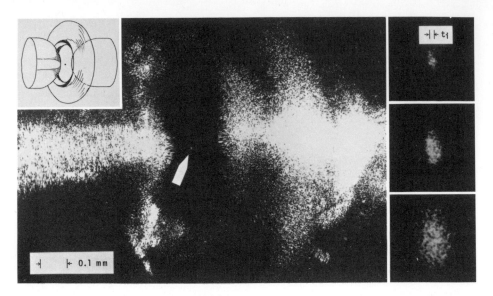

Fig.4 _Micro-photographic images of 1, 2 and 3 trapped Ba⁺ ions._ _The large_
photograph shows the ∿ 2 µm thick image (white arrow) of a single ion inside
the rf quadrupole trap as viewed through the gap between the ring and the left
cap-electrodes (trap structure illuminated by scattered laser light). _A sketch_
of the whole trap structure seen from the same angle is inserted. _The three_
small photos, going from top to bottom, show, 10-fold enlarged, the central
trap region, containing 1, 2 and 3 ions, from [11].

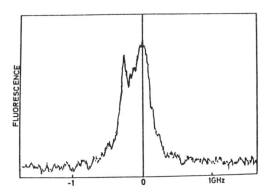

Fig.5 _Resonance fluorescence near_
493 nm from an individual barium
ion vs. scanned frequency ω_a _of_
650 nm laser. _The 493 nm laser fre-_
quency ω _was adjusted to_ $\omega_{sp} - \omega =$
$2\pi \cdot 280$ _MHz to effect side band_
cooling. _The frequency coordinate_
calibration is given for $\omega_a - \omega_{pd}/$
2π. _The sharp two-photon transi-_
tion $6^2S_{1/2} - 6^2P_{1/2} - 5^2D_{3/2}$ _to_
the metastable $5^2D_{3/2}$ _level occur-_
ing for $\omega - \omega_a = \omega_{sp} - \omega_{pd}$ _is visi-_
ble at $\omega_a - \omega_{pd} = -2\pi \cdot 280$ _MHz._
Here ω_{sp}, ω_{pd} _are the respective_
single photon resonance frequen-
cies, from [17].

Ba⁺ which has a lifetime of 17 sec, see Fig. 5. This two-photon transition
is obviously still strongly power-broadened.

7. Conclusion

After a fairly difficult start, mono-ion oscillator spectroscopy now appears to be off and running, with at least five groups all over the world actively participating in such studies. The demonstration of electronic linewidths and long-term frequency stabilities in such passive devices of 1 Hz and less is likely to act as a strong stimulus for the future development of laser sources of improved short-term stability, much closer to the fundamental limit near $\sim 10^{-4}$ Hz and also of new types of laser harmonic generators. Thus the current promise of an atomic line spectral resolution of about 1 part in 10^{18} may be realized in the not too far future.

8. References

1. E.N. Fortson, F.G. Major and H.G. Dehmelt, Phys. Rev. Lett. 16, 221 (1966), and H.A. Schuessler, E.N. Fortson and H.G. Dehmelt, Phys. Rev. 187, 5 (1969).
2. H.G. Dehmelt, Phys. Rev. 103, 1125 (1956).
3. F.G. Major and G. Werth, Phys. Rev. Lett. 30, 1155 (1973).
4. H.G. Dehmelt, Bull. Am. Phys. Soc. 18, 1571 (1973).
5. H.G. Dehmelt, Bull. Am. Phys. Soc. 20, 60 (1975).
6. D. Wineland and H. Dehmelt, Bull. Am. Phys. Soc. 20, 637 (1975).
7. R. Van Dyck, Jr., P. Ekstrom and H. Dehmelt, Nature 262, 776 (1976), and R.S. Van Dyck, Jr., P.B. Schwinberg, and H.G. Dehmelt in "New Frontiers in High-Energy Physics", edited by B. Kursunoglu, A. Perlmutter, and L. Scott, (Plenum, N.Y., 1978).
8. H. Dehmelt and P. Toschek, Bulletin APS 20, 61 (1975).
9. H. Dehmelt and H. Walther, Bulletin APS 20, 61 (1975).
10. W. Neuhauser, M. Hohenstatt, P. Toschek and H. Dehmelt, Phys. Rev. Lett. 41, 233 (1978).
11. W. Neuhauser, M. Hohenstatt, P.E. Toschek and H. Dehmelt, Phys. Rev. A22, 1137 (1980).
12. D.J. Wineland and W.M. Itano, Physics Letters 82A, 75 (1981).
13. H. Dehmelt, private communication (1976) and Morris Loeb Lectures in Physics, Harvard University (1977).
14. H. Dehmelt, Nature 262, 777 (1976); Bull. Am. Phys. Soc. 24, 634 (1979).
15. D.J. Wineland and W.M. Itano, Phys. Rev. A20, 1521 (1979).
16. J. Javanainen, Appl. Phys. 23, 175 (1980).
17. W. Neuhauser, M. Hohenstatt, P.E. Toschek and H.G. Dehmelt, in "Spectral Line Shapes," B. Wende, editor. Walter de Gruyter & Co., Berlin, New York 1981).

Laser Cooling and Double Resonance Spectroscopy of Stored Ions

W.M. Itano and D.J. Wineland
Frequency and Time Standards Group, Time and Frequency Division
National Bureau of Standards, 325 S. Broadway
Boulder, CO 80303, USA

1. Introduction

The use of ion storage techniques for spectroscopy is motivated by the fact that ions can be confined by electric and magnetic fields for long periods of time without suffering the large perturbations which usually accompany other methods of confinement, such as those due to collisions with buffer gas molecules. Linewidths as small as a few Hz and Q's as high as 10^{10} have been observed on ground-state hyperfine transitions of atomic ions stored in rf quadrupole traps [1-3]. The accuracy of these measurements has been limited largely by the second-order Doppler shift. The signal-to-noise ratios have been limited by the small number of ions that can be stored (about 10^5-10^6) and by the difficulty of detecting transitions. Hyperfine and Zeeman transitions have been detected by charge exchange [1], fluorescence [2,3], photodetachment [4], and photodissociation [5].

In this paper we discuss recent work at the National Bureau of Standards (NBS) in this area. The second-order Doppler shift can be reduced by laser (resonant light pressure) cooling. This technique has been demonstrated in experiments on stored ions by our group at NBS [6-8] and also by a group at Heidelberg [9]. Other work at NBS has been directed toward the realization of high-efficiency laser-optical-pumping, double-resonance detection techniques [10]. These techniques have been used to make the first highprecision hyperfine structure measurements of $^{25}Mg^+$ [11]. We anticipate that the laser cooling and double resonance techniques will find practical application in the development of frequency standards based on stored ions [12].

2. Apparatus

A block diagram of the apparatus is shown in Fig.1. The Mg^+ ions are confined in a Penning-style ion trap. They are irradiated by light resonant
with the 3s $^2S_{1/2} \to$ 3p $^2P_{3/2}$ 280 nm transition and also by rf and microwave
radiation resonant with various transitions between ground-state sublevels.
Scattered 280 nm photons are collected by a mirror and counted by a photomultiplier tube, with a net detection efficiency of about 10^{-5}.

A Penning trap confines ions by a combination of a uniform static magnetic field, $\vec{B} = B_0 \hat{z}$, and a quadrupolar electrostatic potential. The magnetic
field causes the ions to move in circular "cyclotron" orbits in the x-y plane.
The electric fields provide a harmonic restoring force for the axial (z)
motion and cause the centers of the cyclotron orbits to move in the x-y plane
in circular "magnetron" orbits around the trap symmetry axis. The angular
frequencies corresponding to the axial, cyclotron, and magnetron motions are
denoted by ω_z, ω_c', and ω_m respectively. For typical operating conditions,
$\omega_z \cong 2\pi \cdot 200$ kHz, $\omega_c' \cong 2\pi \cdot 800$ kHz, and $\omega_m \cong 2\pi \cdot 25$ kHz. Typically,
$B_0 \cong$ 1T, and a potential $V_0 \cong 7$ V is applied across the trap electrodes,
which have inside dimensions of about 1 cm. The axial and cyclotron motions
are thermal and stable. The magnetron motion is nonthermal and unstable,
since an increase in the size of the magnetron orbit leads to a decrease in
the total (kinetic plus potential) energy. However, storage times of about
1 day have been observed, even without laser cooling. The background pressure is typically $\leq 10^{-8}$ Pa.

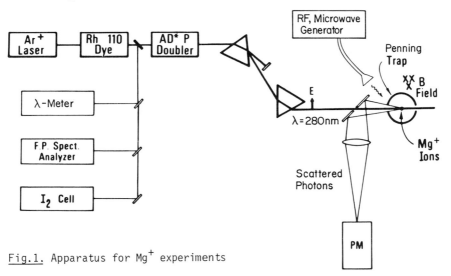

Fig.1. Apparatus for Mg^+ experiments

The 280 nm radiation was produced by frequency doubling the output of a single-mode cw Rhodamine 110 dye laser in a 90° phase-matched AD*P crystal. Between 5 and 30 μW of UV radiation was generated in a bandwidth of about 1 MHz. The dye laser could be long-term stabilized to discrete frequencies to less than 1 MHz by locking it to saturated absorption features in I_2. Fine tuning could be done by varying the trap magnetic field to Zeeman shift the Mg^+ levels. For efficient laser cooling, the UV linewidth must be less than the Doppler width of the transition (3 GHz at 300K), but need not be much less than the natural linewidth (43 MHz).

3. Laser Cooling in a Penning Trap

The basic principle of laser cooling is that light pressure can be used to damp the velocity of an atom or ion if the light frequency is tuned slightly below that of a strong optical transition. The process is easily described for the case of free or harmonically bound atoms [9,13-16]. Atoms whose velocities are directed toward the light source see a frequency Doppler shifted closer to resonance; if their velocities are directed away from the source, the frequency shift is away from resonance. Therefore, the atoms tend to absorb photons when their velocity is directed toward the source. This cools them, since the momentum of the absorbed photons reduces the atomic momentum. (The photons are re-emitted in random directions.) Three-dimensional cooling can be obtained with six laser beams directed along the ± x, ± y, and ± z directions. A single beam suffices to cool all oscillational modes of an atom bound in a three-dimensional harmonic potential, if the modes are nondegenerate and if the beam is not directed along one of the principal axes.

Cooling of the axial and cyclotron modes of an ion in a Penning trap takes place by the process just described for free or harmonically bound atoms, if the light is tuned below resonance [17]. Cooling of the magnetron mode can be accomplished by focusing the light beam so that it is more intense on the side of the trap axis on which the magnetron motion recedes from the light source [6-8,17]. By "cooling" of a mode, we mean the reduction of the average kinetic energy in that mode, regardless of what happens to the potential energy. In order to efficiently cool all modes, the angle between the beam and the z axis should be about 45°; the optimum angle depends on the angular distribution of the scattered photons [17]. We have derived rate equations for the laser cooling of an ion in a Penning trap by a single nonuniform light beam [17]. Solving these equations, we find that it should be possible to cool an ion to a "temperature" T given by $k_B T \cong \hbar \gamma/2$, where k_B is Boltzmann's constant and γ is the natural linewidth of the upper state. The "temperature" is defined in terms of the mean kinetic energy and may be different for the different modes. For Mg^+, this minimum temperature is about 1 mK.

4. Optical Pumping and Double Resonance

The extremely long ground-state relaxation times possible for stored ions make it possible to observe very narrow transition linewidths and to observe very weak optical pumping processes. An ion must scatter about 10^4 optical photons in order to be cooled significantly below room temperature. Therefore, even weak depopulation pumping processes would have an adverse effect on the cooling. Cooling, in turn, influences the optical pumping by reducing the Doppler broadening until it is less than the natural linewidth. This makes it possible for a monochromatic laser to interact strongly with all of the ions and not just with a small velocity class.

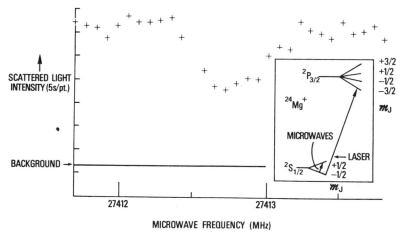

Fig.2. Optical pumping and double resonance of $^{24}Mg^+$

Generally, the $(^2S_{1/2}, M_J = -1/2) \rightarrow (^2P_{3/2}, M_J = -3/2)$ Zeeman component was used for laser cooling of $^{24}Mg^+$, which has I=0 (see Fig.2). This transition does not lead to depopulation pumping of the ground state, since the $M_J = -3/2$ excited state sublevel must decay to the $M_J = -1/2$ ground state sublevel. When this transition is driven near resonance with light polarized perpendicular to the magnetic field, the $(^2S_{1/2}, M_J = -1/2) \to (^2P_{3/2}, M_J = +1/2)$ and $(^2S_{1/2}, M_J = +1/2) \to (^2P_{3/2}, M_J = -1/2)$ transitions are driven weakly in their Lorentzian wings (the Zeeman splitting is much greater than the Doppler broadening). In the steady state, which is achieved in $\leq$ 1s, $16/17 \doteq 94\%$ of the population is in the $M_J = -1/2$ ground-state sublevel [10].

The ground-state Zeeman transition $(^2S_{1/2}, M_J = -1/2 \to +1/2)$ can be detected by a decrease in the fluorescence intensity. This "flop-out" detection method can be very efficient, since a transition due to a single microwave photon interrupts the flow of scattered optical photons until the ion is pumped back to the $M_J = -1/2$ sublevel by weak, off-resonance scattering. Since the decrease in the number of scattered photons per ion can be very large, it is possible to make up for poor light collection and detector quantum efficiency, so that the transition can be detected with nearly unit efficiency. Similar detection methods have been proposed previously 18 .

The $(^2S_{1/2}, M_I = -5/2, M_J = -1/2) \to (^2P_{3/2}, M_I = -5/2, M_J = -3/2)$ hyperfine-Zeeman component was generally used for laser cooling of $^{25}Mg^+$ ($I = 5/2$). In the steady state, about 16/17 of the population is pumped into the $(M_I = -5/2, M_J = -1/2)$ ground-state sublevel. Pumping into the $M_J = -1/2$ manifold takes place by the same mechanism as in $^{24}Mg^+$. Pumping into the $M_I = -5/2$ sublevels takes place because of hyperfine coupling in the excited state [10]. Any transition which decreases the population in the $(M_I = -5/2, M_J = -1/2)$ ground-state sublevel can be detected by a decrease in the fluorescence intensity.

5. Results

In the Penning trap, low temperatures become easier to achieve as the ion density is reduced. This is because the radial electric field due to space charge increases the magnetron velocity. This problem does not exist for a single, isolated ion, and the lowest temperatures were observed for this case [8].

Figure 3 shows the fluorescence from a small number of $^{24}Mg^+$ ions as a function of time. After the ions were cooled and localized at trap center, an oven containing ^{25}Mg (98% isotopic purity) was heated in order to induce the resonant charge exchange reaction $(^{24}Mg^+ + {}^{25}Mg \to {}^{24}Mg + {}^{25}Mg^+)$. The

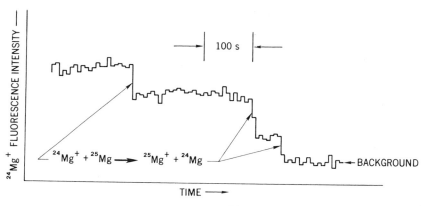

<u>Fig.3.</u> Fluorescence from 3, 2, and 1 ^{24}Mg$^+$ ions

resulting ^{25}Mg$^+$ ions were ejected from the trap by resonant cyclotronmagnetron rf excitation. The three large step decreases in fluorescence are due to loss of the ^{24}Mg$^+$ ions, one at a time, and the last plateau above background is the fluorescence from a single ion. The signal was about 50 photons/s per ion.

The "temperature" of a single ^{24}Mg$^+$ ion was determined from the Doppler width by optical-optical double resonance (see Fig.4). One laser was tuned slightly below the (^{2}S$_{1/2}$, M$_J$ = -1/2) → (^{2}P$_{3/2}$, M$_J$ = -3/2) transition, to provide cooling and fluorescence detection. A low power laser was swept continuously across the (^{2}S$_{1/2}$, M$_J$ = -1/2) → (^{2}P$_{3/2}$, M$_J$ = -1/2) transition, which was detected by a decrease in the fluorescence. The resulting lineshape reflects both the natural and Doppler broadenings. The data points represent 10s integrations; the connecting lines are only for clarity. Simulated curves are shown for temperatures of 0 K and 1000 mK. We estimate that T = 50 ± 30 mK. Since the light was incident at 82° with respect to the z axis, this is essentially a measurement of the cyclotron-magnetron (x-y) temperature. The axial (z) temperature was estimated to be about 600 mK by probing the axial excursions with a focused laser beam. According to our calculations, it should be possible to obtain a cyclotron-magnetron temperature of about 1 mK and an axial temperature of about 11 mK for these conditions [17]. At present, the discrepancy is not understood, but may be due to the presence of impurity ions in the trap.

Ground-state Zeeman and hyperfine transitions were detected by the optical-pumping, double-resonance methods outlined previously. The only transition in ^{24}Mg$^+$ is the electronic spin flip transition (see Fig.2). Several

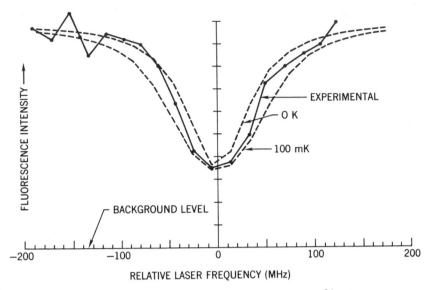

Fig.4. Optical-optical double resonance of a single ^{24}Mg$^+$ ion

transitions in ^{25}Mg$^+$, corresponding to both nuclear and electronic spin flip transitions, were observed. Most transitions were broadened by the instability of B_o, which was, at best, about 1 ppm in a few seconds. The magnetic field derivative of the $(M_I, M_J) = (- 3/2, + 1/2) \leftrightarrow (- 1/2, + 1/2)$ transition goes to zero at $B_o \cong 1.2398$ T. Near this field, the transition was observed with linewidths as narrow as 0.012 Hz (see Fig.5). The oscil-

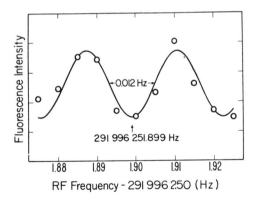

Fig.5. ^{25}Mg$^+$ $(M_I,M_J) = (- 3/2), + 1/2) \leftrightarrow (- 1/2, + 1/2)$ ground-state hyperfine resonance

latory lineshape results from the use of the Ramsey interference method [19], which was implemented by driving the transition with two coherent rf pulses of 1.02 s duration separated by 41.4 s. The center of the resonance can be determined with an uncertainty of about 10^{-11}. Several different transition frequencies were measured in order to obtain separately the hyperfine constant (A = -596.254 376 (54) MHz) and the nuclear-to-electronic g factor ratio (g_I/g_J = 9.299 484 (75) × 10^{-5}) [11].

6. Future Possibilities

Work is being initiated at NBS on the development of a microwave frequency and time standard based on a hyperfine transition in ^{201}Hg$^+$ and of an optical frequency standard based on a two-photon transition to a metastable state in ^{199}Hg$^+$ or ^{201}Hg$^+$. Both proposed standards are based on a cloud of Hg$^+$ ions stored in a Penning trap and use the 6s $^2S_{1/2}$ → 6p $^2P_{1/2}$ 194 nm transition for laser cooling.

The proposed microwave frequency standard is based on the (F, M_F) = (1,1) ↔ (2,1) transition, which is field-independent to first order at $B_o \cong$ 0.534 T, with frequency $\cong$ 25.9 GHz. If B_o can be controlled to slightly better than 0.1 ppm over the ion cloud, the fractional frequency shift can be kept below 10^{-15}. All other systematic shifts, such as those due to the second-order Doppler effect, collisions, the trap electric fields, or thermal radiation, appear to be less than 10^{-15}. It should be possible to observe the transition with a Q of 2.6 × 10^{12} or better, by using optical pumping and detection techniques similar to those demonstrated with ^{25}Mg$^+$. The accuracy of this standard could be as good as 10^{-15}, which is about 100 times better than that of the best frequency standards now available.

The proposed optical frequency standard is based on the two-photon-allowed $5d^{10}$ 6s $^2S_{1/2}$ → $5d^9$ $6s^2$ $^2D_{5/2}$ Hg$^+$ transition, which has a natural Q of 7.4 × 10^{14} [12]. The first-order Doppler effect can be eliminated by driving the transition with counter-propagating 563.2 nm laser beams. Hyperfine-Zeeman components whose magnetic field derivatives vanish at particular values of B_o exist in ^{199}Hg$^+$ and ^{201}Hg$^+$. The two-photon transition can be detected with high efficiency by using the 194.2 nm fluorescence intensity as a probe of the ground state population. Taking full advantage of the high Q transition would require a laser with linewidth less than 1 Hz, which does not exist at present. However, linewidths ≤100 Hz appear feasible and could be used for initial experiments. If the laser linewidth is less than the natural line-width, then the ac Stark shift is about 2×10^{-15} near saturation. All other systematic shifts appear to be less than 10^{-15}.

The method currently being investigated for generating the required 194.2 nm radiation is sum-frequency mixing in a KB5 crystal of the output of a 792 nm single-mode cw ring dye laser and the second harmonic, generated in an ADP crystal, of the output of a 514 nm stabilized, single-mode cw Ar^+ laser. The method has been demonstrated previously with pulsed lasers [21]. Further details of these proposed Hg^+ frequency standards are published elsewhere [12].

Acknowledgments. We wish to acknowledge experimental assistance by Drs. J.C. Bergquist and R.E. Drullinger. This work was supported in part by the Air Force Office of Scientific Research and the Office of Naval Research.

References

1. H.A. Schuessler, E.N. Fortson, H.G. Dehmelt: Phys. Rev. *187*, 5-38 (1969)
2. F.G. Major, G. Werth: Phys. Rev. Lett. *30*, 1155-1158 (1973)
3. R. Blatt, G. Werth: Z. Phys. A *299*, 93 (1981)
4. R.M. Jopson, D.J. Larson: Bull. Am. Phys. Soc. *25*, 1133 (1980)
5. C.B. Richardson, K.B. Jefferts, H.G. Dehmelt: Phys. Rev. *165*, 80-87 (1968)
6. D.J. Wineland, R.E. Drullinger, F.L. Walls: Phys. Rev. Lett. *40*, 1639-1642 (1978)
7. R.E. Drullinger, D.J. Wineland, J.C. Bergquist: Appl. Phys. *22*, 365-368 (1980)
8. D.J. Wineland, W.M. Itano: Phys. Lett. *82*A, 75-78 (1981)
9. W. Neuhauser, M. Hohenstatt, P.E. Toschek, H. Dehmelt: Phys. Rev. Lett. *41*, 233-236 (1978); Appl. Phys. *17*, 123-129 (1978); Phys. Rev. A *22*, 1137-1140 (1980)
10. D.J. Wineland, J.C. Bergquist, W.M. Itano, R.E. Drullinger: Opt. Lett. *5*, 245-247 (1980)
11. W.M. Itano, D.J. Wineland: Phys. Rev. A, to be published
12. D.J. Wineland, W.M. Itano, J.C. Bergquist, F.L. Walls: Proc. 35th Ann. Symp. on Freq. Control (U.S. Army Electronics Command, Fort Monmouth, NJ, 1981)
13. T.W. Hansch, A.L. Schawlow: Opt. Commun. *13*, 68-69 (1975)
14. D.J. Wineland, H. Dehmelt: Bull. Am. Phys. Soc. *20*, 637 (1975)
15. D.J. Wineland, W.M. Itano: Phys. Rev. A *20*, 1521-1540 (1979)
16. J. Javanainen: Appl. Phys. *23*, 175-182 (1980)
17. W.M. Itano, D.J. Wineland: Phys. Rev. A, submitted
18. H. Dehmelt: Bull. Am. Phys. Soc. *20*, 60 (1975)
19. N.F. Ramsey: *Molecular Beams* (Oxford Univ. Press, London 1956) pp.124-134
20. P.L. Bender, J.L. Hall, R.H. Garstang, F.M.J. Pichanick, W.W. Smith, R.L. Barger, J.B. West: Bull. Am. Phys. Soc. *21*, 599 (1976)
21. R.E. Stickel, F.B. Dunning: Appl. Opt. *17*, 981-982 (1978)

Spectroscopy of Trapped Negative Ions

D.J. Larson and R.M. Jopson
Department of Physics, University of Virginia
Charlottesville, VA 22901, USA

Introduction

Negative ions, while similar to neutral atoms and molecules and positive ions in many respects, have properties which set them distinctly apart. The lack of a long range coulomb interaction between the extra electron and the neutral has some important consequences. The number of bound states in atomic negative ions is very small (usually one) much to the spectroscopist's dismay. Electron correlations play a primary role in determining binding energies and other properties and thus negative ions should be a good testing ground for atomic theory. Photodetachment cross sections, at least near threshold, can be described in simple terms.

A dramatic example of the difference between negative ions and neutral atoms can be found in the comparison of our work on photodetachment in a magnetic field to the several studies on photoionization in a magnetic field. [1,2] Because of the short range interaction between atom and de-parting electron, a theory which ignores the final state interaction provides a rather good description of the experimental data for photodetachment while photoionization in a magnetic field is strongly affected by the final state interaction. Thus, while photodetachment and photoionization are analogous processes, the differences due to the different character of the negative ion are substantial.

Despite impressive gains in the last several years, the amount of spectroscopic information about negative ions is small compared to the information available for positive ions and neutral atoms and molecules. This is understandable since negative ions can be difficult to produce and the spectroscopy requires specialized techniques. Much of the information about the structure of these ions has been derived from experiments using photodetachment spectroscopy. [3,4] The spectroscopic information about atomic ions derived from these experiments has been limited to electron affinities and fine structure separations. One goal of our work with trapped ions is to provide new kinds of information about negative ions by developing techniques for direct observation of transitions between bound state sublevels. [5] This goal has been partially realized in experiments on the Zeeman structure of S^- and O^- described below. [6]

Experimental Technique

We have observed the Zeeman resonances in the ground $^2P_{3/2}$ state of trapped $^{32}S^-$ and $^{16}O^-$ ions by exploiting differences in the photodetachment cross sections for the magnetic sublevels. The levels of O^- and of neutral oxygen and the detached electron are shown as a function of magnetic field in Fig. 1. S^- has identical structure except that the fine structure and photodetachment energies are larger. The diagram does not explicitly show the higher lying fine structure levels of the neutral. It also does not show the continuum of states due to motion of the electron in the direction of the magnetic field.

A photodetachment transition to a particular state of the neutral atom and a particular spin and cyclotron state of the electron has a cross section which is sharply peaked near threshold due to the density of states for a single degree of freedom. [1] The Zeeman structure of ion and neutral gives a separation of about $1/6 \mu_0 B$ between the thresholds for different magnetic sublevels. For ion cloud temperatures of $< 10K$ and fields of $> 1T$ the various thresholds would be resolved and frequency selective detachment could be used to create and probe population differences. In our experiments the ion temperatures are as much as 900 K and the individual thresholds are not resolvable due to motional broadening. However, one can exploit polarization dependent differences in the photodetachment rates since the $|m_J| = 3/2$ states photodetach much more rapidly than the $|m_J| = 1/2$ states when illuminated with π polarized light at a frequency above the thresholds

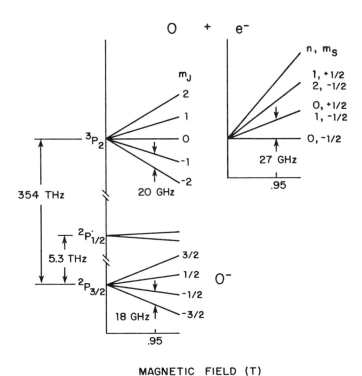

MAGNETIC FIELD (T)

Fig. 1. The energy levels of O^-, the lowest fine structure state of O, and the spin and cyclotron states of the detached electron are shown as a function of magnetic field. The figure does not show the continuum of electron states due to motion along the magnetic field.

for both types of states. For S⁻ we used light about 200 GHz above the
threshold and for O⁻ we used light which was about 25 THz above threshold.

To observe a Zeeman transition, we can use the light to eliminate most of
the ions with $|m_J|$ = 3/2 and leave a population of mostly the $|m_J|$ = 1/2
states. If one drives the m_J = +1/2 to m_J = +3/2 transition during or
between pulses of light, those ions in the m_J = +1/2 state will detach via
the m_J = +3/2 state, leaving mostly ions in the m_J = -1/2 state in the trap.
Thus driving the Zeeman transition during photodetachment will reduce the
number of ions remaining in the trap. One can detect the m_J = -1/2 to
m_J = -3/2 transition in a similar manner. These transitions are not at
the same frequency but are split due to the partial decoupling of L and S
in the magnetic field. Using the polarization dependence of the cross
section is simpler than using the frequency dependence in that the light
frequency does not need to be precisely controlled, but it does not allow
observation of the m_J = +1/2 to m_J = -1/2 transition for the geometry
where the light propagation direction is perpendicular to the magnetic
field. Thus the observable spectrum is two separated lines corresponding
to the $|m_J|$ = 1/2 to $|m_J|$ = 3/2 transitions.

The use of ion traps adds several dimensions to work with negative ions.
[8] The manipulation and observation of populations in ion sublevels is
simplified by the long confinement times and detection sensitivity provided by an ion trap. For microwave transitions, Doppler broadening
effectively disappears since the Dicke criterion of confinement to a region
less than half a wavelength in size can be satisfied. Long confinement
times provide the possibility of narrow lines and thus permit probing of
transitions with high resolution. In addition the possibility of cooling
of the internal and external degrees of freedom can aid substantially in
reducing motional contributions to linewidths as well as in achieving
large ground state populations and in identifying transitions, especially
in molecular ions. BRAUMAN and coworkers have used photodetachment of
negative ions trapped in an ion cyclotron resonance spectrometer to study
a number of molecular negative ions. [9]

This work used a Penning style ion trap both for observing the Zeeman
transitions in trapped ions and for measurement of the magnetic field
using trapped electrons. A partial schematic of the experimental apparatus
is shown in Fig. 2. To measure the Zeeman transitions, the ions are created
in the trap by dissociative attachment using a few hundred nanoampere
electron beam with an energy of 1 to 2 volts incident on carbonyl sulfide
(for S⁻) or nitrous oxide (for O⁻) at about 2×10^{-8} torr. The gas is
then removed, leaving on the order of 10^4 ions trapped in a magnetic field
of about 1T with the trap endcaps at -1.7 volts relative to the ring electrode. After waiting several seconds, a few milliwatts of π polarized
laser light photodetaches the ions. Simultaneously, 18 GHz radiation with
power less than 140 µW drives a Zeeman transition in the ion ground state.
After 1 to 3 seconds of photodetachment and Zeeman driving, the number
of ions remaining is recorded. The number of ions is determined by
driving their axial motion and measuring the current induced on the trap's
ring electrode. The procedure is repeated for each microwave frequency,
scanning over the resonance. A diagram of the timing sequence for each
data point with parameters characteristic of S⁻ is shown in Fig. 3.

To measure the magnetic field, the trap is filled with electrons by
directing a 12 eV electron beam into it with the endcaps biased at -10
volts relative to the ring. After waiting for several seconds, radiation

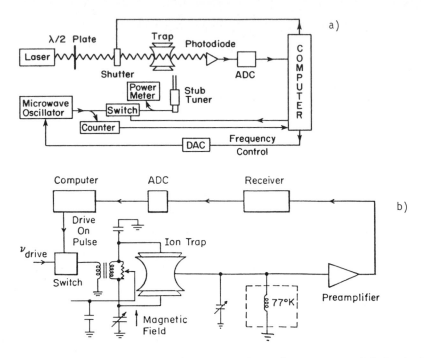

Fig.2. Schematic diagram of the experimental apparatus. (a) Optical and Microwave System, (b) Ion Detection Electronics

at about 27 GHz with power of less than 1 μW drives electrons out of the trap by exciting their cyclotron motion and then the number of electrons left is recorded. This procedure is repeated as the frequency is scanned over the resonance. A data set typically consists of a scan of each Zeeman resonance with measurements of the magnetic field before and after. A picture of a Zeeman resonances in O^- is shown in Fig. 4.

Results and Analysis

The resonances in O^- are three to five times smaller than those obtained in S^-. This is probably because the 790 mm light used for O^- allows detachment to all fine structure states of the neutral atom and the difference in detachment rates between the $|m_J| = 3/2$ and $|m_J| = 1/2$ states is not as large as the difference when only the transition to the $^3P_2$ state is energetically allowed. However there are other differences between the two species which could play a role, including the fact that O^- is lost due to reactions with background gas at a much higher rate. Under our typical conditions, the lifetime of S^- in the trap is larger than a minute and the whole data cycle takes a few seconds.

The O^- resonance shown in Fig. 4, while of limited signal to noise, appears to have the flattened bottom typical of resonances we have observed

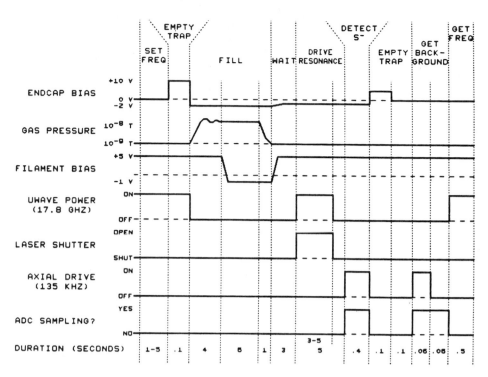

Fig. 3. The sequence of events used to measure a single point of the Zeeman resonance is shown. The parameters used for S⁻ are given.

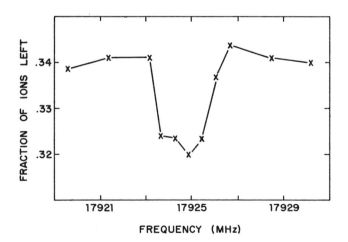

Fig. 4. A Zeeman resonance in O⁻ at a field of about 0.95T. Each point is the average of six separate measurements. The data points are connected by straight lines.

many times in S⁻. This occurs when the microwave power is increased above the minimum level necessary to see the transition. This shape can be understood with the use of a simple rate equation model which includes the effects of photodetachment and Zeeman transitions on the various levels. [10] Some results of this model are shown in Fig. 5. The fraction of ions left after simultaneous photodetachment and Zeeman driving is plotted as a function of the Zeeman transition rate for three different photodetachment times. The parameters used are appropriate for the conditions under which the S⁻ data was taken. The differences in signal as a function of Zeeman transition rate saturate at or a little below the point where the Zeeman rate is equal to the photodetachment rate for the $|m_J| = 3/2$ states. Thus for microwave power levels where the Zeeman transition rate at the center of the line is greater than the photodetachment rate, the line will appear to have a flattened bottom. We assume that the major contribution to the linewidth is magnetic field inhomogeneities. All ions sample different magnetic field regions due to their magnetron and axial motions in the trap, so that nearly all the ions can participate in the transition.

The average of the two observed Zeeman frequencies in each ion gives a value for the g_J factor when compared with the electron cyclotron frequency. (The cyclotron frequency is corrected for the approximately 5 ppm shift due to electric fields in the trap). If we assume L·S coupling, the splitting of the two resonances gives a value for the fine structure separation. The results are

$$S^-: \quad g_J = 1.333\ 984\ (28), \quad \nu_{f.s.} = 488\ (5)\ cm^{-1}$$

$$O^-: \quad g_J = 1.333\ 934\ (56), \quad \nu_{f.s.} = 177\ (10)\ cm^{-1}.$$

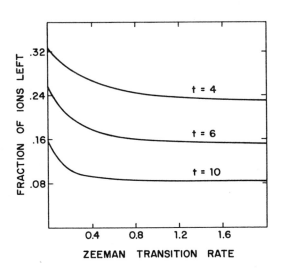

Fig.5. The fraction of ions left after illuminating with both light and microwaves is plotted as a function of Zeeman transition rate for different illumination times. The Zeeman rate is measured in units of the photodetachment rate for the $m_j = 3/2$ state. The time is in units of the inverse of the photodetachment rate. These curves were calculated from a simple rate equation model using parameters appropriate to detachment near threshold. The curves show the saturation of the signal as the Zeeman transition rate is increased.

Many separate determinations of the S$^-$ frequencies have been made. The statistical accuracy of the line centers on lines which are 60 ppm wide is typically on the order of a few ppm. A search for systematic effects suggested a possible problem due to magnetic field inhomogeneities. [6] The error estimate given for g_J is two to three times the size of changes actually observed. The splitting of the resonances should not be as strong a function of magnetic field problems as the average frequency. However, it is more sensitive to small asymmetries in the lines of the type observed in highly inhomogeneous fields. The uncertainty quoted for the S$^-$ fine structure is about twice the statistical error and reflects our estimate of the possible systematic effects. The numbers for O$^-$ are the result of very recent data and must be considered preliminary, pending further measurements. We have arbitrarily assigned uncertainties twice the size for S$^-$.

The value for the fine structure in O$^-$ is in agreement with an earlier result from laser photodetached electron spectrometry. [11] The result for S$^-$ may be in slight disagreement with the results from electron spectrometry and from laser threshold photodetachment. [12]

The values for the g-factors provide new information about the structure of the negative ions. No calculation of these parameters has yet been reported but we can compare the numbers measured to those observed and calculated in the isoelectronic atoms F and Cl. [13,14] In all cases the Lande value for the g factor is 4/3. The largest correction to the Lande value comes from the anomalous moment of the electron. It contributes a shift of +773 x 10^{-6} from the Lande value. The other shifts depend upon the atomic structure and are presumably dominated by relativistic and diamagnetic contributions. The difference between the Lande value corrected for the anomalous moment and the observed values is shown below.

$$O^-: -172 \ (56) \times 10^{-6} \qquad F: -246 \times 10^{-6}$$
$$S^-: -122 \ (28) \times 10^{-6} \qquad Cl: -179 \times 10^{-6}$$

The numbers appear to form a reasonable pattern with the corrections definitely smaller in the negative ions than in the isoelectronic neutral atom.

Conclusions

We have demonstrated a new technique for direct measurements of transitions between sublevels of negative ions. The technique has been applied to observation of the Zeeman structure of S$^-$ and O$^-$, yielding information about g-factors and fine structure separations. The technique can be readily extended to the stable states of a variety of negative ions.

Acknowledgements

This work has been supported by the U.S. Office of Naval Research and by the National Science Foundation.

Refenences

1. W. A. M. Blumberg, R. M. Jopson, and D. J Larson, Phys. Rev. Lett. 40, 1320 (1978). W. A. M. Blumberg, Wayne M. Itano, and D. J. Larson, Phys. Rev. A 19, 139 (1979).

2. M. L. Zimmerman, J. C. Catro, and D. Kleppner, Phys. Rev. Lett. 40, 1083 (1978). K. T. Lu, F. S. Tomkins, H. M. Crosswhite and H. Crosswhite, Phys. Rev. Lett. 41, 1034 (1978). R. R. Freeman, N. P. Economou, G. C. Bjorkland and K. T. Lu, Phys. Rev. Lett. 41, 1463 (1978).

3. H. Massey, Negative Ions, 3rd ed. (Cambridge U. Press, New York, 1976).

4. H. Hotop and W. C. Lineberger, J. Phys. Chem. Ref. Data 4, 539 (1975).

5. R. M. Jopson and D. J. Larson, Optics Lett. 5, 531 (1980).

6. R. M. Jopson and D. J. Larson, to be published.

7. W. Happer, Rev. Mod. Phys. 44, 169 (1972).

8. H. G. Dehmelt, in Advances in Atomic and Molecular Physics, D. R. Bates and I. Esterman, eds. (Academic, New York, 1967 and 1969), vols. 3 and 5.

9. A. H. Zimmerman and J. I. Brauman, J. Chem. Phys. 66, 5823 (1977).

10. R. M. Jopson, Thesis, Harvard University, 1981.

11. F. Breyer, P. Frey and H. Hotop, Z. Physik A 286, 133 (1978).

12. W. C. Lineberger and B. W. Woodward, Phys. Rev. Lett. 25, 424 (1970).

13. V. Beltran-Lopez, and H. G. Robinson, Phys. Rev. 123, 161 (1961).

14. V. Beltran-Lopez, E. Ley Koo, N. Segovia, and E. Blaisten, Phys. Rev. 172, 44 (1968).

Atomic Deceleration, Monochromatization and Trapping in Laser Waves: Theory and Experiments

V.S. Letokhov and V.G. Minogin

Institute of Spectroscopy, USSR Academy of Sciences,
142092 Moscow Trotzk Podol'skii Rayon, USSR

1. Introduction

Great interest has been shown recently in the problem of effect of resonant laser radiation on the translational state of atomic particles. There are evidently two main causes which are responsible for rapid progress in this field of laser physics. First, these studies are motivated by a natural wish to understand the laws of atomic motion in laser fields new for atomic physics and spectroscopy. The second serious cause is a quest to develop effective and technological methods of deep cooling of atoms and the ways of permanent localization of cooled atoms in light fields on the basis of resonant light pressure [1,2]. Such atoms are not only interesting subjects of investigation but, also, promising applications have been predicted for them in precision spectroscopic studies including frequency standards and in atomic physics.

   In the present report, according to today's state of the problem, we shall focus our attention on analysis of some points of atomic motion in laser fields directly related to radiative atomic cooling and localization of cold atoms in restricted space region.

2. Fundamentals of the Theory

Generally speaking, different configurations of light fields are of interest for applications. Nevertheless, since any field can be presented as a superposition of plane travelling waves it is the equation describing the atomic motion in a plane resonant travelling wave that is of basic interest for theory. The question on the form of this equation was first raised in [3], the explicit form of equation was found in [4,5]. We describe briefly the method of derivation of the equation considering, for simplicity, that an atom has two levels and assuming that the lower level is the ground one and the upper level is subjected to radiative decay into the ground one with the total decay rate 2γ. The natural line width 2γ is assumed, as it usually takes place, to exceed the recoil energy $r = \hbar k^2/2M$, where $\hbar k$ is the photon momentum.

The starting point for deriving an atomic motion equation is the Wigner density matrix equation describing both the internal and translational atomic states. For a two-level atom this operator equation coincides with a system of four integro-differential equations for the density matrix components $\rho_{ij}(\vec{r},\vec{p},t)$. In the case concerned $\gamma \gg r$ we may get rid of describing the evolution of internal atomic state and consider the equations on a time scale $\Delta t \gg \gamma^{-1}$. In this case, according to the principle of reduced description [6], the density matrix equations can be transformed by the Bogolyubov method into the equation for atomic distribution function w (r,p,t). The latter is the infinite Fokker-Plank equation with coefficients determined by all possible processes of momentum exchange between an atom and light wave. In an approximation sufficient for all practical purposes the third- and high-order derivatives can be omitted in this equation and in this case the equation has the form (for a wave propagating in the positive direction of the z axis)

$$\frac{\partial w}{\partial t} + \bar{v}\,\frac{\partial w}{\partial \vec{r}} = -\frac{\partial}{\partial p_z}\,(F_z w) + \sum_{i=x,y,z}\frac{\partial^2}{\partial p_i^2}\,(D_{ii}w) \tag{1}$$

where the light pressure force and the momentum diffusion tensor are determined by the relation [4,5]

$$F_z = \hbar k\gamma\,\frac{G}{1 + G + (\Omega - kv_z)^2/\gamma^2} \tag{2}$$

$$D_{ii} = \frac{1}{2}\,\hbar k F_z\left(d_{ii} + \delta_{3i}\left\{1 + \frac{F_z}{\hbar k\gamma}\,[(\Omega - kv_z)^2/\gamma^2 - 3]\right\}\right) \tag{3}$$

in which G is the saturation parameter, $\Omega = \omega - \omega_0$ is the difference between the wave and atomic transition frequencies, and coefficients α_{ii} depend on wave polarization [4].

Equation (1) determines the evolution of the distribution function w both for single atom and for an ensemble of noninteractive atoms. Below we shall apply it to the problem of radiative deceleration of an atomic beam by a counter-propagating laser wave.

3. Radiative Deceleration of Atomic Beams

Radiative deceleration of a thermal atomic beam is a direct way to produce the cold atoms with the temperature down to 10^{-2} K. Let us consider the peculiarities of atomic velocity distribution evolution in a beam assuming that the atoms are moving in the positive direction of the z axis ($V = V_z$) and the plane light wave propagates counter to the beam. In this case the wave frequency is assumed to be red-shifted by the value of Doppler shift $k\bar{v}$ so that the wave is in exact resonance with atoms having the average thermal velocity $\bar{v}$

$$\omega = \omega_0(1 - \bar{v}/c) \quad . \tag{4}$$

Under condition (4) the light pressure force

$$F_z = -\hbar k\gamma \frac{G}{1 + G + k^2(\bar{v} + v)^2/\gamma^2} \tag{5}$$

effectively slows down a narrow atomic velocity group near the resonant velocity $\bar{v} = -\Omega/k$ and with velocity width $\Delta v \sim \Gamma/k$ determined by the absorption linewidth $\Gamma = 2\gamma(1 + G)^{\frac{1}{2}}$. Simultaneously the force due to its nonlinear velocity dependence leads to velocity monochromatization forming a narrow peak from the decelerated atoms (Fig.1). The role of velocity (momentum) diffusion in this case consists in velocity distribution broadening.

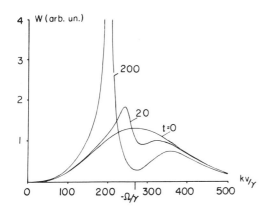

Fig.1. The deformation of velocity distribution for a spatially uniform atomic beam calculated from eq. (1) for $\bar{v} = 220$ γ/k, $\Omega/\gamma = -270$, $G = 10^3$. The value $(ku)^{-1}$ is taken as a time unit, where $u = \hbar k/M$ is recoil velocity. For the transition 3S-3P of sodium atoms, for example, the time unit equals 3.2 μs

Figure 1 illustrates the ideal case of a spatially uniform atomic beam for which the distribution function w does not depend on coordinate. For a real atomic beam w is a function of both velocity and coordinate and the form of

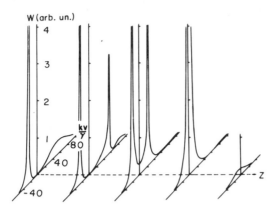

W(arb. un.)

Fig.2. The velocity distribution dependence on the interaction length z for stationary atomic beam at $\Omega/\gamma = -70$, $G = 10$. The distributions from left to right correspond to dimensionless coordinates $z = 0$; $5 \cdot 10^3$; 10^5; $2 \cdot 10^5$, $4 \cdot 10^5$, where length unit is $\gamma/k^2 u$, $u = \hbar k/M$

velocity distribution changes with coordinate z, i.e., with the length of atom-wave interaction. Due to the velocity decreasing along the beam motion direction the stationary atomic beam irradiated by counter-propagating light wave is split in velocity space (Fig.2). While one part of the atoms moves in a positive direction, the slow atoms change their velocity direction and move due to acceleration in the direction of wave propagation.

From a practical point of view, of main interest are the macroscopic characteristics of an atomic beam near the turning point where a major part of atoms have zero velocity. This task can be solved by reducing eq. (1) to hydrodynamic equations for atomic density n, average velocity <v> or atomic flux j = n<v> and temperature T. The principal basis for such a transformation of eq. (1) is the fact that before the turning point the local velocity distribution is established in the beam. The results of macroscopic analysis [7] enable us to determine both the evolution of hydrodynamic variables along the z axis and the value of the deceleration length. Specifically, Fig.3 shows the character of variation in the atomic flux j and longitudinal temperature $T_0 = <(v - <v>)^2>/2$ which characterizes the width of longitudinal velocity distribution of decelerated atoms (the width of the velocity peak on Fig.2)

It should be stressed that macroscopic description can be applied only at a sufficiently long interaction time between atoms and light wave when the atomic velocity distribution has a stationary form. The quantitative criterion for validity of the macroscopic equations is: $\Delta t >> r^{-1}$, where Δt is a time scale for hydrodynamical quantities. In coordinate space this criterion establishes the characteristic length $\Delta z = \bar{v} r^{-1}$ from the atomic beam source starting with which the macroscopic approach may be used. For typical atomic

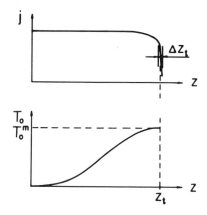

Fig.3. The dependences of atomic flux j = n<v> and longitudinal temperature T_0 on coordinate z

values this characteristic length has an order 0.1 cm. Below we shall see that this value is considerably less than the atomic deceleration length.

The path of decelerated atoms up to the turning point equals

$$Z_t = \frac{\lambda}{12G}\left(\frac{\gamma}{ku}\right)\left(\frac{\Omega}{\gamma}\right)^4 , \tag{6}$$

the characteristic length of turning region is

$$\Delta Z_t = \frac{\lambda(1 + \alpha_{zz})}{4G}\left(\frac{|\Omega|}{\gamma}\right)^3 , \tag{7}$$

and the longitudinal and transverse temperatures near the turning region (for a parallel initial beam) equal

$$T_z = T_0^m = \frac{\hbar(1 + \alpha_{zz})|\Omega|}{4k_B} \tag{8}$$

$$T_i = \frac{4\alpha_{ii}}{1 + \alpha_{zz}}T_0^m \tag{9}$$

where $\alpha_{ii} = \alpha_{xx}, \alpha_{yy}$. At a moderate intensity of light wave ($\sim$ W/cm^2) the deceleration length has a value reasonable for experiments with atomic beams. For example, in case of a thermal beam of calcium atoms irradiated by a wave resonant to the transition 4S - 4P (λ = 423 nm), with G = 10, parameters (6) - (9) are: Z_t = 110 cm, ΔZ_t = 2 $\cdot 10^{-3}$ cm, T_0 = 10^{-2} K.

Radiative deceleration of thermal beams of sodium atoms has been studied in experiments [8-10]. Without going into the details of the experimental schemes we point out that cyclicity of interaction between atoms and resonant laser radiation on the transition 3S - 3P was achieved either through optical

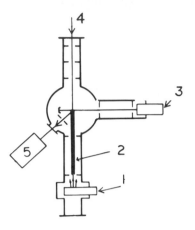

Fig.4. Experimental scheme for deceleration and velocity monochromatization of sodium atomic beam:
1 - atomic beam source,
2 - interaction region,
3 - the source for perpendicular atomic beam which is used for frequency scaling,
4 - laser beams,
5 - photomultiplier [8-10]

orientation of atoms or with the use of two-frequency laser radiation. In the second case one of the laser frequencies was close to the frequency of the transition $3S_{1/2}(F=1) - 3P_{3/2}$ which provided long time resonant atom-field interaction. To observe the velocity distribution profile the frequency turning low intensity probe laser wave is used which excites the resonance fluorescence. The latter is detected by photomultiplier (Fig.4).

The typical experimental profile of velocity distribution modified by resonance-radiation pressure is shown in Fig.5. Left peak in Fig.5 is due to the velocity monochromatization.

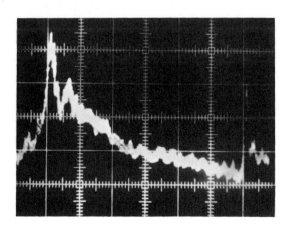

Fig.5. Fluorescence intensity versus frequency of the probe laser wave [10]. One division of the horizontal scale corresponds to 930 MHz or $5.5 \cdot 10^4$ cm/sec

4. Optical Trap for Cold Atoms

The principal possibility to decrease the temperature of an atomic beam to the value $T_0^m \sim 10^{-2}$ K makes it really possible to restrict the motion of cold atoms in nonuniform light fields. Several schemes of optical traps for cold atoms have been suggested recently [2,11,12]. The general analysis of the problem of optical traps [13] shows that there are a lot of light field configurations providing atomic localization. The simplest atomic trap can be formed by six divergent laser beams propagating along the axes of a rectangular coordinate system in the directions $\pm x$, $\pm y$, $\pm z$ and creating a central-symmetry field. To explain the necessity of setting up just such a field we shall discuss the basic principles of trap designing.

Let, for example, the motion of atoms be limited along the z axis. Then it is necessary that two counter-running beams should propagate along the z axis. In order that each beam decelerating the atom moving towards it not accelerate the atom moving in the direction of the beam, it is necessary that the beam frequencies must be red-shifted relative to the atomic transition frequency $\omega_0 (\omega < \omega_0)$. Then, in order that atoms with a low but positive velocity be decelerated when $z > 0$ and atoms with a low negative velocity be decelerated when $z < 0$, it is necessary that at $z > 0$ the intensity of the beam propagating in the -z direction must be higher and at $z < 0$ the intensity of the beam propagating in the +z direction must be higher. This requirement enables us to conclude: the beams must diverge to the centre of the trap as shown in the upper part of Fig.5.

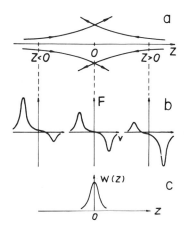

Fig.6. The principal scheme of an optical one-dimensional trap (a), the light pressure force as a function of atomic velocity for three different coordinates z (b) and profile of atomic distribution along z axis (c)

Fixing the frequences of the laser beams and their shapes in the way discussed above one may calculate the light pressure force and velocity diffusion as a functions of velocity and coordinate z. The general character of the force dependence on atomic velocity v along the z axis is shown in the middle part of Fig.5. This figure shows that the evolution of an atomic ensemble under the action of light pressure force is determined by narrowing of the atomic velocity distribution due to nonlinear force-velocity dependence and drift of the atoms to the centre of the trap due to a velocity-average value of force. Atomic momentum diffusion in any trap section only broadens the velocity distribution. On the whole, due to the damping of atomic motion under the action of the friction part of the light pressure force the Maxwell velocity distribution sets in with the local temperature $T(z)$. The average force increasing from the trap axis causes Boltzmann density distribution to set in [13]

$$n(z) = n(0) \exp[-(z/\bar{z})^2]$$ (10)

with the characteristic dimension of the localization region

$$\bar{z} = \frac{1}{4} \left(\frac{b}{l} + \frac{1}{B}\right)^{\frac{1}{2}} (1 + G + \Omega^2/\gamma^2)(|\Omega|/\gamma G)^{\frac{1}{2}} q_0$$ (11)

where b is the laser radiation invariant, 2l is the distance between the beam focuses, $q_0 = \frac{1}{2} (b\lambda/\pi)^{\frac{1}{2}}$ is the radius of the beams in focuses.

The minimum value of $\bar{z}$ can be attained at $\Omega = -\gamma$ and the saturation parameter $G \sim 1$:

$$\bar{z}_{min} = \sqrt{2} q_0 \quad .$$ (12)

Under these conditions the minimum temperature can be attained in the centre of the trap

$$T_{min} = \hbar\gamma/k_B \quad .$$ (13)

A three-dimensional trap gives nothing new in qualitative respect to the results presented. Moreover, since in a one-dimensional trap the localization region is smaller than the beam diameter it follows from the above considerations that it is sufficient to use three mutually perpendicular pairs of counter-running beams to localize atoms in a real three-dimensional space.

Thus, by injecting atoms cooled to temperature $T_0^m \sim 10^{-2}$ K into an optical trap one can accumulate atoms with $T_{min} \sim 10^{-3} - 10^{-4}$ K in the region $\bar{z}_{min} \sim 10^{-2} - 10^{-1}$ cm. The localization time for a single atom in the trap is limited, of course, by collisions with the residual gas and evidently cannot be longer than several hours. Therefore, to maintain a stationary density in the trap one should keep all the time the flow of incoming atoms at a certain level.

5. Conclusion

Six years after the first proposals on radiative cooling of atoms and ions the main progress in experimental studies has been connected with cooling ions. Specifically, in one of the recent experiments the ion temperature 10^{-2} K [14] was achieved whereas atomic deceleration and cooling is undergoing just the first test experiments. Even though there is a considerable gap in the level of experiments with atoms and ions, the monochromatization of an atomic beam and obtaining cold atoms remaining an important problem in atomic physics and spectroscopy. One may hope that cooling and localization of atoms in laser fields will lead to progress in solving such problems with further development of frequency standards and precision spectroscopy of single atoms.

References

1. T.W. Hänsch, A.L. Schawlow: Optics Comm. *13*, 68 (1975)
2. V.S. Letokhov, V.G. Minogin, B.D. Pavlik: Zh. Eksp. Teor. Fiz. *72*, 1328 (1977)
3. E.V. Baklanov, B.Ya. Dubetsky: Opt. Spectrosc. *41*, 1 (1976)
4. V.G. Minogin: Zh. Eksp. Teor. Fiz. *79*, 2044 (1980)
5. R.J. Cook: Phys. Rev. A22, 1078 (1980)
6. R.L. Liboff: *Introduction to the Theory of Kinetic Equations* (Wiley, New York 1969)
7. T.V. Zuyeva, V.S. Letokhov, V.G. Minogin: Zh. Eksp. Teor. Fiz. *81*, 84 (1981)
8. V.I. Balykin, V.S. Letokhov, V.I. Mishin: Zh. Eksp. Teor. Fiz. *78*, 1376 (1980)
9. V.I. Balykin, V.S. Letokhov, V.G. Minogin: Zh. Eksp. Teor. Fiz. *80*, 1779 (1981)
10. S.V. Andreyev, V.I. Balykin, V.S. Letokhov: To be published
11. A. Ashkin: Phys. Rev. Lett. *40*, 729 (1978)
12. J.P. Gordon, A. Ashkin: Phys. Rev. A*21*, 1606 (1980)
13. V.G. Minogin: To be published
14. W. Neuhauser, M. Hohenstatt, P.E. Toschek, H. Dehmelt: Phys. Rev. A*22*, 1137 (1980)

Part IX
Surface and Solid State

FM Spectroscopy and Frequency Domain Optical Memories

G.C. Bjorklund, W. Lenth[1], M.D. Levenson, and C. Ortiz[2]

Dept. K46/282, IBM Research Laboratory, 5600 Cottle Road
San Jose, CA 95193, USA

1. Introduction

There has been considerable recent interest in spectroscopic techniques involving the use of frequency modulated (FM) laser radiation to measure the absorption or dispersion associated with narrow spectral features. In this paper, we review the basic principles of FM spectroscopy, present new results on FM excitation spectroscopy, and describe the application of these techniques to frequency domain optical memories.

2. FM Spectroscopy

FM spectroscopy is a new method of optical heterodyne spectroscopy capable of sensitive and rapid detection of absorption or dispersion features with the full spectral resolution characteristic of cw dye lasers [1]. Figure 1a shows a typical experimental arrangement. The output of a single-axial-mode dye laser oscillating at optical frequency ω_c is passed through a phase modulator driven sinusoidally at radio frequency ω_m to produce a pure FM optical spectrum consisting of a strong carrier at frequency ω_c with two weak sidebands at frequencies $\omega_c \pm \omega_m$. A key concept is that ω_m is large compared to the width of the spectral feature of interest, so that, as illustrated in Fig. 1b, the spectral feature can be probed by a single isolated sideband. Both the absorption and dispersion associated with the spectral feature can be separately measured by monitoring the phase and amplitude of the rf heterodyne beat signal that occurs when the FM spectrum is distorted by the effects of the spectral feature on the probing sideband. Since single-mode dye lasers have little noise at radio frequencies, these beat signals can be detected with a high degree of sensitivity. Furthermore, the entire lineshape of the spectral feature can be scanned by tuning either ω_c or ω_m.

It should be noted that the use of sidebands for laser spectroscopy is not new. Standard (nonheterodyne) absorption spectroscopy with sidebands has been performed by Corcoran et al. [2], Mattick et al. [3], and Magert et al. [4]. Heterodyne spectroscopy with amplitude modulated (AM) sidebands has been accomplished by Szabo [5] and Erickson [6]. Wavelength modulation spectroscopy with lasers, as has been done by

---

[1]Permanent address: Institut für Angewandte Physik, Universität Hamburg, Jungiusstr. 11, D-2000 Hamburg 36, Fed. Rep. of Germany.

[2]Permanent address: Instituto de Optica, Serrano 121, Madrid, Spain.

This work was partially supported by the Office of Naval Research.

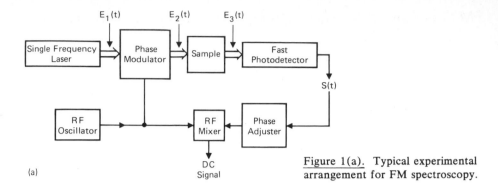

Figure 1(a). Typical experimental arrangement for FM spectroscopy.

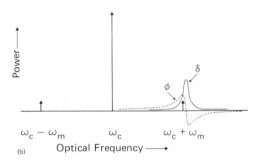

Figure 1(b). Frequency domain illustration of FM spectroscopy.

Hinckley et al. [7] and Tang et al. [8] can be viewed as heterodyne spectroscopy with very closely spaced FM sidebands. Pound [9] and Harris et al. [10] have recognized that the appearance of rf beats provides a sensitive indication of FM distortion. However, the case of widely separated FM sidebands for optical spectroscopy with attendant advantages of zero background signal, rapid response to transients, and laser limited resolution has only recently been accomplished [1,11,12].

The FM optical field $E_2(t)$ incident on the sample is given by

$$E_2(t) = E_0\left[-\frac{M}{2}\cos(\omega_c-\omega_m)t + \cos\omega_c t + \frac{M}{2}\cos(\omega_c+\omega_m)t\right] \quad (1)$$

where E_0 is the electric field amplitude of the original laser beam and M is the modulation index. The time varying intensity $I_3(t)$ of the distorted FM spectrum emerging from the sample is given by

$$I_3(t) = \frac{cE_0^2 M}{8\pi} \cdot \exp[-2\delta(\omega_c)]$$

$$\cdot\left\{[\phi(\omega_c-\omega_m)-2\phi(\omega_c) + \phi(\omega_c+\omega_m)]\sin\omega_m t\right.$$

$$\left. + [\delta(\omega_c-\omega_m) - \delta(\omega_c+\omega_m)]\cos\omega_m t\right\}, \quad (2)$$

where $\phi(\omega)$ is the optical phase shift and $\delta(\omega)$ the amplitude absorption introduced by the sample. The in-phase (cos $\omega_m t$) component of the beat signal is thus proportional to the difference in loss experienced by the upper and lower sidebands, whereas the quadrature (sin $\omega_m t$) component is proportional to the difference between the phase shift experienced by the carrier and the average of the phase shifts experienced by the sidebands. The null signal that results when the FM spectrum is not distorted can be thought of as arising from a perfect cancellation of the beats between each of the sidebands and the carrier.

The lineshape theory of (2) was experimentally tested using a Fabry-Perot resonator in reflection to provide a Lorentzian absorption line with a full width at half-maximum of 298 MHz [1]. Figure 2 shows experimental and theoretical in-phase and quadrature signals obtained when the sidebands were scanned through the resonance by tuning the laser frequency ν_c with ν_m held constant at 925 MHz. As the laser is scanned toward increasing frequency, the in-phase data shows a negative going absorption lineshape followed at a spacing of $2\nu_m$ by a positive going absorption lineshape. The quadrature data show three overlapping dispersion curves which occur when the upper sideband, the carrier, and the lower sideband sweep through the resonance.

We have obtained similar data for saturation holes in I_2 vapor [13]. The output of a tunable dye laser at frequency ν_c was split into a pump and probe beam. The probe beam was FM modulated and then transited an I_2 vapor cell to fall upon a fast photodetector. The pump beam was chopped at 30 kHz and was passed through the I_2 cell in the direction opposite the probe. The FM saturation spectroscopy signal was extracted by lock-in detection of the 30 kHz signal produced by the RF mixer. Figure 3a shows a typical 15 peak I_2 hyperfine pattern obtained when ν_c is swept through the I_2 transition located 2 GHz below the lower frequency sodium D line with ν_m held constant at

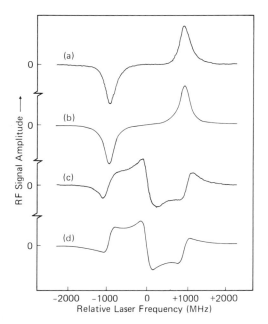

Figure 2. Experimental and theoretical lineshapes for FM spectroscopy of a Fabry-Perot resonance: (a) experimental in-phase signal; (b) theoretical in-phase signal; (c) experimental quadrature signal; and (d) theoretical quadrature signal.

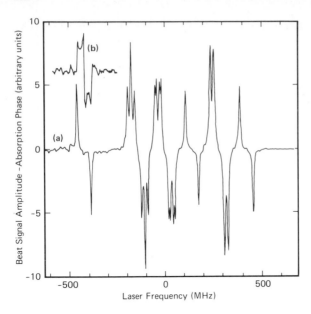

Figure 3. (a) Hyperfine splitting of a single rovibronic I_2 vapor line resolved by FM saturation spectroscopy with the absorption phase detected. Either the positive-going or the negative-going peaks reproduce the hyperfine spectrum; (b) Dispersion phase lineshape due to the lowest frequency hyperfine component.

71 MHz and the in-phase (absorption) signal detected. Resonances occur when $\nu_c = \nu_0 \pm \nu_m/2$ where ν_0 is the frequency of each transition in the molecular rest frame. Figure 3b shows the lineshape obtained when the quadrature (dispersion) signal was detected.

In principle, when sufficient laser power is utilized, the sensitivity of FM spectroscopy is limited only by shot noise and the minimum detectable absorption $\delta_{min} = \delta(\nu_c + \nu_m) - \delta(\nu_c - \nu_m)$ that can be detected in integration time τ is given by $\delta_{min} = 2[\eta M^2 (P_0/h\nu_c)\tau]^{-1/2}$, where η is the photodetector quantum efficiency and P_0 is the total incident laser power [14,15]. Assuming that $\eta = 1$, $P_0 = 5 \times 10^{-3}$W, and $M = 0.1$, a value of $\delta_{min} = 1.5 \times 10^{-7}$ should be detectable for $\tau = 1$ sec, and $\delta_{min} = 0.005$ should be detectable for $\tau = 10^{-9}$ sec.

The ability to rapidly perform absorption measurements was demonstrated using short pulses of FM light [15]. An acousto-optic modulator was utilized to break up the FM light into trains of microsecond duration probes with arbitrary separation. The laser frequency ν_c was scanned over a Doppler broadened I_2 absorption line and the in-phase (absorption) signal monitored with ν_m held constant at 80 MHz. Figure 4(a) shows the standard transmission spectrum of the I_2 line as the laser was scanned slowly over 5 GHz. Figure 4(b) shows the in-phase (absorption) signal obtained by FM spectroscopy when the acousto-optic modulator was driven with a 1:1 duty cycle and a lock-in amplifier used for signal averaging and Figure 4(c) shows the pulsed FM signal displayed directly on a storage oscilloscope when the laser was scanned over 5 GHz in 250 msec and the acousto-optic modulator was set to produce 1 μsec long laser pulses separated by 5 msec.

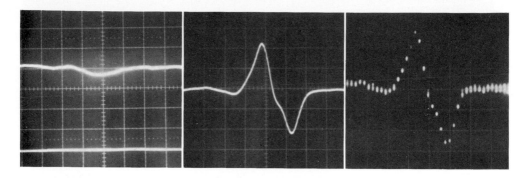

Figure 4. (a) Transmission spectrum of an I_2 vapor line, laser scan range 5 GHz; (b) FM spectroscopy spectrum of the same I_2 vapor line utilizing a lock-in amplifier for signal averaging; (c) Trace of FM spectroscopy signals obtained from 1 μsec long FM-laser pulses when tuning the laser over 5 GHz in 250 msec.

3. FM Excitation Spectroscopy

We have recently completed an experimental and theoretical investigation of the nature of the time dependence of the intensity of the incoherent fluorescence emitted by a three-level system pumped with FM laser radiation. The system is assumed to consist of a ground state $|g>$ and excited states $|f>$ and $|i>$. States $|g>$ and $|f>$ and states $|f>$ and $|i>$ are assumed to be dipole coupled. The FM light pumps the $|g> \rightarrow |f>$ transition and fluorescent emission occurs from $|f>$ down to an intermediate state $|i>$ which then rapidly relaxes to $|g>$.

Since the intensity of the fluoresence out of level $|f>$ is always proportional to the excited state population ρ_{ff}, it is sufficient to utilize second-order perturbation theory to calculate $\rho_{ff}^{(2)}(t)$. The $|g> \rightarrow |f>$ transition lineshape is assumed to be a homogeneously broadened Lorentzian with center frequency ω_{gf} and full width of half maximum of 2 Γ_{gf}, where Γ_{gf} is the transverse relaxation rate. The rate of decay of excited state population in $|f>$ is given by Γ_{ff}. When the system is pumped with FM light described by (1), $\rho_{ff}^{(2)}(t)$ has terms at ω_m given by

$$\rho_{ff}^{(2)}(t) = \left[\frac{\mu_{fg}^2 E_0^2 M}{8\hbar} \right] \cdot \left[\frac{1}{\omega_m^2 + \Gamma_{ff}^2} \right]^{1/2}$$

$$\cdot \left\{ \left[- D(\omega_c - \omega_m) + 2D(\omega_c) - D(\omega_c + \omega_m) \right] \cos(\omega_m t + \phi) \right.$$

$$\left. + \left[A(\omega_c + \omega_m) - A(\omega_c - \omega_m) \right] \sin(\omega_m t + \phi) \right\} \qquad (3)$$

where μ_{fg} is the dipole matrix element between $|f>$ and $|g>$, $\phi = \text{Arc } \tan(\Gamma_{ff}/\omega_m)$, $D(\omega) = (\omega - \omega_{fg})/[(\omega - \omega_{fg})^2 + \Gamma_{fg}^2]$, and $A(\omega) = 1/[(\omega - \omega_{fg})^2 + \Gamma_{fg}^2]$. The quantities $D(\omega)$ and $A(\omega)$ correspond to the dispersive and absorptive parts of the linear susceptibility.

It is interesting to compare the FM excitation result of (3) with the FM absorption result of (2). It can be seen that, aside from an overall phase factor, the functional dependence on ω_c is identical for both cases. Thus, lineshapes obtained by scanning ω_c with ω_m constant can be made the same for both FM excitation and FM absorption spectroscopy, provided that the phase of the signal detection electronics is properly adjusted in each case. The functional dependence on ω_m differs by the factor of $[\omega_m^2 + \Gamma_{ff}]^{1/2}$, which arises from a roll off of the response of the three-level system as ω_m becomes large compared to the characteristic relaxation rate.

A series of experiments was conducted to verify the results of (3) and to directly compare the lineshapes and signal strengths obtained by FM excitation spectroscopy to FM absorption spectroscopy. Figure 5 shows the experimental set-up. The three-level system was derived from N_1 aggregate color centers in a NaF host crystal at 2K. An approximate 200 MHz wide permanent photochemical hole burned into the N_1 zero-phonon line at 6070Å served as the Lorentzian absorption line [16]. An FND-1000 photodiode followed by phase sensitive rf detection electronics was utilized both to measure the strength of the modulation of the intensity of the Stokes-shifted fluorescence and to detect the heterodyne beat signal in the transmitted laser beam.

Figure 6 shows typical experimental lineshape, obtained by scanning ν_c with ν_m held constant at 90 MHz and with the phase of the rf detection electronics adjusted to optimize the $A(\omega_c+\omega_m)-A(\omega_c-\omega_m)$ signal for FM excitation and the $\delta(\omega_c-\omega_m)-\delta(\omega_c+\omega_m)$ signal for FM absorption. It can be seen that apart from the (arbitrary) difference in polarity, the FM excitation and FM absorption lineshapes are in

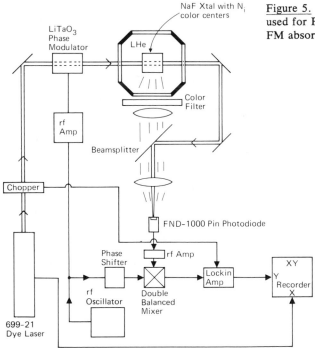

Figure 5. Experimental arrangement used for FM excitation spectroscopy and FM absorption spectroscopy.

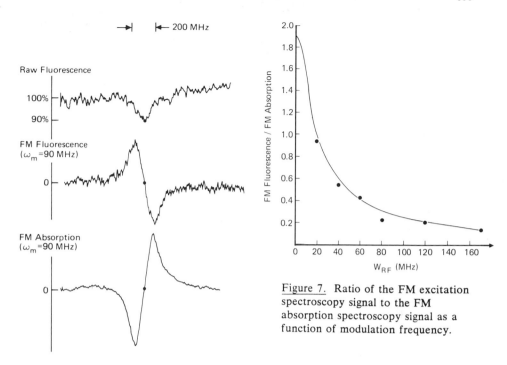

Figure 6. Photochemical hole burned into the N_1 center zero-phonon line of NaF and detected by standard excitation spectroscopy, FM excitation spectroscopy, and FM absorption spectroscopy.

Figure 7. Ratio of the FM excitation spectroscopy signal to the FM absorption spectroscopy signal as a function of modulation frequency.

reasonable agreement. This experiment was repeated for several discrete values of ν_m between 20 and 170 MHz. Figure 7 shows the ratio of the FM excitation signal to the FM absorption signal as a function of ν_m. The agreement to the $[\omega_m^2 + \Gamma_{ff}^2]^{1/2}$ falloff is quite good for $\Gamma_{ff} = 8.3 \times 10^7$ sec^{-1}.

4. Frequency Domain Optical Memories

The spatial storage density of conventional optical memories is limited by diffraction to a maximum of 10^8 bits/cm^2 for planar geometries. The phenomenon of photochemical hole burning allows optical frequency to be utilized as an additional dimension for the organization of an optical memory [17]. Certain low temperature solid state materials such as aggregate color centers or molecules in solution have spectra which exhibit relatively sharp inhomogeneously broadened absorption lines with widths $\Delta\omega_I$ varying from several cm^{-1} in crystalline solids to over 100 cm^{-1} in glasses and polymers. Each such inhomogeneous line is the result of the superposition of many much narrower homogeneous lines of width $\Delta\omega_H$ whose center frequencies are shifted by the random local environments in the crystal. The value of $\Delta\omega_H$ is determined by intrinsic molecular relaxation processes and at 2K can be as narrow as 0.00067 cm^{-1} (20 MHz) in crystals and 0.1 cm^{-1} in glasses. Thus, $\Delta\omega_H$ is typically 10^3-10^4 times narrower than $\Delta\omega_I$.

The phenomenon of photochemical hole burning (PHB) occurs when a permanent photochemical process is induced by a laser beam tuned to a particular frequency within the inhomogeneous line. The laser radiation interacts only with that subset of molecules, whose local environments are such that the laser wavelength is contained within their homogeneous absorption lines. Since the photochemistry depletes the population of this subset, the absorption coefficient at the exact laser wavelength ω_L is reduced. In this way, a permanent "hole" or dip is produced in the inhomogeneous line profile. For shallow holes, the hole width is equal to twice $\Delta\omega_H$. Thus, 10^3 resolvable holes can be burned in a given inhomogeneous line. As shown in Fig. 8a, 10^3 bits of information could be recorded at each spatial storage location by the presence or absence of holes at specific frequency locations within the inhomogeneous line. A spatial storage density of $10^3 \times 10^8$ bits/cm^2 or 10^{11} bits/cm^2 is thus ultimately possible.

At present, the most promising materials for optical memories based upon PHB are the aggregate color centers in alkali halide crystal hosts [18,19]. Figure 8b shows a string of binary information recorded by 150 MHz wide holes burned into a portion of the 5749Å zero-phonon line of an aggregate color center in NaF. In addition to possessing good hole-burning properties, these materials meet the practical requirements of stability at room temperature, good optical quality, reversibility, potential compatibility with GaAlAs lasers, and capability of being produced in thin films [19].

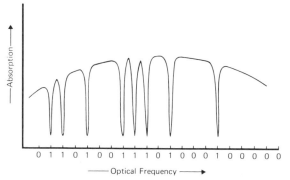

0 1 1 0 1 0 0 1 1 1 0 1 0 0 0 1 0 0 0 0 0

Figure 8(a). Illustration of information storage by photochemical hole burning. Bits of information are encoded by the presence or absence of photochemical holes at various frequency locations of an inhomogeneously broadened absorption line.

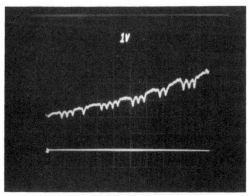

0 111 010 0111100111 001100111 0

Figure 8(b). Sequence of 150 MHz wide photochemical holes burned into a 15 GHz wide portion of the 5749Å zero-phonon line of an aggregate color center in NaF. Holes were detected by standard excitation spectroscopy.

Information would be written into a PHB optical memory by focusing a tunable laser beam to a particular spatial location in the recording medium and then tuning the laser across the inhomogeneous absorption band while modulating the intensity to write "ones" or "zeroes" at each frequency location. Reading of the information is accomplished using the focused tunable laser together with FM spectroscopy. The attenuated output of the laser, at frequency ν_c, would be passed through a phase modulator driven at ν_m on the order of 50 MHz, and then focused to the desired spatial location. The transmitted light would be monitored with a fast photodetector and rf detection electronics set to detect the absorption (in-phase) signal. Assuming a nominal hole width of 100 MHz, the FM spectroscopy signal would provide a differentiated spectrum of the absorption band and holes. This effectively suppresses the slowly varying background absorption and enhances the signals from the holes. Data rates of 30 Mbits/sec per laser beam should be achievable for reading.

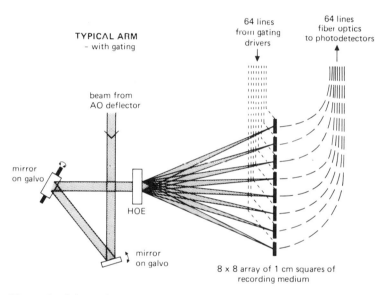

Figure 9. Schematic illustration of a writing and reading scheme for a frequency domain optical storage system. An acousto-optic (AO) deflector can be used for directing the laser beam to 100 different arms.

Figure 9 shows a schematic of one arm of a large direct access storage system based on this design. The mirrors on the galvos would be capable of addressing 10^6 spatial spots in an xy pattern with a random access time of 3 μsec. The holographic optical element (HOE) would serve to split the laser beam into 64 spatially multiplexed beams focused to a 10 μm spot size. Each 1 cm square of recording material would contain 10^6 spatial locations and would thus contain 10^9 bits of information. The entire arm thus contains 6.4×10^{10} bits of information. The proposed storage system would contain on the order of 100 arms or 6.4×10^{12} bits.

References

1. G.C.Bjorklund, Opt. Lett. $\underline{5}$, 15 (1980)
2. V.J.Corcoran, R.E.Cupp, J.J.Gallagher, and W.T.Smith, Appl. Phys. Lett. $\underline{16}$, 316 (1970)
3. A.T.Mattick, A.Sanchez, N.A.Kurnit, and A.Javen, Appl. Phys. Lett. $\underline{23}$, 675 (1973)
4. G.Magerl, E.Bonek, and W.A.Kreiner, Chem. Phys. Lett. $\underline{52}$, 473 (1977)
5. A.Szabo, Phys. Rev. B$\underline{11}$, 4512 (1975)
6. L.E.Erickson, Phys. Rev. B$\underline{16}$, 4731 (1977)
7. E.D.Hinckley and P.L.Kelley, Science $\underline{171}$, 635 (1971)
8. C.L.Tang and J.M.Telle, J. Appl. Phys. $\underline{45}$, 4503 (1974)
9. R.V.Pound, Rev. Sci. Instr. $\underline{17}$, 490 (1946)
10. S.E.Harris, M.K.Oshman, B.J.McMurtry, and E.O.Amman, Appl. Phys. Lett. $\underline{7}$, 185 (1965)
11. J.L.Hall et al., to be published
12. S.Ezekiel et al., to be published
13. G.C.Bjorklund and M.D.Levenson, Phys. Rev. A$\underline{24}$, 166 (1981)
14. G.C.Bjorklund, W.Lenth, M.D.Levenson, and C.Ortiz, SPIE $\underline{286}$ (1981)
15. W.Lenth, C.Ortiz, and G.C.Bjorklund, Opt. Lett. $\underline{6}$, 351 (1981)
16. M.D.Levenson, R.M.Macfarlane, and R.M.Shelby, Phys.Rev.B$\underline{22}$, 4915(1980)
17. D.Haarer, SPIE $\underline{177}$, 97 (1979)
18. R.M.Macfarlane and R.M.Shelby, Phys. Rev. Lett. $\underline{42}$, 788 (1979)
19. C.Ortiz, R.M.Macfarlane, R.M.Shelby, W.Lenth, and G.C.Bjorklund, Appl. Phys. $\underline{25}$, 87 (1981)

Spectroscopy of Very Weakly Absorbing Condensed Media

C.K.N. Patel and E.T. Nelson

Bell Telephone Laboratories, 600 Mountain Ave.
Murray Hill, NJ 07974, USA

and

A.C. Tam

IBM Research Laboratory
San Jose, CA 95193, USA

1. INTRODUCTION

During the past three years opto-acoustic (OA) spectroscopy of condensed phase materials has been applied with considerable success to spectroscopic studies of a variety of weakly absorbing liquids, solids and powders. The technique which has been shown to have exceptional advantages in these studies uses a) a pulsed tunable dye laser as the source of radiation, b) an immersed (in the case of liquids) or attached (in the case of solids) piezo-electric transducer for converting the acoustic signal into electrical output signal, and c) gated detection for discriminating against interfering signals in the time domain. A convenient acronym derived for the scheme is PULPIT (*PU*lsed *L*aser *PI*ezo-electric *Tr*ansducer). The PULPIT opto-acoustic spectroscopy relies upon the nonradiative relaxation of the optical radiation absorbed by the liquid or the solid sample leading to local transient temperature rise which results in an acoustic impulse travelling outwards from the illuminated region.

The idea and the first demonstration of acoustic detection of optical radiation by materials dates back to 1880 when A. G. Bell[1] showed that sunlight dispersed with a prism and absorbed by different materials gave different amounts of sound, as measured by a listener's ear. The technique of OA spectroscopy (optical radiation at any wavelength absorbed and acoustic signal detected) was used intermittently between 1880 and the late 1960's when the development of high power tunable lasers led the revival of the technique and helped delineate its enormous potential for spectroscopy of weakly absorbing gaseous, liquids or solid samples. Atwood and Kerr[2] used a capacitance manometer to study OA spectroscopy and called their experimental technique "spectrophone". Kreuzer[3] was the first to use a sensitive microphone together with a fixed frequency He-Ne laser at 3.39 μm to detect CH_4 for exploring the OA detection idea. Kreuzer and Patel[4] were the first to carry out OA spectroscopy (measuring wavelength dependent absorption of a sample) of dilute mixtures of NO and H_2O in air using a tunable spin-flip Raman laser[5]. Subsequent progress in gas-phase OA spectroscopy has been very rapid (present minimum detection capability $\alpha\ell \approx 10^{-10}$[6], and has led to its application to pollution detection[4,7,8], excited state studies[9], chemical kinetics[10], multiphonon absorption kinetics[11], etc. The extension of gas phase OA technique to condensed phase sample was carried out in 1973 by Harshberger and Robin[12] and by Rosencwaig[13] where they detected optical absorption in condensed phase samples by measuring acoustic signals using a microphone which is included in a cell filled with a nonabsorbing gas enclosing the condensed phase smple. This scheme, called photo-acoustic spectroscopy, is generally not suitable for measuring ultrasmall optical absorptions ($(\alpha\ell)_{min} \approx 10^{-3}$–$10^{-4}$) and suffers from undesirable chopping frequency dependence[14]. Nonetheless, because the photo-acoustic spectroscopy depends on relative sizes of optical absorption depth and thermal diffusion length in the condensed phase sample, it is seen to be useful for the study of thin films[15], and for depth profiling studies[16]. Piezo-electric detection of transient acoustic pulse due to pulsed laser light being absorbed in a condensed phase sample[17], the subject of present studies, has turned out to be an extremely sensitive technique for low loss measurements.

In this paper we will give a brief description of the fundamental signal generation process[18] which will be followed by a discussion of physical phenomena that set a lower limit to the ultimate achievable sensitivity and also provide a comparison between alternate schemes for ultralow absorption detection. Next we describe the experimental techniques. At present we have demonstrated an absorption measurement capability of $(\alpha \ell)_{min} \sim 10^{-7}$[19], determined by experimental problems rather than intrinsic limit set by electrostriction[18]. The ability to measure these small $\alpha \ell$ using PULPIT opto-acoustic spectroscopy has allowed a number of exciting linear and nonlinear absorption studies of liquids[17-23], solids[24], powders[25], and thin liquid films[26]. Further, because the piezo-electric transducer has no obvious low temperature operational limit (as long as the temperature is below the Curie temperature of the piezo-electric transducer material) low temperature PULPIT opto-acoustic spectroscopy is clearly possible. We show that for a number of solid state materials, such as quartz, PULPIT opto-acoustic spectroscopy is considerably more sensitive at low temperatures than at room temperature[27]. Recently, we have obtained PULPIT opto-acoustic spectra of liquid CH_4 (94K)[28], liquid C_2H_4 (113K)[29,30], and solid H_2 (13K)[31] in the 550 nm to 1.6 μm region of wavelengths. We will conclude by pointing out some future applications of PULPIT opto-acoustic spectroscopy.

2. PULPIT OPTO-ACOUSTIC SIGNAL GENERATION

A condensed phase sample (liquid or solid) having an absorption coefficient $\alpha(\nu)$ is illuminated by a pulse of light at a frequency ν, of duration τ_p, and energy E (see Fig.1 of Ref. 18). Diameter of the illuminated volume is 2R. Absorption of optical radiation and subsequent decay of the excitation in the medium through nonradiative processes causes a transient increase in the temperature of the illuminated volume and consequential increase in the volume which launches a radially outward propagating acoustic pulse. The duration of the pressure pulse is determined by the relative size of the laser pulse duration, τ_p or the transit time of an acoustic wave $\tau_a = R/v_a$ across the illuminated cylindrical volume of the material. Here v_a is the sound velocity. The pressure wave consists of a compression pulse followed by a rarefaction pulse. The approximate duration of the pressure pulse, τ_{pp}, is $\approx \tau_p$ for $\tau_p > \tau_a$, and it is $\approx \tau_a$ for $\tau_p < \tau_a$. This pressure pulse when intercepted by a piezo-electric transducer will generate an output voltage V_{oa} which is proportional to the amplitude of the pressure pulse. It has been shown earlier that this voltage is proportional to the absorption coefficient of the medium. Without going through details, the final expressions for the amplitude of the voltage pulse induced in the transducer having a response time of τ_r is [18,27,32],

$$\text{For } \tau_p > \tau_a, \tau_r, \quad \frac{V_{oa}}{E} = K_p \frac{\beta v_a}{C_p} \alpha(\nu) \tag{1}$$

$$\text{For } \tau_r > \tau_a \text{ and } \tau_a > \tau_p, \quad \frac{V_{oa}}{E} = K_a \frac{\beta v_a^2}{C_p} \left[\frac{Z_{abs}}{Z_{abs} + Z_{pzt}} \right] \alpha(\nu) \tag{2}$$

$$\text{For } \tau_a > \tau_p, \tau_r, \quad \frac{V_{oa}}{E} = K_r \frac{\beta v_a^2}{C_p} \alpha(\nu) \tag{3}$$

where β is the volumetric expansion coefficient, C_p is the specific heat at constant pressure, and Z_{abs} and Z_{pzt} are the acoustic impedances of the optical absorber and of the piezo-electric transducer, respectively. K_p, K_a, and K_r are constants determined by the specific geometry, and by the details of the transducer sensitivity, resonances, etc. Here we have assumed a weak absorption case, i.e., $\alpha \ell \ll 1$.

In order to obtain a quantitative measure of $\alpha(\nu)$ as a function of ν we need to know the various constants in Eqs. (1)-(3). While β, v_a and C_p for most materials are known with sufficient accuracy, the parameters that determine K's are either poorly determined or often not determined at all. Thus it is convenient to determine the K's from comparison with a "doped" sample [18,19]. Three observations are in order. First, that none of the constants, K's or β, v_a and C_p have any dependence on the optical frequency. Second, the K's do not depend upon the specific material once the geometry is fixed. And third, apart from the temperature dependence of $\alpha(\nu)$, the material parameters β, v_a and C_p generally exhibit strong temperature dependence. Keeping these three observations in mind, we can transfer the calibration of PULPIT opto-acoustic spectroscopy from one material to another by appropriate scaling through known β, v_a and C_p, as prescribed by appropriate relation of the Eqs. (1), (2) or (3).

For most liquids, the temperature dependence of ν_a, β and C_p results in only small corrections as the temperature is varied. For solids, however, this is not the same since both β and C_p approach rapidly to zero (although, fortunately not at the same rates). Thus for liquids at the usual temperatures at which PULPIT opto-acoustic measurements are carried out, the temperature dependence is not important except in certain pathologic cases. One example of such a pathologic liquid is water at 4° C where the thermal expansion coefficient of water goes to zero and hence the opto-acoustic signal, too, goes to zero[21] as expected from Eqs. (1)-(3). For solids, on the other hand, we have shown[27] that in general room temperature may be the less desirable for PULPIT opto-acoustic measurements compared to lower temperatures.

3. EXPERIMENTAL DETAILS

A generic schematic diagram of the experimental apparatus used in almost all of the PULPIT opto-acoustic spectroscopy studies is described in Ref. 18. It consists of the following:

a. *Pulsed tunable dye laser.* We have used two types of dye lasers. The first one is a flash lamp pumped dye laser which is capable of tuning over a region from ~450 nm to ~700 nm (by using a variety of dyes) and produces pulses of duration ~1-2 μsec, peak power ~1 kW (i.e. pulse energy ~1-2 mJ) at a repetition rate of ~10 Hz. The second laser is a dye laser pumped by the second harmonic of 1.06 μm radiation from a Q-switched Nd:YAG laser. This laser (so far) has been tuned from ~550 nm to 750 nm and produces pulse energies of ~50 mJ in pulses of duration ~7 nsec at 10 Hz repetition rate. Because of the high peak powers possible with this system, we have used stimulated Raman scattering in a high pressure hydrogen gas cell to down-shift the fundamental of the tunable dye laser output to wavelengths as long as 1.6 μm for long wavelength PULPIT opto-acoustic spectroscopy [30].

b. *Opto-acoustic cell and the piezo-electric transducer* assembly which have been described in detail in our earlier papers[17,21,25,26,33]. In all of these, the common factor is the piezo-electric transducer assembly shown in (shown in Fig. 2 of Ref. 18) which is constructed out of stainless steel and houses the lead zirconate-titanate piezo-electric transducer. The reasons for use of the stainless steel enclosure, as opposed to a bare pzt element are detailed in Ref. 33. Briefly, the all stainless steel construction minimizes the danger of contamination in the case of liquid opto-acoustic studies, the highly polished stainless steel diaphragm facing the sample minimizes interference signals arising from light scattered from the bulk of the optical sample being absorbed by the transducer, and finally the all metallic enclosure reduces RFI pickup problems associated with pulsed/Q-switched lasers operating in the near vicinity.

4. SENSITIVITY OF PULPIT OPTO-ACOUSTIC SPECTROSCOPY

By carefully studying the sixth harmonic absorption due to the C-H vibration in C_6H_6 [17-20] we have ascertained that for a pulse energy of ~1 mJ at a repetition rate of 10 Hz we can measure an absorption coefficient as small as $\alpha_{min} \sim 10^{-7}$ cm^{-1}. This ability corresponds to a minimum absorbed energy measurement of $E_{min} \sim 10^{-10}$ Joules. Notice that because PULPIT opto-acoustic spectroscopy is a calorimetric technique, the quantity of importance is $E_{min} = E\alpha\ell$. We can measure even smaller absorption coefficients by increasing E (keeping ℓ constant). However, how far one can go depends upon other processes which may conspire to give pulsed acoustic signals which can interfere with that arising from optical absorption. Three such sources of interference are: (1) window absorption (in the case of liquids) or surface absorption (in the case of solids), (2) bulk elastic light scattering due to particulate matter in liquids, and imperfections, etc. in solids, and (3) electrostriction. The first two do not represent intrinsic limitation to α_{min} since by careful sample and window preparation, the acoustic signals arising from the first two are minimized. Further, time-gating of the transient signal output from the piezo-electric transducer can further reduce the contamination of data caused by spurious signals[18]. The electrostriction signal which arises due to a density perturbation induced by the high intensity laser radiation cannot be eliminated since the form of the transient signal caused by electrostriction mimics that arising due to optical absorption (except for the sign of the signal which depends on whether the material has a positive or negative polarizability). Thus it is instructive to critically evaluate the estimates of electrostriction limited α_{min}(es) (i.e. when the acoustic signal amplitude from electrostriction equals that from optical absorption). Brueck et al[34] claim to have estimated and measured $\alpha_{min}(es)$ in liquid nitrogen using a pulsed CO_2 laser as the source of radiation. They report $\alpha_{min}(es)_{Brueck} \approx 4 \times 10^{-5}$ cm^{-1} for a CO_2 laser beam focused down to ~ 200 μm. Since $\alpha_{min}(es)_{Brueck} \propto$ (beam diameter)$^{-1}$, $\alpha_{min}(es)_{Brueck}$ is not a universal number. For linear PULPIT opto-acoustic spectroscopy, laser beam diameters smaller than ~ 2 mm is rarely used. Thus $\alpha_{min}(es)$ now becomes $\sim 4 \times 10^{-6}$ cm^{-1}. Careful analysis of absorption data due to sixth harmonic of C_6H_6 for various dilutions in CCl_4 [19], and of the 6th, 7th and 8th harmonics of C_6H_6 [20], has clearly indicated that we have no difficulty measuring α's as small as 10^{-7} cm^{-1}. If there is a constant background signal present, it is smaller than $\sim 10^{-7}$ cm^{-1}. The Brueck et al. analysis[34] is not supported by experimental data on C_6H_6, which show that $\alpha_{min}(es) < 10^{-7}$ cm^{-1}. Notice that this can further be reduced by making the laser beam diameter larger.

Further we have carried out two photon absorption studies in C_6H_6 using the PULPIT opto-acoustic spectroscopy which utilized the flash lamp pumped dye laser ($\tau_p \sim 1-2$ μsec) beam focused down to ~ 100 μm. Without going through details [22], we can estimate $\alpha_{min}(es)$ for C_6H_6 and 100 μm beam diameter $\lesssim 10^{-6}$ cm^{-1}.

One final comment on ultimate limitation imposed by electrostriction is in order. Electrostriction signal is largely wavelength independent and thus it represents a flat background (at whatever level of α it becomes observable). If in the PULPIT opto-acoustic spectroscopy we are looking for wavelength dependent peaks, it should be possible to electronically subtract the background signal. Subtraction of constant signals as large as 100 times wavelength dependent peaks are possible. Thus even the observed (for C_6H_6) $\alpha_{min}(es)_{obs} < 10^{-7}$ cm^{-1} is not a lower limit for spectroscopy.

5. COMPARISON BETWEEN PULPIT OPTO-ACOUSTIC SPECTROSCOPY AND PHOTOREFRACTIVE SPECTROSCOPY

Another technique closely related to the PULPIT opto-acoustic technique is the photorefractive (PR) method. In PR measurements, two laser beams are usually used: (i) a strong "excitation beam" that may be pulsed or CW, and generates a refractive-index gradient that may diffuse or propagate outwards depending respectively on whether the refractive index change is due to a temperature gradient that diffuses outwards, or due to a propagating acoustic wave; (ii) a weak CW "probe beam" that measures the refractive index gradient that deflects or distorts the probe beam (the probe must be weak enough so that itself does not generate any refractive index gradient). In most PR methods, the refractive index gradient due to the diffusive temperature gradient only is probed. If the excitation and probe beams coincide, the PR method is usually called a "thermal lensing" (TL) method, which was developed by Albrecht and coworkers[35] as a sensitive spectroscopic tool. (Of course, even in the absence of the probe beam, some degree of self-thermal-lensing is also present, as first reported by Leite et al[36] who suggested it as a sensitive spectroscopic tool[37].) In general, the probe beam need not be coincident or even parallel to the excitation beam. Boccara et al[38] pointed out that sensitivity of PR spectroscopy may be larger for noncoincident PR detection compared to TL; this is possible because in the TL method, the probe beam is situated in a distribution of refractive-index gradient (produced by the excitation beam) while in the non-coincident PR method, the probe can be positioned at the maximum refractive-index gradient.

PR methods and the PULPIT OA methods for spectroscopy of weak absorption lines in condensed matter are seen to have competitive sensitivities, according to the experimental data in the published literature. Our PULPIT OA detection method has been demonstrated to be capable of measuring an absorption coefficient α of 10^{-7} cm^{-1} for a 1-mJ laser. The TL method of Albrecht and coworkers[35] is seen to have sensitivity approaching our method; however, alignment of the two beams, pinhole, and lenses in the TL or PR methods are usually quite delicate, and interpretation of results may not be straightforward. This can be readily seen if we note that the initial TL results of Albrecht et al[35] indicate that the absorption coefficient of liquid C_6H_6 at 607 nm, α (607) was 6×10^{-4} cm^{-2}, while later on, it was realized that a thick lens formalism was needed [39] for data interpretation, and the result was revised to be α (607) = 2.3×10^{-3} cm^{-1}, which is the same as the already published OA result[19]. Thus Jackson et al[40] claimed that their noncoincident PR spectroscopy method can detect absorption coefficient of 10^{-7} cm^{-1} for a 1 mJ excitation laser. However, alignment difficulties are pointed out by them, and also, their spectrum[38] of the 607 nm absorption line of neat C_6H_6 do not precisely agree with our data[17,19] nor with the data of Stone[41] in the red wing, leading one to question the possibility of systematic errors in the PR methods.

A rigorous comparison of PR methods and OA methods was claimed to have been made by Brueck et al[34] although they have only considered the coincident PR case (i.e., TL). They claimed that the measured OA detection limit (using a PULPIT OA detection scheme as ours) is 4×10^{-5} cm^{-1} in liquid N_2, while the corresponding detection limit in their TL detection is 6×10^{-7} cm^{-1}. They further claimed that the theoretical detection limit (in principle) is even more in favor of TL methods. According to them[34] electrostriction limits the sensitivity of the OA technique, and that also of the TL technique to a less extent. We believe that Brueck et al[34] claims and statements are at best only partially correct as discussed in Sec. 4. As noted by Bebchuck et al [42] the fractional volume change $(\delta V/V)_{es}$ due to electrostriction is proportional to I/B, where I is the intensity of the excitation beam and B is the compressibility of the material. Hence, electrostrictive effects can be reduced if I is reduced by using an excitation beam of large cross-sections (e.g., ~ 2 mm diameter).

The inescapable conclusion is that PR methods are not superior to PULPIT OA methods because of increased sensitivity, as claimed by Brueck et al[34]. However, PR methods do have one distinct advantage when transducers cannot be attached to the sample (e.g., for a micro-sample, or for a highly corrosive gas sample[43]); in this case, the noncontact remote-sensing feature of the PR technique is a major advantage. On the other hand, PULPIT OA methods seem to have the following advantages over PR methods when transducers can be attached to the condensed sample (coupling through a quartz couvette[44] is also possible if the condensed sample is corrosive): a) Simple alignment and straightforward data interpretation (in the PR case, alignment is complex, and data interpretation requires thick lens formalism[39] and corrections for all sources of optical distortion effects, and so errors are more liable to creep in); b) Fast data taking is limited by the transducer ringing time of typically less than 100 μsec (in the PR case, the time taken for a thermal lens to diffuse and vanish is typically 0.1 sec, and thus data taking is some 10^3 times slower compared to the OA case).

6. EXPERIMENTAL RESULTS

The demonstrated ability to measure absorption coefficients as small as $\sim 10^{-7}$ cm^{-1} makes PULPIT opto-acoustic spectroscopy a powerful tool for the study of absorption spectra of weakly absorbing condensed phase materials. We will comment briefly on some of the large number of studies and treat the more recent results in some detail.

a. Linear Spectroscopy at Ambient Temperatures:

Aromatic hydrocarbons: Absorption spectra of 6th, 7th and 8th harmonics of C_6H_6 and sixth harmonics of a number of substituted benzenes have been obtained[17-20,44].

Water: Accurate absorption spectra of liquid H_2O and D_2O obtained through PULPIT opto-acoustic spectroscopy[21] are at present the most reliable data on these two extremely important solvents.

Thin Liquid Films: Measurements on thin liquid films (1 μm - 5 μm thick) show that films as thin as monolayers can be measured for strongly absorbing liquids[26].

Powders: An ingeneous experimental arrangement [25] has allowed PULPIT opto-acoustic spectroscopy of powdered materials which are highly scattering. Spectra obtained[25] of Ho_2O_3, Er_2O_3 and Dy_2O_3 indicate that this capability may turn out to be crucial in exploratory studies[45] of low loss materials for long wavelength optical fibers.

b. Nonlinear Spectroscopy:

Two Photon Spectroscopy: By increasing the laser intensity two photon PULPIT opto-acoustic spectra of the $^1A_{1g}-^1B_{2u}$ transition of C_6H_6 have been obtained. (See also the discussion in Section 4.) First accurate measurements of two photon absorption cross-section were reported[22]. Smallest two photon absorptions that can be measured are $\sim 10^{-7}$ cm MW^{-1} corresponding to two photon absorption corss-sections of $\sim 10^{-53}$ cm^4 sec $photon^{-1}$ mol^{-1} for presently used lasers.

Raman Gain Spectroscopy: By having two laser radiations whose frequencies are separated by a Raman allowed frequency of the medium, PULPIT opto-acoustic Raman gain spectroscopy (OARS) of condensed phase materials such as C_6H_6, 1,1,1 trichloroethane, acetone, toluene and n-hexane have been carried out for presently used lasers[23]. Raman gains as small as 10^{-5} cm^{-1} can be measured.

c. Low Temperature Spectroscopy:

Liquids: We have recently succeeded in obtaining the first high quality absorption spectra of liquid CH_4 (96 K)[28] and liquid C_2H_4 (113 K)[29,30] using the PULPIT opto-acoustic spectroscopy. The reasons for the choice of these gases lie in their importance in the planetary spectra [28,46,47]. The OA cell was modified so that it can be cooled down to temperatures as low as 1.2K using a Janis Varitemp dewar [28,31].

Fig. 1 shows an absorption spectrum of liquid CH_4 (94 K) obtained[28] using the PULPIT opto-acoustic spectroscopy technique.

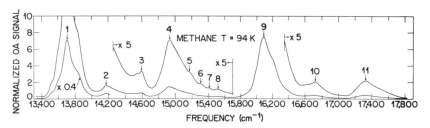

Fig. 1. PULPIT opto-acoustic absorption spectrum of methane.

The observed band center, band widths, strengths, and asymmetries of the marked peaks are summarized in Ref. 28. It can be ascertained from the signal-to-noise ratio that we have the ability to measure absorption coefficients $< 10^{-6}$ cm^{-1} for cryogenic liquids. Significant differences exist between gas and liquid CH_4 absorption features to allow a search for existence of liquid CH_4 droplets in the atmospheres of the giant planets as observed through ground based optical astronomy [28].

We have obtained similar absorption spectra for liquid C_2H_4 (113 K). Further, using the process of stimulated Raman scattering as described in Section 3, we have extended the C_2H_4 spectra to wavelengths ~ 1.6 μm[30]. Even at long wavelengths and at cryogenic temperatures, we clearly have retained the capability of PULPIT opto-acoustic spectroscopy to measure absorption coefficients smaller than 10^{-6} cm^{-1}.

Solids: In cryogenic solids, the band spectra of the type seen for CH_4 and for C_2H_4 are expected to sharpen up, allowing accurate determination of line centers. Recently, we have started the study of liquid and solid H_2 to look for its weak overtone spectra in the visible and the near IR region[48]. From Ref. 48 we can obtain the expected collision induced absorption coefficients for liquid H_2 for the overtone spectra.

Figs. 2 and 3 show absorption spectra of solid H_2 at 13K in the 1.2 μm and 0.8 μm regions, respectively, obtained using PULPIT opto-acoustic spectroscopy [31]. Sharp, well-defined, narrow (linewidth ~ 15 cm^{-1}) lines are seen and these are correlated reasonably well with the high pressure gas phase data for 0-2 and 0-3 transitions, respectively, of McKellar and Welsh[48] and the single molecule calculated line positions[49] for single transitions where during a collision one H_2 molecule is excited to v=2 (or 3) level and for double transition where during a collision, both of the H_2 molecules are excited. It is interesting to note that McKellar and Welsh data were obtained using a (density)^2 × path length ($\rho^2 \ell$) product of $\sim (6.1-19) \times 10^5$ (amagat)^2 meter while our preliminary data on solid H_2 are obtained using a $\rho^2 \ell \approx 3.6 \times 10^3$ (amagat)^2 m. The 0-2 solid H_2 spectra have been seen earlier[50] using a 12 cm path length. The 0-3 solid H_2 have never been reported previously. Even at this preliminary stage, the differences in the high pressure gas and solid H_2 data are striking when compared with data in Ref. 48. Tentative identifications are: for the 0-2 band -

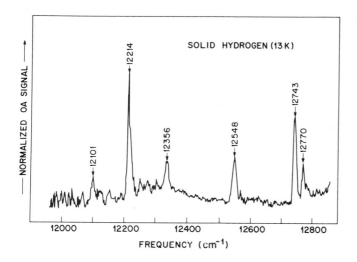

Figs. 2.

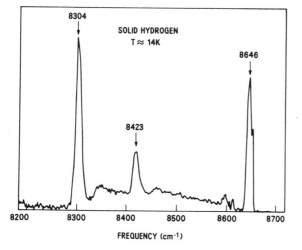

Fig. 3.

Figs. 2 and 3. Overtone absorption spectra of solid hydrogen.

$\nu = 8304$ cm^{-1}, $\quad$ Q$_1$(1) + Q$_1$(1,0), $\quad$ $\nu = 8423$ cm^{-1}, $\quad$ Q$_2$(1,0) + S$_0$(0), $\quad$ $\nu = 8646$ cm^{-1}, Q$_1$(1,0) + S$_1$(1); and for the 0-3 band - $\nu = 12101$ cm^{-1}, Q$_3$(1,0) + S$_0$(0); $\nu = 12214$ cm^{-1}, Q$_2$(1) + Q$_1$(1,0); $\quad$ $\nu = 12356$ cm^{-1}, $\quad$ Q$_3$(1,0) + S$_0$(1); $\quad$ $\nu = 12548$ cm^{-1}, $\quad$ S$_2$(0) + Q$_1$(1,0); $\nu = 12743$ cm^{-1}, S$_2$(1) + Q$_1$(1,0); $\nu = 12770$ cm^{-1}, Q$_2$(1,0) + S$_1$(1). We notice that double transitions definitely predominate as compared with high pressure data indicating a need for appropriate new theory. Further, the line positions are ~10–20 cm^{-1} too low for the observed solid H$_2$ 0–2 and 0–3 transitions from the calculated single molecule transitions (indicating a small but finite binding energy of the bound H$_2$ complexes).

7. CONCLUSION

From the above it is clear that we are at the threshold of applying PULPIT opto-acoustic spectroscopy to a variety of exciting scientific studies and to important technological problems. The presently demonstrated capability of measuring absorption coefficients as small as 10^{-7} cm^{-1} (or $\alpha\ell \approx 10^{-7}$) with pulse energies of 10^{-3} J over a broad wavelength region can not but lead to a major revolution in the spectroscopy of transparent liquids and solids. Where resolution is not an important consideration, we expect to replace the pulsed tunable dye laser with a pulsed xenon arc lamp in conjunction with a monochromator which will yield the needed ~1 mJ energy per pulse in a ~100 cm^{-1} bandwidth. Such a development will make PULPIT opto-acoustic spectroscopy a versatile tool of everyday use. Finally, we close by listing some of the future directions that PULPIT opto-acoustic spectroscopy may follow: 1) Studies of higher order Raman processes; 2) Materials testing; 3) Trace and impurity detection; 4) Studies of forbidden transitions; 5) Studies of excited states; 6) Monolayers; 7) Opto-acoustic microscopy; and 8) Extensions to electron beam and X-ray excitation of the medium.

REFERENCES

1. A. G. Bell, *Proc. Am. Assoc, Adv. Sci.* **29,** 115 (1880).

2. E. L. Kerr and J. G. Atwood, *Appl. Opt.* **7,** 915 (1968).

3. L. B. Kreuzer, *J. Appl. Phys.* **42,** 2934 (1971).

4. L. B. Kreuzer and C. K. N. Patel, *Science* **173,** 45 (1971).

5. C. K. N. Patel and E. D. Shaw, *Phys. Rev. Lett.* **24,** 451 (1970). See also the review articles: C. K. N. Patel in "Laser Spectroscopy" ed. A. Mooradian and R. G. Brewer (Plenum Press, 1974), pp. 471-491; C. R. Pidgeon and S. D. Smith, *Infrared Physics* **17,** 515 (1977).

6. C. K. N. Patel and R. J. Kerl, *Appl. Phys. Lett.* **30,** 578 (1977).

7. L. B. Kreuzer, C. K. N. Patel and N. D. Kenyon, *Science* **177,** 347 (1972); R. J. H. Voorhoeve, C. K. N. Patel, L. E. Trimble, and R. J. Kerl, *Science* **190,** 149 (1975).

8. C. K. N. Patel, E. G. Burkhardt, and C. A. Lambert, *Science* **184,** 1173 (1974); E. G. Burkhardt, C. A. Lambert, and C. K. N. Patel, *Science* **188,** 1111 (1975); C. K. N. Patel, *Optical and Quantum Electronics* **8,** 145 (1976).

9. C. K. N. Patel, R. J. Kerl, and E. H. Burkhardt, *Phys. Rev. Lett.* **38,** 1204 (1977); C. K. N. Patel, *Phys. Rev. Lett.* **40,** 535 (1978).

10. L. A. Farrow and R. E. Richton, in *Proceedings of the SPIE Conference on Photoacoustic Spectroscopy,* (April 20,24 1981), to be published.

11. V. N. Bragatashvilli, I. N. Knyazev, V. S. Letokhov, and V. V. Lobko, *Opt. Comm.* **18,** 525 (1976).

12. W. R. Harshbarger and M. B. Robin, *Acc. Chem. Res.* **6,** 329 (1973).

13. A. Rosencwaig, *Optics Comm.* **7,** 305 (1973).

14. A. Rosencwaig and A. Gersho, *J. Appl. Phys.* **47,** 64 (1976); L. C. Aamodt, J. C. Murphy, and J. G. Parker, *J. Appl. Phys.* **48,** 927 (1972); F. A. McConald and G. C. Wetsel, Jr., *J. Appl. Phys.* **49,** 2313 (1978).

15. See for example A. Rosencwaig in "Optoacoustic Spectroscopy and Detection" ed. Y-H. Pao (Academic Press, 1977), pp. 194-239.

16. M. J. Adams and G. F. Kirkbright, *Analyst* **102,** 678 (1977).

17. C. K. N. Patel and A. C. Tam *Appl. Phys. Lett.* **34,** 467 (1979).

18. C. K. N. Patel and A. C. Tam, *Rev. Mod. Phys.* **53,** 517 (1981).

19. A. C. Tam, C. K. N. Patel and R. J. Kerl, *Optics Lett.* **4,** 81 (1979).

20. C. K. N. Patel, A. C. Tam, and R. J. Kerl, *J. Chem. Phys.* **71,** 1470 (1979).

21. C. K. N. Patel and A. C. Tam, *Nature* **280,** 302 (1979); A. C. Tam and C. K. N. Patel, *Appl. Opt.* **18,** 3348 (1979).

22. A. C. Tam and C. K. N. Patel, *Nature* **280,** 304 (1979).

23. C. K. N. Patel and A. C. Tam, *Appl. Phys. Lett.* **34**, 760 (1979).

24. C. L. Sam and M. L. Shand, *Opt. Comm.* **31**, 174 (1979).

25. A. C. Tam and C. K. N. Patel, *Appl. Phys. Lett.* **35**, 843 (1979).

26. C. K. N. Patel and A. C. Tam, *Appl. Phys. Lett.* **36**, 7 (1980).

27. E. T. Nelson and C. K. N. Patel, *Optics Letters* **6**, 354 (1981).

28. C. K. N. Patel, E. T. Nelson and R. J. Kerl, *Nature* **286**, 368 (1980).

29. E. T. Nelson and C. K. N. Patel, *PNAS* **78**, 702 (1981).

30. E. T. Nelson and C. K. N. Patel, *Appl. Phys. Lett.* (to be published).

31. C. K. N. Patel, E. T. Nelson and R. J. Kerl, (to be published).

32. K. A. Naugol'nykh, *Sov. Phys. Acoust.* **23**, 98 (1977).

33. C. K. N. Patel in Proceedings of Bad Honnef Photoacoustic Spectroscopy Workshop (to be published).

34. S. R. J. Breuck, H. Kildal and L. J. Belanger, *Optics Comm.* **34**, 199 (1980).

35. R. L. Swofford, M. E. Long and A. C. Albrecht, *J. Chem. Phys.* **65**, 179 (1976).

36. R. C. C. Leite, R. S. Moore, and J. R. Whinnery, *Appl. Phys. Lett.* **5**, 141 (1964).

37. D. Solimini, *Appl. Optics* **5**, 1931 (1966).

38. A. C. Boccara, D. Fournier, W. Jackson, and N. M. Amer, *Optics Lett.* **5**, 377 (1980).

39. H. L. Fang and R. L. Swofford, *J. Appl. Phys.* **50**, 6609 (1979).

40. W. B. Jackson, N. B. Amer, A. C. Boccara, and D. Fourrier, *Appl. Opt.* **20**, 1333 (1981).

41. J. Stone, *Appl. Optics,* **17**, 2876 (1978).

42. A. S. Bebchuk, V. M. Mizin and N. Ya. Salova, *Opt. Spectros.* **44**, 91 (1978).

43. A. C. Tam, W. Zapka, K. Chiang and W. Inaino, Paper MB1 in the Technical Digest Second International Conference on Photoacoustic Spectroscopy, Berkeley, June 22-25, 1981 (unpublished).

44. A. C. Tam and C. K. N. Patel, *Opt. Lett.* **5**, 27 (1980).

45. C. K. N. Patel in Proceedings of the SPIE Symposium on Infrared Optical Fibers, Feb. 12, 1981 (to be published).

46. L. P. Giver, *J. Quant Spectrosc. Radiat. Transfer* **19**, 311 (1978).

47. K. R. Ramaprasad, J. Caldwell, and D. S. McClure, *Icarus,* **35**, 400 (1978).

48. For an excellent review of earlier studies in high density H_2 see A. R. W. McKellar and H. L. Welsh, *Proc. Roy. Soc. (London)* **A322**, 421 (1971).

49. J. Foltz, D. D. Rank, and T. A. Wiggins, *J. Mol. Spectry* **21**, 203 (1966).

50. E. J. Allin, H. P. Gush, W. F. J. Hare, J. L. Hunt, and H. L. Welsh, in proceedings of "Colloques Internationaux du Centre National de La Recherches Scientifique" No. LXVII, July 1-6, 1954, Bellevue (published by CNRS, France, 1959) pp. 21-39.

Optical Hole-Burning and Ground State Energy Transfer in Ruby

P.E. Jessop and A. Szabo

Division of Electrical Engineering, National Research Council of Canada
Ottawa, Ontario K1A 0R8, Canada

1. Introduction

A recent application of homogeneous linewidth laser spectroscopy of optical solids is in energy transfer studies[1,2]. This paper describes a high resolution hole-burning study of the R_1 line in ruby giving information on the energy transfer mechanism which determines the saturated lineshape. As is well known, the zero phonon R_1 lines in ruby at liquid helium temperatures are inhomogeneously broadened due to crystal defects. When the $^4A_2(\pm\frac{1}{2}) \to \bar{E}$ transition ($R_1(\frac{1}{2})$ hereafter) is saturated by a narrowband laser, the result shown in Figure 1 occurs[3].

As expected, a hole in the absorption lineshape appears at the laser frequency, however an unexpected effect was the large decrease in the off-resonant absorption (ORA). Such a decrease unequivocally indicates the presence of energy transfer. We ask (1) what is the interaction producing the transfer, and (2) does the transfer occur in the ground or excited states?

Consider first the effect of non-resonant phonon-assisted transfer, in the excited state, on the ORA when the $R_1(\frac{1}{2})$ line is hole-burnt. We expect the ORA of both the $R_1(\frac{1}{2})$ and $R_1(3/2)$ lines to <u>decrease</u> because of population in the common $\bar{E}$ level. However, if ground state transfer occurs between $\pm\frac{1}{2}$ and $\pm3/2$ spin levels (via magnetic dipole induced spin-spin cross relaxation

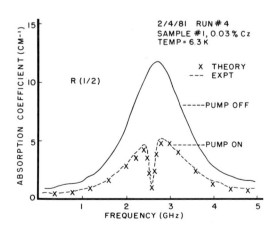

Fig. 1 Hole-burning of $R(\frac{1}{2})$ line in ruby in zero magnetic field.

[4]), then a population argument shows that while the $R_1(\frac{1}{2})$ ORA will <u>decrease</u>, that for the $R_1(3/2)$ will <u>increase</u>. Also a hole, rather than an antihole will appear for $R_1(3/2)$ when the $R_1(\frac{1}{2})$ line is hole-burnt.

2. Experiment and Results

The apparatus is shown in Fig. 2. A CW, fixed frequency ruby laser (<5 MHz linewidth[3]) saturated the $R_1(\frac{1}{2})$ line at 693.4 nm. A scanning, single frequency (1 MHz linewidth) CW dye laser probes the R_1 lineshape. 300 μwatts, focussed to a density of ∿15 W/cm^2, was used to hole-burn and ∿1 μwatt used for probing. The probe beam diameter inside the 1.6 mm thick sample was 3X smaller than the pump to minimize edge effects. All results are for zero magnetic field.

Fig. 3 shows the complete R_1 lineshape when $R_1(\frac{1}{2})$ is pumped and we observe that the $R_1(3/2)$ ORA <u>increases</u>. For ground state transfer, a simple population argument says that the relative changes of the ORA should be

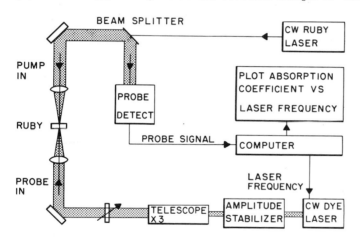

Fig. 2 Experimental set-up.

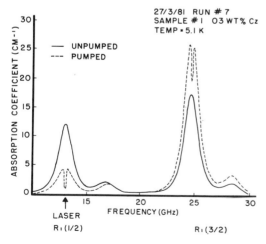

27/3/81 RUN #7
SAMPLE #1 03 WT% Cz
TEMP = 5.1 K

Fig. 3 Effect of $R_1(\frac{1}{2})$ hole-burning on the $R_1(\frac{1}{2})$ and $R_1(3/2)$ absorption lineshapes.

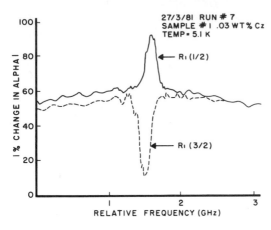

Fig. 4 Comparison of the absolute % change in the $R_1(\tfrac{1}{2})$ and the $R_1(3/2)$ absorption coefficients vs line position when the $R_1(\tfrac{1}{2})$ is hole-burnt as in Fig. 3

nearly equal for the two lines (at 5K). Computer processing of the data gives the results shown in Fig. 4. The ORA changes are seen to be equal within $\sim 5\%$ quantitatively supporting the ground state transfer mechanism. Also from Fig. 4, we see that the ORA change is independent of frequency indicating that all ions come to the same spin temperature.

3. Discussion and Conclusions

A 3 level rate equation analysis similar to [3] gives for the absorption lineshape

$$\alpha(\nu) = \frac{2\alpha_u(\nu)}{1+f}\left[1 - \frac{\varepsilon\delta(2+f)(w_h+\delta)}{(w_h+\delta)^2 + 4(\nu-\nu_0)^2}\right] \tag{1}$$

when the line is pumped at ν_0. In (1), $\alpha_u(\nu)$ is the unsaturated coefficient, f is related to the $^4A_2$ spin temperature, T_s by $f = \exp h\nu'/kT_s$; $\nu' = 11.5$ GHz, and w_h is the homogeneous linewidth (FWHM). δ is a power broadening parameter given by

$$\delta = w_h[S(f+2)(f+1)^{-1} + 1]^{\tfrac{1}{2}} \tag{2}$$

where the saturation parameter $S = 2P\tau\sigma_0/\pi w_h$, τ is the $\bar{E}$ lifetime, P is photons cm^{-2} sec^{-1}, $\sigma_0 = 4.9 \times 10^{-9}$ cm^2 sec^{-1} for $R_1(\tfrac{1}{2})$ and $E\perp c$ and $\varepsilon = [(2+f)+(1+f)/S]^{-1}$. Assuming a constant T_s for all ions leads to a relationship between f and S

$$(f-1)^2[S(f+2)(f+1)^{-1}+1] = (KS)^2 \tag{3}$$

where $K = 3(\pi \ln 2)^{\tfrac{1}{2}}(\tau_s w_h/5\tau w_i)$, τ_s is the spin lattice time of 0.2 sec and w_i is the inhomogeneous width. A fit of the theory to the data is shown in Fig. 1 giving experimentally derived parameters which are compared with calculated ones in Table 1. As observed, a rather large discrepancy exists in the homogeneous linewidth values. We speculate that the 65 MHz value obtained by CW methods may be broadened by spectral diffusion due to slow rearrangements of the Al nuclear spins. If $w_h = 26$ MHz is used in the calculations, the values T_s and S come into closer agreement adding support to this idea.

In conclusion, ground state energy transfer is found to be the dominant mechanism determining the optical saturation behaviour of the dilute samples studied so far. Moreover, the observed frequency independent spin temperature suggests little correlation between the optical frequency and spatial distribution, i.e. microscopic broadening exists[3]. Finally, the effects reported can have important consequences for device technology. For an application such as a hole-burning optical memory[5], spin-spin transfer will reduce the contrast ratio. In another application, the laser-pumped maser[6,7], such transfer is beneficial and low-noise maser operation up to 300 GHz may be feasible.

Table 1   Comparison of experimental and calculated lineshape parameters

PARAMETER		EXPT	CALC
Spin temperature	T_s	0.48 K	0.31 K
Saturation	S	20.0	9.4
Homogeneous width	w_h	26 MHz	65 MHz[1]

[1]From low power fluorescence line narrowing and hole-burning.

Acknowledgement

We wish to acknowledge the excellent technical assistance of R.E. Monnon.

References

1. P.M. Seltzer, D.L. Huber, B.B. Barnett and W.M. Yen, Phys. Rev. B17, 4979 (1978)
2. P.E. Jessop and A. Szabo, Phys. Rev. Lett. 45, 1712 (1980); S. Chu, H.M. Gibbs, S.L. McCall and P. Passner, ibid, 1715 (1980)
3. P.E. Jessop, T. Muramoto and A. Szabo, Phys. Rev. B21, 926 (1980)
4. N. Bloembergen, S. Shapiro, P.S. Pershan and J.O. Artman, Phys. Rev. 114, 445 (1959)
5. A. Szabo, Proceeding Lasers '80, New Orleans, Dec. 15-19, 1980 (in press); ibid, G.C. Bjorklund
6. D.P. Devor, IEEE Trans. MTT 11, 251 (1963)
7. A. Szabo, J. Appl. Phys. 39, 5425 (1968)

Surface-Enhanced Nonlinear Optical Effects and Detection of Absorbed Molecular Monolayers

Y.R. Shen, C.K. Chen, T.F. Heinz, and D. Ricard

Department of Physics, University of California
Berkeley, CA 94720, USA

and

Materials and Molecular Research Division, Lawrence Berkeley Laboratory
Berkeley, CA 94720, USA

In recent years, we have been interested in the study of nonlinear optical effects at interfaces. Our motivation is twofold. First, while nonlinear optics in a bulk medium is now well developed, that at an interface has not received much attention since the early investigations of BLOEMBERGEN and co-workers on second-harmonic reflection from a surface [1]. Second, with the understanding of surface nonlinear optics, we hope to use it for the study of surface properties, in particular, for the study of molecular adsorption and overlayers on surfaces. There already exist various techniques for studies of this nature, such as photoemission, LEED, Auger, infrared, and Raman spectroscopy [2], but laser optical techniques are intrinsically better in spectral resolution.

A few recent developments in surface nonlinear optics in relation to surface study are of interest. We have found that surface coherent anti-Stokes Raman scattering with picosecond pulses can have the sensitivity of detecting submonolayers of molecules on surfaces [3]. Then, HERITAGE [4] and LEVINE et al. [5] have demonstrated the possibility of obtaining Raman spectra of very thin films and molecular monolayers with picosecond Raman gain spectroscopy. In another area, it has been found that Raman scattering of adsorbed molecules on a rough surface of some metals can be enhanced by five or six orders of magnitude [6], enabling the study of adsorbed molecular submonolayers on those surfaces. We have realized that such surface enhancement should also occur in other nonlinear optical processes [7]. In this paper, we discuss the observation of a number of surface-enhanced nonlinear optical effects. We also show the feasibility of using second-harmonic generation to detect the adsorption of molecular monolayers on a metal surface in an electrolytic solution [8].

For surface-enhanced Raman scattering (SERS), it is now believed by most researchers that the enhancement arises mainly from the large local fields present near rough metal structures [9], although it is perhaps also partially due to the interaction of molecules with the metal. The strong local-field effect has its origin in the resonant excitation of local surface plasmons on the rough metal structures. We can, in general, write the local field as $E_{loc}(\omega) = L(\omega)E(\omega)$, with E being the incoming optical field and L the local-field correction factor. On the surface of a metal spheroid, if the diameter of the spheroid is much less than the wavelength $\lambda = 2\pi c/\omega$, it is easily shown that [9]

$$L(\omega) \propto [\in_m(\omega) - \in(\omega)]/[\in_m(\omega) + 2\in(\omega)] , \qquad (1)$$

where $\in_m(\omega)$ and $\in(\omega)$ are the dielectric constants of the metal and the surrounding dielectric medium, respectively. At the surface plasmon resonance,

$Re[\epsilon_m(\omega) + 2\epsilon(\omega)] = 0$, and $L_{max}(\omega) \propto Im[\epsilon_m(\omega) + 2\epsilon(\omega)]^{-1}$ is resonantly enhanced. This is possible for metals because $Re(\epsilon_m)$ is negative below the bulk plasmon resonance. More generally, the local field on a particular surface area of a small metal aggregate of a certain shape can be written as $L(\omega) \propto 1/f(\epsilon)$, and the surface plasmon resonance is excited when $Re[f(\epsilon)] = 0$.

A nonlinear optical process is usually governed by the nonlinear polarization [10]

$$P^{(n)}(\omega) = \chi^{(n)}(\omega = \omega_1 + \cdots + \omega_n)E_1 \cdots E_n \tag{2}$$

with $\chi^{(n)} = L(\omega)L(\omega_1) \cdots L(\omega_n)N\alpha^{(n)}$, where $\alpha^{(n)}$ and $\chi^{(n)}$ are the nth-order polarizability and susceptibility, respectively, and N is the density of atoms or molecules. We note that if $L(\omega)$, $L(\omega_1)$, ..., $L(\omega_n)$ are all resonantly enhanced, then $\chi^{(n)}$ and the corresponding nonlinear optical effect would be greatly magnified. In the case of Raman scattering, the effective Raman cross-section σ_{eff} is proportional to $\chi^{(3)}(\omega_s = \omega_\ell - \omega_\ell + \omega_s)$. Therefore, we have

$$\sigma_{eff} = L^2(\omega_\ell)L^2(\omega_s)\sigma . \tag{3}$$

If $L(\omega_\ell) = L(\omega_s) = 20$, which is a conservative estimate on silver, one finds $\sigma_{eff} = 1.6 \times 10^5 \sigma$. On top of this local-field enhancement, the Raman cross-section σ can, of course, be further enhanced through the interaction of the molecules with the metal.

Eq.(2) suggests that other nonlinear optical processes should also experience such a local-field enhancement. In particular, second-harmonic generation is proportional to $|\chi^{(2)}(2\omega)|^2$ and therefore has a local-field enhancement factor $\eta_{loc}(2\omega) = L^4(\omega)L^2(2\omega)$. Similarly, third-harmonic generation has $\eta_{loc}(3\omega) = L^6(\omega)L^2(3\omega)$, and coherent anti-Stokes Raman scattering has $\eta_{loc}(\omega_a) = L^4(\omega_\ell)L^2(\omega_s)L^2(\omega_a)$. If we assume, appropriate for rough silver, $L(\omega) = L(\omega_\ell) = L(\omega_s) = L(\omega_a) = 20$ and $L(2\omega) = L(3\omega) = 1$, we find $\eta_{loc}(2\omega) = 1.6 \times 10^5$, $\eta_{loc}(3\omega) = 6.4 \times 10^7$, and $\eta_{loc}(\omega_a) = 2.6 \times 10^{10}$. These are huge enhancement factors, suggesting that even if only a monolayer of atoms or molecules contributes to the nonlinear optical effects, they could be observable.

We have indeed observed surface-enhanced second-harmonic generation (SHG) from a roughened silver surface [7]. In this case, SHG from a metal surface is believed to come from the first one or two surface layers of metal atoms. The enhancement can be directly measured by comparing the signal with the second-harmonic reflection from a smooth silver film. A Q-switched Nd:YAG laser at 1.06 μm was used in the experiment to provide the pump beam. The sample was roughened by an electrolytic oxidation-reduction process and then rinsed in distilled water and dried by nitrogen. The second-harmonic (SH) reflection from the rough surface was highly diffuse. Its integrated power over the nearly isotropic angular distribution was found to be $\sim 1 \times 10^4$ times larger than the collimated second-harmonic signal from the smooth surface. The results are shown in Fig.1. That the signals were indeed from SHG is confirmed by the quadratic dependence of the signals on the input laser power, together with a measurement of the spectrum. To find the actual local-field enhancement, we realize that essentially all the enhanced SHG came from the silver aggregates on the roughened surface. Electron micrographs showed that the surface consists of ~ 500 Å Ag aggregates separated by $1500 - 3000$ Å. We therefore estimate that the aggregates occupied a fraction of roughly 5% of the total surface. Then, if we assume that the

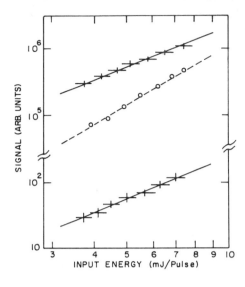

Fig.1 Power-law dependences of the nonlinear signal on silver. The upper and lower solid curves show the quadratic dependence of the diffuse SH signal from the rough bulk sample and of the collimated SH signal from the smooth film, respectively; the dashed curve shows the cubic dependence of the diffuse anti-Stokes signal from the rough bulk sample

coherent dimension of surface second-harmonic generation is approximately equal to the size of the aggregates, the actual local field enhancement of SHG from the aggregates is $n_{\ell oc}(2\omega) = 10^4/0.05 = 2 \times 10^5$. This may still be an underestimate, since not all aggregates give the maximum local-field enhancement. Since ω is very different from 2ω, $L(\omega)$ and $L(2\omega)$ cannot both be at surface plasmon resonances. For the maximum enhancement, we expect $L(\omega)$ to be appreciable, but $L(2\omega) \sim 1$. Then, from $n_{\ell oc}(2\omega) = L^4(\omega)L^2(2\omega)$, we infer $L(\omega) \sim 20$, which seems reasonable.

We have used, in a separate experiment [11], a pump beam at 0.53 and 0.68 μm and found that the surface enhancement of SHG was only $\sim 10^2$. This is expected because $L(\omega)$ is roughly proportional to $\in_m(\omega)/Im[\in_m(\omega)]$ and the latter decreases as ω approaches the bulk plasmon resonance of the metal. We have also found that the surface enhancement of sum-frequency generation with ω_1 at 1.06 μm and ω_2 at 0.53 μm is about three orders of magnitude smaller than that of SHG with ω at 1.06 μm, which can be understood from the relation $n_{\ell oc}(\omega_1 + \omega_2)/n_{\ell oc}(2\omega) = L^2(\omega_1 + \omega_2)/L^2(\omega)$. Surface-enhanced SHG from copper and gold has also been seen. The enhancement is at least one order of magnitude weaker than that for silver and can be explained by the smaller L resulting from the heightened values of $Im\in$ for copper and gold [7].

Among the surface-enhanced nonlinear optical effects, BOYD and YU [11] in our laboratory have observed third-harmonic generation from a roughened gold surface with a pump beam at 1.06 μm. For a 10 nsec pulse, with peak intensity of 5 MW/cm^2 in an area of .2 cm^2, a rough estimate suggests that the third-harmonic signal from a smooth gold surface is about 3 orders of magnitude weaker than the second-harmonic signal in the absence of any surface enhancement. With surface enhancement, it should be about 1 order of magnitude weaker, in general agreement with our measurements. We have also observed surface-enhanced luminescence from roughened surfaces of silver, copper, and gold with one- and two-photon excitation. Luminescence with one-photon excitation from a smooth silver surface was rather weak, and that with two-photon excitation is not observable. We can consider luminescence with n-photon excitation as an (n + 1)-photon process, but the surface enhancement factor in this case is more difficult to estimate because the radiation efficiencies from smooth and rough surfaces may be very different. Hyper-Raman scattering is another three-photon process. It is rather weak and not easily observable even in bulk media. With surface enhancement, however, it even becomes possible to observe hyper-Raman scattering from molecules adsorbed on surfaces, as has recently been reported by CHANG for SO_2 on silver powder [12]. The local-field enhancement factor in this case is $[n_{loc}]_{HR} = L^4(\omega_\ell)L^2(\omega_S) = 1.6 \times 10^5$, if $L(\omega_\ell) = 20$ and $L(\omega_S) = 1$. Surface-enhanced coherent anti-Stokes Raman scattering (CARS) is especially interesting, since, as we mentioned earlier, the local-field enhancement factor is as high as 2.6×10^{10} for $L(\omega) = 20$. In addition, interactions between the molecules and metal can further enhance the process. One would then expect that it would be possible to observe surface-enhanced CARS from molecules adsorbed on silver. Unfortunately, discrimination of the anti-Stokes signal against elastic scattering of the pump beams from the rough surface is difficult. Work to detect surface-enhanced CARS is presently still in progress.

Since one can observe SHG from one or two layers of metal atoms on a metal surface, one would expect that it is also possible to observe SHG from a monolayer of adsorbed molecules. This is confirmed by a theoretical estimate using the equation [8]

$$S = \frac{256\pi^3\omega}{\hbar c^3} |N_a\alpha^{(2)}|^2 I_1^2 AT \quad \text{photons/pulse,} \tag{3}$$

where S is the SHG signal, N_a is the surface density of molecules, $\alpha^{(2)}$ is the second-order polarizability, I_1 is the intensity of a p-polarized pump beam incident at $45°$, A is the beam cross-section, and T is the pulsewidth, all in cgs units. If $\alpha^{(2)} \sim 1 \times 10^{-29}$ esu, $N_a \sim 4 \times 10^{14}$ cm^{-2}, $I_1 \sim 1$ MW/cm^2 at 1.06 μm, $A = 0.2$ cm^2, and $T = 10$ nsec, we find that $S \sim 1.5 \times 10^3$ photons/pulse. Thus, even without surface enhancement, SHG from a monolayer of adsorbed molecules with $|N_a\alpha^{(2)}| \gtrsim 10^{-15}$ esu should be detectable if the pump beam does not induce surface damage.

By preparing a monolayer of p-nitrobenzoic acid adsorbed on an aluminum oxide surface, we have obtained direct evidence for the detectability of a molecular monolayer by SHG in the absence of any surface enhancement effects. In this case, the adsorbed monolayer produced ~ 5 times as much second-harmonic signal as the substrate alone. In another case of interest, we have studied SHG from a smooth Ag electrode in an electrolytic cell, which consisted of a glass cell containing 0.1M KCl in doubly distilled water. In addition to the Ag electrode, a Pt working electrode and a saturated calomel electrode were submerged in the electrolytic cell, with a bias voltage applied between the Pt and Ag electrodes. The effect of the double charge layer in the solution at the Ag surface on SHG can be readily seen. The observed reflected second-harmonic signal as a function of the voltage drop between Ag and the reference saturated calomel electrode (SCE), V_{Ag-SCE}, appears in Fig.2. This curve can be understood as the result of interference between the SHG from Ag and the field-induced SHG from the double charge layer. In fact, the data can be fit very well by an effective nonlinear susceptibility

$$\chi_{eff}^{(2)} = \chi_{Ag}^{(2)} + \chi_{DC}^{(2)} \tag{4}$$

with $\chi_{DC}^{(2)} = aE_0 + bE_0^2$, where E_0 is the DC electric field in the double charge layer and is assumed to be linear in V_{Ag-SCE} [13]. We assume that $\chi_{Ag}^{(2)}$ is independent of V_{Ag-SCE}.

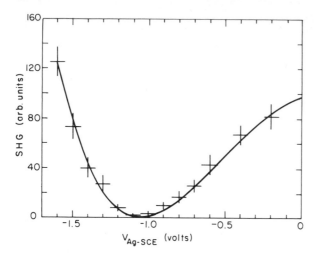

Fig.2 SH from a smooth silver film in 0.1M KCl versus V_{Ag-SCE}

With surface enhancement, SHG from adsorbed molecules is even more easily observable [8]. This can be shown by monitoring the diffuse SH from the Ag electrode submerged in the electrolytic cell during an oxidation-reduction cycle. The first one or two electrolytic cycles roughen the Ag surface and presumably activate the adsorption sites. The SHG signal from the roughened Ag electrode is then monitored during a subsequent cycle. A typical result is shown in Fig.3. At the beginning of the oxidation cycle, the signal rises sharply, indicating the formation of AgCl on the surface. The signal levels

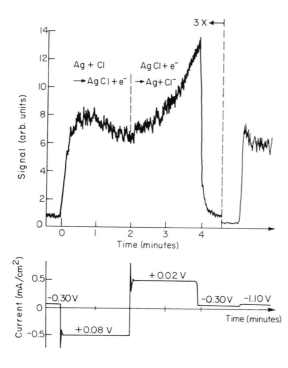

Fig.3 Current and diffuse SH as a function of time during and after an electrolytic cycle. The voltages listed in the lower curve are V_{Ag-SCE}. Pyridine (.05M) was added to the 0.1M KCl solution following the completion of the electrolytic cycle

off after an average of 3 or 4 monolayers of AgCl are formed, as judged from
the amount of charge transfer that occurred. This confirms the picture that
only the surface layers of molecules contribute to SHG. In the present case,
the field-induced SHG in the solution is presumably averaged out at the rough
surface, and the signal from AgCl in Fig.3 is rather insensitive to V_{Ag-SCE}.
At the end of the reduction cycle, the signal drops suddenly as the last 3 or
4 layers of AgCl are reduced. The rise in the signal near the end of the
cycle can probably be explained by the increase of surface roughness through
redeposition of silver.

A striking demonstration of the capability of SHG for detecting a monolayer of adsorbed molecules is provided by the case of pyridine adsorbed on the
silver electrode in an electrolytic solution. As shown in Fig.3, after the
reduction cycle is completed, if pyridine (0.05M) is added to the solution
and a bias corresponding to $V_{Ag-SCE} = -1.1$ v is applied, the second-harmonic
signal increased by 25 — 50 times. This signal arises from the adsorption of
a monolayer of pyridine on the Ag surface, which is known to depend on
V_{Ag-SCE} [6]. The SHG data in Fig.4 show that pyridine begins to be adsorbed
at $V_{Ag-SCE} \sim -0.6$ v and reaches a monolayer at ~ -0.9 v.

We can use SHG from adsorbed pyridine to obtain the adsorption isotherm,
i.e., the surface density of adsorbed pyridine molecules versus pyridine concentration in the electrolytic solution. We assume that the second harmonic
signal can be expressed by $P(2\omega) = (A + BN_a)^2$, where the BN_a term is from adsorbed pyridine, with N_a being the surface density of pyridine molecules, and
the A term coming from the Ag background. Fig.5 gives the measured $(\sqrt{P(2\omega)} -$
A) $\propto N_a$ versus the pyridine concentration ρ in the electrolytic solution for
$V_{Ag-SCE} = -1.0$ v. It happens that the result can be fit nicely with the
simple Langmuir equation [14]

$$N_a = \frac{\rho}{K + \rho} N_{as} ,$$

where N_{as} is the saturated value of N_a and K (in mole/ℓ) is related to the
adsorption free energy ΔG by

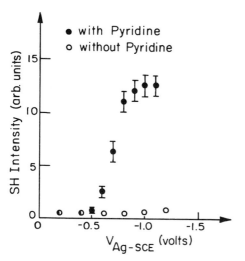

Fig.4 Diffuse SH signal versus
V_{Ag-SCE} following an electrolytic
cycle, with .05M pyridine and 0.1M
KCl dissolved in water

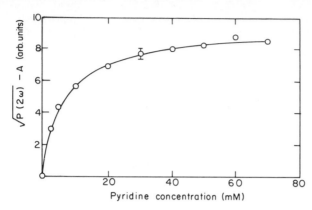

Fig.5 Equilibrium ($\sqrt{P(2\omega)}$-A) versus bulk pyridine concentration. The solid curve is a theoretical fit to the experimental data using the Langmuir model

$$K = 55 \exp(- \Delta G/RT) \; .$$

From the theoretical fit, we find ΔG = 5.1 kcal/mole for pyridine on silver in 0.1M KCl aqueous solution. From surface-enhanced Raman scattering, we have deduced a very similar adsorption isotherm, differing only in that the initial slope of the curve is somewhat higher, with a corresponding ΔG = 5.7 kcal/mole.

Other molecules adsorbed on Ag, including CN, cyanopyridines, pyrimidine, and pyrazine, have also been studied. In all these cases, adsorption and desorption of a molecular monolayer by control of the bias voltage can be observed. The case of pyrazine is of special interest. While the pyridine molecule ⬡N is not centrosymmetric, the pyrazine molecule N⬡N does have inversion symmetry. Therefore, the former has a nonvanishing $\alpha^{(2)}$, but the latter has a vanishing $\alpha^{(2)}$. If, however, the molecules are chemically adsorbed on silver, then both will have nonvanishing $\alpha^{(2)}$ because the interaction between molecules and metal breaks the symmetry. The SHG signal from

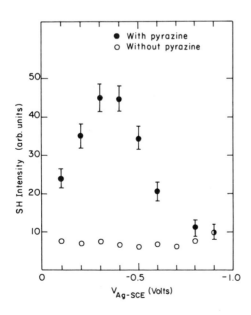

Fig.6 Diffuse SH signal versus $V_{Ag\text{-}SCE}$ following an electrolytic cycle in 0.1M KCl, with and without .05M pyrazine present

the adsorbed pyrazine versus V_{Ag-SCE} is presented in Fig.6. In comparison with Fig.4, it is seen that the maximum signal from pyrazine is only $4-5$ times smaller than that from pyridine. This means that $\alpha^{(2)}$ for pyrazine is only ~ 2 times smaller than $\alpha^{(2)}$ for pyridine. This result confirms the fact that pyrazine is chemically adsorbed on Ag with a strong metal-adsorbate interaction, as previously suggested by the Raman measurements [15].

Detection of molecular monolayers by SHG unfortunately lacks spectral selectivity. The technique can, however, be extended with sum- and difference-frequency generation. As one or both frequencies approach resonances, the signal is expected to be greatly enhanced. The resonant enhancement feature should then allow us to carry out spectroscopic studies of adsorbed molecules. With pulsed lasers, transient spectroscopy of adsorbed molecules may also become possible.

Acknowledgement T.F.H acknowledges partial support as an NSF graduate fellow. This work was supported by the Director, Office of Energy Research, Office of Basic Energy Sciences, Materials Science Division of the U.S. Department of Energy under Contract Number W-7405-ENG-48.

References

1. N. Bloembergen and P. S. Pershan, Phys. Rev. 128, 606 (1962); N. Bloembergen, R. K. Chang, S. S. Jha, and C. H. Lee, Phys. Rev. 174, 813 (1968)
2. See, for example, G. A. Somorjai, M. A. Van Hove, Adsorbed Monolayers on Solid Surfaces, Structure and Bonding, Vol.38 (Springer, Berlin, Heidelberg, New York 1979)
3. C. K. Chen, A. R. B. de Castro, Y. R. Shen, and F. DeMartini, Phys. Rev. 43, 946 (1979)
4. J. P. Heritage and D. L. Allara, Chem. Phys. Lett. 74, 507 (1980)
5. B. F. Levine, C. G. Bethea, A. R. Tretola, and M. Korngor, Appl. Phys. Lett. 37, 595 (1980)
6. M. Fleischmann, P. J. Hendra, and A. J. McQuillan, Chem. Phys. Lett. 26, 163 (1974); D. L. Jeanmaire and R. P. Van Duyne, J. Electroanal. Chem. 84, 1 (1977)
7. C. K. Chen, A. R. B. de Castro, and Y. R. Shen, Phys. Rev. Lett. 46, 145 (1981)
8. C. K. Chen, T. F. Heinz, D. Ricard, and Y. R. Shen, Phys. Rev. Lett. 46, 1010 (1981)
9. See J. Gersten and A. Nitzan, J. Chem. Phys. 73, 3023 (1980) and references therein
10. See, for example, N. Bloembergen, Nonlinear Optics (Benjamin, New York, 1964)
11. G. T. Boyd and Z. H. Yu (unpublished)
12. R. K. Chang (to be published)
13. For weak static fields, see B. F. Levine and C. G. Bethea, J. Chem. Phys. 63, 2666 (1975)
14. See, for example, M. J. Rosen, Surfactants and Interfacial Phenomena (J. Wiley and Sons, New York, 1978), Chap.2
15. R. Dornhaus, B. M. Long, R. E. Benner, and R. K. Chang, Surf. Sci. 93, 240 (1980)

Surface Enhanced Raman Scattering from Lithographically Prepared Microstructures

P.F. Liao

Bell Telephone Laboratories, Rm. 4F-433,
Holmdel, N.J. 07733, USA

Observations of large (10^6) enhancements of Raman cross sections for molecules adsorbed onto suitably roughened silver surfaces have generated much interest. The enhancement allows easy detection of submonolayer coverage of molecules on surfaces and presents the opportunity to obtain detailed information about molecules on surfaces by means of Raman spectroscopy. Recently many other optical processes such as second harmonic generation, dye luminescence and two-photon absorption have also been shown to be enhanced on such silver surfaces. In this paper we discuss the mechanism for the enhancement and describe experiments which utilize surfaces having "controlled roughness". These surfaces consist of arrays of isolated submicron silver particles which are uniform in shape and size. Modern techniques of microlithography are used to produce these surfaces. With the help of these surfaces we are able to elucidate the mechanism responsible for Surface Enhanced Raman Scattering (SERS) and can design the surface for maximum enhancement. We have, to date, found[1] enhancements of 10^7 on such surfaces.

The observations that surface roughness is generally required for the existence of large enhancement and the observations[2] of enhancement for layers of molecules beyond the first layer has led to electromagnetic models for SERS in which plasmon resonances of the microscopic bumps on the surface act to increase the local field at the molecules and to amplify the re-radiated field of the Raman active molecule.

The electromagnetic theory was first proposed by Moscovits[3] and has been elaborated on by several groups[3-5]. The enhancing surface is generally modeled as covered with a random array of metal spheroids. The main features of these theories are easily seen by considering a single dielectric ellipsoid with an external laser field, E_L directed along the principal axis of the ellipsoid, and a nearby molecule also located on the principal axis. If the ellipsoid major (a) and minor (b) axis dimensions are such that (a, b $\ll \lambda$) the problem can be solved in an electrostatic approximation. To further simplify the problem we replace the ellipsoid by a point dipole of magnitude[6] $\vec{\mu}_E = \alpha_E \vec{E}$ with

$$\alpha_E = \frac{-ab^2\varepsilon_0}{3} \frac{1-\varepsilon/\varepsilon_0}{1-(1-\varepsilon/\varepsilon_0)A} \quad . \tag{1}$$

Here ε and ε_0 are the dielectric constants of the ellipsoid material and of the surrounding medium respectively; A is a depolarization factor given by

$$A = \frac{ab^2}{2} \int \frac{ds}{(s+a^2)^{3/2}(s+b^2)} \quad .$$

Equation (1) contains a resonance denominator and at resonance the dipolar field of the ellipsoid becomes very large and induces a large Raman molecular polarization. The resulting molecular dipole moment μ_m oscillating at the Stokes frequency ω_S is given by $\mu_m(\omega_S) = 2\alpha R\alpha E(\omega_L)E_L/\epsilon_0 r^3$. Here ω_L is the laser frequency and r is the distance from the center of the ellipsoid to the molecule. The field of the molecular dipole in turn polarizes the ellipsoid to produce an ellipsoid dipole at the Stokes frequency, $\mu_E(\omega_S)=2\alpha_E(\omega_S)\mu_m(\omega_S)/\epsilon_0 r^3$, which is larger than the usual Raman molecular dipole by the factor

$$f = \frac{4}{9} \frac{[1-\epsilon(\omega_s)/\epsilon_0]}{[1-(1-\epsilon(\omega_s)/\epsilon_0)A]} \frac{[1-\epsilon(\omega_L)/\epsilon_0]}{[1-(1-\epsilon(\omega_L)/\epsilon_0)A]} \left[\frac{ab^2}{r^3}\right]^2 . \tag{2}$$

The net enhancement of the Raman intensity is given by $|f|^2$. This expression is essentially the same as that given in reference 5. The major difference between the result given in equation (2) and the solution to the correct boundary value problem for the case of an ellipsoid is the enhancement caused by the concentration of the field around the tips of the ellipsoid by what Gersten and Nitzan have referred to as the "lightening rod" effect. Recently LIAO and WOKAUN have shown that this effect can be expressed as a simple factor

$$\gamma^2 = \left\{\frac{3}{2}\left(\frac{a}{b}\right)^2(1-A)\right\}^2$$

with which equation (2) should be multiplied. This "lightening rod" effect is an important one. For a 3:1 aspect ratio ellipsoid it is responsible for $\gamma^4 = 2\times10^4$ of the total enhancement for molecules located at the ellipsoid tips.

The dielectric properties of silver are such that the resonance at $\epsilon \simeq \epsilon_0(1-1/A)$ occurs in or near the visible region of the spectrum with the exact value dependent on the particle shape through the factor A. For $\epsilon_0 = 1$, a silver sphere would have a resonance at 3.5eV i.e., in the ultraviolet part of the spectrum near 350nm. A 3:1 ellipsoid is resonant at $\epsilon = -8.25\epsilon_0$ which would imply the resonance is shifted to $\sim$ 5000Å for such silver ellipsoids when $\epsilon_0 = 1$. The resonances are quite intense since the imaginary part of ϵ (i.e., ϵ_2) is near zero in the visible region of the spectrum. These examples show that as the ellipsoid is made more needlelike, the particle plasmon resonance corresponding to electron motion along the major axis is shifted toward longer wavelengths. Increasing ϵ_0, for example by dipping the surface into inert liquids, will also shift the resonance to longer wavelengths. Both of these predictions are borne out by our lithographically produced samples.

In order to obtain a good comparison with the particle plasmon theory we use microlithographic techniques to fabricate regular arrays of isolated uniformly sized and shaped silver particles of 100nm dimension.

The fabrication steps are shown in Fig. 1. A silicon wafer with a 500nm thick thermally grown oxide layer is coated with 300Å of chrome and 1000Å of photoresist as shown in Fig. 1a. The photoresist is patterned by exposing it to a 300nm period interference pattern formed by two 325nm He-Cd laser beams. Two exposures are required, in between which the sample is rotated by 90°, to create a crossed grating pattern. After development, the resulting array of resist posts (Fig. 1b) is used as a mask for argon ion milling a chrome pattern. Argon ions of 500eV energy sputter away chrome which is not protected by photoresist. Chrome makes a durable mask for the final highly directional reactive plasma etching[8] of the SiO_2 in a $CH F_3$ low pressure plasma.

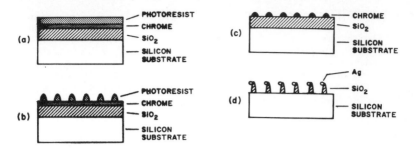

Fig. 1 Lithographic steps for production of microstructure surface

The resulting substrate pattern consists of an array of tall SiO_2 posts on a silicon wafer. By evaporating silver at grazing incidence onto this substrate we obtain isolated silver particles on the tops of the posts. The aspect ratio of the silver particles can be varied by adjusting the angle of evaporation.

After silver is deposited onto the substrate, a monolayer of CN molecules is adsorbed onto the silver by exposing the sample to HCN vapor. Raman light, Stokes shifted by the 2144 cm^{-1} CN stretch mode, is then excited with 20-100mW of laser light from either an argon ion or Rh6G dye laser and measured through a double monochromator, photon counting system set for 8 cm^{-1} resolution. The samples are mounted at 60^0 to the horizontally polarized incident laser beam.

In Fig. 2 the variation of the measured normalized peak Raman intensity versus excitation photon energy is shown for the sample in a nitrogen atmosphere. The normalization eliminates the ω^4 density of states factor in the expression for the Raman cross section, the laser intensity and the detector sensitivity. Data is shown for two different aspect ratio particles. The open circles correspond to the $\sim$ 3:1 aspect particles, while the solid points were taken with $\sim$ 2:1 aspect ratio particles. In each case one clear reso-

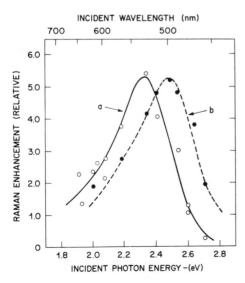

Fig. 2 Dependence of the Raman signal on the aspect ratio of the silver ellipsoids. The normalized Raman intensity of the CN (2144 cm^{-1}) vibration in nitrogen is shown as a function of incident photon energy.
(a) 3:1 aspect ratio ellipsoids.
(b) 2:1 aspect ratio ellipsoids.

nance is observed. Evaluation of the complete expression for the enhancement, including the lightning rod effect, would predict two narrow peaks separated by ω_{vib} = 0.27eV. One peak corresponds to the resonance of the incident light and one to the resonance of the emitted Raman light. However, if one broadens the theoretical resonances by increasing ε_2 to 2.0, or if one assumes our particles have a slight distribution (±12%) of aspect ratios, the two peaks coalesce into one, and the width, location and shift of the peak can be fit to the data yielding an aspect ratio of 3.9:1 for the open circle data and 3.2:1 for the solid points. These values compare favorably with the observed ratios, especially considering that we have neglected the effects of inter-actions between particles, the presence of the SiO_2 posts, and deviations from ellipsoidal shape.

Calibration of the absolute detection sensitivity of our apparatus against the known output of a standard tungsten lamp indicates a $\sim 10^7$ enhancement of the CN Raman cross section at the resonant peak (assuming a Raman cross section of 3 x 10^{-30} cm^2 and one monolayer coverage[4] of 10^{15} cm^{-2}).

A further test of the particle plasmon model can be made by varying the die-lectric constant of the surrounding medium. As we discussed earlier, if ε_0 is increased, the resonance should shift to the red. In Fig. 3 we see that this behavior is indeed verified as the sample was immersed in either water (ε_0=1.77) or cyclohexane (ε_0=2.04). Unfortunately our dye laser could not be tuned sufficiently to completely resolve the resonances. The predicted resonance for 3.9:1 aspect ellipsoids with a 12% distribution would be at 1.9eV in water and 1.5eV in cyclohexane.

The general utility of these surfaces is immediately evident. Clearly one can design them to enhance the particular wavelengths of interest. One can "tune" the particle resonances to increase the Raman efficiency. Gersten and Nitzan have predicted enhancements as large as 10^{11} for properly sized and shaped particles. Because the spacings between the particles are well con-trolled, these samples are also ideal for the study of inter-particle inter-

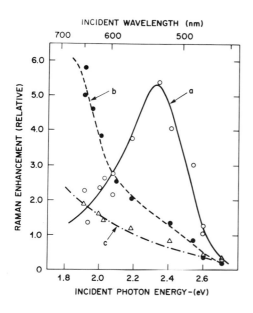

Fig. 3 Dependence of the CN Raman enhancement on the dielectric con-stant (ε_0) of the surrounding medium, for 3:1 ellipsoids. (a) Nitrogen (ε_0=1), (b) H_2O (ε_0=1.77), (c) cyclo-hexane (ε_0=2.04)

actions. Such studies are underway. Other electromagnetic processes are also enhanced by the intense fields near the particles, and for these processes lithographic samples have unique properties. We have found enhancement of luminescence, two-photon absorption, and second harmonic generation. In the case of second harmonic generation, the grating like character of particle arrays resulted in generation of second harmonic radiation in directions into which no fundamental radiation is diffracted.

Acknowledgements

This work has been the result of the efforts and collaboration of many people. At Bell Laboratories J. G. Bergman, D. S. Chemla, A. Glass, J. P. Heritage, C. V. Shank, T. H. Wood and A. Wokaun made especially important contributions. Microfabrication advice was received from L. D. Jackel, E. Hu and R. E. Howard of Bell Laboratories. The microstructures were made during a visit by the author to the microstructures laboratory at M.I.T. where J. Melngailis, A. Hawryluk and N. P. Economou directly contributed to the final results.

References

1. P. F. Liao, J. G. Bergman, D. S. Chemla, A. Wokaun, J. Melngailis, A. M. Hawryluk and N. P. Economou, Chem. Phys. Lett. to be published, 1981.
2. J. E. Rowe, C. V. Shank, D. A. Zwemer and C. A. Murray, Phys. Rev. Lett. 44:1770 (1980).
3. M. Moscovits, J. Chem. Phys. 69:4159 (1978).
4. S. L. McCall, P. M. Platzman and P. A. Wolff, Phys. Lett. A 77:381 (1980); M. Kerker, D. S. Wang, H. Chen, Appl. Optics 19:4159 (1980); C. Y. Chen and E. Burstein, Phys. Rev. Lett. 45:1287 (1980).
5. J. I. Gersten and A. Nitzan, J. Chem. Phys. 73:3023 (1980).
6. See for example, C. J. F. Bottcher, Theory of Electric Polarization, Vol. 1 (Elsevier Sci. Publ. Co., NY 1973) p. 79.
7. P. F. Liao and A. Wokaun, to be published.
8. H. W. Lehmann and R. Widmer, Appl. Phys. Lett. 32:163 (1978).

Investigation of Molecule–Surface Interaction by Laser-Induced Fluorescence

F. Frenkel, J. Häger, W. Krieger, and H. Walther

Max-Planck-Institut für Quantenoptik
D-8046 Garching, Fed. Rep. of Germany

C.T. Campbell, G. Ertl, H. Kuipers, and J. Segner

Institut für Physikalische Chemie der Universität München
D-8000 München 2, Fed. Rep. of Germany

The study of molecule-surface scattering gives important information on the kinetics of adsorption and desorption, on molecule-surface potentials, and in addition on the mechanisms of catalytic reactions. In principle several scattering processes are possible: elastic and inelastic scattering, trapping/desorption, and adsorption. In most of the experiments reported so far the scattering process has been studied by measuring the angular and the velocity distribution of the scattered particles. In order to get a complete description, however, a knowledge of the internal state distribution is also important. As in the case of gas phase scattering the internal state distribution of surface-scattered molecules can be successfully measured by laser-induced fluorescence. This technique has been applied to measure the rotational distribution of surface-scattered NO molecules [1,2]. Here we report on one of these experiments [1], where in addition the angular distribution of the scattered molecules was obtained.

The experimental setup is shown in Fig. 1. Its central part is the scattering chamber with a rotatable Pt(111) surface. A supersonic molecular beam is produced by expanding 300 Torr NO through a small nozzle. During the experiments a pressure of $\cong 10^{-10}$ Torr could be maintained in the scattering chamber. A rotatable quadrupole mass filter was used for determining the angular distribution of scattered particles. The molecules were excited by the frequency-doubled pulses of an excimer-pumped dye laser. The pulse length was

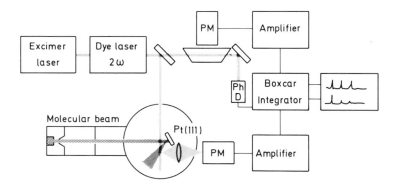

Fig. 1   Schematic drawing of the experimental arrangement (PM = photomultiplier tube, PhD = photodiode)

about 5 ns. About 10 µJ of pulse energy was available at the excitation wave-length of 226 nm. The laser beam could be crossed either with the incoming molecular beam or with the scattered molecules. The fluorescence light from the molecules was focused onto a photomultiplier tube whose signals were amplified, averaged by a boxcar integrator, and recorded on a strip chart recorder. The fluorescence signals from an NO-filled cell were used to cor-rect the signals from the beam for changes of the laser intensity and to facilitate identification of the spectral lines.

The NO molecules were excited from the X $^2\Pi_{1/2}(v'' = 0)$ ground state to the A $^2\Sigma^+(v' = 0)$ state. Both states consist of rotational levels with quan-tum numbers J'' and J', respectively, whose energy separations are well known. The measured fluorescence intensity $I_{J'J''}$ is proportional to the Hönl-London factor $S_{J'J''}$ and to the population density $N_{J''}$ of the corres-ponding ground state level:

$$I_{J'J''} \sim [S_{J'J''}/(2J''+1)]N_{J''}.$$

Measured line intensities therefore yield direct information about the popu-lation densities $N_{J''}$.

In thermal equilibrium the population densities $N_{J''}$ follow a Boltzmann distribution:

$$N_{J''} \sim (2J''+1) \exp(E_{J''}/kT_{rot}),$$

where $E_{J''}$ is the energy of the J''th level and T_{rot} the rotational tempera-ture. In a plot of $\ln[N_{J''}/(2J''+1)]$ such a distribution yields a straight line whose slope allows T_{rot} to be determined.

The measurements consisted of taking spectra of the rotational lines by varying the laser frequency. The spectra measured in the incoming molecular beam yielded a Boltzmann distribution with a rotational temperature of 32 K as shown in the inset of Fig. 2. This indicates that in the supersonic beam the rotational degrees of freedom are in thermal equilibrium.

Two sets of data measured in the reflected molecular beam are shown in Fig. 2. They correspond to different surfaces. The upper set of points (Fig. 2(a)) was obtained with a Pt(111) surface kept at a temperature of 290 K. At this temperature an adlayer of NO is formed on the substrate, and the incident NO molecules are scattered at this adsorbed layer. Within the error bars the measured points allow a fit with a straight line, lead-ing to a rotational temperature of the molecules of (280 ± 30) K. This shows that the rotational motion of the molecules is fully accommodated to the surface temperature, indicating that the scattering process is domina-ted by a trapping/desorption mechanism. This interpretation is supported by the angular distribution measurements, which give an almost perfect cosine distribution [3].

The lower set of data (Fig. 2(b)) was obtained with a Pt(111) surface that was covered by a carbon overlayer. The surface temperature was 330 K. Taking into account the smaller error bars the measured points deviate from a

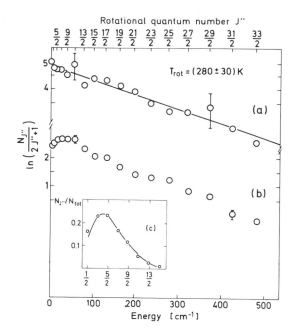

Fig. 2 Plot of $\ln[N_{J''}/(2J''+1)]$ versus energy of the rotational states (a) after scattering from an NO-covered Pt(111) surface (290 K), (b) after scattering from a carbon-covered Pt(111) surface (330 K), (c) rotational energy distribution of incident molecular beam

straight line, especially at small rotational quantum numbers. This effect is attributed to inelastic scattering processes, leading to direct transformation of translational into rotational energy. Theoretical calculations by Nichols and Weare [4] based on a classical two-dimensional model predict a pronounced maximum in the distribution of the scattered molecules at small rotational energies. This maximum can be identified qualitatively with the deviations from a Boltzmann distribution observed in the experiment.

The angular distribution of the molecules scattered at the carbon covered Pt(III) surface is shown in Fig. 3. The broad lobe of scattered particles in the specular direction is a further indication that direct inelastic scattering events are predominant.

Recently also investigation of the NO scattering from pure graphite crystals has been started. The preliminary results show - within the limits of error - a Boltzmann-distribution corresponding to a rotational temperature lower than the surface temperature. This agrees qualitatively with the results obtained for the carbon-covered surface so that it is quite probable that also in the case of the pure graphite crystal direct inelastic scattering events are dominant.

The measurements show that laser-induced fluorescence is applicable to study molecule-surface scattering. The technique offers promising prospects for further investigations on the dynamics of gas-solid interactions. The measurements can easily be extended to include the vibrational degrees of

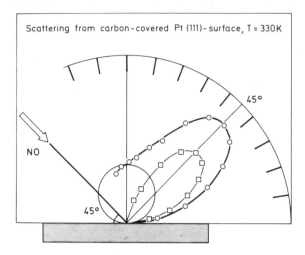

Fig. 3 Angular distribution of NO molecules scattered from a carbon-covered Pt(111) surface (circles). The squares are obtained by subtracting a cosine distribution

freedom as well as changes of the electron angular momentum upon scattering at a surface. In principle measurements of internal energy distributions of catalytic reaction products should also be possible.

References

1. F. Frenkel, J. Häger, W. Krieger, H. Walther, C.T. Campbell, G. Ertl, H. Kuipers, and J. Segner, Phys. Rev. Lett. 46, 152 (1981).

2. G.M. McClelland, G.D. Kubiak, H.G. Rennagel, and R.N. Zare Phys. Rev. Lett. 46, 831 (1981)

3. C.T. Campbell, G. Ertl, and J. Segner, to be published.

4. W.L. Nichols and J.H. Weare, J. Chem. Phys. 66, 1075 (1977)

Hyperfine Studies of Rare Earth Ions Dilute in Optical Solids

L.E. Erickson and K.K. Sharma

Division of Electrical Engineering, National Research Council
Ottawa, K1A OR8, Canada

We would like to describe some recent studies of the hyperfine structure (hfs) of Pr^{3+} dilute in a single crystal of $LiYF_4$. In this work we have observed nqr in an excited state [1], a zero field ground state magnetic resonance line in which resolved structure is due to the magnetic dipole-dipole interaction between the Pr nucleus and neighboring F nuclei, and extra nqr transitions in the presence of an external magnetic field in which both Pr and F nuclei are seen to participate [2,3].

The trivalent Pr ion possesses two equivalent 4f electrons in addition to a closed "Xenon" shell. The Hund ground state is $^3H_4$ and an excited state $^1D_2$ is happily one Rh6G photon higher in energy. In the $LiYF_4$ host crystal, Pr replaces a Y ion and sees a crystal electric field of tetragonal symmetry S_4 which is a small perturbation of the "free" ion levels and partially lifts the 2J+1 degeneracy. The lowest states of $^3H_4$ and $^1D_2$ are non-degenerate. Pr has only one naturally occurring isotope and it has a nuclear spin of 5/2. The Hamiltonian for this system was given by TEPLOV [4]

$$H=D[I_z^2 - 1/3\, I\,(I+1)]+E\,[I_x^2 - I_y^2] + h\sum_i \gamma_i H_i I_i \ , \quad i=x,y,z \qquad (1)$$

where D and E represent the sum of three interactions that due to (1) the second order magnetic hyperfine interaction, (2) the quadrupole moment of Pr and the field gradient produced by the 4f electrons and (3) the quadrupole moment of Pr and the field gradient of the lattice charges. In S_4 symmetry, E=0 and each state will consist of three hyperfine levels of separation 2D and 4D in the absence of an external static magnetic field. That is two nqr lines should be observed, one at twice the frequency of the other. The second order magnetic hyperfine interaction contributes an anisotropic term to the gyromagnetic ratio tensor [4].

The experiment is simple in concept and execution. A single crystal of $LiYF_4$ doped with 0.1 at .% Pr is cooled to 4.5K and illuminated with a single-mode frequency-stabilized CW dye laser beam which is resonant with the ground state and the lowest state of $^1D_2$. The crystal fluorescence quickly falls to about 1/20 of its initial value. The optical pumping process "burns" a hole in the inhomogeneously broadened absorption line, by selectively pumping a hyperfine state of an ion resonant with the laser. Other hyperfine states are pumped but they correspond to different ions in different strain fields of the crystal. The excitation is distributed between the pumped hf state of the lowest excited state $^1D_2$ and other hf states of the ground state. The result is a very large polarization of the hf levels. If the crystal is bathed in an rf magnetic field ($\sim$0.1 Gauss rms), whose frequency is swept slowly, the crystal luminescence is observed to decrease slightly at 5123 kHz (Fig. 1a) and 10245 kHz, and to increase substantially at 9488 kHz (Fig. 1b) and 18949 kHz. The former is nqr in the excited $^1D_2$ state, and the latter is nqr of ground state. This optical-rf method is so sensitive that you may observe nuclear resonance of 10^{10} spins by looking at the crystal with your eye in a lighted room, and noting the change in fluorescence as the rf magnetic field is turned off and on.

The measured nqr frequencies were used to obtain D and E for the two electronic levels. These are given in Table 1.

Table 1 Best Fit Hamiltonian Parameters for Pr^{3+} in $LiYF_4$.

	D [kHz]	E [kHz]	$\gamma_z/2\pi$ [kHz/G]	$\gamma_x/2\pi$ [kHz/G]
Ground state	4738.3 ±0.2	52 ±13	-1.924 ±0.005	-8.22 ±0.2
Excited State	2561.2 ±0.6	12 ± 8	-1.653 ±0.020	-1.480 ±0.020

As expected for S_4 symmetry, E=0. One may also study these transactions in a weak static magnetic field. The Hamiltonian (1) was fitted to such measurements, made in a 30 - 50 Gauss steerable magnetic field, and the results are also given in Table 1. The z-axis of the axial gyromagnetic ratio tensor was found to coincide with the crystal c-axis within the accuracy of the experiment (±1 deg.).

The data for the excited state were used together with the crystal field wavefunctions of MORRISON et al [5], and the Hamiltonian (1) to derive a value for the hyperfine constant A_J=616±51 MHZ for the excited $^1D_2$ state. This "measured" value compares to A_J=754 MHz obtained from the formulae of WYBOURNE [6] with corrections for intermediate coupling. The data were also used to obtain the constituents of D [2].

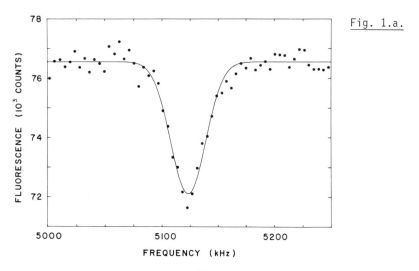

Fig. 1.a.

Fig. 1 Zero magnetic field Pr^{3+} nqr transitions I_z=1/2 - I_z=3/2 for (a) the lowest level of the excited $(^1D_2)$ state and (b) the ground state $(^3H_4)$.

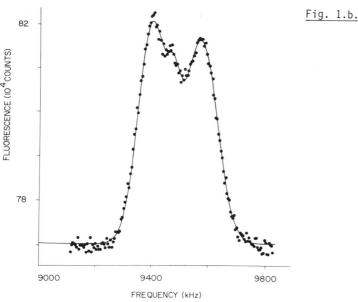

Fig. 1.b.

The low frequency nqr line of the ground state in zero field shows structure, while that of the excited state is featureless and much narrower. The respective high frequency peaks show no structure. This structure is due to the magnetic dipole-dipole interaction of the Pr nucleus with the neighboring F nuclei, which included in the quadrupole Hamiltonian [2,3] gives

$$H = D[S_z^2 - 1/3\,S\,(S+1)] + a S_z I_z - a(\gamma_x/4\gamma_z)(S_+ I_- + S_- I_+) + b[S_z I_+ + (\gamma_x/\gamma_z) I_z S_+]$$

$$+ b^*[S_z I_- + (\gamma_x/\gamma_z) I_z S_-] + c(\gamma_x/\gamma_z) S_+ I_+ + c^*(\gamma_x/\gamma_z) S_- I_-$$

$$- h\left(\frac{\gamma_z}{2\pi} S_z + \frac{\gamma_F}{2\pi} I_z\right) H \cos\theta - h\left(\frac{\gamma_z}{2\pi} S_x + \frac{\gamma_F}{2\pi} I_x\right) H \sin\theta \tag{2}$$

where

$$a = (h^2/r^3)\gamma_z\gamma_F(1 - 3\cos^2\theta') \tag{3}$$
$$b = -(3/2)(h^2/r^3)\gamma_z\gamma_F \cos\theta' \sin\theta' e^{-i\phi'} \tag{4}$$
$$c = -(3/4)(h^2/r^3)\gamma_z\gamma_F \sin^2\theta' e^{-2i\phi'} \tag{5}$$

where S, I are the Pr and F nuclear spin, and θ' and ϕ' give the orientation of the Pr-F vector in the coordinates of the principal axis system. The parameters a, b and c were calculated from the crystallographic data and the measured gyromagnetic ratios. In doing this, two sets of a, b and c were obtained corresponding to two near neighbor F nuclei. The results are shown in Table 2 together with an experimental set which fits Fig. 1a. If we consider the sign of a and the ratios a/b, a/c, it appears that the experimental data correspond to an FII interaction.

Table 2 Dipole-Dipole Coupling Constants for the $LiYF_4:Pr^{3+}$ Ground State

	FI	FII	Zero Field	50 Gauss
a	2.5	-3.0	-10	-2 kHz
b	3.7	2.9	- 7	3 kHz
c	3.8	1.3	- 3.5	- kHz

The absolute magnitude differ by a factor of three, which would indicate the model is too simple for the zero field case.

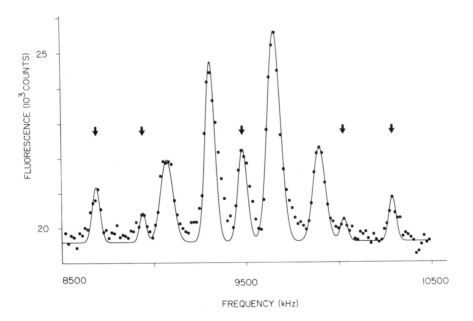

Fig. 2 Ground state nqr 2D transitions in a static external magnetic field of 100 Gauss along the crystal c-axis. The transitions marked with arrows are coupled transitions.

The low frequency transition, observed in a magnetic field of 100 Gauss along the z-axis is shown in Fig. 2. In the absence of dipolar interactions, only the central pair of resonance lines would be observed. The other unmarked pair is normally forbidden for the magnetic field along the z-axis. The extra lines, marked by arrows, are coupled transitions, involving simultaneous "flipping" of Pr and F nuclear spins. Many similar measurements in a field of 50 Gauss are summarized in Fig. 3. Because of the nature of the data, we obtained a fit of the Hamiltonian to the high frequency data using a perturbation approach and using those parameters in the exact Hamiltonian obtained the solid lines shown in Fig. 3. The solid lines appear to describe the experimental situation. These 50 Gauss parameters, shown in Table 2, are similar in magnitude to the calculated a and b for FII. The curve positions are insensitive to c in the presence of a field. Fig. 3 gives a clear demonstration of the dipole-dipole nature of the extra transitions. In Fig. 2, only nine of sixteen transitions are visible because of

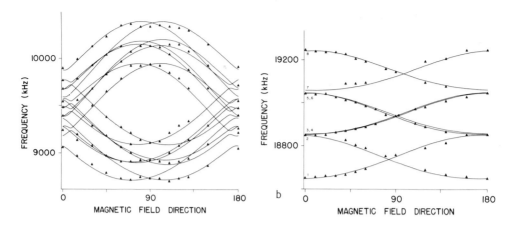

Fig. 3 A summary of ground state nqr measurements in a magnetic field of 50 Gauss as a function of the direction of the field in the xz plane for the (a) 2D and (b) 4D transitions.

insufficient resolution. In a field of 350 Gauss, we have observed thirteen transitions. Clearly, a better fit would be obtained using high field data.

In summary, we have demonstrated that optical-rf studies of rare earth hfs make it possible to study interesting systems which have been inaccessable because the conventional techniques lacked sensitivity.

References

1. K. K. Sharma and L. E. Erickson, Phys. Rev. Lett. 45, 294 (1981)
2. K. K. Sharma and L. E. Erickson, Phys. Rev. B23, 69 (1981)
3. K. K. Sharma and L. E. Erickson, J. Phys. C: Solid State Phys. 14, 1329 (1981)
4. M. A. Teplov, Zh. Eksp. Teor. Fiz. 53, 1510 (1967)(Sov. Phys. JETP 26, 872 (1968))
5. L. Esterowitz, F. J. Bartoli, R. E. Allen, D. E. Wortman, C. A. Morrison and R. P. Leavitt, Phys. Rev. B19, 6442 (1979)
6. B. G. Wybourne, Spectroscopic Properties of Rare Earths, Interscience 1965.

Part X
Vacuum Ultraviolet

Anti-Stokes Scattering as an XUV Radiation Source*

S.E. Harris, R.W. Falcone, M. Gross[†], R. Normandin, K.D. Pedrotti,
J.E. Rothenberg, J.C. Wang, J.R. Willison, and J.F. Young

Edward L. Ginzton Laboratory, Stanford University
Stanford, CA 94305, USA

1. Introduction

Laser induced scattering from atoms stored in a metastable level may be used to produce an XUV radiation source with several unique properties: narrow linewidth, tunability, linear polarization, picosecond pulsewidth, and relatively high peak spectral brightness [1].

In this paper we summarize progress on the use of this radiation source as a unique instrument for XUV spectroscopy, and as a possible flashlamp for a 200 Å laser.

A step in the development of this source was made by ZYCH et al. [2], who demonstrated that the maximum spectral brightness of the source is that of a blackbody with a temperature of the metastable storage level. In their experiments (Fig.1) a 100 ps long pulse of 1.06 µ radiation was incident on a He glow discharge with a metastable population of about 5×10^{11} atoms/cm^3. The peak intensity of the laser emission at 569 Å and 637 Å was compared to that of the resonance lines at 584 Å and 537 Å. Results are shown in Table 1, where it is seen that the peak count rate and peak spectral brightness of the anti-Stokes emission at 569 Å are 17 times and 139 times greater than those of the 584 Å resonance line.

The first spectroscopic use of the anti-Stokes radiation source was demonstrated by FALCONE et al. [3], who used it to resolve the isotopic splitting of the $2s\,^1S_0$ levels of ^{3}He and ^{4}He. The experiment was done in an internal configuration, with a known target level of Ne acting as a narrow band detector of generated anti-Stokes radiation from each of the He isotopes.

2. Absorption Spectroscopy of the $3p^6$ Shell of Potassium

In the last several months we have used the anti-Stokes process as an external source of radiation for high resolution absorption spectroscopy of the $3p^6$ shell of K [4]. A schematic of the apparatus is shown in Fig.2. Inci-

* This work was jointly supported by the Office of Naval Research, the Air Force Office of Scientific Research, the Army Research Office, and the National Aeronautics and Space Administration.

† Laboratoire de Spectroscopie Hertzienne, Ecole Normale Superieure, Paris, France.

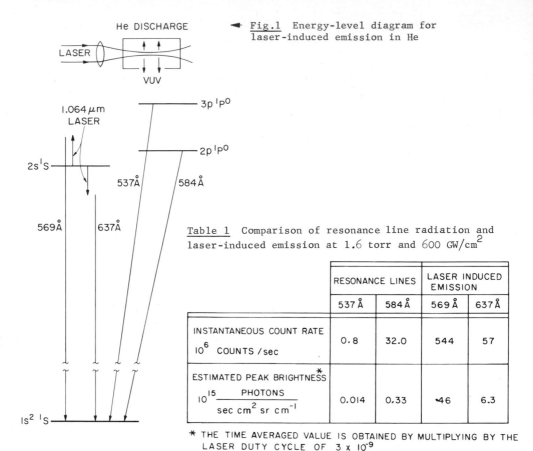

He DISCHARGE

LASER

VUV

◄ Fig.1 Energy-level diagram for laser-induced emission in He

1.064 μm LASER

3p $^1P^0$

2p $^1P^0$

2s ^{1}S

537Å   584Å

569Å   637Å

1s^2 ^{1}S

Table 1 Comparison of resonance line radiation and laser-induced emission at 1.6 torr and 600 GW/cm^2

	RESONANCE LINES		LASER INDUCED EMISSION	
	537Å	584Å	569Å	637Å
INSTANTANEOUS COUNT RATE 10^6 COUNTS /sec	0.8	32.0	544	57
ESTIMATED PEAK BRIGHTNESS* 10^{15} $\frac{PHOTONS}{sec\ cm^2\ sr\ cm^{-1}}$	0.014	0.33	46	6.3

* THE TIME AVERAGED VALUE IS OBTAINED BY MULTIPLYING BY THE LASER DUTY CYCLE OF 3 x 10^{-9}

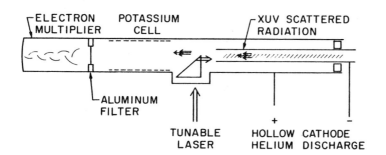

ELECTRON MULTIPLIER   POTASSIUM CELL   XUV SCATTERED RADIATION

ALUMINUM FILTER

TUNABLE LASER   HOLLOW CATHODE HELIUM DISCHARGE

Fig.2 Schematic of apparatus used for absorption spectroscopy of potassium

dent dye laser radiation tuned from 5990 Å to 4880 Å results in generated anti-Stokes radiation in the 182,975 cm^{-1} — 186,769 cm^{-1} spectral region. The generated radiation passes through a K absorption cell, a thin aluminum filter, and is incident on an electron multiplier. For an incident laser pulse energy of 50 mJ, at 6000 Å, about 10^8 XUV photons per pulse are produced in the 60 cm discharge length. The solid angle of the detector reduces the effective flux to about 800 photons per pulse, of which about 10% pass through the aluminum filter. At ten pulses per second our measured signal, assuming a 10% detector quantum efficiency, corresponds to about 200 photons per second to the detector.

A typical absorption scan of K is shown in Fig.3. The level of the background emission from the plasma with the tunable laser blocked is shown to the left of the dashed line. In most cases the absorption of the observed K features brought the transmission back down to this background level, indicating that any other laser induced emission from the plasma is not of significant amplitude.

The broader lines in Fig.3 have been observed earlier by MANSFIELD [5]. The narrower lines have not been previously reported. The narrowest of these has a width of 1.9 cm^{-1}, which is not much larger than the theoretical linewidth of the XUV radiation source. This width corresponds to an autoionizing time of greater than 3 ps, indicating a transition to an upper level which in the approximation of LS coupling is forbidden to autoionize.

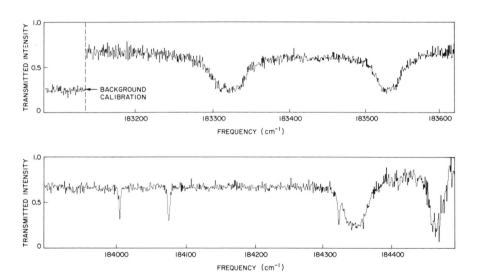

<u>Fig.3</u> Absorption scans of potassium. Vapor pressure is 10^{15} atoms/cm^3, and cell length is 5 cm.

3. Emission Spectroscopy of Lithium

In Li, LS coupling holds sufficiently well that doubly excited inner shell levels which are forbidden by Coulombic selection rules from autoionizing may have radiative branching ratios which approach unity. XUV transitions which originate from doubly excited even parity ^2P levels and odd parity $^2D^o$ levels are good candidates for lasers in the 200 Å spectral region [6]. One of the strongest of these transitions, $1s2p^2\ ^2P \rightarrow 1s^22p\ ^2P^o$, occurs at 207 Å (Fig.4). It has a calculated oscillator strength, Einstein A coefficient, and branching ratio of $f_{12} = 0.13$, $A_{21} = 1.97 \times 10^{10}$ sec^{-1}, and 0.83, respectively [7]. The terminal level of the transition ($1s^22p^2\ ^2P^o$) may be emptied by ionization by an incident laser pulse.

The emission spectra of a microwave heated Li plasma was obtained with a heat-pipe waveguide, as shown schematically in Fig.5. The plasma was driven with a pulsed magnetron operating at 9.4 GHz with a peak power of about 200 kW, a 2 μs pulsewidth, and a 448 Hz repetition rate. The observed spectra are shown in Fig.6 [8]. Strong transitions of the type $1s2p^2\ ^2P \rightarrow 1s^2np\ ^2P^o$, as well as radiative autoionization $1s2p^2\ ^2P \rightarrow 1s^2\ ^1S + \hbar\omega + e^-$ is observed.

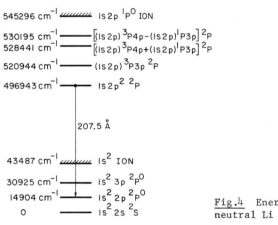

Fig.4 Energy level diagram of neutral Li (from [7])

LITHIUM HEAT–PIPE WAVEGUIDE DETAIL

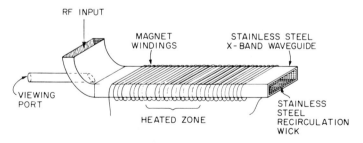

Fig.5 Schematic of heat-pipe waveguide

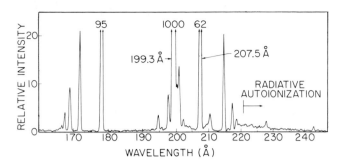

Fig.6 Emission spectrum of Li

A confirmation of the sufficiency of the coulombic selection rules is obtained by noting the absence of the $1s2p^2$ ^2P level as observed by ejected electron spectroscopy. By contrast, the $1s2p^2$ ^2S and $1s2p^2$ ^2P levels, though of the same configuration and parity, are allowed to autoionize and so are seen in the ejected-electron spectra, but not observed here.

In a recent experiment, J. R. WILLISON has observed anti-Stokes scattering from the $Li^+(1s2s)$ level. The apparatus consists of a microwave excited, ridged waveguide with a side viewing port, an aluminum filter, and an electron multiplier. A tuning range of about ± 100 cm^{-1}, centered about the $Li^+(1s2s - 1s2p)$ transition at 9581 Å is observed.

4. 200 Å Laser — Metastable Storage

The strongest of the potential laser transitions discussed in the preceding section have radiative lifetimes of about 50 ps. At a Doppler width of 4 cm^{-1} their gain cross section is ~ 5×10^{-14} cm^2. An excited level atom density of 10^{12} atoms/cm^3 will produce a gain of 5%/cm or e^{15} in a length of 3 m. To attain an excited state density of 10^{12} atoms/cm^3 in less than 50 ps appears prohibitively difficult. Our proposals are based on obtaining this density in a metastable level of Li or Li$^+$ and then using an intense picosecond laser pulse to transfer the population to the selected target level. The same laser or an additional laser would be used to deplete the terminal level.

Metastable levels which are appropriate for storage are the low lying quartet levels of neutral Li, for example $(1s2s2p$ $^4P^o_{5/2})$, and the singlet and triplet levels of Li$^+$. A discussion of storage and transfer from quartet levels of the alkali atoms is given in [6] and [9]; here we limit the discussion to radiative transfer from the ion.

An example of radiative (anti-Stokes) flashlamp pumping of a 199.7 Å laser is shown in Fig.7. The flashlamp is fired with a 1.06 μ pulse, which has a detuning of 1038 cm^{-1} from the $1s2s$ ^1S - $1s2p$ ^1P Li$^+$ transition. Assuming an excited $1s2s$ ^1S ion density of 3×10^{12} ions/cm^3 and a ground level ion density of 3×10^{14} ions/cm^3, the two-photon blackbody limit for the flashlamp power density in a 4 cm^{-1} bandwidth is 3.8×10^4 W/cm^2. Optical frequency shifts will probably limit the allowable 1.06 μ power density to 4×10^{10} W/cm^2. For a 3 mm beam radius, and therefore a peak 1.06 μ power of about 10^{10} W, the XUV flashlamp intensity is 9.3×10^3 W/cm^2.

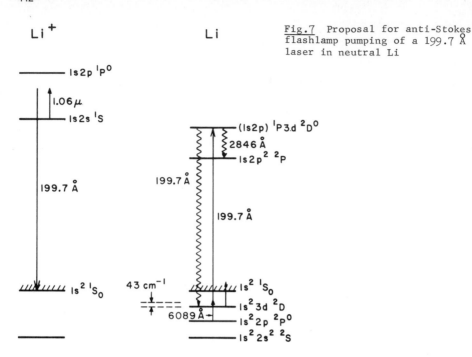

Fig.7 Proposal for anti-Stokes
flashlamp pumping of a 199.7 Å
laser in neutral Li

The flashlamp is absorbed by two photon pumping of the $(1s2p)$ ^1P $2d$ $^2D^o$ level [10] from (discharge produced) population in the upper level of the Li resonance line. The required laser wavelength is 6089 Å. Due to the small detuning (43 cm^{-1}), the power and energy requirements on this laser are sufficiently modest that a long pulsed dye laser may be used. For a $1s^2 2p$ ^2P population of 8×10^{15} atoms/cm^3, a 2 mm diameter beam with a pulse width of 5 ns requires a power and energy of 2.5×10^5 W and 1.2 mJ.

For the conditions of the last two paragraphs, and a 100 ps long 1.06 μ pumping pulse, we obtain an upper level laser atom density of about 5×10^{11} atoms/cm^3. Assuming an empty ground level, this corresponds to a gain of exp (10) in 3 m.

We now envision that the laser pulse which is used to empty the ground $(1s^2 3d$ $^2D)$ level will follow the anti-Stokes pumping pulse. There are two reasons for this: One may make use of radiative trapping which is expected on the laser transition and thus pump for several times longer than the radiative lifetime. Also, the possibility of reducing the blackbody temperature of the anti-Stokes radiation from the Li ion (as the 3d level is dumped) is avoided. The wavelength window which allows emptying the terminal laser level, without emptying the upper level, is 8194 Å - 7554 Å. A possibility is to Stokes shift 5320 Å, first in N$_2$, then in H$_2$ to obtain 8120 Å. The cross section for photoionization of 3d atoms at this wavelength is 1.8×10^{-17} cm^2 [11]. To reduce the 3d population by a factor of exp $(- 10)$, in a time of 50 ps, therefore requires a power density of 2.7×10^9 W/cm^2. To decisively overcome 2p-3d electron excitation may require a power density which is an order of magnitude larger.

A 199.7 Å laser of this type would probably have an output energy and power of about 0.1 mJ and 10^6 W, and could readily operate at 10 pps.

5. Hollow Cathode Discharge

A key step toward the realization of the discharge conditions in the preceding section has been made by R. W. FALCONE [12]. A pulsed, low induc-
tance hollow cathode discharge with an active length of about 10 cm has been
built and studied (Fig.8). The discharge operates as a heat pipe with a
ground level Li density of about 1 torr. Operating voltage and current are
about 3 kV and 1.5 kA. Lithium populations are measured by absorption of a
laser probe beam. A typical absorption scan of the Li^+ 1s2s $^3S \rightarrow$ 1s2s ^3P
transition is shown in Fig.9. A summary of measured discharge conditions
is given in Table 2.

HOLLOW CATHODE DISCHARGE

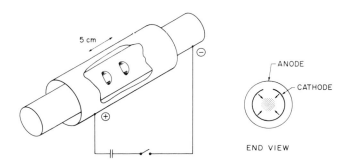

Fig.8   Schematic of pulsed hollow cathode

Li^+ (Is 2s ^3S) ABSORPTION SCAN

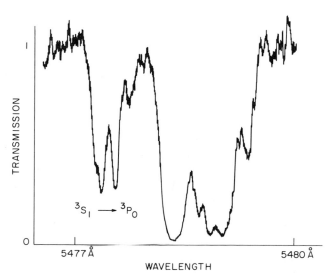

Fig.9 Absorption scan
of triplet population
in Li

Table 2   FALCONE Results

$Li(1s^2 2s)$	2×10^{16} atoms/cm^3
$Li(1s^2 2p)$	8×10^{15}
$Li^+(1s^2)$	2×10^{14}
$Li^+(1s2s\,^1S)$	3×10^{11}
$Li^+(1s2s\,^3S)$	4×10^{12}

As it pertains to the example of the previous section, the blackbody temperature, i.e., the ratio of $Li^+(1s2s\,^1S)$ to $Li^+(1s^2)$ atoms, is about an order of magnitude lower than required.

6. Laser Designation

To reach the proposed upper laser level $(1s2p)\,^1P\,3d\,^2D^o$ from ground requires excitation of two electrons. Normally, fluorescence at 2846 Å (Fig.7) is not observed in our discharge. By tuning a laser to the $1s^2 2p\,^2P^o$ - $1s^2 3d\,^2D$ transition at 6105 Å, a population of about 10^{15} atoms/cm^3 is produced in the $1s^2 3d\,^2D$ level. Single electron excitation (1s - 2p) is then effective in populating $(1s2p)\,^1P\,3d\,^2D^o$; we then see fluorescence at 2846 Å.

7. Summary

Anti-Stokes radiation has potential as a powerful tool for high resolution spectroscopy in the XUV spectral region. By employing discharge techniques such as that of Section 5, and using other storage species, extension to selected regions in much of the 100 Å to 1000 Å wavelength range may be possible.

Considerable work on the discharge will be necessary before a laser can be constructed with this technology.

Acknowledgements

The authors thank C. F. BUNGE for preprints of his papers. His spectroscopic calculations [7,10] are invaluable to our work. Helpful discussions with T. LUCATORTO, S. MANSON, and A. WEISS are also gratefully acknowledged.

References

1. S. E. Harris, Appl. Phys. Lett. 31, 498 (1977).
2. L. J. Zych, J. Lukasik, J. F. Young, and S. E. Harris, Phys. Rev. Lett. 40, 1493 (1978).
3. R. W. Falcone, J. R. Willison, J. F. Young, and S. E. Harris, Optics Lett. 3, 162 (1978).
4. Joshua E. Rothenberg, J. F. Young, and S. E. Harris, Optics Lett. (to be published, August 1981).

5. M. W. D. Mansfield, Proc. R. Soc. Lond. A 346, 539 (1975).
6. S. E. Harris, Optics Lett 5, 1 (1980).
7. C. F. Bunge, Phys. Rev. A 19, 936 (1979).
8. J. R. Willison, R. W. Falcone, J. C. Wang, J. F. Young, and S. E. Harris, Phys. Rev. Lett. 44, 1125 (1980).
9. Joshua E. Rothenberg and Stephen E. Harris, IEEE J. Quant. Elect. QE-17, 418 (1981).
10. R. Jauregui and C. F. Bunge, Phys. Rev. A 23, 1618 (1981).
11. S. Manson (personal communication).
12. R. W. Falcone and K. D. Pedrotti, "Pulsed Hollow Cathode Discharge for XUV Lasers" (in preparation).

The Study of Atomic and Molecular Processes with Rare-Gas Halogen Lasers

H. Egger, M. Rothschild, D. Muller, H. Pummer, T. Srinivasan,
J. Zavelovich, and C.K. Rhodes

Department of Physics, University of Illinois at Chicago Circle, P.O.Box 4348
Chicago, IL 60680, USA

Abstract

A discussion of the use of ultraviolet rare-gas halogen lasers for the study
of high-lying electronic states is provided. Specific examples involving
(1) the examination of collisionally assisted nonlinear radiative processes
and (2) the configuration of ultrahigh-spectral-brightness sources are given.

I. Introduction

Atomic and molecular spectroscopy in the vacuum ultraviolet (VUV) region
(λ < 200 nm) provides important data on the properties of highly excited
electronic states. This includes the study of perturbed spectra, processes
leading to dissociation and ionization, and collisional effects. In recent
years, ultraviolet rare-gas halogen (RGH) lasers have proved to be useful
tools for performing studies of this kind on a variety of atoms and molecules.

Since commercially available RGH lasers readily generate output powers
in excess of 10 MW at several ultraviolet frequencies [1], the consideration
of multiquantum amplitudes considerably extends the applicability of these
light sources for the examination of high-lying atomic and molecular states [2].
For example, the nonlinear technique of excitation has been utilized [3,4,5]
to obtain both spectroscopic and kinetic data on states in the 10- to 13-eV
range for H_2, Kr, and CO. In addition, similar methods have been applied to
the study of cryogenic liquids [6] and collisionally assisted multiphoton
absorption [7].

Considerably enhanced flexibility for spectroscopic purposes is achieved
in configurations which optimize the spectral brightness of the RGH systems.
For both KrF* (248 nm) and ArF* (193 nm), it has been possible to generate
essentially transform-limited outputs [8,9] at power levels exceeding 10 MW.
In both cases, the improvement in spectral brightness, as compared to a free-
running oscillator of comparable output power, was greater than 10^9.

In this article, we will describe (1) observations of collision-induced
nonlinear absorption in rare-gas media and (2) ultrahigh-spectral-brightness
RGH sources, two aspects of our current activity.

II. Pressure-Induced Multiphoton Ionization of Rare Gases

In recent studies [7], observations of multiphoton pressure-induced absorption in rare gases have been made using KrF* (248 nm) and XeF* (351 nm) sources. In one set of experiments, a sample consisting of ∿100-900 torr of Kr was irradiated with the focused output of a conventional KrF* (248 nm) laser (Lambda Physik EMG 101). As a consequence of this irradiation, the sample emitted a broadband fluorescence centered at ∿147 nm, which was detected in a direction perpendicular to the 248-nm laser beam. This fluorescence arises from the transition between the lowest excited states of Kr_2^*, 1_u and 0_u^+, and the repulsive part of the ground-state (0_g^+) potential [10]. The integrated Kr_2^* fluorescence was found to be proportional to the third power of the laser intensity, as shown in Fig. 1, and to the square of the Kr pressure, as illustrated in Fig. 2. The observed intensity dependence clearly implies that the 147-nm fluorescence originates from a three-photon process. Indeed, the energy of three 248-nm photons is above the ionization limit [11] of Kr; and the energy pathways through which the resulting Kr^+ evolves to form Kr_2^* efficiently are well established [1,12].

The first two excited states of the Kr atom, $5s[3/2]_2^0$ and $5s[3/2]_1^0$, lie within ∿600 cm^{-1} of the energy of two 248-nm photons (see Fig. 3). For the free atom, however, this two-photon resonance condition does not enhance the amplitude for the three-photon ionization process, since the electric dipole matrix elements connecting the ground state to the 5s intermediate states in the two-quantum amplitude have zero value, because of parity considerations. Interatomic perturbations, of course, can affect the symmetry of the atomic states, enabling the corresponding matrix elements to acquire a non-vanishing value. From the viewpoint of molecular states, the two-quantum resonance acquires an allowed character. Consequently, the amplitude for three-photon ionization becomes sensitive to the details of the molecular structure, including both the position of the molecular curves and the variation of the matrix elements with internuclear separation.

The observed quadratic pressure dependence of the 147-nm fluorescence illustrated in Fig. 2 can be understood in light of the molecular picture [13]. On this basis it is expected that the two-quantum resonant enhancement contributes appreciably to the amplitude at internuclear separations R ≤ 5 A, values for which the molecular states deviate significantly from the corresponding atomic asymptotes (see Fig. 3). Thus, the excitation occurs principally in Kr_2 molecules, either bound or free, rather than in Kr atoms. Furthermore, the energy mismatch of ∿600 cm^{-1} in the Kr atoms is considerably reduced at R ≈ 4 A, by the combined effect of upward curvature of the 1_g and 2_g states [14] and the van der Waals well in the ground state [15].

If the collision partner of the Kr is a dissimilar system, such as Ar or Ne, several aspects of the two-quantum resonant enhancement are expected to change, specifically (1) the variations in well depth and equilibrium position [15,16] of the van der Waals ground-state potential, (2) the shift in the position of the excited states, (3) the alterations in the participating matrix· elements, and (4) the relaxation of the g↔u selection rule operative in the homonuclear system Kr_2. Depending on the relative importance of these factors, the resulting three-photon ionization rate in Kr/Ar and Kr/Ne mixtures can be either greater or less than that characteristic of krypton alone. Experimentally, compared to the ionization rate of the Kr_2 system, we observed an enhancement of ∿1.6 in Kr/Ar and a reduction at least five-fold in Kr/Ne.

Pressure-induced multiquantum phenomena of this nature are not unique to the particular case of Kr at 248 nm. It is expected that a large variety of systems will exhibit similar or closely related effects. For instance, in similar studies, pressure-induced four-photon ionization of Xe by a XeF* (351 nm) laser has also been observed. In this case, the four-photon process was enhanced by a three-photon near-resonant molecular Xe_2 state which originates from the $7s[3/2]_1^0$ atomic level.

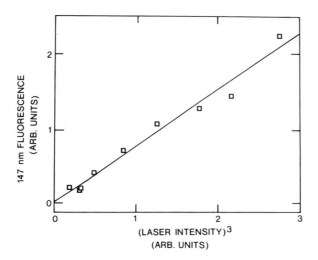

Fig. 1 The integrated 147-nm Kr_2* fluorescence as a function of the third power of the laser intensity. The abscissa is in units of approximately 3.4×10^{29} $(W/cm^2)^3$.

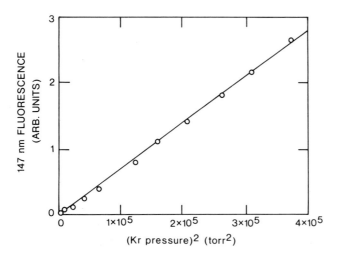

Fig. 2 The integrated 147-nm Kr_2* fluorescence as a function of the square of the Kr pressure

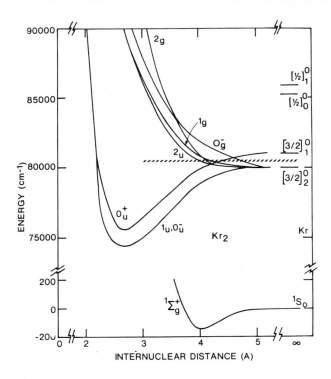

Fig. 3   The energy potentials of Kr_2 and Kr.   The shaded horizontal line corresponds to the energy of two 248-nm photons.

III. High-spectral-brightness Tunable RGH Systems

Although the experiments outlined above were all conducted with free-running RGH oscillators, considerable spectroscopic refinement is both achievable and useful in a wide range of atomic and molecular studies. It has been shown [8,9] that in a proper configuration, it is possible to combine tunability ($\sim$100 cm^{-1}), high output power (>10 MW), a narrow spectral width ($\sim$200 MHz), and a low beam divergence ($\sim$30 μrad), in a single RGH system. Indeed, a spectral brightness very close to the fundamental bound has been demonstrated [8,9] for both KrF* (248 nm) and ArF* (193 nm). Naturally, such improved performance vastly enhances the experimental usefulness of existing excimer lasers, especially in the area of nonlinear multiphoton processes. For example, precise tuning to a particular final state, or through a specific intermediate resonance, sharply alters the excitation rate and reveals considerable spectroscopic information regarding high-lying states. As an example of the utility of such sources for the direct study of molecular states in the XUV range, spectrally sharp and tunable 83-nm radiation, arising from third-harmonic generation [17] of KrF* (248 nm), has been used to analyze quantitatively [18] weakly predissociated levels in H_2, HD, and D_2.

The details of the ArF* (193 nm) system serve to illustrate the nature of the procedures used to achieve high-brightness operation. The overall system is outlined in Fig. 4. The output of a single-frequency, cw dye laser at 580 nm was pulse-amplified in a three-stage XeF* pumped amplifier. The $\sim$10-nsec, 20-mJ visible pulses were focused into a Sr heat pipe [19], making possible the generation of third-harmonic (193 nm) pulses with $\sim$5 nsec duration and 200 mW peak power. These pulses were then amplified in two discharge-pumped ArF* laser amplifiers, to produce $\sim$5-nsec, 30-mJ pulses at 193 nm. The linewidth of these pulses was determined, with the aid of a 1.5-GHz free-spectral-range Fabry-Perot interferometer, to be $\sim$0.01 cm^{-1}. Similarly, the beam divergence was nearly diffraction-limited, $\sim$5 x 15 μrad. In addition, the absolute frequency of the dye laser was known to within 0.02 cm^{-1} by interferometric comparison with a stable HeNe laser [20]. Therefore, tunability within the ArF* gain profile is achieved by setting the dye laser at the appropriate frequency. Special attention was given to suppression of the broadband amplified spontaneous emission, both in the dye amplifier and in the ArF* amplifiers. Adequate discrimination was accomplished by the appropriate use of spatial filtering, and an added grating in the dye amplifier system. In recent experiments, the addition of a final power amplifier stage (Lambda Physik EMG 200) has yielded the generation of pulses with output energies in excess of 300 mJ. Finally, frequency tripling to 64 nm in Kr, H_2, and Ar has been observed.

IV. Summary

Ultraviolet RGH lasers can readily be applied to the measurement of a wide range of electronic physical processes. Since very high peak powers are available, nonlinear amplitudes are particularly useful for the study of states lying in the VUV range.

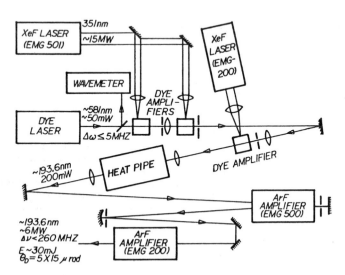

Fig. 4 A schematic outline of the narrow-bandwidth, tunable ArF* laser system

V. Acknowledgments

Support for special equipment for these studies was provided by the Department of Energy under contract no. DE-AC02-79ER10350. Additional support was furnished by the Air Force Office of Scientific Research under grant no. AFOSR-79-0130, the National Science Foundation under grant no. NSF 78-27610, the Dept. of Energy under contract no. CO-280-ET-33065, and the Office of Naval Research. The assistance of J. Wright and S. Vendetta in performing these studies is gratefully acknowledged.

VI. References

1. C. A. Brau, in Excimer Lasers, ed. by C. K. Rhodes, Topics in Applied Physics, Vol.30 (Springer, Berlin, Heidelberg, New York 1979), p. 87

2. W. K. Bischel, J. Bokor, D. J. Kligler, and C. K. Rhodes, IEEE J. Quantum Electron. QE-15, 380 (1979)

3. D. J. Kligler and C. K. Rhodes, Phys. Rev. Lett. 40, 309 (1977); D. J. Kligler, J. Bokor, and C. K. Rhodes, Phys. Rev. A 21, 607 (1980)

4. J. Bokor, J. Zavelovich, and C. K. Rhodes, Phys. Rev. A 21, 1453 (1980)

5. J. Bokor, J. Zavelovich, and C. K. Rhodes, J. Chem. Phys. 72, 965 (1980)

6. M. Wilson and D. Muller, private communication

7. M. Rothschild, W. Gornik, J. Zavelovich, and C. K. Rhodes, J. Chem. Phys. (to be published)

8. R. T. Hawkins, H. Egger, J. Bokor, and C. K. Rhodes, Appl. Phys. Lett. 36, 391 (1980)

9. H. Egger, T. Srinivasan, K. Hohla, H. Scheingraber, C. R. Vidal, H. Pummer, and C. K. Rhodes, Appl. Phys. Lett. (to be published)

10. C. K. Rhodes and P. W. Hoff in Laser Spectroscopy, edited by R. G. Brewer and A. Mooradian (Plenum Press, New York, 1974), p. 113

11. C. E. Moore, Atomic Energy Levels, vol. II, NSRDS-NBS 35 (1971)

12. C. W. Werner, E. V. George, P. W. Hoff, and C. K. Rhodes, IEEE J. Quantum Electron. QE-13, 769 (1977)

13. J. P. Colpa in Physics of High Pressures and Condensed Phases, ed. by A. van Itterbeek (North-Holland Publishing Co., Amsterdam, 1965), p. 490; S. E. Harris, J. F. Young, W. R. Green, R. W. Falcone, J. Lukasik, J. C. White, J. R. Willison, M. D. Wright, and G. A. Zdasiuk in Laser Spectroscopy IV, ed. by H. Walther, K. W. Rothe, Springer Series in Optical Sciences, Vol.21 (Springer, Berlin, Heidelberg, New York 1979), p. 349

14. O. Vallee, N. Tran Minh, and J. Chappelle, J. Chem. Phys. 73, 2784 (1980)

15. Y. Tanaka, K. Yoshino, and D. E. Freeman, J. Chem. Phys. 59, 5160 (1973)

16. C. Y. Ng, Y. T. Lee, and J. A. Barker, J. Chem. Phys. 61, 1996 (1974)

17. H. Egger, R. T. Hawkins, J. Bokor, H. Pummer, M. Rothschild, and C. K. Rhodes, Opt. Lett. 5, 282 (1980)

18. M. Rothschild, H. Egger, R. T. Hawkins, J. Bokor, H. Pummer, and C. K. Rhodes, Phys. Rev. A 23, 206 (1981)

19. C. R. Vidal, Appl. Opt. 19, 3897 (1980)

20. F. W. Kowalski, R. T. Hawkins, and A. L. Schawlow, J. Opt. Soc. Am. 66, 965 (1976)

Optical Second Harmonic Generation by a Single Laser Beam in an Isotropic Medium

R.R. Freeman, J.E. Bjorkholm, R. Panock

4C-330, Bell Laboratories
Holmdel, NJ 07733, USA

and

W.E. Cooke

Department of Physics, University of Southern California
Los Angeles, CA 90007, USA

We have observed optical second harmonic generation (SHG) in sodium vapor, in the absence of any external fields, using a single input laser beam tuned near either the allowed 3s-4d two photon transition or the dipole-forbidden 3s-5p two-photon transition. In both cases a tunable doubling of the laser was observed with surprisingly large conversion (10^{-5} to 10^{-6}). When the laser was tuned to the allowed 3s-5s two-photon transition, no doubling was observed. In this paper we present the results of our experimental investigations of these phenomena, and discuss the results in the context of proposed theoretical models for the effect.

The observation of second harmonic generation in a gaseous vapor is not without precedence. Although several proofs of the impossibility of SHG in an isotropic material have been given, they are restricted to collinear plane waves. In his comprehensive paper on nonlinear interactions, PERSHAN [1] discussed the possibility of using higher order multipole moments to observe SHG in gases by employing non-collinear plane waves. Later, BETHUNE, SMITH, and SHEN [2] demonstrated optical sum-frequency generation in gases via this technique, and several groups have demonstrated SHG in atomic vapors by applying magnetic or electric fields to the vapor. [3]

Only recently there have been reports of SHG in gaseous vapors from a single focused laser beam without the application of external fields. In 1978, MOSSBERG, FLUSBERG, and HARTMANN, reported SHG in thallium vapor [4], and in 1979 MIYAZAKI, SATO, and KASHIWAGI reported weak SHG production in sodium vapor using high intensity (10^{12}W/cm^2) mode-locked 1.06 light [5]. In 1980, HEINRICH and BEHEMBURG reported SHG from a single beam in atomic barium [6], OKADA, FUKUDA and MATSUOKA observed SHG in Li [7] and BOKOR, FREEMAN, PANOCK and WHITE demonstrated the doubling of 2800 Å to 1400 Å with conversions of 10^{-6} in mercury vapor [8].

Our measurements were all made in sodium vapor. A pulsed dye laser was spatially filtered, linearly or circularly polarized, and loosely focused into a cell containing sodium vapor. The output was filtered to remove the fundamental, and spectrally analyzed with a spectrometer. In Fig. 1 the relevant energy levels of the sodium atom are shown. When the laser was tuned near the dipole-allowed two-photon 3s-4d transition at 5787 Å, or near the dipole-forbidden two-photon 3s-5p transition at 5707 Å, doubling was observed. The laser intensities were varied from approximately 10^5W/cm^2 to as high as 10^9W/cm^2. The linewidth of the laser was less than 0.3 cm^{-1}. The doubling was found to be widely tunable about the resonance enhancements (see Fig. 2); at maximum input powers and optimal tunings, conversions greater than 10^{-5} were observed. The SHG output was collimated, polarized, and, to within our

spectrometer resolution (0.1 Å), exactly 1/2 of the input wavelength. We
also observed several other emissions from the cell. A strong coherent
beam at 3302 Å corresponding to the 4p-3s transition was observed, especially
under conditions of pumping the 3s-4d two-photon transition. In addition,
a broad band, incoherent fluorescence was also observed, the exact spectral
nature of which depended upon the details of the detuning from the resonance
enhancements. When we attempted to observe doubling by tuning to the two-
photon allowed 3s-5s transition at 6024 Å, we again observed strong coherent
emission at 3302 Å, with a broad background fluorescence, but no SHG.

In Fig. 2, the relative output power of the SHG as a function of wavelength
is shown. Notice how the output is enhanced asymmetrically about the resonances. In the 3s-5p case there is virtually no output on the short wavelength side of the resonance. The shapes of these tuning curves was found
to be a sensitive function of the vapor density. This dependence and the
shapes of the curves convince us that the observed SHG arises from a bulk
interaction (i.e., bulk dispersion plays a role in the process).

We have made measurements of the intensity dependence of the SHG signal:
for the 3s-4d case we were restricted at low intensities by the scattered
light in the spectrometer from the 3302 Å parametric output. In this case,
at the lowest detectable signal, the output scaled as the square of the
input intensity, saturating to sublinear at the highest intensities. In the
3s-5p case we had less scattered 3302 Å light to contend with and were able
to record output at much lower intensities. In this case we found an initial
dependence greater than I^2 which rapidly reduced to I^2, and then to linear,
with increasing intensity.

The mode of the SHG output was examined in the far field for its spatial
dependence and polarization. The spatial mode was monitored by scanning a
pinhole across the output beam just before the lens used to focus the light
into the spectrometer. In the 3s-4d case, when the input beam was linearly
polarized the output was bimodal in the direction of the input polarization,
and linearly polarized in this direction. When circularly polarized input
light was used, a donut output mode was observed with roughly circular
polarization. In the 3s-5p case, when the input beam was linearly polarized, the output mode was a donut, with linear polarization in the direction of the input polarization. In both cases, the modes of the SHG had
their maxima well outside the mode pattern of the fundamental.

We have been tempted to analyze our results in terms of coupling via quadrupole moments of the sodium atom. For example, the 3s-4d case would proceed via dipole-dipole excitation followed by radiation by the 4d-3s quadrupole. In the 3s-5p case, the process would be dipole-quadrupole with radiation by the 5p-3s dipole. In the 3s-5s case, there is no quadrupole (or
higher order) moment connecting the s-states so that the process is not expected to radiate, as observed. We have calculated the power radiated from
an isotropic medium by an induced quadrupole moment at the second harmonic
frequency, as for the proposed 3s-4d process. The isotropy of the medium
was rigorously invoked. For a focused incident beam (paraxial approximation) linearly polarized in the x-direction and propagating along z, the
second harmonic field at $\vec{r}$ (far-field) radiated from the incremental volume
dV_0, located at $\vec{r}_0$, is

$$d(E_{2\omega}) \propto \left[\frac{x}{z} E_x^2 + E_x E_z \right] e^{i\vec{k}_{2\omega} \cdot (\vec{r} - \vec{r}_0)} dV_0 \quad . \tag{1}$$

The second term in this expression arises from the longitudinal field that a laser beam of finite transverse extent must have. If one neglects this term, and integrates (1) over the interaction region, an expression for the output power, P_0, is obtained that is in close agreement with our results. However, when both terms of (1) are used, the calculated output power is reduced by the factor

$$P_{2\omega}/P_0 = \left| \left(2\vec{k}_\omega - \vec{k}_{2\omega} \right)/2k_\omega \right|^2 \quad ; \tag{2}$$

this result is anywhere from 10^{-6} to 10^{-8} smaller than our observed power.

Recently, BETHUNE [9] has analyzed the various possible mechanisms for SHG in a single beam, and has specifically examined the mechanism proposed by MIYAZAKI [10]. He concluded that MIYAZAKI et al [11] observed SHG due to electric fields arising from charge separation following multiphoton ionization. In a recent paper, BETHUNE [12] has examined how unintentional polarization scrambling or higher order spatial mode patterns in a single beam can lead to quadrupole coupled SHG in a single beam. Recently, MATSUOKA et al. [7] have concluded from their SHG studies in Li that electric fields induced from ionization are responsible for the doubling.

We have performed many experimental tests of the above models for our case and find that none of them explain our results. For example, we have irradiated the sodium with an additional laser at 1.06μ and monitored the increased production of ions while doubling on the 3s-4d transition and attempting to double on the 3s-5s transition. Even with greatly increased ionization due to the intense 1.06μ irradiation, no change in the doubling efficiency was observed on 3s-4d, and no doubling whatsoever was observed on 3s-5s. We have purposely introduced higher order mode structure in our input laser, and have produced completely random polarization modes as well. In all cases we found no change in our output efficiencies for any of the transitions. Finally, as best evidenced in the 3s-5p transition case, the polarization of the output SHG is not along the intensity gradients of the input beam, in direct opposition to the model proposed by MATSUOKA et al. [7].

We are currently working on a mechanism to explain our results that relies upon ionization and quadrupole coupling. The multiphoton ionization that inevitably occurs is proposed to play a role that is distinct from that of creating electric fields due to charge separation. Here a spatial variation of the density of ground state atoms is produced by the combined effects of multiphoton ionization and the spatial variation of the laser beam intensity. Under these conditions, the isotropy of the material is destroyed, and the integral over the elemental contributions in (1) does not give the greatly attenuated result of (2); rather it yields a result that is in rough qualitative agreement with our observations.

In conclusion, we have presented the broad outlines of an experimental study of SHG in isotropic materials with a single laser beam without the imposition of external fields. The process is shown to be efficient (on the order of allowed third-order processes), tunable, and may well find practical applications in the production of deep VUV light. For example, doubling of 2800 Å light to 1400 Å light in mercury has been obtained in our laboratory [8]; JAMROZ, LAROCQUE, and STOICHEFF have observed doubling of 3202 Å to 1601 Å in zinc vapor [13]; and it should be possible to double approximately 1930 Å (ArF* excimer laser) to 965 Å in Kr vapor.

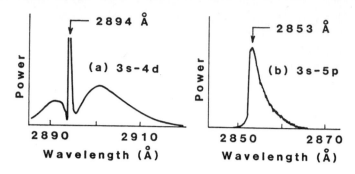

Fig.1 Energy levels of the sodium atom relevant for our experiments. The upward arrows denote the input light at the fundamental frequency and the broad downward arrows indicate the observed second-harmonic radiation. Note that no SHG is observed on the 3s-5s transition.

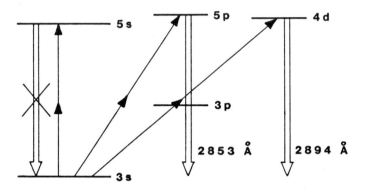

Fig.2 Typical tuning curves for SHG; the second-harmonic output power as a function of the second-harmonic wavelength. Curve (a) is for the 3s-4d transition and is a smoothed version of point-by-point data; continuous scanning of the input wavelength could not be used because of the strong background (3302 Å and fluorescence). Curve (b) is for the 3s-5p transition and was obtained with a continuous scan. The exact shapes of the curves were strongly dependent on the sodium density and laser intensity.

References

1. P. S. Pershan, Phys. Rev. 130, 919 (1963)
2. D. S. Bethune, R. W. Smith, and Y. R. Shen, Phys. Rev. Lett. 37, 431 (1976); Phys. Rev. A 17, 297 (1978)
3. A. Flusberg, T. Mossberg and S. R. Hartmann, Phys. Rev. Lett. 38, 694 (1977); Phys. Rev. Lett. 38, 59 (1977) M. Matsuoka, N. Nakatsuka, H. Uchiki, M. Mitsunaga Phys. Rev. Lett. 38, 894 (1977)
4. T. Mossberg, A. Flusberg and S. R. Hartmann, Optics Comm. 25, 121 (1978)

5. K. Miyazaki, T. Sato, H. Kashiwagi, Phys. Rev. Lett. <u>43</u>, 1154 (1979)
6. J. Heinrich and W. Behmenburg, Appl. Phys. <u>23</u>, 333 (1980)
7. J. Okada, Y. Fukuda and M. Matsuoka, J. Phys. Soc. Japan (to be published)
8. J. Bokor, R. Freeman, R. Panock and J. White, Opt. Lett. <u>6</u>, 182 (1981)
9. D. S. Bethune, Phys. Rev. A <u>23</u>, 3139 (1981)
10. K. Miyazaki, Phys. Rev. A <u>23,</u> 1350 (1981)
11. K. Miyazaki, T. Sato, and <u>H.</u> Kashiwagi, Phys. Rev. A <u>23</u>, 1358 (1981)
12. D. S. Bethune, Opt. Lett. <u>6</u>, 289 (1981)
13. B. P. Stoicheff, person communication

Quantitative VUV Absorption Spectroscopy of Free Ions

T.J. McIlrath

National Bureau of Standards, Washington, D.C. 20234 and
Inst. for Phys. Sci. and Tech., University of Maryland
College Park, MD 20742, USA

and

T.B. Lucatorto

National Bureau of Standards
Washington, D.C. 20234, USA

In this paper we present some results of our studies of Quantitative Vacuum Ultraviolet Absorption Spectroscopy of Free Ions using Laser Driven Ionization. Details of the experiments have been published elsewhere [1]. A metal vapor is contained in a heat-pipe oven with a known density and column length. The resonance line of the atoms is excited with a long pulse ($\approx 1 \mu s$) dye laser at $\approx 1MW/cm^2$ powers. Collisional processes involving free electrons lead to nearly complete ionization ($95\pm5\%$) of the vapor in a time scale which is too short for any migration out of the line of sight. Absorption spectra are obtained photographically in the 10-60nm region with a 3m grazing incidence spectrograph. If the ion has a resonance line which can be reached with an appropriate laser, e.g. Ba, Ca, La, etc., then the first ion can subsequently be pumped to produce complete conversion to the second stage of ionization and so forth. Only one stage of ionization is generally present at one time if the driving laser saturates the resonance transition. The electron excitation temperature depends on the ionization temperature of the species being ionized and is generally $\lesssim 1ev$. Fig.1 shows the absorption spectra of Li during the long pulse excitation as the vapor evolves from entirely neutral Li to $95\pm5\%$ helium-like Li^+. From these data we obtained a photoionization cross section for $Li^+ \rightarrow Li^{++}$ at the series limit (16.39nm) of $2.0\pm1.0Mb$ [2].

Recently we have made studies on Ba where excitation at 553.7nm provides a column of Ba^+ ions and subsequent excitation at 493.5nm provides Ba^{++} ions. The electronic structure of the neutral is $(Kr)4d^{10} 5s^2 5p^6 {}^1S_0$ and the Ba^+ and Ba^{++} have one and two of the $6s^2$ electrons removed respectively. Figure 2 shows the absorption of Ba^{++} between 32 and 38nm where the $5p^6 \rightarrow 5p^5$ (ns,nd) series go to the ${}^2P_{3/2}$ and ${}^2P_{1/2}$ limits of Ba^{+++}. This series can be compared directly with Xe **spectra** for the equivalent series [3] and it is found that the Ba^{++} autoionizing widths between the two limits are much smaller than the Xe widths and the channel interaction is considerably reduced in Ba^{++}. Details of this analysis are being prepared for publication [4].

The excitation of the $4d^{10}$ electrons can be studied in the 10 to 15nm region and fig.3 shows the absorption cross section for Ba, Ba^+ and Ba^{++}. Transitions out of the $4d^{10}$ shell in the rare earths, including Ba, have been of considerable interest for several years to solid state and surface physics researchers because the core electrons are deep enough that the absorption features are essentially atomic in character. The features are easily identified but the photo emission spectra are dependent on the valent state of the atom in the solid and thus these transitions can be used to monitor the atomic valent state in both metals and insulators [5]. Different valent states in solids correspond to chemical removal of valence electrons but the ionic core still interacts with the metallic conduction electrons. In our experiments

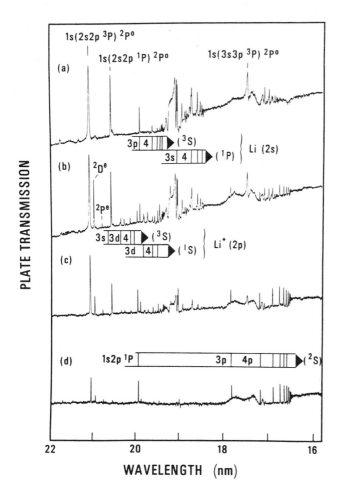

Fig.1 Li absorption spectra during excitation by 900ns laser pulse. (a) Before laser initiation, pure Li (b) 200ns after laser initiation (c) 400ns after laser initiation (d) 600ns after initiation, 95±5% Li⁺. Broad persistant structure from 17 to 18nm is a source feature. Resolved feature at 19.1nm used to monitor neutral density. Bar graphs indicate series with inner shell ($1s^2$) excitation in Li($1s^22s$), Li*($1s^22p$) and Li⁺($1s^2$) going to the Li⁺ and Li⁺⁺ limits respectively.

the valence electrons are removed in the gas phase and the absorption spectra of the free ions studied as a reference for the solid state absorption.

In the case of Ba neutral there are weak absorptions below the $4d^9$ limit which have been identified as transitions to $4d^9$ $4f^3P_1$ and $^3D_1$ levels and to $4d^96p$ levels, but the bulk of the oscillator strength lies above the ionization limit in what has been called a *delayed, giant* or *collective effect* resonance . This has been identified as a transition to $4d^9\varepsilon f^1P$ states and in general it is found that in all cases the bulk of the oscillator strength involves transitions to $n,\varepsilon f^1P$ states. The fact that the discrete transitions

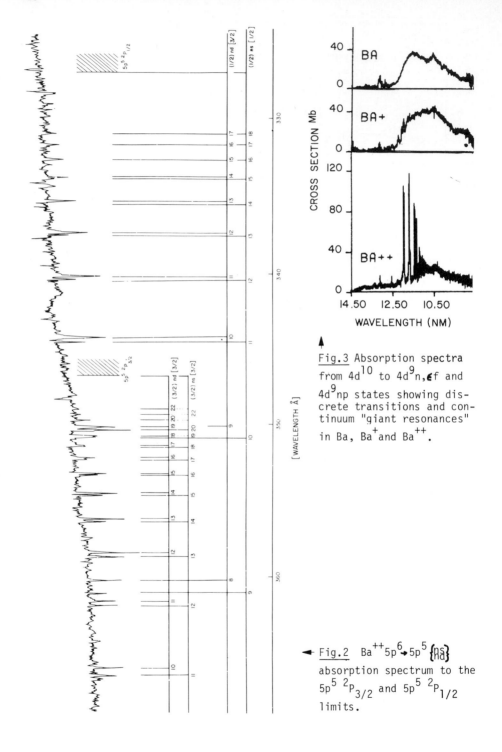

Fig.3 Absorption spectra from 4d^{10} to 4d^{9}n,ϵf and 4d^{9}np states showing discrete transitions and continuum "giant resonances" in Ba, Ba^{+} and Ba^{++}.

Fig.2 Ba^{++}5p^{6}→5p^{5} $\{$ns nd$\}$ absorption spectrum to the 5p^{5} ^{2}P$_{3/2}$ and 5p^{5} ^{2}P$_{1/2}$ limits.

are weak reflects the fact that the bound $nf\,^1P$ states are all concentrated in the outer well of the double well potential and there is little spatial over-lap with the 4d wave functions. Above the ionization threshold the ϵf wave-functions also have little overlap with the 4d wavefunctions until ϵ becomes large enough to surmount the potential barrier and allow penetration into the inner well. Hence the observation of a *delay* in the continuum.

In Ba^{++} the situation is dramatically different. In this case the removal of the shielding valence electrons causes a partial collapse of the $nf\,^1P$ wave functions into the inner well so there is a strong overlap with the 4d electron and extremely strong discrete absorption features with a corresponding reduc-tion of the continuum absorption.

There are a large number of strong discrete features and calculations of the Ba^{++} energy levels lead to the conclusion that the spectrum shows inter-mediate coupling with approximately L-S coupling only for the 4f terms, while at the same time the radial functions are strongly term dependent.

The experimental technique of resonant laser driven ionization has pro-vided the opportunity to study the absorption spectrum of ionic species and to follow changes in core electron transitions as the valence electrons are sequentially removed. The quantitative nature of the experiments allow ac-curate comparisons of the strengths of autoionizing and continuum transitions.

The authors acknowledge the assistance of J. SUGAR, A. WEISS and S. YOUNGER in interpreting the spectra and W.T. HILL III in obtaining spectra.

References

1. T.B. Lucatorto and T.J. McIlrath, Appl. Opt. <u>19</u>, 3948 (1980) and refer-ences therein.

2. T.J. McIlrath and T.B. Lucatorto, Phys. Rev. Lett. 38, 1390 (1977)

3. J. Berkowitz, <u>Photoabsorption</u>, <u>Photoionization</u> <u>and</u> <u>Photoelectron</u> <u>Spect-roscopy</u>, Academic Press, New York, 1979, p. 181.

4. W.T. Hill III, T.B. Lucatorto, T.J. McIlrath and J. Sugar, to be published.

5. L.I. Johansson, J.W. Allen, I. Lindan, M.H. Hecht and S.B.M. Hagström, Phys. Rev. B <u>21</u>, 1408 (1980): T.C. Chiang, D.E. Eastman, F.J. Himpsel, G. Kiandl and M. Aono, Phys. Rev. Lett. <u>23</u>, 1846 (1980).

Laser-Assisted Collisional Energy Transfer Between Rydberg States of Carbon Monoxide

J. Lukasik and S.C. Wallace*

Laboratoire d'Optique Quantique du C.N.R.S., Ecole Polytechnique
F-91128 Palaiseau, France

1. Introduction

Laser-field effects in collisions now constitute a well documented class of phenomena in atomic media [1-3]. We report the first observation of laser-assisted collisions in a molecular system, namely in carbon monoxide (CO). The measured cross-sections for intermolecular energy transfer processes involving high energy ($E \sim 11eV$) Rydberg states are larger than $5 \times 10^{-16}cm^2$ at laser intensities of $\sim 10^{10}W/cm^2$, in spite of the fact that the rovibronic energy levels in a molecule spread out or "dilute" the electronic transition probabilities. Thus, these cross-sections are large enough to clearly demonstrate the potential feasibility of "transition state" spectroscopy in both photodissociation and reactive scattering processes utilizing such laser-field effects.

The types of laser-assisted energy transfer processes which we studied in CO are as follows :

$$CO(1)\{A^1\Pi, v_1 = 5\} + CO(2)\{X^1\Sigma^+, v_2 = 0\} + \hbar\omega \rightarrow$$

$$\rightarrow CO(1)\{X^1\Sigma^+, v_1 = 1\} + CO(2)\{B^1\Sigma^+, v_2 = 0\} \qquad (1a)$$

$$\rightarrow CO(1)\{X^1\Sigma^+, v_1 = 0\} + CO(2)\{B^1\Sigma^+, v_2 = 1\} \qquad (1b)$$

and these are indicated schematically in Fig.1(a) and 1(b). A laser pulse at t=0 creates a population ($>10^{13}cm^{-3}$) in a rovibronic level (v=5, J=12) of the $A^1\Pi_o$ state via two photon absorption of a frequency doubled dye laser ($\lambda = 2784\text{Å}$) After a 12nsec delay the transfer laser prepares a dressed state [4-5] (indicated by the dotted line) which can transfer its energy to another CO molecule if it *simultaneously* makes a collision. This may also be described as a virtual absorption in the presence of a real collision [4]. Using the known oscillator strengths [6], Franck-Condon factors [6-7] and rotational line strengths for the B←A and B→X transitions in CO we estimate that we are in the strong field regime [3a] for our laser intensities and therefore the cross-section is given by [1,3a]

$$\sigma = 1.4 \frac{3\pi}{\sqrt{3}} \left(\frac{\mu_1 \ \mu_2 \ \mu_3}{\hbar^2 \ \bar{v} \ \Delta\omega} \right) E \qquad (2)$$

*Permanent address: Department of Chemistry, University of Toronto, Toronto, Ontario M5S 1A1, Canada

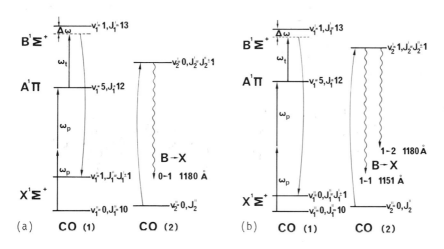

Fig.1 Energy level schemes for laser-assisted intermolecular energy transfer in CO. Fig.(a) and (b) correspond to processes (1a) and (1b) described in text

where μ_1, μ_2 and μ_3 are matrix elements corresponding to B←A (1,5)(CO(1)), B→X (CO(1)) and B←X (CO(2)) transitions respectively, $\Delta\omega$ is the detuning from a real rovibronic level in the B state and E is the transfer laser field strength. Allowing for a $10 cm^{-1}$ detuning from a rovibronic level in the $B^1\Sigma^+$ state and a factor of 1/3 for the orientational average [3a] with respect to the linearly polarized transfer laser (P/A = $8 \times 10^9 W/cm^2$), we estimate $\sigma = 17 \times 10^{-16} cm^2$ for (1a) and $\sigma = 3 \times 10^{-16} cm^2$ for (1b).

2. Experimental

The laser system for these studies was designed to produce two independently tunable, and synchronized in time, dye laser pulses. This was achieved by using a single Nd:Yag oscillator and then dividing the output into two beams which were further amplified and used to pump the two dye lasers. The initial excitation pulse at 2784 Å was 8nsec in duration with a pulse energy of 6mJ and a linewidth (FWHM) of $0.8 cm^{-1}$. The transfer laser at $\lambda \sim 5800$ Å was 10nsec in duration with a maximum pulse energy of 25 mJ and a linewidth (FWHM) of $<0.06 cm^{-1}$ and was delayed by 12nsec with respect to the pump pulse. The pump and transfer laser beams were counter propagating and focused to a beam waist of 200µm inside a stainless steel cell with LiF windows. Fluorescence was detected at 90° through a 1 meter Seya-Namioka VUV monochomator with a solar blind EMI photomultiplier with a C_{SI} cathode (1mm MgF_2 window) and the absolute sensitivity calibration of this detection system has already been described [8].

3. Results and Discussion

At low intensities of the transfer laser ($10^5 W/cm^2$), we observe only the direct (intramolecular) excitation of the $B^1\Sigma^+(v'=1)\leftarrow A^1\Pi(v''=5)$ transitions, with line positions as observed by conventional spectroscopy [9]. As can be seen from the portion of the R branch shown in the lower trace of Fig.3 (p_{CO}=30 torr) we are populating primarily J=12 in the $A^1\Pi$ state with a small contribution from J=13, but with no J=11 (both J=12 and 13 are excited simultaneously because the bandwidth of the pump laser is comparable to their separation at the S branch A←X band head). Even for pressures as high as 100 torr, there is no evidence from these excitation spectra for any rotational relaxation in this vibrational level of the $A^1\Pi$ state. Measurement of the B→X fluorescence spectrum originating from v'=1 gave a branching ratio for (1,1), (1,2) and (1,3) emission of 4:1:0.3, which is in good agreement with the calculated ratio of Franck-Condon factors [7] of 4:1.3:0.4 derived from the RKR potential energy curves for CO.

Under conditions of high transfer laser intensity (8×10^9W/cm^2) we observe a dramatic change in the excitation spectrum. If we detect only the emission from v'=1 of the B state ((1,1) at 1151 Å), then the sharp lines observed at low intensity become *assymetrically* broadened ($\Delta\nu$=3.7 ± 0.3cm^{-1}) as illustrated in Fig.2. Although we do indeed observe a red wing to the line, it is probably premature to interpret this as being directly analogous to the quasi-static wing observed in atomic media [2].

In Fig.3, we show the excitation spectrum taken while simultaneously monitoring the fluorescence from both the v'=0 and v'=1 levels of the $B^1\Sigma^+$ state. This choice was made because (0,1) emission lies too close ($\Delta\lambda$=0.5Å) to the (1,2) emission to be resolved with our monochromator and because we also wish to emphasize that both processes (1a) and (1b) are occuring simultaneously. As well as the assymetric lineshape around the real transition noted above, there is also an additional set of peaks of lower intensity but *reproducible* in position as for example indicated by the arrow in Fig.3. Furthermore the emission spectra obtained when exciting at 5776 Å and 5778.2Å give different intensity ratios at 1151 Å ((1,1) emission) and 1181 Å ((0,1) +(1,2) emission) namely 4:1 vs. 2:1, corroborating the additional presence of (0,1) emission resulting from process (1a).

A detailed theoretical analysis of the rotational state dependence (taking into account the Boltzman distribution of acceptor states of CO(2)) of proces-

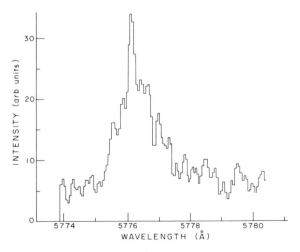

INTENSITY (arb units)

WAVELENGTH (Å)

Fig.2   Excitation spectrum of CO (scanning the transfer laser λ_t) at high intensity (P/A=8×10^9W/cm^2), detecting B→X (1,1) emission at 1151 Å. The transfer laser is delayed by 12nsec with respect to the pump p_{CO}=30 torr)

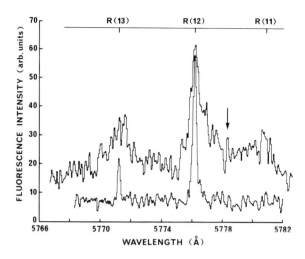

Fig.3 Excitation spectrum of CO, detecting all B→X emission ((0,1)+(1,1)+(1,2)). Lower trace: P/A=10^5W/cm^2 and upper trace: P/A= 8×10^9W/cm^2. The signals registered between the R lines correspond to the background level observed without the transfer laser (see lower trace)

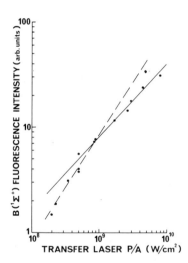

Fig.4 Intensity dependence of the switched collision process, p_{CO}= 100 torr and λ_t=5778.2Å. Solid and broken lines represent slopes 1/2 and 1 respectively

ses (1a) and (1b) reveals an extremely dense spectrum of switched collisions with $\Delta\omega$ extending from 0 up to 200 cm^{-1}. We have in fact observed similar quasi-continuous spectral features, well above the background level, tuning the high intensity transfer laser as far as 150cm^{-1} from any real B←A (1,5) transition. The intensity dependence of the observed fluorescence for λ_t= 5778.2Å (indicated by the arrow in Fig.3) is given in Fig.4. It is initially linear in P/A and then goes over to a (P/A)$^{1/2}$ dependence at high intensities as is predicted by theory in eq.(2). This behaviour was observed at several different wavelengths, except in a 0.4cm^{-1} region around the intramolecular transitions shown in the low intensity scan in Fig.3.

An intramolecular process, which could contribute to the observed emission is the production of anti-Stokes radiation induced by the transfer laser. Several reasons lead us to believe that an anti-Stokes process can be eliminated.

Firstly, the frequency dependence of anti-Stokes radiation must *smoothly* follow the intramolecular excitation as shown in the lower trace of Fig.3. Secondly, it must be symmetric with respect to the frequency detuning. Finally, the emission spectrum (vibrational intensities) must be independent of the wavelength of the transfer laser. All of these properties are in direct contrast to what is actually observed.

It is possible to make an estimate of the cross-section observed experimentally for the sum of processes (1a) and (1b). We calculate that 2 % of the $A^1\Pi$ state is transferred to the $B^1\Sigma^+$ state via this laser assisted collisional process at 8×10^9W/cm. This corresponds to a total cross-section of $\sigma_{a+b}=5\times10^{-16}$cm^2 as compared to our initial estimate for $\sigma_{a+b}=20\times10^{-16}$cm^2. Agreement is reasonable considering the total uncertainties in the input parameters.

This work shows the important role that laser-assisted collisional phenomena can play in laser spectroscopy studies at high laser intensities. As the first observation of switched collisions to achieve inelastic energy transfer in a molecular system, it is an important step in developping and understanding models for laser field effects in molecular dynamics and provides the basis for exploring the rich field of dressed state chemistry in other areas such as photofragmentation and even reactive scattering. Finally, the large cross-sections achieved in these experiments establish the feasibility of future applications, as for example exciting VUV laser transitions via laser-induced collisions.

The authors gratefully acknowledge helpful discussions with W.R. Green and G. Grynberg as well as the experimental help of J. Robert. They thank D.L. Albritton for supplying his unpublished Franck-Condon factors for the B-X transition in CO. Technical assistance of M. Bierry and M. Chateau is also greatly appreciated.

4. References

1. W.R.Green, J. Lukasik, J.R. Willison, M.D. Wright, J.F. Young and S.E. Harris, Phys.Rev.Lett., 42, 970 (1979)
2. (a) R.W. Falcone, W.R. Green, J.C. White, J.F. Young and S.E. Harris, Phys.Rev.A15, 1333 (1977) ; (b) W.R. Green, M.D. Wright, J. Lukasik, J.F. Young and S.E. Harris, Opt.Lett., 4, 265 (1979) ; (c) P. Cahuzac, and P.E. Toschek, Phys.Rev.Lett., 40, 1087 (1978) ; (d) C. Brechignac, P. Cahuzac and P.E. Toschek, Phys.Rev., A21, 1969 (1980) ; (e) A.V. Hellfeld, J. Caddick and J. Weiner, Phys.Rev.Lett., 40, 1369 (1978)
3. (a) L.I. Gudzenko and S.I. Yakovlenko, Sov.Phys.JETP, 35, 877 (1972) (b) S. Yeh and P.R. Berman, Phys.Rev., A19, 1106 (1979) ; (c) A.M.F. Lau, Phys.Rev.,A16, 1535 (1977) ; (d) J.M. Yuan, J.R. Liang and T.F. George, J.Chem.Phys., 66, 1107 (1977)
4. S.E. Harris, R.W. Falcone, W.R. Green, D.B. Lidow, J.C. White, J.F. Young, in Tunable Lasers and Applications, ed. by A. Mooradian, T. Jaeger, P. Stokseth, Springer Series in Optical Sciences, Vol.3 (Springer, Berlin, Heidelberg, New York 1976)
5. C. Cohen-Tannoudji, Cargèse Lectures in Physics, vol.2, p.347 (Gordon and Breach, New York, 1968)
6. P.H. Krupenie, The Band Spectrum of CO, NSRDS-NB55 (1966)
7. D.L. Albritton, Franck-Condon factors, personal communication
8. W.R. Green and J. Lukasik, Optics Letters, 5, 531 (1980)
9. J. Danielak, R. Kepa, K. Ojczyk, M. Rytel, Acta.Phys.Polon., A39, 29 (1971)

Part XI
Progress in New Laser
Sources

Optically Pumped Semiconductor Platelet Lasers

M.M. Salour*

Department of Electrical Engineering and Computer Science

and

Research Laboratory of Electronics, Massachusetts Institute of Technology
Cambridge, MA 02139, USA

Lasers exist in many shapes and forms. Yet in a field so young, the search for new types of lasers remains a leading research field in itself. The new type of laser under discussion is based upon optical pumping in semiconductors. Various individual features of such a device have already been realized [1] and the attractiveness of these individual features potentially combine pleasingly in the newly developed laser [2,3]. Semiconductor lasers have been around since 1962. The years have seen such advances in them that they have now been operated even at room temperatures [4]; by electron beam [5], optical [6] or diode carrier injection pumping [7]; at a variety of frequencies, including the visible regions; and with a large variety of materials. They can be made smaller than any type of laser to date. With increased development in microstructure technology, one is therefore impressed by the possibility of the smallness conceivable with a semiconductor laser.

Semiconductors have already been optically pumped [1]. Optical pumping greatly increases the variety of semiconductors that can be used, compared to the common scheme of injection pumping. They have the advantage over diode and dye lasers that virtually any direct-band-gap semiconductor can be used, thereby increasing the available spectral range. To design a semiconductor laser, one should know what distinguishes it from other types of gain media and have at one's fingertips typical numbers for properties relevant to laser operation. Lasers vary greatly in power, operating wavelength, cavity design, method of pumping, mode discipline (mode-locking, single-frequency, or chaotic operation). Yet the single most frequent way in which the laser worker identifies his equipment is by the type of gain medium, since the medium will strongly influence, if not dictate, the other considerations of laser design.

The most singular feature of semiconductor lasers is that one is not dealing with gain centers (atoms, ions, molecules, complexes) sparsely distributed in a passive medium or empty space, but rather with the problem of inverting the atoms in an entire block of solid, unlike any other kind of laser. Since the absorption and gain lengths in a semiconductor laser are very small indeed compared to other lasers, this central fact bears on the choice of pumping scheme: active medium, heatsinking, cavity design, and sample geometry as semiconductors are crystals and their ordering implies spatial anisotropy (selection rules), and polarization, since polarization effects all depend on crystal orientation in general.

A single bulk semiconductor is the simplest amplifier medium: it is cheap, readily available, and requires far less processing than heterostructure.

---

*Alfred P. Sloan Fellow.

To date, most optically pumped semiconductor lasers have used either crystal faces [1] or closely attached mirrors [8] as the cavity reflectors; this prevents the insertion of tuning elements into the cavity and lowers its optical quality. We recently reported the first cw optically pumped semiconductor laser in an external cavity [2]. The lasing medium was a cadmium sulfide platelet pumped longitudinally by an Ar$^+$ laser. The laser had an output power of 9 mW and a linewidth of 0.1 nm. The power conversion efficiency was 10%. The output is both prism- and temperature-tunable over a range of 9 nm.

The crystals used were very thin (<5 μm) cadmium sulfide platelets chosen for flatness and parallelism by observing under a microscope the interference patterns created by a sodium lamp. The best crystals varied in thickness by less than 3 wavelengths over an area of 5 mm by 5 mm. The crystals were mounted onto a piece of sapphire using a thin film of silicone oil (Fig.1). The same side of the sapphire was dielectrically coated with a maximum reflectivity mirror. The pump beam was focused onto the crystal to a spot size of ∿5 μm by a 10X microscope objective, which also served to collimate the crystal fluorescence. This beam was separated from the pump beam by a polarizing beamsplitter which transmitted 98% of the CdS emission. In order to make this possible, the c-axis of the CdS crystal was vertically oriented. Then its fluorescence, which primarily has $E \perp c$, was polarized perpendicularly to the vertically polarized Ar$^+$ laser beam.

The semiconductor laser beam passed through a prism and could be tuned by rotating the output mirror about the vertical axis [2]. The sapphire mirror was held by copper pieces which were attached to a liquid nitrogen chamber using copper braid. It could also be tilted in two directions to align this end of the cavity. This arrangement allowed sample temperatures as low as 94 K. The microscope objective was also held in the vacuum, which was sealed by two AR-coated windows.

The spontaneous emission and laser spectra without the prism are shown in Fig.2. The spontaneous emission is characterized by Fabry-Perot modes due to interference between the direct emission and that reflected by the sapphire mirror. As the pump power is increased for 40 μW to 1 W, the spontaneous emission shifts to the red, consistent with local heating or high exciton density effects [9].

The spectrum changes dramatically, however, if the external mirror is then brought into alignment (Fig.2c). An increase in power by a factor of up to 2000 is observed, and the linewidth narrows to less than 0.1 nm. The lasing occurs near the peak of one of the Fabry-Perot modes observed in the spontaneous emission pattern.

The output beam had a divergence of less than 1 mrad in the TEM$_{00}$ mode with a cavity length of 20 cm. If the cavity alignment was slightly changed, lasing occurred in higher order modes.

With a prism inserted into the cavity with the 99.5% output mirror, the laser could be tuned to three different Fabry-Perot modes, covering the range between 495 and 501 nm. Within the strongest of these three modes it could be tuned over 1.8 nm with a 0.1 nm bandwidth. The wavelengths between these modes could be reached by moving the crystal laterally so that the crystal Fabry-Perot length changed slightly. In addition, when the temperature of the sample was raised from 95 K to 140 K the laser wavelength increased from 497 to 504 nm. The wavelength could be varied continuously over this range by a combination of temperature and prism tuning. At temperatures above

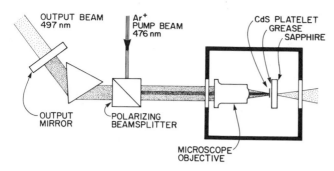

Fig.1 Cavity design for experiments with prism tuning

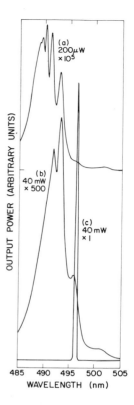

Fig.2 Emission spectra of CdS platelet at 95 K.
(a) 200 μW pump power, (b) 40 mW pump power with output mirror misaligned, (c) same pump power as (b) but
with output mirror aligned. Note the differences in
vertical sensitivity. The linewidth shown in part (c)
is detector-limited; the actual linewidth of 0.1 nm
was determined by increasing the monochromator resolution

140 K the damage threshold was comparable to the laser threshold. Lasing could
be accomplished using any of the 488-, 476-, 473-, or 458-nm lines of the Ar+
laser as a pump.

This work demonstrated the first prism-tunable optically pumped semiconductor laser. We have extended this technique to CdSe and CdSSe crystals,
using a 514-nm Ar+ pump. Then several crystals were mounted adjacent to each
other on the same mirror, yielding a laser which is easily tunable from 500
to 700 nm. Crystals of AlGaAsP or ZnCdS could also be used. Heating problems
in the crystal might be further reduced by the use of epitaxial layers of
CdSe grown on sapphire [10].

For synchronous pumping the Ar+ laser was actively mode-locked to produce
100-psec pulses on the 476-nm line, and the cavity was lengthened to 1.8 m.
The pulse width was measured in an autocorrelator using a lithium formate
crystal. Typical spots gave stable 4-10 psec (assuming a single-sided exponential shape; see Fig.3) pulses with 3-5 mW output power and 50 mW of pump
power. Up to 26 mW of output power could be maintained for a short while.

The cavity length can be changed by 500 μm without adversely affecting
the pulse shape (Fig.4). This is in sharp contrast to synchronously pumped
dye lasers, where changes of only microns can drastically alter the pulse
shape. This indicated that passive pulse shaping may be taking place. The
output spectrum is much broader than that encountered in cw operation, and
often lasing in more than one mode of the crystal Fabry-Perot occurs. We

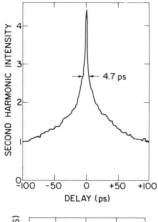

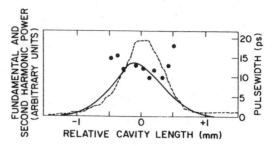

Fig.4 Dependence of the laser output-power (solid line) and peak second harmonic power (dashed line) on the relative cavity length. The dots refer to the output pulse width at various cavity lengths as measured by auto-correlation assuming a single-sided exponential

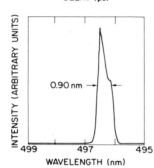

Fig.3 (a) Autocorrelation trace taken from laser with 3.2 mW output power. A pulse width of 4.7 ps is calculated if a single-sided exponential shape is assumed but the large signals in the tails of the pulse indicate an 8 ps pulse width. (b) Simultaneous spectrum, taken with 0.08 nm resolution

have demonstrated that antireflection coating of the crystal face eliminates these modes, increases the cavity bandwidth, and results in shorter pulses[11].

These lasers differ from dye lasers in that no jet fluctuations are present, eliminating a very strong source of noise. They can be operated completely in a vacuum, eliminating atmospheric pressure fluctuations present in dye laser cavities. In addition, the spontaneous emission spectrum is narrower than those of dyes, allowing a stabilized single-frequency laser to operate with fewer wavelength-selecting elements while tuning can be done by varying the temperature. We believe that lasers of this type have the capability for single-frequency operation tunable throughout most of the visible and near IR.

The author thanks C.B. Roxlo, D. Bebelaar, R. Putnam, D.A. Johnson, D.C. Reynolds, A. Mooradian, H.A. Haus, and E.P. Ippen. This work was supported by the U.S. Air Force Office of Scientific Research.

References

[1] M.R. Johnson and N. Holonyak, J. Appl. Phys. $\underline{39}$, 3977 (1968); S.R. Chinn, J.A. Rossi, C.M. Wolfe, and A. Mooradian, IEEE J. Quant. Electron. $\underline{QE-9}$, 294 (1973); N. Menyuk, A.S. Pine, and A. Mooradian, IEEE J. Quant. Electron. $\underline{QE-11}$, 477 (1975).

[2] C.B. Roxlo, D. Bebelaar, and M.M. Salour, Appl. Phys. Lett. $\underline{38}$, 307 (1981).

[3] C.B. Roxlo and M.M. Salour, Appl. Phys. Lett. $\underline{38}$, 738 (1981).

[4] J.A. Rossi, S.R. Chinn, and A. Mooradian, Appl. Phys. Lett. $\underline{20}$, 84 (1974).

[5] C.E. Hurwitz, Appl. Phys. Lett. $\underline{8}$, 121 (1966).

[6] T.C. Damen, M.A. Dugay, J.M. Wiesenfeld, J. Stone, C.A. Burris, in Picosecond Phenomena II, ed. by R. Hochstrasser, W. Kaiser, C.U. Strank, Springer Series in Chemical Physics, Vol.14 (Springer, Berlin, Heidelberg, New York 1980), p. 38.

[7] E.P. Ippen, D.J. Eilenberger, R.W. Dixon, in Picosecond Phenomena II, ed. by R. Hochstrasser, W. Kaiser, C.U. Strank, Springer Series in Chemical Physics, Vol.14 (Springer, Berlin, Heidelberg, New York 1980), p. 21

[8] J. Stone, C.A. Burris, and J.C. Campbell, J. Appl. Phys. $\underline{51}$, 3038 (1980).

[9] R.F. Leheny and J. Shah, Phys. Rev. Lett. $\underline{37}$, 871 (1976).

[10] V.N. Martynor, S.A. Modvedev, L.L. Aksenova, and Tu. D. Avchukhov, Sov. Phys. Crystallogr. $\underline{24}$, 743 (1979).

[11] W.L. Cao, A.M. Vaucher, and Chi H. Lee, Appl. Phys. Lett. $\underline{38}$, 306 (1981); $\underline{38}$, 653 (1981)

Short Wavelength Laser Design*

P.L. Hagelstein

Special Studies Group, Lawrence Livermore Laboratory
Livermore, CA 94550, USA

Introduction

The field of EUV and soft X-ray lasers has suffered from a shortage of
working laboratory laser sources to study ever since the conception of
the field in the mid to late 1960's [1]. Admittedly, a great deal of
work has been done over the last decade and a half [2]; nonetheless,
the scientific world still awaits the researcher who, in his laboratory,
will push back the frontier and claim the EUV and soft X-ray spectral
regions as conquered domains for the laser physicists of the quantum
electronics community. It is not yet obvious to what uses scientists
shall put these territories once they have been won. To push the analogy
further, these new lands appear not to be overly laden with scientific
gold and jewels; however, in time workers should reap the fruits of their
labor in the areas of holography, spectroscopy, non-LTE kinetics, plasma
diagnostics, and perhaps some day in the area of photo-lithography [3].

At the Lawrence Livermore Laboratory, there is a distinct emphasis on
the development of computational physics resources with which to attack
difficult physics and engineering problems [4]. The laboratory also has a
proclivity for bringing sheer brute force to bear on its more obstinate
problems, and as a result is endowed with (among other facilities) one of
the planet's more powerful laser systems (SHIVA). The combination of these
and other factors has led to a theoretical effort during the past half
decade, one goal of which has been to create computational tools for the
investigation of high intensity laser-pumped short wavelength laser systems,
as well as laser systems pumped by other means.

One result of this research has been the development of a general short
wavelength laser design code XRASER [5]. This code has been applied to the
design of several proposed lasers using a variety of schemes (photoioniza-
tion-, photoexcitation-, and recombination-pumping schemes). Documentation
of this work is underway; however, as much of the quantum electronics
community is not yet familiar with the general approach, we shall restrict
our discussion largely to certain aspects of the simplest of the schemes --
the Ne II 27 eV photoionization-pumped laser scheme. In particular, we shall
examine the behavior of a proposed laser target under different pump
intensities for 1.06μ light and for frequency tripled (0.35μ) light. This
topic is of interest as several laser fusion laboratories are contemplating

*Work performed under the auspices of the U.S. Dept. of Energy by Lawrence
Livermore Nat. Lab. under Contract W-7405 Eng-48 and supported by the Fannie
and John Hertz Foundation.

converting to frequency tripled radiation to reduce preheat due to suprathermal electrons. We shall find that the shorter wavelength pump radiation
results in improved operation of the Ne II laser due primarily to the higher
absorption of the blue light.

Motivation

Before delving into the details of the Ne II scheme, some explanation is in
order as to why one should be interested in a scheme of this sort. For
example, coherent light has been generated via frequency multiplication of
1.06μ light at energies highter than 27 eV, and with an efficiency [6] higher
than appears to be possible with designs that have been examined so far.
As a laser, such a device would at best be of academic interest. In defense
of the scheme, there are two points which might lead one to take an interest:
1. The laser scheme is quite simple and the atomic physics of the levels
involved are well known such that one may calculate the dynamical behavior
of the laser and its output and hope to obtain agreement with experimental
results.
2. The small signal gain can be quite high (in excess of 50/cm), which is
important since SHIVA and other large laser facilities are unable to focus
over a distance longer than 4 mm (were there a cavity built with lossy mirrors around the laser medium, then the gain requirements would be decreased;
nonetheless, the occurence of mirrors on a laser target at SHIVA in the near
term is unlikely).
These two features of the scheme, that of being readily calculable, and that
of being readily realizable, are features currently shared by very few laser
schemes in this and higher energy spectral regimes. Indeed, there are very
few schemes at all which show sufficiently high gain under the conditions
which are available at SHIVA such that a short-wavelength laser designer
might be willing to proffer it to his highly critical and unforgiving conservative experimentalist colleagues which sometimes frequent the larger
laser facilities. The basic problem is that SHIVA and other laser-fusion
laser facilities were designed for fusion experiments and not for short
wavelength laser experiments. We have requested that SHIVA NOVA be modified
so as to be more compatible with short wavelength laser requirements, and we
hope that future large laser facilities might be designed with these requirements in mind.

We view the Ne II laser scheme as the simplest of a sequence of laser
schemes leading to ever higher laser transition energies. The physics,
design, and experimental issues are more complex for the higher energy laser
schemes which we have examined (which include resonantly-pumped photoexcita-
tion schemes between 50 eV and 180 eV, and electron collisional excitation
pumped schemes below 100 eV [5]). Even were other schemes with higher energy
and efficiency readily available (that could be carried out today on SHIVA),
it would still be an attractive experiment to do in order to compare with the
numerical results which are available.

It should be noted that the Ne II scheme does not require the full power of
SHIVA to pump it. The power requirements are directly related to the pulse
length, color, and focusing capabilities of the candidate pump laser system.
The issues of providing clean (spatially homogeneous) cylindrically focused
(with high aspect ratio) intense infrared or visible radiation has not been
explored within the laser fusion community. The current scheme for using
SHIVA employs no cylindrical focusing; rather, the side-by-side placement of

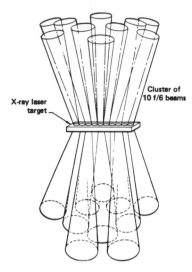

Figure 1 Possible beam configuration for a short wavelength laser experiment on SHIVA.

circular spots along the length of the target as shown in Figure 1. It is apparent that the Ne II laser can be pumped with an incident pulse of peak power well under a terawatt were a good cylindrically focusing system in the blue available.

The Ne II Scheme and Calculations

Removal of a 2s electron from the Ne I $2s^2p^6$ ground state results in the production of a Ne II $2sp^6$ ^2S ion, which lies about 27 eV above the two Ne II $2s^2p^5$ ^2P levels (see Figure 2). Preferential removal of 2s electrons over 2p electrons can be effected by photoionization pumping with X-rays above 250 eV and below 850 eV as can be seen from perusal of the single electron cross sectional data shown in Figure 3. Coupling between the lower and upper laser

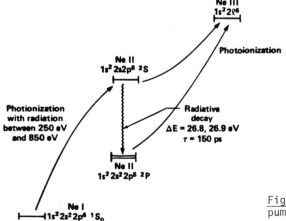

Figure 2 Ne II photoionization-pumped laser scheme.

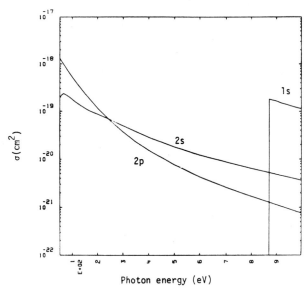

Figure 3 Photoionization cross sections per electron for Ne I.

levels is weak, and all one must worry about is that the neon density be low enough such that the electron avalanche which occurs with the onset of photoionization pumping does not spoil the inversion prematurely, as electron collisional processes favor population of the lower laser state. As the scheme is self terminating, there is no recovery once the lower laser states become filled.

Table 1 Optimized gain for selected pump pulses for 1μ and $1/3\mu$ light

Pump Pulse	Peak pump rate/2s electron		Gain (1/cm) and neon density $(1/cm^3)$	
	1μ	$1/3\mu$	1μ	$1/3\mu$
90 ps 5×10^{13}W/cm^2	4.0×10^8	7.8×10^8	46 1.3×10^{17}	89 1.9×10^{17}
1×10^{14}W/cm^2	1.2×10^9	9.0×10^8	75 2.0×10^{17}	103 2.1×10^{17}
1.5×10^{14}W/cm^2	2.4×10^9	9.5×10^8	103 2.7×10^{17}	107 2.7×10^{17}
50 ps 5×10^{13}W/cm^2	2.3×10^8	5.8×10^8	57 2.7×10^{17}	110 3.1×10^{17}
1×10^{14}W/cm^2	6.0×10^8	1.0×10^9	125 3.6×10^{17}	122 2.7×10^{17}
1.5×10^{14}W/cm^2	1.0×10^9	1.2×10^9	130 3.6×10^{17}	144 2.7×10^{17}

The peak pump rate per 2s electron is computed in the limit of a flashlamp and filter of infinite extent. The kinetics were computed using pump rates less by a geometrical factor of 0.43. The target specifications are as follows: 250 Å Mo, .1μ CH, 1μ LiF, and .1μ CH.

In order to produce X-rays in the 500 eV range, we have chosen thin-film flashlamp and filter systems to convert the incident optical pump laser light into X-rays. At 10^{14} W/cm^2 incident intensity, molybdenum is a reasonable choice of flashlamp material by virtue of its strong M-N shell radiation emitted by the dominant M-shell ions present. Low energy radiation (below 250 eV) must be filtered out as it is quite ruinous to the inversion kinetics (see Figure 3) and for this reason we have selected LiF sandwiched by thin (sub 1000 Å) layers of parylene as a filter to remove it. The resulting flashlamp and filter system has four layers (Mo/CH/LiF/CH) and the target which results has these thin films above and below (with Mo facing outwards) with low density (about $2 \times 10^{17}/cm^3)_2$neon inside. Irradiation by a 90 ps pulse of 1.06 µ light at 5×10^{13} W/cm^2 yields a gain pattern which, at the time of peak gain, resembles what is shown in Figure 4.

In Table 1 we show the results of a set of calculations of target dynamics, kinetics, and gain at the center of the laser target (the gain is highest at the target center, although as shown in Figure 4, there is a slow variation of gain in space around the center). The radiation spectrum was calculated from opacity data generated by the laser fusion code LASNEX [4] (see [5] for a discussion of this type of calculation) and was used to drive the laser kinetics in XRASER ([5] also has a discussion of the laser kinetics). The laser gain was optimized over the initial neon density, and the peak gain and associated density are tabulated. Also given are peak photoionization rates per electron for Ne I 2s electrons, which is a measure of the strength of pumping (as the inner filter edge heats up, low energy thermal emission occurs, which causes 2p ionization and leads to spoiling of the gain; hence, the peak 2s pump rate does not tell the entire story). For this calculation we have assumed that the suprathermal electrons penetrate into the laser medium and displace the thermal electrons, leading to a very high electron temperature. For the blue light, we have assumed that no suprathermal electrons are present in the laser medium, which is a good approximation for all but the highest intensity cases, where the gain should be higher by about 20%.

Gain (cm^{-1})

A : 2.16E+01
B : 2.37E+01
C : 2.58E+01
D : 2.79E+01
E : 2.99E+01
F : 3.20E+01
G : 3.41E+01
H : 3.62E+01
I : 3.83E+01
J : 4.04E+01

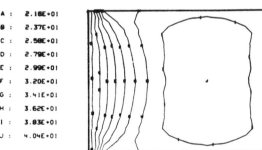

Figure 4 Gain isocontours on a cross section of the laser. Thinfilm flash-lamp and filter systems are above and below the laser medium and cold brass is to either side. The laser is 400µ across and would extend outward from the paper a length corresponding to 4 mm.

Discussion

From the results presented in Table 1 we observe that the proposed laser target design is calculated to have higher small signal laser gain for blue light irradiation by nearly a factor of two at the lower incident intensity, given that one is allowed to vary the density of the neon gas within the laser target. The primary effect is simply that the blue light is absorbed with higher efficiency (we have assumed 30% absorption in the infrared and 70% for the blue light), which leads immediately to stronger pump rates and hence higher gain. Another major effect is the increased or decreased contribution of the electron collisional stripping of the Ne I 2p electrons, depending on whether suprathermal electrons dominate the electron distribution of the laser medium. Fewer and colder hot electrons are generated by the blue pump radiation, which is unfortunate because the colder thermal electrons are more efficient in removing the 2p electrons from Ne I. This is not necessarily true above 1.5×10^{14} W/cm^2, where there are enough hot electrons to make a difference; however, at this high intensity, one is driving the target much too hard (the peak gain occurs well before the peak of the pulse).

References

1 M. A. Duguay and P. M. Rentzepis, Appl. Phys. Let. 10, 350 (1967)
2 A general review of the different areas of the field is given in P. L. Hagelstein, ."Physics of Short Wavelength Laser Design," LLL UCRL-53100. Also useful are B. Carrigan, "X-ray Lasers (A Bibliography with Abstracts)," NTIS/PS-79/0011 (1979), R. C. Elton, Adv. X-ray Anal. 21, 1 (1978), A. V. Vinogradov, I. I. Sobel'man and E. A. Yukov, J. Physique 39, C4 (1978). An excellent (but by now dated) review is R. W. Waynant and R. C. Elton, Proc. IEEE 64, 1059 (1976)
3 Although there are some applications for laboratory short wavelength lasers of the laser-pumped variety (see S. Jorna, "X-ray Laser Applications Study," Physical Dynamics, Inc. Final Rept. PD-LJ-77-159 (1977) or a more recent study by D. J. Nagel (to be published (1980))), the potential applications have yet to provoke the investment in time, money and effort necessary to develop them. The energy range of interest for X-ray lithography is above 250 eV, H. I. Smith, D. C. Flanders, J. Vac. Sci. Tech. 17, 533 (1980).
4 For an example of a difficult physics problem and the use of both massive computational resources as well as sheer brute force, see the "Laser Program Annual Report - 1979," UCRL-50021-79 and the annual reports of the past several years as well.
5 The code XRASER and much work relating to this paper is discussed in the Ph.D. Thesis of the author, the reference for which is given in [2].
6 J. Reintjes, C. Y. She, R. C. Eckardt, N. E. Karangelen, R. A. Andrews, Appl. Phys. Let. 30, 480 (1977)

Free Electron Laser Experiment on the ACO Storage Ring (Orsay)

P. Elleaume[1], C. Bazin[2], M. Billardon[3], D.A.G. Deacon[4], Y. Farge,
J.M.J. Madey[4], J.M. Ortega[3], Y. Pétroff. K.E. Robinson[4], and M.G. Velghe

LURE, Bâtiment 209 C, Université de Paris-Sud
F-91405 Orsay, France

1. Introduction

Since the first demonstration of the Free Electron Laser (FEL) in Stanford
[1], several experiments have been proposed using various types of relativistic electron beam sources. The purpose of our experiment in Orsay (LURE-
Stanford Collaboration) is to use the electron beam of the ACO storage ring
(140 MeV to 540 MeV) to build a F.E.L. in the 5000 Å range.

2. Undulator

An undulator is a magnet structure
giving a transverse periodic field B
along a trajectory followed by the
electrons (Fig. 1). Such a field
exerts an acceleration on the electrons and is responsible of the so-
called synchrotron radiation emission. The spectrum of the radiations
emitted in a narrow solid angle centered on the electron velocity is
quite different from the broad spectrum emitted by the beam in a uniform field, it is peaked around
harmonics of a resonant wavelength λ
(Fig. 2) :

$$\lambda_i = \frac{\lambda_o}{2i\gamma^2} \left(1 + \frac{k^2}{2} + \gamma^2\theta^2\right) \qquad (1)$$

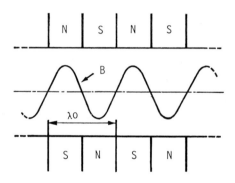

Fig. 1 : Undulator field

1   Département de Physico-Chimie, Service de Photophysique, CEN-Saclay,
     91190 Gif-sur-Yvette, France
2   Laboratoire de l'Accélérateur Linéaire, Bâtiment 200, Université de
     Paris-Sud, 91405 Orsay, France
3   Ecole Supérieure de Physique et Chimie, 10 rue Vauquelin 75231 Paris
     Cedex 05, France
4   High Energy Physics Lab, Stanford University, Stanford CA 94305, USA
5   Laboratoire de Photophysique Moléculaire, Bât. 210, Université de Paris-
     Sud, 91405 Orsay, France

where

λ_o : the undulator period (several cm)

i : integer (harmonic number)

γ : electron energy/mc^2 (several hundreds)

k : proportional to the magnetic field (order of unity)

θ : angle between observation
direction and electron
trajectory

The width of the peaks is about $\frac{1}{N}$
where N is the number of period of
the undulator.

A 23 period superconducting undulator with a 4 cm period and 4 kG
design field [2] has been installed
on a straight section of ACO. Characteristics of the spontaneous
emission were measured both at
240 MeV, the injection energy and
at 150 MeV and are in good agreement with theory [3, 4].

3. Gain

In quantum electrodynamics,
synchrotron radiation is
viewed as a Compton back-
scattering of the virtual
photons from the magnets.
The stimulation of this
two photons process (one
photon processes are for-
bidden for free electrons)
makes possible to operate
an undulator as an ampli-
fier for use in a laser
[5].

An experiment has been
settled [6] to measure the
gain (Fig. 3). It mainly
consists of a double de-
tection of the modulation
(due to absorption or am-
plification by free elec-
trons) of a CW argon laser
probe at the ring rotation

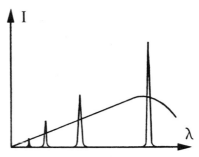

Fig. 2 : Synchrotron radiation
emission spectrum
(Undulator : peaked, uniform
field : broad)

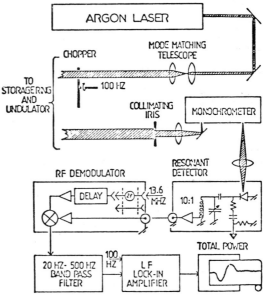

Fig. 3 : Lay-out of the gain measurement
experiment

482

frequency (27 MHz) and at a chopper frequency ($\sim$ 100 Hz). Gain and spontaneous emission at λ = 4880 Å as function of the electron energy are presented in figure 4. The gain is as predicted [7], the derivative of the spontaneous emission, the width is about 40 % larger than predicted which is explained by the field measurement [4]. The largest measured peak gain averaged over the laser mode is about 3×10^{-4} per pass [6].

Effect of the argon laser on the electron bunch has been observed in term of a bunch lengthening of about 4 % for 1.7 Watt power [8]. Bunch lengthening and energy spreading are expected to be the main causes of saturation of the FEL.

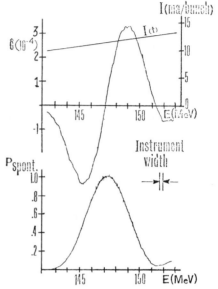

Fig. 4 : Gain and spontaneous emission at λ = 4880 Å v.s. electron energy

4. Future plans

A 5.5 meter long cavity (exactly one fourth of the ring peremeter to synchronize light and electron pulses) has been installed in the vacuum of the ring. Mirrors having a total loss $\lesssim$ .05 % will be used. With two bunches stored in ACO, the threshold gain is about .1 % per pass and is therefore presently too high to achieve oscillation. To increase the gain, we are presently building a permanent magnet optical klystron [10] which consists of two identical undulators separated by a dispersive section. Enhancement of the gain at least by a factor of 3 is expected at high current [10] and should allow investigation of the behaviour of laser near threshold.

We should point out that ACO is an old storage ring designed for e^- - e^+ experiments and is not optimized for FEL operation. The new generation of machines built specially for synchrotron radiation (Aladdin, Brookhaven, Bessy, Super-ACO) will have longer straight sections, higher peak current and lower energy spread allowing gain of several percent using the easy permanent magnet technology. This gives the possibility of operating a sub-nanosecond pulsed laser continuously tunable by changing the electron energy (see Eq. 1) in the range 20000-2500 Å.

5. Acknowledgments

This work has been supported by the DGRST, Contract 79-7-1063, the DRET, Contract 79-073 and the AFOSR, Contract F 49620-80-C-0068.

We would like to thank the technical staff of L.U.R.E. and members of the Linear Accelerator Laboratory.

References

1. D.A.G. Deacon, L.R. Elias, J.M.J. Madey, G.J. Ramian, H.A. Schwettman and T.I. Smith, Phys. Rev. Lett., 38, 892 (1977)

2. C. Bazin, Y. Farge, M. Lemonnier, J. Pérot and Y. Pétroff, Nucl. Inst. & Meth. 172 (1980) 61

3. C. Bazin, M. Billardon, D.A.G. Deacon, Y. Farge, J.M. Ortega, J. Pérot, Y. Pétroff and M. Velghe, Journal de Phys. Lettres 41 (1980) 547

4. M. Billardon et al., to be published

5. J.M.J. Madey, J. Appl. Phys. 42 (1971) 1906

6. D.A.G. Deacon, J.M.J. Madey, K.E. Robinson, C. Bazin, M. Billardon, P. Elleaume, Y. Farge, J.M. Ortega, Y. Pétroff and M.F. Velghe, Proc. 1981 Part. Acc. Conf., IEEE Trans. Nucl. Sc.

7. J.M.J. Madey, Nuovo Cimento, 50 B, 64 (1979)

8. K.E. Robinson et al., to be published

9. J.M. Ortega et al., to be published

10. N.A. Vinokurov, Proc. Xth Intern. Conf. on High Energy Charged Particle Accelerators, Serpukhov, Vol. 2 (1977), p. 454

11. P. Elleaume, to be published in *Physics of Quantum Electronics,* (Addison Wesley)

Simple UV, Visible and IR Recombination Lasers in Expanding Metal-Vapor Plasmas

W.T. Silfvast and O.R. Wood, II

Bell Telephone Laboratories
Holmdel, NJ 07733, USA

The use of the electron-ion collisional recombination process in an expanding, segmented plasma, has led to the development of a new class of metal-vapor lasers in the metals of Mg, Al, Ca, Cu, Zn, Ag, Cd, In, Sn, Pb, and Bi. These lasers operate in the ultraviolet at wavelengths as short as 298nm, in the visible, and in the infrared up to 1.950μm. These extremely simple lasers, which are easily constructed, have long operating life, and operate at repetition rates up to 1 kHz, produce output powers of 1-10 W/pulse for durations of 5-100μs. This type of laser, generated by a row of segmented arcs, is called a Segmented-Plasma-Excitation-Recombination laser or SPER laser.[1] The extension of this type of laser excitation process to one that uses laser-produced plasmas as the excitation source is expected to lead to VUV and XUV lasers.

The production of a recombination laser involves two sequential processes; (1) ionization or plasma formation; and (2) recombination and laser emission. The ionization process, in order to be efficient, requires the production of ions at densities somewhat higher than those allowed during the lasing process. Initial densities of the order of 10^{15}-10^{16} cm^{-3} for single ions (higher for more highly stripped ions) allow a large percentage of the excitation energy to be concentrated at a single ionization stage thus enhancing the potential efficiency and gain of the laser.

Expansion is then required to reduce the density and also to induce the cooling of the electrons so that reattachment or recombination can occur with the ions. Adiabatic expansion into a vacuum produces cooling of the plasma; however, expansion into a background gas has been found to be significantly more effective in inducing rapid cooling which more effectively populates the upper laser level. In addition, segmentation of the plasma into a series of spherical plasmas also greatly enhances the gain as compared with an elongated plasma.

The most favorable laser transitions are those that occur across the large gaps in the energy level spectrum of a given ion stage that are located just below the closely spaced grouping of levels merging with the continuum of the next higher ion stage. The closely spaced levels allow virtually all of the population of the next higher ion stage to move downward, via the collisional recombination process, to the upper laser level, thereby concentrating a large fraction of the ionization energy at a specific energy level. This stored energy can then be extracted by means of stimulated emission to one or more lower lying levels (across the energy level gap referred to earlier).

Sources of expanding plasmas that have been used in making these recombination lasers include laser-produced plasmas and pulsed metal-vapor arcs. The laser-produced plasmas have the advantages of being able to control the plasma formation temperature and thus the specific ion stage by varying the laser intensity, and of controlling the initial density by adjusting the size of the focal region of the laser when focused onto the metal to be vaporized. The disadvantage of the laser-produced plasma technique is the difficulty in making a long gain region to enhance the prospect of obtaining laser action.

The advantage of using a series of pulsed metal-vapor arcs, as used in the SPER laser, is in the simplicity of making a long gain region. This is done by attaching a series of metal strips onto an insulating substrate leaving a 1-mm gap between each strip. When a high-voltage, high-current, 2-5μs duration pulse of electrical energy is applied to the series of strips in the presence of a low-pressure background gas, a row of arc plasmas are formed in the gaps which subsequently expand into the surrounding background gas where recombination and consequently laser action occur.

Eight of the strongest ultraviolet and visible SPER laser transitions provide a broad spectrum of useful laser wavelengths for possible applications. These include 298nm and 301nm in InIII, 468nm in InII, 492nm in ZnII, 534nm and 538nm in CdII, 537nm in PbII and 580nm in SnII.

Isoelectronic scaling of known laser transitions to identical transitions in the next higher ion stage of the next element in the periodic table allows the optimization of recombination lasers at low ion stages, which result in IR and visible lasers (due to the relatively low ionization potential of low ion stages), and then the scaling upward of similar transitions to higher ion stages in order to obtain short-wavelength lasers. The most significant example of this technique[2], to date, involves a 4f-5d transition which produces laser action at 1.84μm in AgI, at 538nm in CdII, and 301nm in InIII.

Presently, the 4f-3d transition at 116.8nm in CIV is being investigated as a possible VUV recombination laser as a result of isoelectronic scaling of a previously observed laser in BeII at 468nm. Additional laser transitions also being studied in CIV include a UV transition at 252nm (where high-reflecting mirrors are available for optimization of the plasma parameters for efficient production of the necessary C^{4+} species), and a much shorter wavelength XUV transition (5f-3d) at 79.9nm.

References

1. W. T. Silfvast, L. H. Szeto and O. R. Wood, II, Appl. Phys. Lett. 36:615 (1980).
2. W. T. Silfvast, L. H. Szeto and O. R. Wood, II, Appl. Phys. Lett. (to be published August 1, 1981).

A Two-Photon Laser

B. Nikolaus, D.Z. Zhang, and P.E. Toschek
Institut für Angewandte Physik 1, Universität Heidelberg
Albert Ueberlestraße 3-5, D-6900 Heidelberg, Fed. Rep. of Germany

1. Introduction

During the past decade, electronic two-photon transitions [1] have been observed and largely utilized in two-photon absorption of laser light by atoms and molecules. In particular, cancellation of the Doppler shift in standing-wave two-photon absorption has turned into a powerful tool for high-resolution spectroscopy [2].

In contrast, demonstrations of pure *two-photon emission* have been rare [3]. Efficient two-photon emission involving a three-level cascade requires conflicting preconditions for the level scheme. This process is related to Raman emission which, however, essentially differs in the role of spontaneous emission. On the other hand, both processes, when resonantly enhanced, are intimately entangled with single-photon transitions [4]. Off-resonance stimulated Raman emission is "pure" in this sense*. Likewise, the demonstration of pure two-photon emission in general requires tuning the light field(s) *far off* the intermediate resonance.

Two-photon emission may alternatively include two quanta of equal or unequal energy (degenerate or non-degenerate emission, respectively). In the latter case, two-frequency emission stimulated by light injection of only one of the frequencies can be achieved. This process has been recognized in the past when observing the radiation from potassium vapour which was simultaneously pumped by ruby laser light and Stokes light generated in various organic liquids [3]. The temperature tuning range of ruby allowed frequency tuning off the intermediate level by $\sim$ 10 cm^{-1}. Other attempts, e.g. to demonstrate stimulated two-photon emission making use of the forbidden oxygen aurora lines proved unsuccessful. However, various schemes and theoretical models for *two-photon lasers* have been proposed [6-9].

With inversion lacking, non-degenerate two-photon emission may involve one *spontaneously* emitted quantum. An incoherent source based upon this principle was recently demonstrated [10].

We have observed, from the inverted 2s ^{2}S $\rightarrow$ 3d ^{2}D two-photon transition of atomic lithium vapour (Fig. 1), coherent two-photon emission in two ways:

---

*But also on-resonance interactions which emerge from virtual transitions in transparent gases [5].

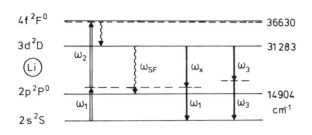

Fig. 1 Level scheme of atomic Li

(*i*) *Non-degenerate* generation of coherent light pulses at frequency ω_x, stimulated by injected light pulses of frequency $\omega_1 \neq \omega_x$, where $\omega_1 + \omega_x = \omega_{31} = \omega(3d - 2s)$.

(*ii*) *Degenerate* single-pass amplification of injected light at frequency $\omega_3 = \omega_{31} / 2$.

2. Exitation Scheme

Li atoms are two-step/two-photon excited by simultaneous inter- action with two light pulses whose sum frequency is close to re- sonance with the $4f\ ^2F$ level. The transient population inversion on the 3d ^2D level is caused by resonant quadrupole (hyper-) Ra- man process [11] accompanied by IR Stokes emission. However, a contribution from collisional redistribution in the closely spaced 4f and 4d levels followed by fast (super-)fluorescent decay can- not be excluded [12].

The 3d population decays by way of superfluorescence - which is easily visible as yellow beam and characterized by large fluc- tuations - into the 2p level. However, the presence of strong pump light close to the 2s - 2p resonance line populates the 2p level and permits two-photon emission on the transition 3d - 2s to compete favourably with the yellow superfluorescence, which may be completely quenched under certain excitation conditions. - The application of a third "probe" light pulse opens yet another channel for decay of the 3d population.

3. Experiments

Two sequences of 4-nsec-long pulses from N_2-laser-pumped dye la- sers at $\lambda_1 667,4$ nm (DCM) and at $\lambda_2 461,8$ nm (Coumarine 1) simul- taneously irradiate, from opposite sides, Li vapour in a heat pipe (s. Fig. 2). Coherent light pulses at ω_x, which show small amplitude fluctuations only, are recorded by monochromator, fast- rise photo-diode, box-car integrator, and XY recorder. Alternati- vely, probe pulses, generated by a third dye laser at $\lambda_3 639.1$ nm, experience gain as revealed by the same photoelectric detection

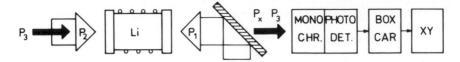

Fig. 2 Scheme of experiment

scheme. Correlations of phase or polarization of the light emitted at ω_x or ω_3 do not exist. This important fact excludes parametric n-wave processes as source of the emission. In addition, the directions of light propagation do not admit phase matching.

4. Results

The output peak power $P_x(\omega_x)$ of the light pulses generated by non-degenerate two-photon emission at ω_x has been measured vs. pump powers P_1 and P_2. Above threshold, P_x depends linearly upon P_2 (Fig. 3a). In contrast, a square-law dependence on P_1 shows up for part of the dynamical range in agreement with P_1 acting both as pump and stimulating light. At high P_1 values, the resonance transition becomes saturated. The superimposed reproducible wiggle seems to indicate Rabi flopping. - As a function of vapour density, P_x depends linearly above threshold (Fig. 3b).

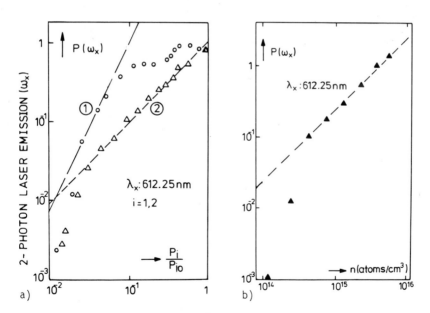

Fig. 3 Non-degenerate two-photon laser emission vs. pump power (a) and vs. vapour density (b)

In Fig. 4, the peak power P_x is shown *vs.* stepwise tuning ω_1 and ω_2 in opposite frequency directions, such that the excitation preserved quasi-resonance with the 4f level. The central minimum occurs from absorption of the pump beams at resonance with the intermediate 2p level. Laser emission is visible even at off-tuning, from this level, as far as 500 cm^{-1}. In the wings, the output power depends on the tuning as $(\Delta\lambda_1)^{-4}$: Pump rate and stimulation by P_1 both contribute a factor $(\Delta\lambda_1)^{-2}$ from the Lorentzian wings of the line shape.

The excited Li vapour acts as an amplifier for light pulses at the probe frequency $\omega_3(\lambda_3 639,15$ nm$)$. This gain competes with and largely suppresses superfluorescence from the 3d level and the non-degenerate oscillation. Fig. 5 shows the output $P_3(z)$ *vs.* input $P_3(0)$. Three series of measurements are plotted: No pump light was applied with sequence 1 which shows probe attenuation by two-photon absorption. This absorption is reduced, when P_1 is on, and a fraction of the Li ground state population is excited by multi-photon absorption and decay into the 3d level (2). When both P_1 *and* P_2 are on, population inversion occurs on the 3d level against the ground state, due to the fast excitation scheme, and P_3 is two-photon amplified.

The corresponding gain depends linearly on the probe light intensity in the unsaturated regime, since

$$dI_3/dz = B_2 \cdot \Delta N \cdot I_3^2 , \tag{1}$$

where ΔN is the inversion density and $B_2 I_3^2$ is the two-photon transition rate per atom. The full lines represent fits according to the two-photon analogue of Lambert-Beer's law,

$$I(z)/I(0) = (1 \pm I(0)/\hat{I})^{-1} , \tag{2}$$

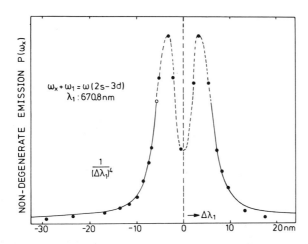

Fig. 4 Non-degenerate two-photon laser emission at ω_x *vs.* detuning of pump light 1 which also stimulates the emission at $\omega_1 = \omega(3d - 2s) - \omega_x$.

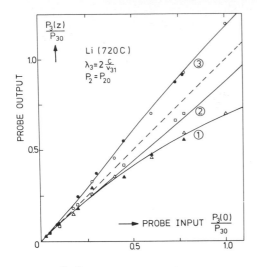

Fig. 5 Output power $P_3(z)$ *vs.* input power $P_3(0)$ of injected probe light pulses at ω_3. (1): No pump light, (2): P_1 present only, (3): P_1 and P_2 generate two-photon gain. - Independent sets of measurements are distinguished.

where $\hat{I}^{-1} = \Delta N \circ \sigma \cdot z$, and σ is the emission cross-section. Saturation has been included in the calculation of curves 2 and 3 in a model where the population density responds fast to variations of light intensity. In contrast, curve 1 does not include saturation since 3d - 2p superfluorescence, following two-photon absorption, suppresses saturation. Upon P_1 excitation, however, this superfluorescence is quenched due to appreciable 2p population.

The two-photon laser which has been demonstrated in these experiments is a quantum-optical device essentially different from (single-photon) laser, parametric oscillator, and stimulated Raman sources. Obvious applications include high-power optical amplifiers whose gain is not prematurely quenched by fluorescence.

References

1 Maria Goeppert-Mayer, Naturwissenschaften 17, 932 (1929); Ann. Phys. (Leipz.) 9, 273 (1931)
2 See, e.g., N. Bloembergen, M.D. Levenson, in High-Resolution Laser Spectroscopy, ed. by K. Shimoda, Topics in Applied Physics, Vol. 13 (Springer, Berlin, Heidelberg, New York 1976) p. 315
3 S. Barak, M. Rodni, and S. Yatsiv, IEEE J. Quantum Electronics QE-5 448 (1969)
4 V.S. Letokhov, V.P. Chebotayev, Nonlinear Laser Spectroscopy, Springer Series in Optical Sciences, Vol. 4 (Springer, Berlin, Heidelberg, New York 1977)
5 P.E. Toschek, in: "Aux Frontières de la Spectroscopie laser", R. Balian, S. Haroche, and S. Liberman, eds., p. 323, North-Holland, Amsterdam 1977
6 P.P. Sorokin and N. Braslau, IBM J. Res. Dev. 8, 177 (1964)
7 A.M. Prokhorov, Science 149, 828 (1965)

8 R.L. Carman, Phys. Rev. A $\underline{12}$, 1048 (1975)
9 L.M. Narducci, W.W. Eidson, P. Furcinitti, and D.C. Eteson, Phys. Rev. A $\underline{16}$ 1665 (1977)
10 S.E. Harris, Appl. Phys. Letters $\underline{31}$, 498 (1977); L.J. Zych, J. Lukasik, J.F. Young, and S.E. Harris, Phys. Rev. Letters $\underline{40}$, 1493 (1978)
11 D. Cotter and M.A. Yuratich, Optics Communic. $\underline{29}$, 307 (1979)
12 For analog superfluorescence see: Ph. Cahuzac, H. Sontag, and P.E. Toschek, Optics Communic. $\underline{31}$, 37 (1979)

List of Contributors

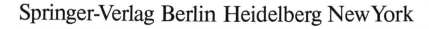

W. Demtröder

Laser Spectroscopy

Basic Concepts and Instrumentation

1981. 431 figures. XIII, 694 pages
(Springer Series in Chemical Physics, Volume 5)
ISBN 3-540-10343-0

This textbook is an introduction to modern techniques and instrumentation in laser spectroscopy. After an elementary discussion of basic subjects such as absorption and dispersion of light, coherence, line broadening effects and saturation phenomena, a detailed outline of spectroscopic instrumentation, such as spectrographs, interferometers and photodetectors is given.
The main part of the book deals with recently developed techniques in laser spectroscopy and the increased sensitivity and spectral resolution they have made possible. Doppler-free methods are particularly emphasized.
This book helps close the gap between classical works on optics and spectroscopy and more specialized publications on modern research in this field. It is addressed to graduate students in physics and chemistry as well as scientists just starting out in this field.

Lasers and Chemical Change

By A. Ben-Shaul, Y. Haas, K. L. Kompa, R. D. Levine

1981. 245 figures. XII, 497 pages
(Springer Series in Chemical Physics, Volume 10)
ISBN 3-540-10379-1

This book deals with one of the most striking applications of the laser: the initiation, control and diagnostics of chemical processes. It covers both chemical lasers and laser chemistry, i. e. the generation of coherent light in the course of chemical reactions and the application of such light to the study of chemical reactions. Touching on both theoretical and experimental aspects, the book is divided into five parts: of the first gives a general introduction, the second a comprehensive theoretical treatment of laser molecule reciprocal action, and the third discusses tools and techniques. The fourth chapter summarizes chemical lasers. The final chapter is **Lasers and Chemical Change** evaluates the results and prospects of laser chemistry provides the first comprehensive treatment of this field on a graduate level.

Laser Spectroscopy III

Proceedings of the Third International Conference, Jackson Lake Lodge, Wyoming, USA, July 4–8, 1977
Editors: J. Hall, J. L. Carlsten

1977. 296 figures. XI, 468 pages
(Springer-Series in Optical Sciences, Volume 7)
ISBN 3-540-08543-2

Laser Spectroscopy IV

Proceedings of the Fourth International Conference, Rottach-Egern, Federal Republic of Germany, June 11–15, 1979
Editors: H. Walther, K. W. Rothe

1979. 411 figures, 19 tables. XIII, 652 pages
(Springer Series in Optical Sciences, Volume 21)
ISBN 3-540-09766-X

Contents: Introduction. – Fundamental Physical Applications of Laser Spectroscopy. Two and Three Level Atoms/High Resolution Spectroscopy. – Rydberg States. – Multiphoton Dissociation, Multiphoton Excitation. – Nonlinear Processes, Laser Induced Collisions, Multiphoton Ionization. – Coherent Transients, Time Domain Spectroscopy. Optical Bistability, Superradiance. – Laser Spectroscopic Applications. – Laser Sources. – Postdeadline Papers. – Index of Contributors.

"... The papers presented at the Conference were of a very high standard in all sections, and this is fully reflected in this volume... I hope that it is obvious from this review how useful this volume will be for those who are interested in the newest methods and advanced applications of the rapidly developing field of laser spectroscopy..."
European Spectroscopy News

Springer-Verlag
Berlin
Heidelberg
New York